# Lecture Notes in Computer Science 13624

## Founding Editors

Gerhard Goos

Juris Hartmanis

## Editorial Board Members

Elisa Bertino, *Purdue University, West Lafayette, IN, USA*

Wen Gao, *Peking University, Beijing, China*

Bernhard Steffen, *TU Dortmund University, Dortmund, Germany*

Moti Yung, *Columbia University, New York, NY, USA*

The series Lecture Notes in Computer Science (LNCS), including its subseries Lecture Notes in Artificial Intelligence (LNAI) and Lecture Notes in Bioinformatics (LNBI), has established itself as a medium for the publication of new developments in computer science and information technology research, teaching, and education.

LNCS enjoys close cooperation with the computer science R & D community, the series counts many renowned academics among its volume editors and paper authors, and collaborates with prestigious societies. Its mission is to serve this international community by providing an invaluable service, mainly focused on the publication of conference and workshop proceedings and postproceedings. LNCS commenced publication in 1973.

Mohammad Tanveer · Sonali Agarwal ·
Seiichi Ozawa · Asif Ekbal · Adam Jatowt
Editors

# Neural Information Processing

29th International Conference, ICONIP 2022
Virtual Event, November 22–26, 2022
Proceedings, Part II

Springer

*Editors*
Mohammad Tanveer
Indian Institute of Technology Indore
Indore, India

Seiichi Ozawa
Kobe University
Kobe, Japan

Adam Jatowt
University of Innsbruck
Innsbruck, Austria

Sonali Agarwal ⓘ
Indian Institute of Information Technology -
Allahabad
Prayagraj, India

Asif Ekbal
Indian Institute of Technology Patna
Patna, India

ISSN 0302-9743          ISSN 1611-3349 (electronic)
Lecture Notes in Computer Science
ISBN 978-3-031-30107-0          ISBN 978-3-031-30108-7 (eBook)
https://doi.org/10.1007/978-3-031-30108-7

# Preface

Welcome to the proceedings of the 29th International Conference on Neural Information Processing (ICONIP 2022) of the Asia-Pacific Neural Network Society (APNNS), held virtually from Indore, India, during November 22–26, 2022.

The mission of the Asia-Pacific Neural Network Society is to promote active interactions among researchers, scientists, and industry professionals who are working in neural networks and related fields in the Asia-Pacific region. APNNS has Governing Board Members from 13 countries/regions – Australia, China, Hong Kong, India, Japan, Malaysia, New Zealand, Singapore, South Korea, Qatar, Taiwan, Thailand, and Turkey. The society's flagship annual conference is the International Conference of Neural Information Processing (ICONIP).

The ICONIP conference aims to provide a leading international forum for researchers, scientists, and industry professionals who are working in neuroscience, neural networks, deep learning, and related fields to share their new ideas, progress, and achievements. Due to the current situation regarding the pandemic and international travel, ICONIP 2022, which was planned to be held in New Delhi, India, was organized as a fully virtual conference.

The proceedings of ICONIP 2022 consists of a multi-volume set in LNCS and CCIS, which includes 146 and 213 papers, respectively, selected from 1003 submissions reflecting the increasingly high quality of research in neural networks and related areas. The conference focused on four main areas, i.e., "Theory and Algorithms," "Cognitive Neurosciences," "Human Centered Computing," and "Applications." The conference also had special sessions in 12 niche areas, namely

1. International Workshop on Artificial Intelligence and Cyber Security (AICS)
2. Computationally Intelligent Techniques in Processing and Analysis of Neuronal Information (PANI)
3. Learning with Fewer Labels in Medical Computing (FMC)
4. Computational Intelligence for Biomedical Image Analysis (BIA)
5. Optimized AI Models with Interpretability, Security, and Uncertainty Estimation in Healthcare (OAI)
6. Advances in Deep Learning for Biometrics and Forensics (ADBF)
7. Machine Learning for Decision-Making in Healthcare: Challenges and Opportunities (MDH)
8. Reliable, Robust and Secure Machine Learning Algorithms (RRS)
9. Evolutionary Machine Learning Technologies in Healthcare (EMLH)
10. High Performance Computing Based Scalable Machine Learning Techniques for Big Data and Their Applications (HPCML)
11. Intelligent Transportation Analytics (ITA)
12. Deep Learning and Security Techniques for Secure Video Processing (DLST)

Our great appreciation goes to the Program Committee members and the reviewers who devoted their time and effort to our rigorous peer-review process. Their insightful reviews and timely feedback ensured the high quality of the papers accepted for publication.

The submitted papers in the main conference and special sessions were reviewed following the same process, and we ensured that every paper has at least two high-quality single-blind reviews. The PC Chairs discussed the reviews of every paper very meticulously before making a final decision. Finally, thank you to all the authors of papers, presenters, and participants, which made the conference a grand success. Your support and engagement made it all worthwhile.

December 2022                                             Mohammad Tanveer
                                                          Sonali Agarwal
                                                          Seiichi Ozawa
                                                          Asif Ekbal
                                                          Adam Jatowt

# Organization

## Program Committee

### General Chairs

M. Tanveer      Indian Institute of Technology Indore, India
Sonali Agarwal      IIIT Allahabad, India
Seiichi Ozawa      Kobe University, Japan

### Honorary Chairs

Jonathan Chan      King Mongkut's University of Technology
     Thonburi, Thailand
P. N. Suganthan      Nanyang Technological University, Singapore

### Program Chairs

Asif Ekbal      Indian Institute of Technology Patna, India
Adam Jatowt      University of Innsbruck, Austria

### Technical Chairs

Shandar Ahmad      JNU, India
Derong Liu      University of Chicago, USA

### Special Session Chairs

Kai Qin      Swinburne University of Technology, Australia
Kaizhu Huang      Duke Kunshan University, China
Amit Kumar Singh      NIT Patna, India

### Tutorial Chairs

Swagatam Das      ISI Kolkata, India
Partha Pratim Roy      IIT Roorkee, India

## Finance Chairs

Shekhar Verma                Indian Institute of Information Technology
                             Allahabad, India
Hayaru Shouno                University of Electro-Communications, Japan
R. B. Pachori                IIT Indore, India

## Publicity Chairs

Jerry Chun-Wei Lin           Western Norway University of Applied Sciences,
                             Norway
Chandan Gautam               A*STAR, Singapore

## Publication Chairs

Deepak Ranjan Nayak          MNIT Jaipur, India
Tripti Goel                  NIT Silchar, India

## Sponsorship Chairs

Asoke K. Talukder            NIT Surathkal, India
Vrijendra Singh              IIIT Allahabad, India

## Website Chairs

M. Arshad                    IIT Indore, India
Navjot Singh                 IIIT Allahabad, India

## Local Arrangement Chairs

Pallavi Somvanshi            JNU, India
Yogendra Meena               University of Delhi, India
M. Javed                     IIIT Allahabad, India
Vinay Kumar Gupta            IIT Indore, India
Iqbal Hasan                  National Informatics Centre, Ministry of
                             Electronics and Information Technology, India

## Regional Liaison Committee

Sansanee Auephanwiriyakul    Chiang Mai University, Thailand
Nia Kurnianingsih            Politeknik Negeri Semarang, Indonesia

Md Rafiqul Islam                   University of Technology Sydney, Australia
Bharat Richhariya                  IISc Bangalore, India
Sanjay Kumar Sonbhadra             Shiksha 'O' Anusandhan, India
Mufti Mahmud                       Nottingham Trent University, UK
Francesco Piccialli                University of Naples Federico II, Italy

## Program Committee

Balamurali A. R.                   IITB-Monash Research Academy, India
Ibrahim A. Hameed                  Norwegian University of Science and Technology
                                     (NTNU), Norway
Fazly Salleh Abas                  Multimedia University, Malaysia
Prabath Abeysekara                 RMIT University, Australia
Adamu Abubakar Ibrahim             International Islamic University, Malaysia
Muhammad Abulaish                  South Asian University, India
Saptakatha Adak                    Philips, India
Abhijit Adhikary                   King's College, London, UK
Hasin Afzal Ahmed                  Gauhati University, India
Rohit Agarwal                      UiT The Arctic University of Norway, Norway
A. K. Agarwal                      Sharda University, India
Fenty Eka Muzayyana Agustin        UIN Syarif Hidayatullah Jakarta, Indonesia
Gulfam Ahamad                      BGSB University, India
Farhad Ahamed                      Kent Institute, Australia
Zishan Ahmad                       Indian Institute of Technology Patna, India
Mohammad Faizal Ahmad Fauzi        Multimedia University, Malaysia
Mudasir Ahmadganaie                Indian Institute of Technology Indore, India
Hasin Afzal Ahmed                  Gauhati University, India
Sangtae Ahn                        Kyungpook National University, South Korea
Md. Shad Akhtar                    Indraprastha Institute of Information Technology,
                                     Delhi, India
Abdulrazak Yahya Saleh Alhababi    University of Malaysia, Sarawak, Malaysia
Ahmed Alharbi                      RMIT University, Australia
Irfan Ali                          Aligarh Muslim University, India
Ali Anaissi                        CSIRO, Australia
Ashish Anand                       Indian Institute of Technology, Guwahati, India
C. Anantaram                       Indraprastha Institute of Information Technology
                                     and Tata Consultancy Services Ltd., India
Nur Afny C. Andryani               Universiti Teknologi Petronas, Malaysia
Marco Anisetti                     Università degli Studi di Milano, Italy
Mohd Zeeshan Ansari                Jamia Millia Islamia, India
J. Anuradha                        VIT, India
Ramakrishna Appicharla             Indian Institute of Technology Patna, India

| | |
|---|---|
| V. N. Manjunath Aradhya | JSS Science and Technology University, India |
| Sunil Aryal | Deakin University, Australia |
| Muhammad Awais | COMSATS University Islamabad, Wah Campus, Pakistan |
| Mubasher Baig | National University of Computer and Emerging Sciences (NUCES) Lahore, Pakistan |
| Sudhansu Bala Das | NIT Rourkela, India |
| Rakesh Balabantaray | International Institute of Information Technology Bhubaneswar, India |
| Sang-Woo Ban | Dongguk University, South Korea |
| Tao Ban | National Institute of Information and Communications Technology, Japan |
| Dibyanayan Bandyopadhyay | Indian Institute of Technology, Patna, India |
| Somnath Banerjee | University of Tartu, Estonia |
| Debajyoty Banik | Kalinga Institute of Industrial Technology, India |
| Mohamad Hardyman Barawi | Universiti Malaysia, Sarawak, Malaysia |
| Mahmoud Barhamgi | Claude Bernard Lyon 1 University, France |
| Kingshuk Basak | Indian Institute of Technology Patna, India |
| Elhadj Benkhelifa | Staffordshire University, UK |
| Sudip Bhattacharya | Bhilai Institute of Technology Durg, India |
| Monowar H Bhuyan | Umeå University, Sweden |
| Xu Bin | Northwestern Polytechnical University, China |
| Shafaatunnur Binti Hasan | UTM, Malaysia |
| David Bong | Universiti Malaysia Sarawak, Malaysia |
| Larbi Boubchir | University of Paris, France |
| Himanshu Buckchash | UiT The Arctic University of Norway, Norway |
| George Cabral | Federal Rural University of Pernambuco, Brazil |
| Michael Carl | Kent State University, USA |
| Dalia Chakrabarty | Brunel University London, UK |
| Deepayan Chakraborty | IIT Kharagpur, India |
| Tanmoy Chakraborty | IIT Delhi, India |
| Rapeeporn Chamchong | Mahasarakham University, Thailand |
| Ram Chandra Barik | C. V. Raman Global University, India |
| Chandrahas | Indian Institute of Science, Bangalore, India |
| Ming-Ching Chang | University at Albany - SUNY, USA |
| Shivam Chaudhary | Indian Institute of Technology Gandhinagar, India |
| Dushyant Singh Chauhan | Indian Institute of Technology Patna, India |
| Manisha Chawla | Amazon Inc., India |
| Shreya Chawla | Australian National University, Australia |
| Chun-Hao Chen | National Kaohsiung University of Science and Technology, Taiwan |
| Gang Chen | Victoria University of Wellington, New Zealand |

| | |
|---|---|
| He Chen | Hebei University of Technology, China |
| Hongxu Chen | University of Queensland, Australia |
| J. Chen | Dalian University of Technology, China |
| Jianhui Chen | Beijing University of Technology, China |
| Junxin Chen | Dalian University of Technology, China |
| Junyi Chen | City University of Hong Kong, China |
| Junying Chen | South China University of Technology, China |
| Lisi Chen | Hong Kong Baptist University, China |
| Mulin Chen | Northwestern Polytechnical University, China |
| Xiaocong Chen | University of New South Wales, Australia |
| Xiaofeng Chen | Chongqing Jiaotong University, China |
| Zhuangbin Chen | The Chinese University of Hong Kong, China |
| Long Cheng | Institute of Automation, China |
| Qingrong Cheng | Fudan University, China |
| Ruting Cheng | George Washington University, USA |
| Girija Chetty | University of Canberra, Australia |
| Manoj Chinnakotla | Microsoft R&D Pvt. Ltd., India |
| Andrew Chiou | CQ University, Australia |
| Sung-Bae Cho | Yonsei University, South Korea |
| Kupsze Choi | The Hong Kong Polytechnic University, China |
| Phatthanaphong Chomphuwiset | Mahasarakham University, Thailand |
| Fengyu Cong | Dalian University of Technology, China |
| Jose Alfredo Ferreira Costa | UFRN, Brazil |
| Ruxandra Liana Costea | Polytechnic University of Bucharest, Romania |
| Raphaël Couturier | University of Franche-Comte, France |
| Zhenyu Cui | Peking University, China |
| Zhihong Cui | Shandong University, China |
| Juan D. Velasquez | University of Chile, Chile |
| Rukshima Dabare | Murdoch University, Australia |
| Cherifi Dalila | University of Boumerdes, Algeria |
| Minh-Son Dao | National Institute of Information and Communications Technology, Japan |
| Tedjo Darmanto | STMIK AMIK Bandung, Indonesia |
| Debasmit Das | IIT Roorkee, India |
| Dipankar Das | Jadavpur University, India |
| Niladri Sekhar Dash | Indian Statistical Institute, Kolkata, India |
| Satya Ranjan Dash | KIIT University, India |
| Shubhajit Datta | Indian Institute of Technology, Kharagpur, India |
| Alok Debnath | Trinity College Dublin, Ireland |
| Amir Dehsarvi | Ludwig Maximilian University of Munich, Germany |
| Hangyu Deng | Waseda University, Japan |

Mingcong Deng                    Tokyo University of Agriculture and Technology,
                                   Japan
Zhaohong Deng                    Jiangnan University, China
V. Susheela Devi                 Indian Institute of Science, Bangalore, India
M. M. Dhabu                      VNIT Nagpur, India
Dhimas Arief Dharmawan           Universitas Indonesia, Indonesia
Khaldoon Dhou                    Texas A&M University Central Texas, USA
Gihan Dias                       University of Moratuwa, Sri Lanka
Nat Dilokthanakul                Vidyasirimedhi Institute of Science and
                                   Technology, Thailand
Tai Dinh                         Kyoto College of Graduate Studies for
                                   Informatics, Japan
Gaurav Dixit                     Indian Institute of Technology Roorkee, India
Youcef Djenouri                  SINTEF Digital, Norway
Hai Dong                         RMIT University, Australia
Shichao Dong                     Ping An Insurance Group, China
Mohit Dua                        NIT Kurukshetra, India
Yijun Duan                       Kyoto University, Japan
Shiv Ram Dubey                   Indian Institute of Information Technology,
                                   Allahabad, India
Piotr Duda                       Institute of Computational
                                   Intelligence/Czestochowa University of
                                   Technology, Poland
Sri Harsha Dumpala               Dalhousie University and Vector Institute, Canada
Hridoy Sankar Dutta              University of Cambridge, UK
Indranil Dutta                   Jadavpur University, India
Pratik Dutta                     Indian Institute of Technology Patna, India
Rudresh Dwivedi                  Netaji Subhas University of Technology, India
Heba El-Fiqi                     UNSW Canberra, Australia
Felix Engel                      Leibniz Information Centre for Science and
                                   Technology (TIB), Germany
Akshay Fajge                     Indian Institute of Technology Patna, India
Yuchun Fang                      Shanghai University, China
Mohd Fazil                       JMI, India
Zhengyang Feng                   Shanghai Jiao Tong University, China
Zunlei Feng                      Zhejiang University, China
Mauajama Firdaus                 University of Alberta, Canada
Devi Fitrianah                   Bina Nusantara University, Indonesia
Philippe Fournierviger           Shenzhen University, China
Wai-Keung Fung                   Cardiff Metropolitan University, UK
Baban Gain                       Indian Institute of Technology, Patna, India
Claudio Gallicchio               University of Pisa, Italy
Yongsheng Gao                    Griffith University, Australia

| | |
|---|---|
| Yunjun Gao | Zhejiang University, China |
| Vicente García Díaz | University of Oviedo, Spain |
| Arpit Garg | University of Adelaide, Australia |
| Chandan Gautam | I2R, A*STAR, Singapore |
| Yaswanth Gavini | University of Hyderabad, India |
| Tom Gedeon | Australian National University, Australia |
| Iuliana Georgescu | University of Bucharest, Romania |
| Deepanway Ghosal | Indian Institute of Technology Patna, India |
| Arjun Ghosh | National Institute of Technology Durgapur, India |
| Sanjukta Ghosh | IIT (BHU) Varanasi, India |
| Soumitra Ghosh | Indian Institute of Technology Patna, India |
| Pranav Goel | Bloomberg L.P., India |
| Tripti Goel | National Institute of Technology Silchar, India |
| Kah Ong Michael Goh | Multimedia University, Malaysia |
| Kam Meng Goh | Tunku Abdul Rahman University of Management and Technology, Malaysia |
| Iqbal Gondal | RMIT University, Australia |
| Puneet Goyal | Indian Institute of Technology Ropar, India |
| Vishal Goyal | Punjabi University Patiala, India |
| Xiaotong Gu | University of Tasmania, Australia |
| Radha Krishna Guntur | VNRVJIET, India |
| Li Guo | University of Macau, China |
| Ping Guo | Beijing Normal University, China |
| Yu Guo | Xi'an Jiaotong University, China |
| Akshansh Gupta | CSIR-Central Electronics Engineering Research Institute, India |
| Deepak Gupta | National Library of Medicine, National Institutes of Health (NIH), USA |
| Deepak Gupta | NIT Arunachal Pradesh, India |
| Kamal Gupta | NIT Patna, India |
| Kapil Gupta | PDPM IIITDM, Jabalpur, India |
| Komal Gupta | IIT Patna, India |
| Christophe Guyeux | University of Franche-Comte, France |
| Katsuyuki Hagiwara | Mie University, Japan |
| Soyeon Han | University of Sydney, Australia |
| Palak Handa | IGDTUW, India |
| Rahmadya Handayanto | Universitas Islam 45 Bekasi, Indonesia |
| Ahteshamul Haq | Aligarh Muslim University, India |
| Muhammad Haris | Universitas Nusa Mandiri, Indonesia |
| Harith Al-Sahaf | Victoria University of Wellington, New Zealand |
| Md Rakibul Hasan | BRAC University, Bangladesh |
| Mohammed Hasanuzzaman | ADAPT Centre, Ireland |

| | |
|---|---|
| Takako Hashimoto | Chiba University of Commerce, Japan |
| Bipan Hazarika | Gauhati University, India |
| Huiguang He | Institute of Automation, Chinese Academy of Sciences, China |
| Wei He | University of Science and Technology Beijing, China |
| Xinwei He | University of Illinois Urbana-Champaign, USA |
| Enna Hirata | Kobe University, Japan |
| Akira Hirose | University of Tokyo, Japan |
| Katsuhiro Honda | Osaka Metropolitan University, Japan |
| Huy Hongnguyen | National Institute of Informatics, Japan |
| Wai Lam Hoo | University of Malaya, Malaysia |
| Shih Hsiung Lee | National Cheng Kung University, Taiwan |
| Jiankun Hu | UNSW@ADFA, Australia |
| Yanyan Hu | University of Science and Technology Beijing, China |
| Chaoran Huang | UNSW Sydney, Australia |
| He Huang | Soochow University, Taiwan |
| Ko-Wei Huang | National Kaohsiung University of Science and Technology, Taiwan |
| Shudong Huang | Sichuan University, China |
| Chih-Chieh Hung | National Chung Hsing University, Taiwan |
| Mohamed Ibn Khedher | IRT-SystemX, France |
| David Iclanzan | Sapientia Hungarian University of Transylvania, Romania |
| Cosimo Ieracitano | University "Mediterranea" of Reggio Calabria, Italy |
| Kazushi Ikeda | Nara Institute of Science and Technology, Japan |
| Hiroaki Inoue | Kobe University, Japan |
| Teijiro Isokawa | University of Hyogo, Japan |
| Kokila Jagadeesh | Indian Institute of Information Technology, Allahabad, India |
| Mukesh Jain | Jawaharlal Nehru University, India |
| Fuad Jamour | AWS, USA |
| Mohd. Javed | Indian Institute of Information Technology, Allahabad, India |
| Balasubramaniam Jayaram | Indian Institute of Technology Hyderabad, India |
| Jin-Tsong Jeng | National Formosa University, Taiwan |
| Sungmoon Jeong | Kyungpook National University Hospital, South Korea |
| Yizhang Jiang | Jiangnan University, China |
| Ferdinjoe Johnjoseph | Thai-Nichi Institute of Technology, Thailand |
| Alireza Jolfaei | Federation University, Australia |

| | |
|---|---|
| Ratnesh Joshi | Indian Institute of Technology Patna, India |
| Roshan Joymartis | Global Academy of Technology, India |
| Chen Junjie | IMAU, The Netherlands |
| Ashwini K. | Global Academy of Technology, India |
| Asoke K. Talukder | National Institute of Technology Karnataka - Surathkal, India |
| Ashad Kabir | Charles Sturt University, Australia |
| Narendra Kadoo | CSIR-National Chemical Laboratory, India |
| Seifedine Kadry | Noroff University College, Norway |
| M. Shamim Kaiser | Jahangirnagar University, Bangladesh |
| Ashraf Kamal | ACL Digital, India |
| Sabyasachi Kamila | Indian Institute of Technology Patna, India |
| Tomoyuki Kaneko | University of Tokyo, Japan |
| Rajkumar Kannan | Bishop Heber College, India |
| Hamid Karimi | Utah State University, USA |
| Nikola Kasabov | AUT, New Zealand |
| Dermot Kerr | University of Ulster, UK |
| Abhishek Kesarwani | NIT Rourkela, India |
| Shwet Ketu | Shambhunath Institute of Engineering and Technology, India |
| Asif Khan | Integral University, India |
| Tariq Khan | UNSW, Australia |
| Thaweesak Khongtuk | Rajamangala University of Technology Suvarnabhumi (RMUTSB), India |
| Abbas Khosravi | Deakin University, Australia |
| Thanh Tung Khuat | University of Technology Sydney, Australia |
| Junae Kim | DST Group, Australia |
| Sangwook Kim | Kobe University, Japan |
| Mutsumi Kimura | Ryukoku University, Japan |
| Uday Kiran | University of Aizu, Japan |
| Hisashi Koga | University of Electro-Communications, Japan |
| Yasuharu Koike | Tokyo Institute of Technology, Japan |
| Ven Jyn Kok | Universiti Kebangsaan Malaysia, Malaysia |
| Praveen Kolli | Pinterest Inc, USA |
| Sunil Kumar Kopparapu | Tata Consultancy Services Ltd., India |
| Fajri Koto | MBZUAI, UAE |
| Aneesh Krishna | Curtin University, Australia |
| Parameswari Krishnamurthy | University of Hyderabad, India |
| Malhar Kulkarni | IIT Bombay, India |
| Abhinav Kumar | NIT, Patna, India |
| Abhishek Kumar | Indian Institute of Technology Patna, India |
| Amit Kumar | Tarento Technologies Pvt Limited, India |

| | |
|---|---|
| Nagendra Kumar | IIT Indore, India |
| Pranaw Kumar | Centre for Development of Advanced Computing (CDAC) Mumbai, India |
| Puneet Kumar | Jawaharlal Nehru University, India |
| Raja Kumar | Taylor's University, Malaysia |
| Sachin Kumar | University of Delhi, India |
| Sandeep Kumar | IIT Patna, India |
| Sanjaya Kumar Panda | National Institute of Technology, Warangal, India |
| Chouhan Kumar Rath | National Institute of Technology, Durgapur, India |
| Sovan Kumar Sahoo | Indian Institute of Technology Patna, India |
| Anil Kumar Singh | IIT (BHU) Varanasi, India |
| Vikash Kumar Singh | VIT-AP University, India |
| Sanjay Kumar Sonbhadra | ITER, SoA, Odisha, India |
| Gitanjali Kumari | Indian Institute of Technology Patna, India |
| Rina Kumari | KIIT, India |
| Amit Kumarsingh | National Institute of Technology Patna, India |
| Sanjay Kumarsonbhadra | SSITM, India |
| Vishesh Kumar Tanwar | Missouri University of Science and Technology, USA |
| Bibekananda Kundu | CDAC Kolkata, India |
| Yoshimitsu Kuroki | Kurume National College of Technology, Japan |
| Susumu Kuroyanagi | Nagoya Institute of Technology, Japan |
| Retno Kusumaningrum | Universitas Diponegoro, Indonesia |
| Dwina Kuswardani | Institut Teknologi PLN, Indonesia |
| Stephen Kwok | Murdoch University, Australia |
| Hamid Laga | Murdoch University, Australia |
| Edmund Lai | Auckland University of Technology, New Zealand |
| Weng Kin Lai | Tunku Abdul Rahman University of Management & Technology (TAR UMT), Malaysia |
| Kittichai Lavangnananda | King Mongkut's University of Technology Thonburi (KMUTT), Thailand |
| Anwesha Law | Indian Statistical Institute, India |
| Thao Le | Deakin University, Australia |
| Xinyi Le | Shanghai Jiao Tong University, China |
| Dong-Gyu Lee | Kyungpook National University, South Korea |
| Eui Chul Lee | Sangmyung University, South Korea |
| Minho Lee | Kyungpook National University, South Korea |
| Shih Hsiung Lee | National Kaohsiung University of Science and Technology, Taiwan |
| Gurpreet Lehal | Punjabi University, India |
| Jiahuan Lei | Meituan-Dianping Group, China |

Pui Huang Leong            Tunku Abdul Rahman University of Management
                          and Technology, Malaysia
Chi Sing Leung            City University of Hong Kong, China
Man-Fai Leung             Anglia Ruskin University, UK
Bing-Zhao Li              Beijing Institute of Technology, China
Gang Li                   Deakin University, Australia
Jiawei Li                 Tsinghua University, China
Mengmeng Li               Zhengzhou University, China
Xiangtao Li               Jilin University, China
Yang Li                   East China Normal University, China
Yantao Li                 Chongqing University, China
Yaxin Li                  Michigan State University, USA
Yiming Li                 Tsinghua University, China
Yuankai Li                University of Science and Technology of China,
                          China
Yun Li                    Nanjing University of Posts and
                          Telecommunications, China
Zhipeng Li                Tsinghua University, China
Hualou Liang              Drexel University, USA
Xiao Liang                Nankai University, China
Hao Liao                  Shenzhen University, China
Alan Wee-Chung Liew       Griffith University, Australia
Chern Hong Lim            Monash University Malaysia, Malaysia
Kok Lim Yau               Universiti Tunku Abdul Rahman (UTAR),
                          Malaysia
Chin-Teng Lin             UTS, Australia
Jerry Chun-Wei Lin        Western Norway University of Applied Sciences,
                          Norway
Jiecong Lin               City University of Hong Kong, China
Dugang Liu                Shenzhen University, China
Feng Liu                  Stevens Institute of Technology, USA
Hongtao Liu               Du Xiaoman Financial, China
Ju Liu                    Shandong University, China
Linjing Liu               City University of Hong Kong, China
Weifeng Liu               China University of Petroleum (East China),
                          China
Wenqiang Liu              Hong Kong Polytechnic University, China
Xin Liu                   National Institute of Advanced Industrial Science
                          and Technology (AIST), Japan
Yang Liu                  Harbin Institute of Technology, China
Zhi-Yong Liu              Institute of Automation, Chinese Academy of
                          Sciences, China
Zongying Liu              Dalian Maritime University, China

| | |
|---|---|
| Jaime Lloret | Universitat Politècnica de València, Spain |
| Sye Loong Keoh | University of Glasgow, Singapore, Singapore |
| Hongtao Lu | Shanghai Jiao Tong University, China |
| Wenlian Lu | Fudan University, China |
| Xuequan Lu | Deakin University, Australia |
| Xiao Luo | UCLA, USA |
| Guozheng Ma | Shenzhen International Graduate School, Tsinghua University, China |
| Qianli Ma | South China University of Technology, China |
| Wanli Ma | University of Canberra, Australia |
| Muhammad Anwar Ma'sum | Universitas Indonesia, Indonesia |
| Michele Magno | University of Bologna, Italy |
| Sainik Kumar Mahata | JU, India |
| Shalni Mahato | Indian Institute of Information Technology (IIIT) Ranchi, India |
| Adnan Mahmood | Macquarie University, Australia |
| Mohammed Mahmoud | October University for Modern Sciences & Arts - MSA University, Egypt |
| Mufti Mahmud | University of Padova, Italy |
| Krishanu Maity | Indian Institute of Technology Patna, India |
| Mamta | IIT Patna, India |
| Aprinaldi Mantau | Kyushu Institute of Technology, Japan |
| Mohsen Marjani | Taylor's University, Malaysia |
| Sanparith Marukatat | NECTEC, Thailand |
| José María Luna | Universidad de Córdoba, Spain |
| Archana Mathur | Nitte Meenakshi Institute of Technology, India |
| Patrick McAllister | Ulster University, UK |
| Piotr Milczarski | Lodz University of Technology, Poland |
| Kshitij Mishra | IIT Patna, India |
| Pruthwik Mishra | IIIT-Hyderabad, India |
| Santosh Mishra | Indian Institute of Technology Patna, India |
| Sajib Mistry | Curtin University, Australia |
| Sayantan Mitra | Accenture Labs, India |
| Vinay Kumar Mittal | Neti International Research Center, India |
| Daisuke Miyamoto | University of Tokyo, Japan |
| Kazuteru Miyazaki | National Institution for Academic Degrees and Quality Enhancement of Higher Education, Japan |
| U. Mmodibbo | Modibbo Adama University Yola, Nigeria |
| Aditya Mogadala | Saarland University, Germany |
| Reem Mohamed | Mansoura University, Egypt |
| Muhammad Syafiq Mohd Pozi | Universiti Utara Malaysia, Malaysia |

| | |
|---|---|
| Anirban Mondal | University of Tokyo, Japan |
| Anupam Mondal | Jadavpur University, India |
| Supriyo Mondal | ZBW - Leibniz Information Centre for Economics, Germany |
| J. Manuel Moreno | Universitat Politècnica de Catalunya, Spain |
| Francisco J. Moreno-Barea | Universidad de Málaga, Spain |
| Sakchai Muangsrinoon | Walailak University, Thailand |
| Siti Anizah Muhamed | Politeknik Sultan Salahuddin Abdul Aziz Shah, Malaysia |
| Samrat Mukherjee | Indian Institute of Technology, Patna, India |
| Siddhartha Mukherjee | Samsung R&D Institute India, Bangalore, India |
| Dharmalingam Muthusamy | Bharathiar University, India |
| Abhijith Athreya Mysore Gopinath | Pennsylvania State University, USA |
| Harikrishnan N. B. | BITS Pilani K K Birla Goa Campus, India |
| Usman Naseem | University of Sydney, Australia |
| Deepak Nayak | Malaviya National Institute of Technology, Jaipur, India |
| Hamada Nayel | Benha University, Egypt |
| Usman Nazir | Lahore University of Management Sciences, Pakistan |
| Vasudevan Nedumpozhimana | TU Dublin, Ireland |
| Atul Negi | University of Hyderabad, India |
| Aneta Neumann | University of Adelaide, Australia |
| Hea Choon Ngo | Universiti Teknikal Malaysia Melaka, Malaysia |
| Dang Nguyen | University of Canberra, Australia |
| Duy Khuong Nguyen | FPT Software Ltd., FPT Group, Vietnam |
| Hoang D. Nguyen | University College Cork, Ireland |
| Hong Huy Nguyen | National Institute of Informatics, Japan |
| Tam Nguyen | Leibniz University Hannover, Germany |
| Thanh-Son Nguyen | Agency for Science, Technology and Research (A*STAR), Singapore |
| Vu-Linh Nguyen | Eindhoven University of Technology, Netherlands |
| Nick Nikzad | Griffith University, Australia |
| Boda Ning | Swinburne University of Technology, Australia |
| Haruhiko Nishimura | University of Hyogo, Japan |
| Kishorjit Nongmeikapam | Indian Institute of Information Technology (IIIT) Manipur, India |
| Aleksandra Nowak | Jagiellonian University, Poland |
| Stavros Ntalampiras | University of Milan, Italy |
| Anupiya Nugaliyadde | Sri Lanka Institute of Information Technology, Sri Lanka |

| | |
|---|---|
| Anto Satriyo Nugroho | Agency for Assessment & Application of Technology, Indonesia |
| Aparajita Ojha | PDPM IIITDM Jabalpur, India |
| Akeem Olowolayemo | International Islamic University Malaysia, Malaysia |
| Toshiaki Omori | Kobe University, Japan |
| Shih Yin Ooi | Multimedia University, Malaysia |
| Sidali Ouadfeul | Algerian Petroleum Institute, Algeria |
| Samir Ouchani | CESI Lineact, France |
| Srinivas P. Y. K. L. | IIIT Sri City, India |
| Neelamadhab Padhy | GIET University, India |
| Worapat Paireekreng | Dhurakij Pundit University, Thailand |
| Partha Pakray | National Institute of Technology Silchar, India |
| Santanu Pal | Wipro Limited, India |
| Bin Pan | Nankai University, China |
| Rrubaa Panchendrarajan | Sri Lanka Institute of Information Technology, Sri Lanka |
| Pankaj Pandey | Indian Institute of Technology, Gandhinagar, India |
| Lie Meng Pang | Southern University of Science and Technology, China |
| Sweta Panigrahi | National Institute of Technology Warangal, India |
| T. Pant | IIIT Allahabad, India |
| Shantipriya Parida | Idiap Research Institute, Switzerland |
| Hyeyoung Park | Kyungpook National University, South Korea |
| Md Aslam Parwez | Jamia Millia Islamia, India |
| Leandro Pasa | Federal University of Technology - Parana (UTFPR), Brazil |
| Kitsuchart Pasupa | King Mongkut's Institute of Technology Ladkrabang, Thailand |
| Debanjan Pathak | Kalinga Institute of Industrial Technology (KIIT), India |
| Vyom Pathak | University of Florida, USA |
| Sangameshwar Patil | TCS Research, India |
| Bidyut Kr. Patra | IIT (BHU) Varanasi, India |
| Dipanjyoti Paul | Indian Institute of Technology Patna, India |
| Sayanta Paul | Ola, India |
| Sachin Pawar | Tata Consultancy Services Ltd., India |
| Pornntiwa Pawara | Mahasarakham University, Thailand |
| Yong Peng | Hangzhou Dianzi University, China |
| Yusuf Perwej | Ambalika Institute of Management and Technology (AIMT), India |
| Olutomilayo Olayemi Petinrin | City University of Hong Kong, China |
| Arpan Phukan | Indian Institute of Technology Patna, India |

| | |
|---|---|
| Chiara Picardi | University of York, UK |
| Francesco Piccialli | University of Naples Federico II, Italy |
| Josephine Plested | University of New South Wales, Australia |
| Krishna Reddy Polepalli | IIIT Hyderabad, India |
| Dan Popescu | University Politehnica of Bucharest, Romania |
| Heru Praptono | Bank Indonesia/UI, Indonesia |
| Mukesh Prasad | University of Technology Sydney, Australia |
| Yamuna Prasad | Thompson Rivers University, Canada |
| Krishna Prasadmiyapuram | IIT Gandhinagar, India |
| Partha Pratim Sarangi | KIIT Deemed to be University, India |
| Emanuele Principi | Università Politecnica delle Marche, Italy |
| Dimeter Prodonov | Imec, Belgium |
| Ratchakoon Pruengkarn | College of Innovative Technology and Engineering, Dhurakij Pundit University, Thailand |
| Michal Ptaszynski | Kitami Institute of Technology, Japan |
| Narinder Singh Punn | Mayo Clinic, Arizona, USA |
| Abhinanda Ranjit Punnakkal | UiT The Arctic University of Norway, Norway |
| Zico Pratama Putra | Queen Mary University of London, UK |
| Zhenyue Qin | Tencent, China |
| Nawab Muhammad Faseeh Qureshi | SU, South Korea |
| Md Rafiqul | UTS, Australia |
| Saifur Rahaman | City University of Hong Kong, China |
| Shri Rai | Murdoch University, Australia |
| Vartika Rai | IIIT Hyderabad, India |
| Kiran Raja | Norwegian University of Science and Technology, Norway |
| Sutharshan Rajasegarar | Deakin University, Australia |
| Arief Ramadhan | Bina Nusantara University, Indonesia |
| Mallipeddi Rammohan | Kyungpook National University, South Korea |
| Md. Mashud Rana | Commonwealth Scientific and Industrial Research Organisation (CSIRO), Australia |
| Surangika Ranathunga | University of Moratuwa, Sri Lanka |
| Soumya Ranjan Mishra | KIIT University, India |
| Hemant Rathore | Birla Institute of Technology & Science, Pilani, India |
| Imran Razzak | UNSW, Australia |
| Yazhou Ren | University of Science and Technology of China, China |
| Motahar Reza | GITAM University Hyderabad, India |
| Dwiza Riana | STMIK Nusa Mandiri, Indonesia |
| Bharat Richhariya | BITS Pilani, India |

| | |
|---|---|
| Pattabhi R. K. Rao | AU-KBC Research Centre, India |
| Heejun Roh | Korea University, South Korea |
| Vijay Rowtula | IIIT Hyderabad, India |
| Aniruddha Roy | IIT Kharagpur, India |
| Sudipta Roy | Jio Institute, India |
| Narendra S. Chaudhari | Indian Institute of Technology Indore, India |
| Fariza Sabrina | Central Queensland University, Australia |
| Debanjan Sadhya | ABV-IIITM Gwalior, India |
| Sumit Sah | IIT Dharwad, India |
| Atanu Saha | Jadavpur University, India |
| Sajib Saha | Commonwealth Scientific and Industrial Research Organisation, Australia |
| Snehanshu Saha | BITS Pilani K K Birla Goa Campus, India |
| Tulika Saha | IIT Patna, India |
| Navanath Saharia | Indian Institute of Information Technology Manipur, India |
| Pracheta Sahoo | University of Texas at Dallas, USA |
| Sovan Kumar Sahoo | Indian Institute of Technology Patna, India |
| Tanik Saikh | L3S Research Center, Germany |
| Naveen Saini | Indian Institute of Information Technology Lucknow, India |
| Fumiaki Saitoh | Chiba Institute of Technology, Japan |
| Rohit Salgotra | Swansea University, UK |
| Michel Salomon | Univ. Bourgogne Franche-Comté, France |
| Yu Sang | Research Institute of Institute of Computing Technology, Exploration and Development, Liaohe Oilfield, PetroChina, China |
| Suyash Sangwan | Indian Institute of Technology Patna, India |
| Soubhagya Sankar Barpanda | VIT-AP University, India |
| Jose A. Santos | Ulster University, UK |
| Kamal Sarkar | Jadavpur University, India |
| Sandip Sarkar | Jadavpur University, India |
| Naoyuki Sato | Future University Hakodate, Japan |
| Eri Sato-Shimokawara | Tokyo Metropolitan University, Japan |
| Sunil Saumya | Indian Institute of Information Technology Dharwad, India |
| Gerald Schaefer | Loughborough University, UK |
| Rafal Scherer | Czestochowa University of Technology, Poland |
| Arvind Selwal | Central University of Jammu, India |
| Noor Akhmad Setiawan | Universitas Gadjah Mada, Indonesia |
| Mohammad Shahid | Aligarh Muslim University, India |
| Jie Shao | University of Science and Technology of China, China |

| | |
|---|---|
| Nabin Sharma | University of Technology Sydney, Australia |
| Raksha Sharma | IIT Bombay, India |
| Sourabh Sharma | Avantika University, India |
| Suraj Sharma | International Institute of Information Technology Bhubaneswar, India |
| Ravi Shekhar | Queen Mary University of London, UK |
| Michael Sheng | Macquarie University, Australia |
| Yin Sheng | Huazhong University of Science and Technology, China |
| Yongpan Sheng | Southwest University, China |
| Liu Shenglan | Dalian University of Technology, China |
| Tomohiro Shibata | Kyushu Institute of Technology, Japan |
| Iksoo Shin | University of Science & Technology, China |
| Mohd Fairuz Shiratuddin | Murdoch University, Australia |
| Hayaru Shouno | University of Electro-Communications, Japan |
| Sanyam Shukla | MANIT, Bhopal, India |
| Udom Silparcha | KMUTT, Thailand |
| Apoorva Singh | Indian Institute of Technology Patna, India |
| Divya Singh | Central University of Bihar, India |
| Gitanjali Singh | Indian Institute of Technology Patna, India |
| Gopendra Singh | Indian Institute of Technology Patna, India |
| K. P. Singh | IIIT Allahabad, India |
| Navjot Singh | IIIT Allahabad, India |
| Om Singh | NIT Patna, India |
| Pardeep Singh | Jawaharlal Nehru University, India |
| Rajiv Singh | Banasthali Vidyapith, India |
| Sandhya Singh | Indian Institute of Technology Bombay, India |
| Smriti Singh | IIT Bombay, India |
| Narinder Singhpunn | Mayo Clinic, Arizona, USA |
| Saaveethya Sivakumar | Curtin University, Malaysia |
| Ferdous Sohel | Murdoch University, Australia |
| Chattrakul Sombattheera | Mahasarakham University, Thailand |
| Lei Song | Unitec Institute of Technology, New Zealand |
| Linqi Song | City University of Hong Kong, China |
| Yuhua Song | University of Science and Technology Beijing, China |
| Gautam Srivastava | Brandon University, Canada |
| Rajeev Srivastava | Banaras Hindu University (IT-BHU), Varanasi, India |
| Jérémie Sublime | ISEP - Institut Supérieur d'Électronique de Paris, France |
| P. N. Suganthan | Nanyang Technological University, Singapore |

| | |
|---|---|
| Derwin Suhartono | Bina Nusantara University, Indonesia |
| Indra Adji Sulistijono | Politeknik Elektronika Negeri Surabaya (PENS), Indonesia |
| John Sum | National Chung Hsing University, Taiwan |
| Fuchun Sun | Tsinghua University, China |
| Ning Sun | Nankai University, China |
| Anindya Sundar Das | Indian Institute of Technology Patna, India |
| Bapi Raju Surampudi | International Institute of Information Technology Hyderabad, India |
| Olarik Surinta | Mahasarakham University, Thailand |
| Maria Susan Anggreainy | Bina Nusantara University, Indonesia |
| M. Syafrullah | Universitas Budi Luhur, Indonesia |
| Murtaza Taj | Lahore University of Management Sciences, Pakistan |
| Norikazu Takahashi | Okayama University, Japan |
| Abdelmalik Taleb-Ahmed | Polytechnic University of Hauts-de-France, France |
| Hakaru Tamukoh | Kyushu Institute of Technology, Japan |
| Choo Jun Tan | Wawasan Open University, Malaysia |
| Chuanqi Tan | BIT, China |
| Shing Chiang Tan | Multimedia University, Malaysia |
| Xiao Jian Tan | Tunku Abdul Rahman University of Management and Technology (TAR UMT), Malaysia |
| Xin Tan | East China Normal University, China |
| Ying Tan | Peking University, China |
| Gouhei Tanaka | University of Tokyo, Japan |
| Yang Tang | East China University of Science and Technology, China |
| Zhiri Tang | City University of Hong Kong, China |
| Tanveer Tarray | Islamic University of Science and Technology, India |
| Chee Siong Teh | Universiti Malaysia Sarawak (UNIMAS), Malaysia |
| Ya-Wen Teng | Academia Sinica, Taiwan |
| Gaurish Thakkar | University of Zagreb, Croatia |
| Medari Tham | St. Anthony's College, India |
| Selvarajah Thuseethan | Sabaragamuwa University of Sri Lanka, Sri Lanka |
| Shu Tian | University of Science and Technology Beijing, China |
| Massimo Tistarelli | University of Sassari, Italy |
| Abhisek Tiwari | IIT Patna, India |
| Uma Shanker Tiwary | Indian Institute of Information Technology, Allahabad, India |

| | |
|---|---|
| Alex To | University of Sydney, Australia |
| Stefania Tomasiello | University of Tartu, Estonia |
| Anh Duong Trinh | Technological University Dublin, Ireland |
| Enkhtur Tsogbaatar | Mongolian University of Science and Technology, Mongolia |
| Enmei Tu | Shanghai Jiao Tong University, China |
| Eiji Uchino | Yamaguchi University, Japan |
| Prajna Upadhyay | IIT Delhi, India |
| Sahand Vahidnia | University of New South Wales, Australia |
| Ashwini Vaidya | IIT Delhi, India |
| Deeksha Varshney | Indian Institute of Technology, Patna, India |
| Sowmini Devi Veeramachaneni | Mahindra University, India |
| Samudra Vijaya | Koneru Lakshmaiah Education Foundation, India |
| Surbhi Vijh | JSS Academy of Technical Education, Noida, India |
| Nhi N. Y. Vo | University of Technology Sydney, Australia |
| Xuan-Son Vu | Umeå University, Sweden |
| Anil Kumar Vuppala | IIIT Hyderabad, India |
| Nobuhiko Wagatsuma | Toho University, Japan |
| Feng Wan | University of Macau, China |
| Bingshu Wang | Northwestern Polytechnical University Taicang Campus, China |
| Dianhui Wang | La Trobe University, Australia |
| Ding Wang | Beijing University of Technology, China |
| Guanjin Wang | Murdoch University, Australia |
| Jiasen Wang | City University of Hong Kong, China |
| Lei Wang | Beihang University, China |
| Libo Wang | Xiamen University of Technology, China |
| Meng Wang | Southeast University, China |
| Qiu-Feng Wang | Xi'an Jiaotong-Liverpool University, China |
| Sheng Wang | Henan University, China |
| Weiqun Wang | Institute of Automation, Chinese Academy of Sciences, China |
| Wentao Wang | Michigan State University, USA |
| Yongyu Wang | Michigan Technological University, USA |
| Zhijin Wang | Jimei University, China |
| Bunthit Watanapa | KMUTT-SIT, Thailand |
| Yanling Wei | TU Berlin, Germany |
| Guanghui Wen | RMIT University, Australia |
| Ari Wibisono | Universitas Indonesia, Indonesia |
| Adi Wibowo | Diponegoro University, Indonesia |
| Ka-Chun Wong | City University of Hong Kong, China |

| | |
|---|---|
| Kevin Wong | Murdoch University, Australia |
| Raymond Wong | Universiti Malaya, Malaysia |
| Kuntpong Woraratpanya | King Mongkut's Institute of Technology Ladkrabang (KMITL), Thailand |
| Marcin Woźniak | Silesian University of Technology, Poland |
| Chengwei Wu | Harbin Institute of Technology, China |
| Jing Wu | Shanghai Jiao Tong University, China |
| Weibin Wu | Sun Yat-sen University, China |
| Hongbing Xia | Beijing Normal University, China |
| Tao Xiang | Chongqing University, China |
| Qiang Xiao | Huazhong University of Science and Technology, China |
| Guandong Xu | University of Technology Sydney, Australia |
| Qing Xu | Tianjin University, China |
| Yifan Xu | Huazhong University of Science and Technology, China |
| Junyu Xuan | University of Technology Sydney, Australia |
| Hui Xue | Southeast University, China |
| Saumitra Yadav | IIIT-Hyderabad, India |
| Shekhar Yadav | Madan Mohan Malaviya University of Technology, India |
| Sweta Yadav | University of Illinois at Chicago, USA |
| Tarun Yadav | Defence Research and Development Organisation, India |
| Shankai Yan | Hainan University, China |
| Feidiao Yang | Microsoft, China |
| Gang Yang | Renmin University of China, China |
| Haiqin Yang | International Digital Economy Academy, China |
| Jianyi Yang | Shandong University, China |
| Jinfu Yang | BJUT, China |
| Minghao Yang | Institute of Automation, Chinese Academy of Sciences, China |
| Shaofu Yang | Southeast University, China |
| Wachira Yangyuen | Rajamangala University of Technology Srivijaya, Thailand |
| Xinye Yi | Guilin University of Electronic Technology, China |
| Hang Yu | Shanghai University, China |
| Wen Yu | Cinvestav, Mexico |
| Wenxin Yu | Southwest University of Science and Technology, China |
| Zhaoyuan Yu | Nanjing Normal University, China |
| Ye Yuan | Xi'an Jiaotong University, China |
| Xiaodong Yue | Shanghai University, China |

Aizan Zafar              Indian Institute of Technology Patna, India
Jichuan Zeng             Bytedance, China
Jie Zhang                Newcastle University, UK
Shixiong Zhang           Xidian University, China
Tianlin Zhang            University of Manchester, UK
Mingbo Zhao              Donghua University, China
Shenglin Zhao            Zhejiang University, China
Guoqiang Zhong           Ocean University of China, China
Jinghui Zhong            South China University of Technology, China
Bo Zhou                  Southwest University, China
Yucheng Zhou             University of Technology Sydney, Australia
Dengya Zhu               Curtin University, Australia
Xuanying Zhu             ANU, Australia
Hua Zuo                  University of Technology Sydney, Australia

## Additional Reviewers

Acharya, Rajul                      Doborjeh, Maryam
Afrin, Mahbuba                      Dong, Zhuben
Alsuhaibani, Abdullah               Dutta, Subhabrata
Amarnath                            Dybala, Pawel
Appicharla, Ramakrishna             El Achkar, Charbel
Arora, Ridhi                        Feng, Zhengyang
Azar, Joseph                        Galkowski, Tomasz
Bai, Weiwei                         Garg, Arpit
Bao, Xiwen                          Ghobakhlou, Akbar
Barawi, Mohamad Hardyman            Ghosh, Soumitra
Bhat, Mohammad Idrees Bhat          Guo, Hui
Cai, Taotao                         Gupta, Ankur
Cao, Feiqi                          Gupta, Deepak
Chakraborty, Bodhi                  Gupta, Megha
Chang, Yu-Cheng                     Han, Yanyang
Chen                                Han, Yiyan
Chen, Jianpeng                      Hang, Bin
Chen, Yong                          Harshit
Chhipa, Priyank                     He, Silu
Cho, Joshua                         Hua, Ning
Chongyang, Chen                     Huang, Meng
Cuenat, Stéphane                    Huang, Rongting
Dang, Lili                          Huang, Xiuyu
Das Chakladar, Debashis             Hussain, Zawar
Das, Kishalay                       Imran, Javed
Dey, Monalisa                       Islam, Md Rafiqul

Jain, Samir
Jia, Mei
Jiang, Jincen
Jiang, Xiao
Jiangyu, Wang
Jiaxin, Lou
Jiaxu, Hou
Jinzhou, Bao
Ju, Wei
Kasyap, Harsh
Katai, Zoltan
Keserwani, Prateek
Khan, Asif
Khan, Muhammad Fawad Akbar
Khari, Manju
Kheiri, Kiana
Kirk, Nathan
Kiyani, Arslan
Kolya, Anup Kumar
Krdzavac, Nenad
Kumar, Lov
Kumar, Mukesh
Kumar, Puneet
Kumar, Rahul
Kumar, Sunil
Lan, Meng
Lavangnananda, Kittichai
Li, Qian
Li, Xiaoou
Li, Xin
Li, Xinjia
Liang, Mengnan
Liang, Shuai
Liquan, Li
Liu, Boyang
Liu, Chang
Liu, Feng
Liu, Linjing
Liu, Xinglan
Liu, Xinling
Liu, Zhe
Lotey, Taveena
Ma, Bing
Ma, Zeyu
Madanian, Samaneh

Mahata, Sainik Kumar
Mahmud, Md. Redowan
Man, Jingtao
Meena, Kunj Bihari
Mishra, Pragnyaban
Mistry, Sajib
Modibbo, Umar Muhammad
Na, Na
Nag Choudhury, Somenath
Nampalle, Kishore
Nandi, Palash
Neupane, Dhiraj
Nigam, Nitika
Nigam, Swati
Ning, Jianbo
Oumer, Jehad
Pandey, Abhineet Kumar
Pandey, Sandeep
Paramita, Adi Suryaputra
Paul, Apurba
Petinrin, Olutomilayo Olayemi
Phan Trong, Dat
Pradana, Muhamad Hilmil Muchtar Aditya
Pundhir, Anshul
Rahman, Sheikh Shah Mohammad Motiur
Rai, Sawan
Rajesh, Bulla
Rajput, Amitesh Singh
Rao, Raghunandan K. R.
Rathore, Santosh Singh
Ray, Payel
Roy, Satyaki
Saini, Nikhil
Saki, Mahdi
Salimath, Nagesh
Sang, Haiwei
Shao, Jian
Sharma, Anshul
Sharma, Shivam
Shi, Jichen
Shi, Jun
Shi, Kaize
Shi, Li
Singh, Nagendra Pratap
Singh, Pritpal

Singh, Rituraj
Singh, Shrey
Singh, Tribhuvan
Song, Meilun
Song, Yuhua
Soni, Bharat
Stommel, Martin
Su, Yanchi
Sun, Xiaoxuan
Suryodiningrat, Satrio Pradono
Swarnkar, Mayank
Tammewar, Aniruddha
Tan, Xiaosu
Tanoni, Giulia
Tanwar, Vishesh
Tao, Yuwen
To, Alex
Tran, Khuong
Varshney, Ayush
Vo, Anh-Khoa
Vuppala, Anil
Wang, Hui
Wang, Kai
Wang, Rui
Wang, Xia
Wang, Yansong

Wang, Yuan
Wang, Yunhe
Watanapa, Saowaluk
Wenqian, Fan
Xia, Hongbing
Xie, Weidun
Xiong, Wenxin
Xu, Zhehao
Xu, Zhikun
Yan, Bosheng
Yang, Haoran
Yang, Jie
Yang, Xin
Yansui, Song
Yu, Cunzhe
Yu, Zhuohan
Zandavi, Seid Miad
Zeng, Longbin
Zhang, Jane
Zhang, Ruolan
Zhang, Ziqi
Zhao, Chen
Zhou, Xinxin
Zhou, Zihang
Zhu, Liao
Zhu, Linghui

# Contents – Part II

## Cognitive Neurosciences

# Cognitive Neurosciences

# Differences in Brain Activation During Physics Problem Solving Across Students with Various Learning Progression: Electrophysiological Evidence Based on Detrended Fluctuation Analysis

Qian Wang[1], Hongan Wang[1], Huihua Deng[1], and Yanmei Zhu[1,2]([✉])

[1] Key Laboratory of Child Development and Learning Science
(Ministry of Education), School of Biological Science and Medical Engineering,
Southeast University, Nanjing 210096, Jiangsu, China
{230179688,230219201,dengrcls,zhuyanmei}@seu.edu.cn
[2] School of Early Childhood Education, Nanjing Xiaozhuang University,
Nanjing 211171, Jiangsu, China

**Abstract.** The Detrended Fluctuation Analysis is a widely used method for analysis of non-stationary time series which has been applied to EEG signals. However, few studies have applied this method to the assessment of cognitive abilities in healthy groups, especially in the context of science education. In this work, for the first time, the DFA method was applied to analyze the EEG time series during physics problem solving. We studied the DFA exponents on brain activation when individuals with different learning progression were solving the physics problems, as well as the relationship between DFA exponents and their performance. Statistical analysis reveals that, excellent groups with the best learning progression demonstrated the higher DFA exponents when compared the other two groups. Since DFA provides correlations between time series in EEG, the correlations are believed to be associated with model dynamical systems which reflect sustained cognitive operations. The results reflected that students in this group have developed the dynamic model systems of physics concepts. They can extract relevant knowledge more accurately and efficiently to build scientific models during problem-solving. The application of DFA method in physics education context may deepen our understanding of the neural basis of problem-solving ability and provide a promising indicator of learning achievement.

**Keywords:** Physics problem solving · Electrophysiological ·
Detrended Fuctuation Analysis

## 1 Introduction

Problem-solving ability is widely regarded as a core skill and key competency in science education. Problem solving is a higher order cognition process involving

M. Tanveer et al. (Eds.): ICONIP 2022, LNCS 13624, pp. 3–12, 2023.
https://doi.org/10.1007/978-3-031-30108-7_1

attention, reasoning, working memory, visual processing, semantic memory and multisensory integration [1]. Despite the impressive amount of research devoted to problem solving research in the field of science education, it is surprising that little is known about the neural processes and brain activation during science problem solving. In this case, research on neural features in science problem solving may have values in revealing neural correlates of knowledge representation and scientific reasoning.

Previous studies on problem solving mainly focused on event-related changes of EEG power, dynamic changes in EEG rhythm and neural representation by function MRI [2–6]. Considering the sustained cognitive operations of the problem-solving process, we presented the temporal correlations in neuronal oscillations. Previous studies on the temporal dependency of neuronal activities have consistently showed that the fluctuations of neuronal signals at many levels of nervous system are controlled by temporal correlations. These results suggest that temporal correlation may represent a compromising indicator of the competing demands of stability and information transmission in neuronal networks [7]. The Detrended fluctuation analysis method (DFA) is a widely used method for analysis of non-stationary time series which has been applied to EEG signals [8,9]. The DFA methods allows quantifying the presence of long- and short-term correlations in time series. Until now, the DFA method has been mostly used to distinguish between healthy and diseased human systems. However, few studies have applied this method to the assessment of cognitive abilities in healthy groups, especially in the context of science education. In this work, for the first time, the DFA method was applied to analyze the EEG time series during physics problem solving. We studied the temporal correlations on brain activation when individuals with different learning progression were solving the physics problems, as well as the relationship between temporal correlations and their problem-solving performance.

## 2 Methods

### 2.1 Participants

Fifty-five graduate students in Southeast University were recruited in the study. According to their academic performance in physics, these students were divided into three groups with various learning progression: excellent group (19 participants; mean age = 20.63; SD = 1.50); moderate group (18 participants; mean age = 21.28; SD = 2.32) and poor group (18 participants; mean age = 22.06; SD = 2.12). All study procedures and research methods were carried out in accordance with the Declaration of Helsinki (1964) by the World Medical Association concerning human experimentation and were approved by the Research Ethics Committee of Affiliated Zhongda Hospital, Southeast University, China.

### 2.2 Stimuli Materials and Procedure

Stimulus in this study were physics problems about the motion of a ball passing through a curved pipe. There were 160 non-repetitive physics problems for the participants to solve. Each trial of a task was shown by two-part stimulus presentation: problem stimulus and answer stimulus including four possible outcomes. As illustrated in Fig. 1, each trial started with the presentation of a central fixation cross on the screen for a random of 1000–1500 ms. Then the problem stimulus of a ball shooting into a curved pipe with certain velocity was presented for 2000 ms. After that, the answer stimulus appeared and remained on the screen for 5000 ms, the participants were required to choose the correct answer from four options by pressing the responding reaction button. All trials were presented in a random order for each participant and the task lasted approximately 25 min. Participants were instructed to respond to answer stimulus as correctly and quickly as possible.

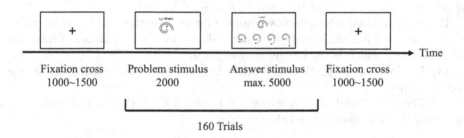

**Fig. 1.** Schematic representation of an experimental trial.

## 2.3 EEG Data Acquisition and Data Preprocess

The EEG activity was recorded from 64 tin electrodes mounted in an elastic cap (NeuroScan Inc., Herndon, Virginia, USA) according to the international 10–20 system. All electrode impedances were maintained below 5 kΩ. The EEG signals were amplified and digitized using a band pass of 0.01–100 Hz and a sampling rate 500 Hz.

After acquisition, EEG data were preprocessed under the EEGLAB and MATLAB platform. EEG signals were re-referenced to the bilateral mastoids, filtered with 30 Hz low-pass filter and a 0.1 Hz high-pass filter. trials were extracted with the epoch of 3500 ms (1000 ms pre-stimulus and 2500 ms post-stimulus intervals with baseline correction. Artifact rejection via independent component analysis (ICA) was performed subsequently for each subject.

## 2.4 Detrended Fluctuation Analysis (DFA) and Scaling Exponents

The complex nature of the electrical brain activity results in a high degree of spatial and temporal fluctuations in the EEG [10]. To understand the EEG activity in a better way, it is important to characterize its fluctuations over different time scales. Detrended fluctuation analysis (DFA) can provide a scaling exponent with a single metric. It is appropriate for nonlinear and non-stationary physiological data such as EEG [9]. In our study, we used DFA to estimate short-range temporal correlations (SRTCs) and long-range temporal correlations (LRTCs) of EEG signals during physics problem solving.

The calculation process of DFA was divided into the following steps:

(1) Time series $x(t)$ with length $N$, $t = 1, 2, \cdots, N$, calculating the cumulative deviation and converted to the new sequence:

$$y(t) = \sum_{i=1}^{t}(x(i) - \bar{x}) \tag{1}$$

where $\bar{x}$ represents mean of time series: $\bar{x} = \frac{1}{N}\sum_{t=1}^{N} x(t)$.

(2) $y(t)$ is divided into m non-overlapping intervals with equal length $n$, where $N$ is the interval length, namely the time scale, and m is the number of intervals (or Windows), which is the integer part of $N/n$.

(3) The local trend $y_n(t)$ was obtained by the least square normal fitting for each sequence.

(4) The local trend of each interval is removed for $y(t)$, and the root mean square of the new sequence is calculated:

$$F(n) = \sqrt{\frac{1}{N}\sum_{t=1}^{N}[y(t) - y_n(t)]^2} \tag{2}$$

(5) Change the size of window length $n$ and repeat steps (2), (3) and (4) to obtain the relationship between different window length $n$ and its corresponding average fluctuation $F(n)$. If there is power law temporal correlation in time series, there is a linear relationship in the logarithmic graph of $F(n)$ and $N$, which is the temporal correlation phenomenon:

$$F(n) \propto n^{\alpha} \tag{3}$$

where $\alpha$ refers to scaling exponent of DFA. In this study, we set a short-range correlation time window length with 0.02 to 0.2 s and a long-range correlation time window length with 0.2 to 1 s. The value of $\alpha$ represents the correlation of time series. Also, $\alpha_1$ refers to short-range correlation and $\alpha_2$ refers to long-range correlation.

## 2.5   Data Analysis

For behavioral data, we considered two behavior measures: (1) Accuracy (ACC), which is the percentage of correct responses phase for physics problems; and (2) Reaction time (RT), which is the time delay from problem onset to key-press for a correct response.

For scaling exponents of EEG data, we both selected electrodes FP1, FPZ, FP2, F3, F1, FZ, F2, F4, FC1, FCZ, FC2, C3, C1, CZ, C2, C4, CP1, CPZ, CP2,

P3, P1, PZ, P2, P4, PO3, POZ, PO4, O1, OZ, O2. Two-way repeated measures analysis of variance (ANOVA) was performed on the $\alpha_1$ and $\alpha_2$, with group (excellent, moderate, poor) as the between-participants factor and with within-participant variable of channel. Furthermore, we set electrode site 1, 3; Z and 2, 4 to represent the left, middle and right hemisphere. We performed a two-way repeated measures analysis of variance with group as the between-participants factor and brain region as the within-participant factor. Greenhouse-Geisser correction was applied to correct for violations of the sphericity assumption when appropriate. We used partial eta squared ($\eta^2$) as the effect size estimate. All post-hoc tests were Bonferroni-corrected. Finally, we performed exploratory correlation analyses, computing Pearson's R between the ACC and scaling exponents $\alpha_2$ of 30 channels.

## 3    Results and Discussions

### 3.1    Behavior Results

The ACCs of responses phase to problems for three group (excellent, moderate, poor) were 96.8% (SD = 0.03), 44.9% (SD = 0.11) and 19.3% (SD = 0.09), respectively. For the RT, excellent group, moderate group and poor group were 1483.98 ms (SD = 412.23), 2520.64 ms (SD = 620.52) and 2507.68 ms (SD = 547.09), respectively. One-way ANOVA revealed that the ACCs of responses was significantly affected by participant type, $F(2, 52) = 405.31$, $p < .00$. Post hoc multiple comparisons showed that there were significant pairwise differences between the three groups ($ps < .00$, adjusted). For the RT, the ANOVA revealed a significant difference of three groups, $F(2, 52) = 23.36, p < .00$. Post hoc results showed that there were significant differences between the excellent group and the moderate group, the excellent group and the poor group ($ps < .00$, adjusted). Our behavior findings indicated that there was a significant difference among three groups of participant in terms of ACC and RT during physics problem solving, and the higher the ACC of participants, the shorter the RT.

### 3.2    Short-Range Temporal Correlations (SRTCs: $\alpha_1$)

The group × channel ANOVA using $\alpha_1$ revealed a significant effect of channel, $F(3, 156) = 7.62$, $p = .0001$, $\eta^2 = .0023$, but the group effect failed to reach statistical significant, $F(2, 52) = .23$, $p = .79$, $\eta^2 = .007$. Specifically, the mean, standard deviation and coefficient of variation of $\alpha_1$ were 1.31±.026 (1.98%) for excellent group, 1.32±.035 (2.65%) for moderate group, and 1.32±.023 (1.74%) for poor group. There was no significant interaction effect, $F(6, 156) = .70$, $p = .64$, $\eta^2 = .004$.

The group × brain region ANOVA showed a significant effect of brain region, $F(2, 86) = 13.04, p < .0001, \eta^2 = .003$. There was no group effect, $F(2, 52) = .24$, $p = .79$, $\eta^2 = .008$, and no significant interaction effect, $F(3, 86) = 1.75, p = .16$, $\eta^2 = .0007$. Post hoc paired t-test results showed that in the group of poor and moderate, the $\alpha_1$ of the right hemisphere and midline was significantly higher than left hemisphere ($t\ value: -3.16 \sim -2.37$, $ps < 0.05$, $Cohen's\ d: -.243 \sim -.037$, adjusted), but in the group of excellent, the $\alpha_1$ of the right hemisphere was significantly lower than midline ($t = 2.31$, $p = .044$, $Cohen's\ d = .03$, adjusted), as shown in Fig. 2. For the $\alpha_1$, we observed no significant difference in group. While, the three groups of participants differed on scaling exponents in different brain regions, and the right hemisphere and midline were higher than the left hemisphere.

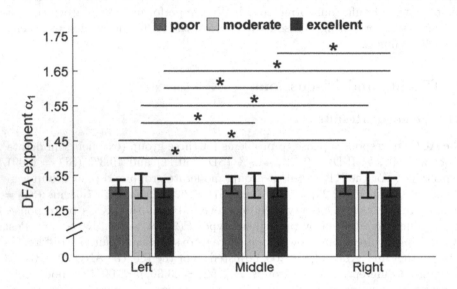

**Fig. 2.** Differences in short-range correlations ($\alpha_1$) among three groups of participants in different brain regions. $* : p < 0.05$.

## 3.3   Long-Range Temporal Correlations (LRTCs: $\alpha_2$)

The group $\times$ channel ANOVA using $\alpha_2$ revealed a significant effect of group, $F(2, 52) = 3.82$, $p = .028$, $\eta^2 = .010$. Specifically, the mean, standard deviation and coefficient of variation of $\alpha_2$ were $1.093\pm.040$ ($3.66\%$) for excellent group, $1.060\pm.054$ ($5.09\%$) for moderate group, and $1.053\pm.061$ ($5.79\%$) for poor group. The effect of channel was not significant, $F(3, 177) = .85$, $p = .47$, $\eta^2 = .002$, and there was no significant interaction effect, $F(6, 177) = .72$, $p = .64$, $\eta^2 = .004$. Post hoc unpaired multiple comparisons showed that the $\alpha_2$ of excellent group was higher than that of the poor and moderate groups in most channels in the midline and right hemisphere, and some of channels in the left hemisphere

($ps$ < 0.05, adjusted), as shown in Fig. 3. Besides, there were more channels with significant differences in the poor group compared to the moderate group, including three channels in the frontal lobe (FPZ, FP2, F2) and the O1 in the occipital lobe.

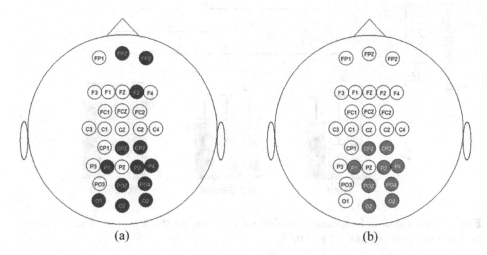

(a)                                        (b)

**Fig. 3.** Differences in long-range correlations ($\alpha_2$) on different channels. Color marked channels indicated a significant difference ($p$ < 0.05). (a) Differences in $\alpha_2$ exponent between excellent and poor was reflected in the channels marked blue color; (b) Differences in $\alpha_2$ exponent between excellent and moderate was showed in the channels with red color. (Color figure online)

The group × brain region ANOVA revealed a significant effect of group, $F(2, 52) = 3.85$, $p = .027$, $\eta^2 = .012$. There was no significant effect of brain region, $F(2, 77) = .87$, $p = .38$, $\eta^2 = .0002$, and no significant interaction effect, $F(3, 77) = .77$, $p = .50$, $\eta^2 = .0003$. Post hoc unpaired multiple comparisons showed that the $\alpha_2$ of excellent group was higher in both midline and right hemisphere than moderate and poor groups ($ps$ < 0.05, adjusted), while the difference between the poor and moderate groups was not significant ($ps$ > 0.05, adjusted), as shown in Fig. 4. Similar to the results of group × channel, the differences in LRTC exponents were mainly in the regions of right hemisphere and midline. Previous researches has demonstrated that attention and inhibitory functions during problem solving were showed on EEG oscillations in the regions of occipital-parietal and central region [11]. This interpretation was consistent with our results.

## 3.4   Correlation Between DFA Exponents and Accuracy

The Pearson correlation test showed that there was a significant positive correlation between the participants' accuracy (ACC) and the long-range correlation

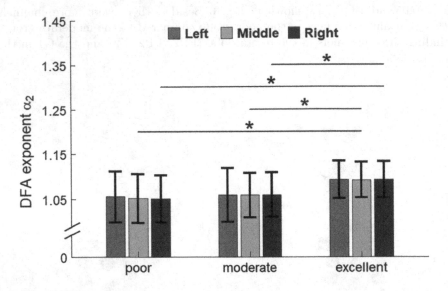

**Fig. 4.** Differences in long-range correlations ($\alpha_2$) among three groups of participants in different brain regions. $*: p < 0.05$.

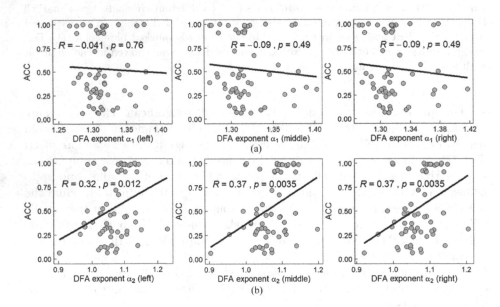

**Fig. 5.** Correlations between DFA exponents and accuracy in different brain regions. The results in order from left to right was: left hemisphere, midline and right hemisphere. (a) Correlation between the short-range correlation exponent ($\alpha_1$) and accuracy; (b) Correlation between the short-range correlation exponent ($\alpha_2$) and accuracy.

$(\alpha_2)$ of 30 channels, as well as the long-range correlation $(\alpha_2)$ of three brain regions. While there was no significant correlation with the short-range correlation $(\alpha_1)$ of 30 channels and the three brain regions. In Fig. 5 we gave the specific results of correlation tests for the $\alpha_1$ and $\alpha_2$ of three brain regions and ACC. The correlation coefficient between the $\alpha_1$ and ACC was close to 0, while the correlation coefficient R between the $\alpha_2$ and ACC ranged from 0.32 to 0.37, $p < .05$. The midline and right hemisphere was higher than left hemisphere. According the correlation, we speculated that LRTC may predict the ACC of physics problem solving for participants.

## 4    Conclusions

In the study, we applied the DFA exponents of EEG signals to investigate the brain activation when students with different learning progression were solving physics problem. By calculating DFA exponents at short and long scales of observation, we explored the scale properties of the EEG time series across the different groups, as well as underlying cognitive mechanism. The present study demonstrates that both of the DFA exponents in participants of three groups were greater than 1 and less than 1.5, regardless of their various learning progression. These results showed that the EEG in three groups exhibited SRTC and LRTC with power-law behavior. Previous researches suggested that a power-law exponent was a sign of critical state, and a brain near a critical state may operate in a more flexible and adaptive way for information transfer, processing, and storage [12]. In our study, this indicator reflected the more flexible extraction and processing of information during physics problem solving correspondingly.

Furthermore, the results showed that excellent groups demonstrated differences in brain activity with a higher LRTC when compared the other two groups. Thus, the excellent group showed a positive temporal dependence (i.e., autocorrelation) within the signal during physics problem-solving. The long-range temporal correlations were believed to be associated with model dynamical systems that show efficiency in learning, rapid information transfer and conception network organization [13,14]. The results reflected that students with excellent learning progression have developed dynamic model systems of physics concepts. They can extract relevant knowledge more accurately and build scientific models efficiently during problem-solving. Moreover, the excellent group presented higher DFA exponents in the regions of occipital-parietal and central region. It implied that more attention and inhibition were inputted during physics problem solving for the excellent group. Meanwhile, we also find that a significant positive correlation between LRTC exponents and ACC. It indicated that higher LRTC exponents was advantageous for effective extraction and transmission of information, thereby supporting solving physics problem correctly. This study suggested that the application of DFA method in physics education context may deepen our understanding of the neural basis of problem-solving ability and provide a promising indicator of learning achievement.

**Acknowledgments.** This work is supported by the Natural Science Foundation of China (Grant No. 62077013).

# References

1. Zhu, Y.M., Wang, Q., Zhang, L.: Study of EEG characteristics while solving scientific problems with different mental effort. Sci. Rep. **11**, 23783 (2021)
2. Zhu, Y.M., Zhang, L., Leng, Y., Pang, R.D., Wang, X.L.: Event-related potential evidence for persistence of an intuitive misconception about electricity. Mind Brain Educ. **13**, 80–91 (2019)
3. Lai, K., et al.: Encoding of physics concepts: concreteness and presentation modality reflected by human brain dynamics. PLoS ONE **7**, e41784 (2012)
4. Zhu, Y.M., Qian, X., Yang, Y., Leng, Y.: The influence of explicit conceptual knowledge on perception of physical motions: an ERP study. Neurosci. Lett. **541**, 253–257 (2013)
5. Mason, R.A., Just, M.A.: Neural representations of physics concepts. Psychol. Sci. **27**, 904–913 (2016)
6. Foisy, L.B., Potvin, P., Riopel, M., Masson, S.: Is inhibition involved in overcoming a common physics misconception in mechanics? Trends Neurosci. Educ. **4**, 26–36 (2015)
7. Jia, H., Yu, D.: Attenuated long-range temporal correlations of electrocortical oscillations in patients with autism spectrum disorder. Dev. Cogn. Neurosci. **39**, 100687 (2019)
8. Smit, D.J., et al.: Scale-free modulation of resting-state neuronal oscillations reflects prolonged brain maturation in humans. J. Neurosci. **31**, 13128–13136 (2011)
9. Peng, C.K., Buldyrev, S.V., Havlin, S., Simons, M., Stanley, H. E., Goldberger, A.L.: Mosaic organization of DNA nucleotides. Phys. Rev. Atom. Mol. Opt. Phys. **49**(2), 1685–1689 (1994)
10. Hwa, R.C., Ferree, T.C.: Scaling properties of fluctuations in human electroencephalogram. Phys. Rev. E **66**, 021901 (2001)
11. Nikulin, V.V., Jönsson, E.G., Brismar, T.: Attenuation of long-range temporal correlations in the amplitude dynamics of alpha and beta neuronal oscillations in patients with schizophrenia. Neuroimage **61**, 162–169 (2012)
12. Beggs, J.M., Timme, N.: Being critical of criticality in the brain. Front. Physiol. **3**, 163 (2012)
13. Botcharova, M., Farmer, S.F., Berthouze, L.: Markers of criticality in phase synchronization. Front. Syst. Neurosci. **8**, 176 (2014)
14. Seleznov, I., et al.: Detrended fluctuation, coherence, and spectral power analysis of activation rearrangement in eeg dynamics during cognitive workload. Front. Hum. Neurosci. **13**, 270 (2019)

# A Dynamic, Economical, and Robust Coding Scheme in the Lateral Prefrontal Neurons of Monkeys

Kazuhiro Sakamoto[1,2]($\boxtimes$), Naohiro Saito[2], Shun Yoshida[2],
and Hajime Mushiake[2]

[1] Department of Neuroscience, Faculty of Medicine, Tohoku Medical and
Pharmaceutical University, 1-15-1 Fukumuro, Miyagino-ku, Sendai 983-8536, Japan
sakamoto@tohoku-mpu.ac.jp
[2] Department of Physiology, Tohoku University School of Medicine,
2-1 Seiryo-machi, Aoba-ku, Sendai 980-8575, Japan

**Abstract.** The lateral prefrontal cortex (lPFC) plays crucial roles in executive functions, including working memory and behavioral planning. The functions of lPFC require conservation of its limited neuronal resources. Herein, we examined lPFC neuronal activities in monkeys during a path-planning task that required behavioral planning and working memory. We analyzed the coding dynamics of final-goal neurons, and found selective and sustained activities toward the final goal, reflecting working memory. Putative excitatory pyramidal neurons shifted their scheme from discrete to collective coding during the preparatory period of the task, whereas inhibitory interneurons used a collective coding scheme.

**Keywords:** lateral prefrontal cortex · monkey · dynamic coding · economical coding · robust coding

## 1 Introduction

The lateral prefrontal cortex (lPFC) plays crucial roles in executive functions, including working memory and behavioral planning [1–3]. To integrate information from other cortical areas and make behavioral decisions in complex, ever-changing environments, the limited neuronal resources of lPFC should be conserved. However, this has not yet been demonstrated in neurophysiological studies.

Early unit studies reported that the firing of lPFC neurons in monkeys reflects working memory processes. In particular, neurons show persistent activity for specific memories, such as the location of a displayed cue, even after it has

This work was supported by JSPS KAKENHI Grant Number JP16H06276 (Platform of Advanced Animal Model Support), 17K07060, 20K07726 (Kiban C), MEXT KAKENHI Grant Number 20H05478, 22H04780 (Hyper-Adaptability) and Japan Agency for Medical Research and Development (AMED) under Grant Number JP18dm0207051. The authors declare no competing financial interests.

M. Tanveer et al. (Eds.): ICONIP 2022, LNCS 13624, pp. 13–24, 2023.
https://doi.org/10.1007/978-3-031-30108-7_2

disappeared [4–6]. Other studies reported that, based on their firing rate, lPFC neurons encode for planned behaviors [7–9]. With advances in computing power and analytical methods, several studies have demonstrated dynamic coding in lPFC neurons; for example, neuronal activity indicates a representational shift from behavioral goals to specific actions [10–12]. Furthermore, our previous study found that dynamic coding neurons also encode specific actions in a resource-saving manner, via "axis coding" of action direction [13].

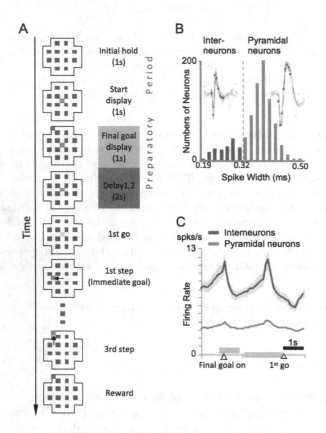

**Fig. 1.** Behavioral task and classification of neurons. A. Path-planning task; each panel indicates a single event. Green, red, and yellow squares indicate the current cursor position, final-goal position, and movement initiation signal, respectively. The delay period was divided into first (delay 1) and second (delay 2) halves. The path may be blocked during the latter half. B. Distribution of spike width for all neurons. Interneurons and pyramidal neurons were divided at the bottom of the distribution. C. Mean firing rates of putative interneurons and pyramidal neurons; shaded areas indicate standard error of mean (SEM). (Color figure online)

In the present study, we evaluated whether lPFC neurons involved in working memory encode dynamic and resource-saving coding schemes. We found that putative excitatory pyramidal neurons shift their scheme from discrete to collective coding during the preparatory period of the path-planning task. In comparison, inhibitory interneurons continue to operate under a collective coding scheme during the preparatory period. These results suggest that working memory neurons in the lPFC encode memories in a dynamic, resource-saving, and robust manner.

# 2    Methods

## 2.1    Subjects

The experiments were performed using two adult male Japanese monkeys (*Macaca fuscata*). The experimental protocols were approved by the Animal Care and Use Committee of Tohoku University (permit # 20MeA-2) and adhered to the National Institutes of Health guidelines for the care and use of laboratory animals, as well as with the recommendations of the Weatherall Report.

## 2.2    Behavioral Task and Neuronal Recording

The details of the experimental and analytical procedures have been described previously [7,11–14].

The monkeys were trained to perform a path-planning task that required the planning of multiple cursor movements, controlled using manipulanda, to reach a goal within a maze (Fig. 1A). To begin the trial, the animals were required to hold the two manipulanda in a neutral position for 1 s (initial hold). Subsequently, a cursor was presented at the center of the maze (start display). One second later, the position of goal cursor was presented for 1 s (final goal display). After a delay (delay 1 or 2), the color of the cursor changed from green to yellow, which served as the initiation signal (1st go). After a 1-s hold period, the next go signal was presented (2nd go). When the cursor reached the final goal position, animals received a reward (reward). To dissociate arm and cursor movements, arm-cursor assignments were altered on completion of a block of 48 trials. In > 89% of trials, monkeys reached the final goal within the minimum number of steps (i.e., three).

We used conventional electrophysiological techniques to obtain in vivo single-cell recordings from the lPFC region of the right hemisphere.

## 2.3    Data Analysis

The present study examined whether neuronal activities observed during a 4-s preparatory period (i.e., start display, final goal display, delay 1, and delay 2) were associated with selectivity for final or immediate goals. For this purpose, multiple linear regression analysis of the spike counts in each 100-ms time window was conducted using the following formula:

$$\text{firing rate} = \alpha + \beta \times (\text{final or immediate goals}), \qquad (1)$$

where $\alpha$ is the intercept and $\beta$ is the set of coefficients. Categorical factors for the final goals were the four final-goal positions presented during the final goal display period, while those for the immediate goals were the four positions of the cursor at the first step. Therefore, three dummy variables were used for each final and immediate goal. Accordingly, $\beta$ included three coefficients. The analyses of final and immediate goals were conducted separately. The F-value at each time point was normalized to the significance level of the $F$-value ($p = 0.05$), and is referred to as the normalized goal selectivity.

Neurons with immediate-goal selectivity higher than the significance level and final-goal selectivity in a certain time period were regarded as having significant immediate-goal selectivity, and defined as immediate-goal neurons. Final goal neurons were defined as those with higher selectivity for final than immediate goals through-out the preparatory period.

To explore the mechanisms underlying neuronal coding, putative interneurons and pyramidal neurons were classified on the basis of the waveforms of their action potentials [15–17]. Pyramidal neurons had a spike width of $> 0.32$ ms, whereas interneurons had a spike width of $< 0.32$ ms (Fig. 1B). The mean firing rate was significantly higher for interneurons than pyramidal neurons during the preparatory period (Fig. 1C).

**Table 1.** Numbers of recorded neurons classified according to goal selectivity and cell type.

|  | Total | Task-related | (Final | Immediate) |
|---|---|---|---|---|
| Pyramidal neurons (% of PN) | 731 (82%) | 322 (79%) | 212 (82%) | 110 (74%) |
| Interneurons (% of IN) | 156 (18%) | 85 (21%) | 47 (18%) | 38 (26%) |
| Total | 887 | 407 | 259 | 148 |

# 3   Results

## 3.1   Database

We recorded neuronal activities in the lPFC of the right hemisphere of two macaque monkeys during a path-planning task. The analyses only included neuronal activity data recorded during correct trials completed within the minimum number of steps. Using linear regression analysis, 887 well-isolated single units were classified as final- or immediate-goal neurons based on the aforementioned criteria. In addition, units were classified as interneurons or pyramidal neurons based on the spike shape (Table 1, Fig. 1B). During the preparatory period, 148 immediate-goal neurons exhibited significant selectivity for immediate goals, defined on the basis of the direction of cursor movements made by each monkey during the first cursor movement period. In addition, 259 final-goal neurons exhibited significant selectivity for final, but not immediate, goals during the same period. We focus on the detailed characteristics of final-goal neurons ($n = 259$).

## 3.2    Example Neurons

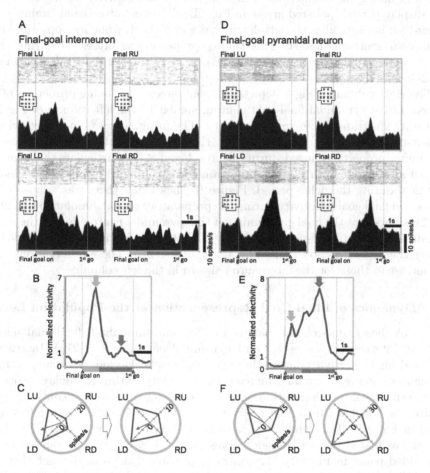

**Fig. 2.** Representative examples of putative final-goal interneurons (A–C) and pyramidal neurons (D–F). A. Raster plots and spike density histograms of the neuronal activity of the interneuron for the four final goals. LU, left-up; RU, right-up; LD, left-down; RD, right-down. B. Time course of the final-goal selectivity of the neuron. C. Final-goal directional tunings at the times indicated by the red and blue arrows in B, respectively. Red arrow, preferred direction and amplitude of final-goal selectivity. (Color figure online)

The left column of Fig. 2 shows a typical example of the activity of interneurons. In this example, the neuronal firing rate was strongly modulated by the direction of the final goal displayed during the final goal display period, but the neuron showed selective activity in the direction of the final goal during delay 2 (i.e., the latter half of the delay period in which some paths may be blocked). This neuron exhibited the highest firing rate for the left-down final goal during the preparatory period (Fig. 2A). These properties were evaluated using regression analysis (Fig. 2B). The left-sided panel of Fig. 2C shows the firing rates associated with four final goals at the time of highest final-goal selectivity during the final goal display period (pale-red arrow in Fig. 2B). The goal-directional tuning was vector-like; in particular, the left-down goal was highest, while its opposite, i.e., right-down goal, was lowest. This tuning property was preserved during the delay 2 period (pale-blue arrow in Fig. 2B), as shown in the right-sided panel of Fig. 2C.

The right column of Fig. 2 depicts the goal-directional tuning properties of a representative pyramidal final-goal neuron, similar to the left column of Fig. 2. Figure 2D illustrates the neuronal activity. The firing rate was increased when the left-up final goal was presented during the final goal display period. However, the tuning property changed during delays 1 and 2. In particular, the neuron showed an increase in firing rate and changed its tuning to favor the left-down final goal during the delay period. Figure 2E shows that there was a sustained increase in final goal selectivity during the preparatory period. Similar to Fig. 2D, Fig. 2F shows that the final goal tuning of this pyramidal neuron during the final goal display period (pale-red arrow in Fig. 2E) changed from left-up to left down during the delay 2 period (pale-blue arrow in Fig. 2E). These observations are in contrast to those for the interneuron shown in the left column.

### 3.3   Dynamics of Final-Goal Representation at the Population Level

Figure 3A shows time-related changes in the mean normalized final goal selectivity of interneurons ($n = 47$) and pyramidal neurons ($n = 212$). Interneurons had high selectivity during the final goal display period, whereas pyramidal neurons showed a gradual increase in selectivity during the delay periods. For interneurons, the distribution of preferred final-goal was significantly biased toward the left-down final goal ($n = 36$, $p = 0.021$, binominal test; left-sided panel in Fig. 2B). This tendency was preserved during delay 2, i.e., the distribution was biased toward left-sided goals ($n = 20$, $p = 0.021$, binominal test; right-sided panel in Fig. 2B). By contrast, pyramidal neurons did not show a biased distribution in terms of the preferred final-goal during the final goal display period ($n = 128$, $p = 0.23$, binominal test; left-sided panel in Fig. 2C), whereas they exhibited significantly high and low distributions of the left-down and right-up goals, respectively ($n = 134$, $p = 3.4 \times 10^{-} - 5$ for LD, p = 0.020 for RU, binominal test; right-sided panel in Fig. 2C).

Analysis of the mean LD-component of firing rate verified the aforementioned observations (Fig. 3D). The LD-component is defined as the LD-RU axis component of the vector in the final-goal directional tuning plot, as shown in Fig. 2 C and F. The LD-component was calculated for each 100-ms time window and neuron. Figure 3D demonstrates that interneurons exhibit biased firing toward the left-down goal compared to the pyramidal neurons during the final goal display period.

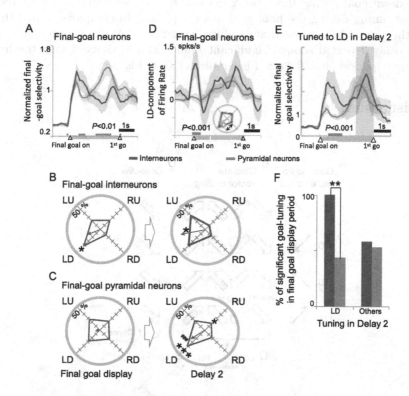

**Fig. 3.** Dynamics of final-goal tuning in populations of interneurons and pyramidal neurons. A. Time-development of mean normalized final goal selectivity. B,C. Distribution of the preferred final goal of interneurons (B) and pyramidal neurons (C) at the most significant time during the final goal display and delay 2 periods, respectively. The type and degree of distribution bias are indicated by arrow direction and length, respectively. D. The mean LD-component of the firing rate over time. Inset, definition of the LD-component of firing rate. E. The mean normalized final-goal selectivity of LD-preferring neurons during the delay 2 period (shaded area). F. Percentage of neurons showing significant final-goal selectivity during the delay 2 and final-goal display periods. The time periods of the significant differences are indicated by gray bars in (A,D,E). ***, $p < 0.001$; **, $p < 0.01$; *, $p < 0.5$. (Color figure online)

Figure 3E and F presents the analysis of neurons that significantly preferred the left-down goal during delay 2 (interneurons, $n = 8$; pyramidal neurons, $n = 55$). The analysis of mean normalized final-goal selectivity demonstrated strong final-goal selectivity of interneurons during the preparatory period, while pyramidal neurons did not show strong final-goal selectivity during the final goal display period. However, pyramidal neurons exhibited an increase in final-goal selectivity during the delay periods. The interneurons that significantly preferred the left-down goal during the delay 2 period ($n = 8$) also exhibited significant final-goal tuning during the final goal display period. In comparison, less than half of the pyramidal neurons that exhibited significant left-down preference during the delay 2 period showed significant final-goal selectivity during the final goal display period ($p = 0.0048$, Fisher's exact test; Fig. 3F).

## 4    Discussion

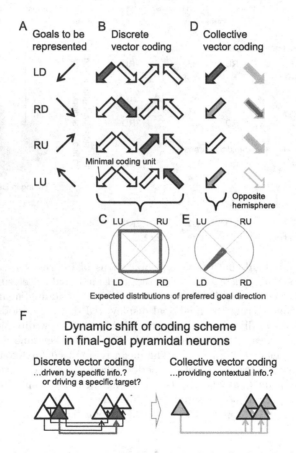

**Fig. 4.** Coding scheme representing the direction of the final goal. A. The four final goals. B,C. Discrete vector coding scheme and corresponding preferred direction distribution. D,E. Collective vector coding scheme and preferred direction distribution. F. Pyramidal neurons dynamically shift coding schemes during the preparatory period. Details are provided in text.

We investigated the tuning dynamics of final-goal-selective neurons in the lPFC of monkeys during a path-planning task. Neuronal firing was sustained during the preparatory period, and was modulated by the direction of the final goal. Putative interneurons showed a biased distribution in terms of the preferred final-goal during the preparatory period, toward the left-down direction. By contrast, the preferred final-goal distribution of putative pyramidal neurons was distributed evenly among the four directions during the final goal display period, whereas it was biased toward the left-down final goal during the delay 1 and 2 periods. The dynamic tuning and biased preferred final goal of these neurons presumably contributes to the robust and economical coding achieved by lPFC, as discussed below.

Final-goal neurons should express are the four directions of the final goal (Fig. 4A). The simplest coding scheme for these four directions is probably the discrete vector coding shown in Fig. 4B, where each direction is encoded by the on/off activity of a certain neuronal population, hereinafter referred to as the minimal coding unit. If all these minimal coding units are in the right hemisphere lPFC from which we recorded neuronal activities, the distribution of the preferred final goals is equal among the four directions, as shown in Fig. 4C. The four final goals are represented by the activities of each minimal coding unit. On the other hand, if the distribution of the preferred final goal is biased, and particularly if it is biased toward a single goal direction (e.g., lower left in our experiment), the coding scheme of the four directions is complicated and requires certain assumptions to be met (Fig. 4D, E). First, the minimal coding unit does not simply encode information in an on/off manner. That is, it is assumed that the maximum response is in the lower left direction (i.e., the preferred direction), and the minimum response is in the upper right direction (i.e., the anti-preferred direction), and the intermediate response is in the other two directions. This is a natural assumption based on the tunings of the neuronal examples shown in Fig. 2C and F. Second, it is assumed that there is another minimal coding unit. Importantly, we recorded neuronal activity from the right hemisphere only. Therefore, it was hypothesized that neurons that prefer the lower right goal predominate in the left hemisphere. For the two-dimensional information related to four final goals, it is natural to assume the existence of a minimal coding unit that encodes a direction at an angle of 90° relative to the lower left. In the present study, the coding scheme suggested by the biased preferred direction distribution is referred to as collective vector coding. As shown in Fig. 4D, in this scheme the four goals are represented by the activity patterns of two minimal coding units.

The final-goal selective pyramidal neurons have a discrete vector coding scheme early in the preparatory period, and a collective vector coding scheme late in the preparatory period (Fig. 4F). Discrete vector coding may reflect the situation where a minimal coding unit is directly driven by other cell groups, and has the advantage of directly driving other cell groups (left panel in Fig. 4F). However, several minimal coding units (four in the current study) are required, and therefore do not contribute to the conservation of neural resources. On the other hand, collective vector coding conserves neural resources and only two minimal coding units were required in the present study. This coding scheme may not be suitable for driving other neuronal groups. However, it may be sufficient if contextual information is provided to other neurons (right-sided panel in Fig. 4F).

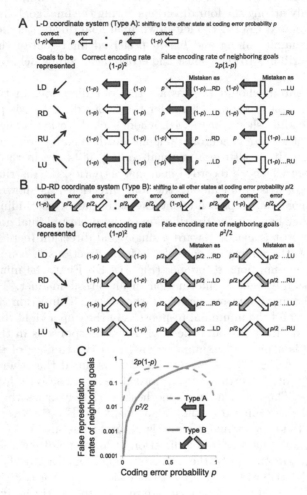

**Fig. 5.** Collective coding schemes and influence of coding error on the final-goal representation. A. L-D coordinate system. B. LD-RD coordinate system shown in Fig. 4D. C. Coding error probability vs. the false representation rate of neighboring goals. The details are presented in the text.

Although lPFC final-goal selective neurons have collective vector coding, it is necessary to determine why it does not use two-dimensional representations of up-down and left-right axes. In Fig. 5A, we assumed the existence of minimal coding units that prefer left and downward directions, respectively (hereinafter referred to as the L-D coordinate system). The four final goals are expressed as the combination of the on/off states of the activities. This coding scheme is simpler and more natural than that used for the lPFC neurons, which is presented schematically in Figs. 4D and 5B (hereinafter referred to as the LD-RD coordinate system).

For the purpose of this discussion, we considered the coding error $p$ of minimal coding units. There is variability in the firing activity of real neurons even under similar experimental conditions, so it is important to consider coding errors. We also considered the probability that the neighboring final goals were mis-encoded, to distinguish between similar behaviors. In the case of the L-D coordinate system, the probability that neighboring final goals are accidentally encoded, i.e., the false representation rate of neighboring goals, is $2p(1 - p)$, as shown in Fig. 5A. In the case of LD-RD coordinate system, there are variations in how coding errors occur. However, we considered a case where two out of three possible error states of a minimal coding unit have a probability of $p/2$, as shown in Fig. 5B. In this case, the rate of false representations, i.e., the rate at which neighboring goals are accidentally encoded, is $p^2/2$. There was no significant qualitative change in the formula for the false representation rate of neighboring goals in the other variants. Figure 5C shows how the false representation rate of neighboring goals changes with the coding error rate of the minimal coding unit. If the coding error rate of the minimal coding unit is low, the false representation rate of neighboring goals would be lower in the LD-RD compared to L-D coordinate system. These results suggest that the coding scheme adopted by lPFC neurons robustly encode the information of final goals.

Previous studies showed that lPFC neurons not only exhibit dynamic representations [10–13], but also use abstract and resource-saving coding schemes [13,18] The present study showed that the coding scheme of lPFC neurons is dynamic, resource-saving, and robust. The neuronal properties are useful for decision-making and behavioral planning, because lPFC is responsible for adapting to complex and ever-changing environments despite limited neuronal resources.

# References

1. Duncan, J.: An adaptive coding model of neural function in prefrontal cortex. Nat. Rev. Neurosci. **2**, 820–829 (2001)
2. Tanji, J., Hoshi, E.: Role of the lateral prefrontal cortex in executive behavioral control. Physiol. Rev. **88**, 37–57 (2008)
3. Passingham, R.E., Wise, S.P.: The Neurobiology of the Prefrontal Cortex: Anatomy, Evolution, and the Origin of Insight. Oxford University Press, Oxford (2012)
4. Funahashi, S., Bruce, C.J., Goldman-Rakic, P.S.: Mnemonic coding of visual space in the monkey's dorsolateral prefrontal cortex. J. Neurophysiol. **61**, 331–349 (1989)
5. Miller, E.K., Erickson, C.A., Desimone, R.: Neural mechanisms of visual working memory in prefrontal cortex of the macaque. J. Neurosci. **16**, 5154–5167 (1996)

6. Romo, R., Brody, C.D., Hernandez, A., Lemus, L.: Neuronal correlates of parametric working memory in the prefrontal cortex. Nature **399**, 470–473 (1999)
7. Saito, N., Mushiake, H., Sakamoto, K., Itoyama, Y., Tanji, J.: Representation of immediate and final behavioral goals in the monkey prefrontal cortex during an instructed delay period. Cereb. Cortex **15**, 1535–1546 (2005)
8. Genovesio, A., Brasted, P.J., Wise, S.P.: Representation of future and previous spatial goals by separate neural populations in prefrontal cortex. J. Neurosci. **26**, 7305–7316 (2006)
9. Genovesio, A., Tsujimoto, S., Wise, S.P.: Encoding goals but not abstract magnitude in the primate prefrontal cortex. Neuron **74**, 656–662 (2012)
10. Machens, C.K., Romo, R., Brody, C.D.: Flexible control of mutual inhibition: a neural model of two-interval discrimination. Science **307**, 1121–1125 (2005)
11. Sakamoto, K., et al.: Discharge synchrony during the transition of behavioral goal representations encoded by discharge rates of prefrontal neurons. Cereb. Cortex **18**, 2036–2045 (2008)
12. Sakamoto, K., Katori, Y., Saito, N., Yoshida, S., Aihara, K., Mushiake, H.: Increased firing irregularity as an emergent property of neural-state transition in monkey pre-frontal cortex. PLoS ONE **8**, e80906 (2013)
13. Sakamoto, K., Saito, N., Yoshida, S., Mushiake, H.: Dynamic axis-tuned cells in the monkey lateral prefrontal cortex during a path-planning task. J. Neurosci. **40**, 203–219 (2020)
14. Mushiake, H., Saito, N., Sakamoto, K., Itoyama, Y., Tanji, J.: Activity in the lateral prefrontal cortex reflects multiple steps of future events in action plans. Neuron **50**, 631–641 (2006)
15. Wilson, F.A., O'Scalaidhe, S.P., Goldman-Rakic, P.S.: Functional synergism between putative gamma-aminobutyrate-containing neurons and pyramidal neurons in prefron-tal cortex. Proc. Natl. Acad. Sci. U.S.A. **91**, 4009–4013 (1994)
16. Rao, S.G., Williams, G.V., Goldman-Rakic, P.S.: Isodirectional tuning of adjacent interneurons and pyramidal cells during working memory: evidence for microcolumnar organization in PFC. J. Neurophysiol. **81**, 1903–1916 (1999)
17. Constantinidis, C., Goldman-Rakic, P.S.: Correlated discharges among putative pyramidal neurons and interneurons in the primate prefrontal cortex. J. Neurophysiol. **88**, 3487–3497 (2002)
18. Shima, K., Isoda, M., Mushiake, H., Tanji, J.: Categorization of behavioural sequences in the prefrontal cortex. Nature **445**, 315–318 (2007)

# Dynamic Characteristics of Micro-state Transition Defined by Instantaneous Frequency in the Electroencephalography of Schizophrenia Patients

Daiya Ebina[1]([✉]), Sou Nobukawa[1,2], Takashi Ikeda[3], Mitsuru Kikuchi[3,4], and Tetsuya Takahashi[3,5,6]

[1] Graduate School of Information and Computer Science, Chiba Institute of Technology, 2-17-1 Tsudanuma, Narashino, Chiba 275-0016, Japan
s1831021MR@s.chibakoudai.jp, nobukawa@cs.it-chiba.ac.jp
[2] National Center of Neurology and Psychiatry, National Institute of Mental Health, Tokyo, Japan
[3] Research Center for Child Mental Development, Kanazawa University, Ishikawa, Japan
[4] Department of Psychiatry and Behavioral Science, Kanazawa University, Ishikawa, Japan
[5] Department of Neuropsychiatry, University of Fukui, Fukui, Japan
[6] Uozu Shinkei Sanatorium, Toyama, Japan

**Abstract.** Recently proposed dynamic functional connectivity (dFC) approach, which focuses on the degree of spatial-temporal variability in regional-pair-wise functional connectivity (FC), is able to detect neural network alternations as the core neural basis of schizophrenia (SZ). Moreover, from the perspective of "emergence" in complex network science, the importance of the establishment of a method to evaluate the neural interactions in the whole brain network, not separating each pair-wise interaction, is emphasized. We proposed the micro-state approach based on the whole-brain instantaneous frequency distribution as one of these methods; this approach opens a new avenue as an evaluation method to detect cognitive function impairment and pathology. Thus, we hypothesized that the application of this micro-state approach to neural activity could detect the other aspects of brain network alterations for SZ previously elucidated in conventional FC and dFC. We applied the micro-state approach to electroencephalography (EEG) signals of SZ patients and healthy controls. The results revealed the alternation of dynamical leading phase transitions between the frontal and occipital regions and right and left hemispheric regions at the beta and gamma bands. This alternation suggested the corpus callosum impairments and abnormal enhancement of functional hub structure at the fast bands as topology of whole brain functional network. Thus, our proposed micro-state approach succeeded in detecting the state transition alternations concerning SZ pathology. This approach might contribute in elucidating new aspects of dFC, thereby resulting in the discovery of a biomarker for SZ.

M. Tanveer et al. (Eds.): ICONIP 2022, LNCS 13624, pp. 25–36, 2023.
https://doi.org/10.1007/978-3-031-30108-7_3

**Keywords:** Schizophrenia · Instantaneous Frequency · Dynamic
Functional Connectivity · Electroencephalography

# 1   Introduction

Schizophrenia (SZ) is a severe, chronic, and intractable mental disorder characterized by distinct positive symptoms, such as delusions, hallucinations, and
thought disorder, and negative symptoms, such as decreased or lost motivation,
emotional flatness, which impair cognitive function [1,2]. Early therapeutic inter-
vention has a significant impact on the prognosis [1]. Therefore, in addition to
the conventional diagnosis based on the medical interview, devising objective
and quantitative biomarkers is desirable to facilitate an early diagnosis. Disconnection of neural networks in SZ is an effective candidate for a biomarker of
SZ, since the disconnection of neural networks in brain results in SZ symptoms
according to the widely known disconnection hypothesis [3].

Studies based on electroencephalograms (EEGs) and functional magnetic resonance imaging (fMRI) have revealed that abnormalities in the interactions of
neural activity between brain regions, which is measured by functional connectivity (FC) defined as synchronization/information flow of neural activity, reflect
the pathology of SZ [4] (reviewed in [5]). Particularly, the combination of EEG
signals with a high temporal resolution and phase synchronization approach
using the high spatial-temporal resolution by reducing the volume conduction
demonstrated the frequency-band and region-specific abnormality of FC in SZ,
i.e., SZ patients have reduced beta band FCs centered in the frontal region and
gamma band FCs throughout a wide-range of brain regions [6].

While, the temporal complexity of local neural activity in individual brain
regions strongly correlates with the global network structure of FC, such as
node degree and centrality [7,8]; therefore the complexity reflects the global
interaction of neural activity among a wide range of brain regions [6,9]. In SZ,
significant increases in complexity at slow temporal scales were identified in
the front-center-temporal regions [10]. This abnormal high complexity may be
attributed to the temporal disorganization of neural activity [10] (reviewed in
[11]). This complexity is also affected by age, symptoms, and medication [11].

In addition to FC and local regional complexity in neural activity, recent
findings of dynamic functional connectivity (dFC), defined by the degree of variation in the FC strength, demonstrated that dFC more accurately reflects the
cognitive abilities than does static FC in SZ [12]. Additionally, by combining
clustering and graph analysis methods with FC transitions in a time window,
spatial and temporal patterns of network dynamics in the whole brain have been
revealed [5,13]. In dFC evaluation by fMRI, temporal changes in the coherence
of blood oxygen level dependent (BOLD) signals in brain regions using sliding
time window analysis have been used [14]. Moreover, studies on dFC of EEG
and magnetoencephalogram (MEG) with high temporal resolution have been
conducted[15]. Considering the application of dFC to EEG and MEG, although
EEG and MEG have a higher temporal resolution than fMRI, the detection of

instantaneous dynamics in the order of several milliseconds is challenging owing to the sliding window process used in the dFC estimation methods [15]. To tackle this issue, recent studies using a method called dynamical phase synchronization (DPS), which is defined by the temporal complexity of instantaneous phase difference of neural activity between regions, successfully detected moment-to-moment network dynamics in dFC during aging by decomposing the EEG signal [9].

The FC approach, including dynamic FC, focuses on pair-wise neural interactions among complex brain interactions [6,9]. In the SZ, the degree of FC at the resting state alternates; especially, in a default mode network (DMN) exhibiting intrinsic activities, this alternation becomes significant [16]. However, brain functions are produced in global and multiple neural activities, and not only in local neural activity [17]. Several recent studies highlighted the importance of evaluating the brain from the perspective of "emergence" of complex systems, where new functions are created through the interaction of multiple elements [18]. "Emergence" research in complex network science have revealed that merely capturing pair-wise neural interaction is insufficient for evaluating complex neural network structures characterized as the brain [19]. Therefore, the need to incorporate an integrated approach for capturing neural interactions of the entire brain has been noted [18]. Considering one of these approaches, in our recent research, from the viewpoint of "emergence", a new micro-state method based on the instantaneous frequency (called as IF micro-state) dynamics was proposed, which enables the evaluation of the neural activity of the entire brain rather than the pair-wise neural interaction between brain regions [20]. Subsequently, in a study on AD patients, this approach revealed that maintaining the occipital leading phase was more difficult in the AD group than in the healthy control group and its degree exhibits the correlation with cognitive function in AD [20].

Thus, considering the fact that cognitive dysfunction is also a core symptom in SZ, we hypothesized that our proposed IF micro-state approach, which can evaluate the integration of the instantaneous frequency dynamic of the whole brain without disassembling pair-wise neural interactions, would provide a new understanding of the relationship between the network dynamics of SZ. To validate our hypothesis, we define EEG signals as micro-states based on instantaneous frequencies in various frequency bands, and evaluated the dynamics of state transitions.

## 2    Materials and Methods

### 2.1    Participants

This study employed the same participants as those in our previous study concerning the abnormality of FC in SZ [6]. The SZ group consisted of 21 right-handed participants. The age-and sex-matched healthy control group (HC) consisted of 31 right-handed healthy participants. The demographic characteristics

**Table 1.** Demographic characteristics of healthy controls (HC) and schizophrenia (SZ) patients

|  | HC | SZ | $p$-value |
|---|---|---|---|
| Male/female | 16/15 | 11/10 | 0.9566 |
| Age, year | 27.9 (8.2) | 28.1 (10.1) | 0.9262 |
| Duration of the illness, months | NA | 24.2 (36.2) | NA |
| BPRS score | NA | 52.6 (13.2) | NA |

Values represent mean (SD).
Abbreviation: BPRS, Brief Psychiatric Rating Scale; SD, standard deviation.

of the two groups are summarized in Table 1. Participants with major neurological diseases, previous electroshock or significant head trauma, or a history of drug or alcohol dependence were excluded. The HC participants were recruited from among staff members at the Kanazawa University Hospital and their family members. They had no personal or family history of psychiatric or neurological diseases, as confirmed by both a self-reported past history and a psychiatric examination of the present mental state using the axis I criteria of the Diagnostic and Statistical Manual-Fourth Edition (DSM-IV). Patients in the SZ group were recruited from the outpatients of Kanazawa University Hospital and met the criteria for DSM-IV SZ at the time of the study, and were subsequently diagnosed with SZ by a specialized clinical psychiatrist. The patients were diagnosed with SZ by an expert clinical psychiatrist. No patient had been treated with neuroleptics before the EEG recording. The Brief Psychiatric Rating Scale (BPRS) was used to assess the patient's symptoms on the day of the EEG recording. The data were approved by the Ethics Committee of Kanazawa University and were conducted in accordance with the aims of the Declaration of Helsinki. Moreover, all the participants agreed to participate in the study upon understanding the study. Informed consent was obtained from the all the participants.

## 2.2 EEG Recordings

EEG data were recorded from 16 electrodes in accordance with the International 10-20 System: Fp1, Fp2, F3, F4, Fz, F7, F8, C3, C4, P3, P4, Pz, T5, T6, O1, and O2. The reference electrode was placed in the linked earlobes, and eye movements during the EEG recording were monitored using an additional electrooculogram. The impedance at each electrode was maintained below $5k\Omega$. EEG was obtained at a sampling frequency of 200 Hz with a band pass filter of 1.5-60 Hz, and stored for offline analysis using an 18-channel system (EEG-44189; Nihon Kohden; Tokyo, Japan). The participants were instructed to lie down with their eyes closed in a soundproof, light-controlled, and electrically shielded recording room; EEG was recorded for 10–15 min in a resting-state condition. Artifacts (e.g., muscle activities, eye movements, or blinks) were visually identified and carefully excluded. One continuous 60-s artifact-free epoch was extracted from every participant. A PLI study in SZ demonstrated network abnormalities in beta and gamma bands [6]. Therefore, we focused on the whole brain IF micro-state of the beta and gamma bands. For each epoch, band-pass filtering was performed to isolate the conventional frequency bands as follows: beta (13-30 Hz), and gamma (30-60 Hz) bands. The first and last 5-s epochs were eliminated from the analysis to avoid interference related to the band pass filtering procedure.

### 2.3    Estimation of the Dynamic State Based on the Instantaneous Frequency Distribution

We defined the state of brain activity by studying the dynamics of the instantaneous frequencies of EEG signals. Using the time series of the instantaneous frequencies of the EEG signals; the following process [20] was used to estimate the state of brain activity (Fig.1).

The Hilbert transform was used to estimate the instantaneous phase $\theta(t)$ $(-\pi \leq \theta \leq \pi)$ for each frequency band. This instantaneous phase has a phase noise called phase slip, which is a large deviation from the set frequency range. Therefore, by performing median filtering, we obtained continuous instantaneous frequencies $IF(t)$ $(-\infty \leq IF(t) \leq +\infty)$ without phase slip. This method was employed in our previous study [9]. Against this instantaneous frequencies $IF_i(t)$ ($i$: electrode position), the frequency was Z-scored among all the electrodes. The z-scored $dIF_i(t)$ of the HC and SZ groups were classified into $k$ clusters by the $k$-means method. Here, the center of the cluster was determined by the z-scored $dIF_i(t)$ of HC. In this study, we set the cluster size to $k = 2, 3, 4$. We set these cluster sizes to approximate the number of clusters generally used in micro-states (reviewed in [21,22]).

### 2.4    Statistical Analysis

To evaluate the dynamic characteristics of state transitions classified by $k$-means, we used the emergence probabilities of each state and its state transition probability. A $t$-test was used to evaluate the difference in the emergence probabilities and the state transition probabilities between the HC and SZ groups. A Benjamini-Hochberg false discovery (FDR) correction was applied to the $t$-values for multiple comparisons of transition probabilities (size of $p$-values: $k \times k$ state transitions $\times$ frequency bands), and for the emergence probabilities of IF micro-state (size of $p$-values: $k$ state $\times$ 2 frequency bands) with $q < 0.05$.

## 3    Results

The parts of (a) for Figs. 2, 3, and 4 demonstrate the group average for the temporal mean $dIF_i(t)$ among the evaluation duration at the beta and gamma bands, in the cases with cluster size $k = 2, 3, 4$, respectively, in the HC and SZ groups. In the parts of (b) for Figs. 2, 3, and 4, the emergence probabilities and

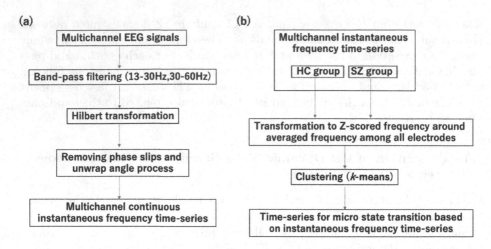

**Fig. 1.** (a) Estimation process for instantaneous frequency time-series of electroencephalography (EEG) signals. (b) Process to determine the IF micro-state time series of EEG signals.

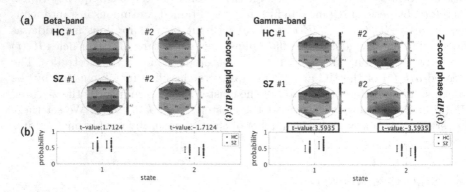

**Fig. 2.** (a) Topographs of the mean values of the $Z$-scored frequency $dIF_i(t)$ in the healthy controls (HC) and schizophrenia (SZ) groups based on the duration for each state (#1 and #2) (cluster size $k = 2$, left and right parts exhibit beta and gamma bands, respectively). In both groups, the region-specific leading phase, such as the frontal and occipital leading, was confirmed. (b) Emergence probability of state #1 and #2. The significant large difference between the HC and SZ groups for the emergence probabilities were confirmed ($q < 0.05$), which is squared by a solid blue circle. (Color figure online)

the $t$-values for these probabilities between the HC and SZ groups are shown. Although common region-specific topological features between the HC and SZ groups were observed, significant differences between the HC and SZ groups in the emergence probability were identified. Particularly, in the case of cluster size $k = 2$ (see Fig. 2), at both the beta and gamma bands, the states for the frontal and occipital leading phases were confirmed. The emergence probability

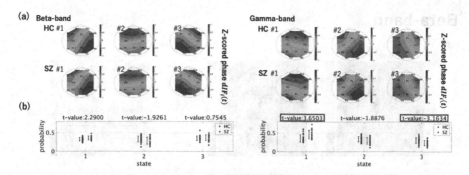

**Fig. 3.** (a) Topographs of the mean values of $Z$-scored frequency $dIF_i(t)$ in the HC and SZ groups based on the duration for each state (#1, #2, and #3) (cluster size $k = 3$, left and right parts exhibit beta and gamma bands, respectively). In both the groups, region-specific leading phase, such as the right and left frontal and occipital leading, was confirmed. (b) Emergence probability of states #1, #2, and #3. The significant large difference between the HC and SZ groups for the emergence probabilities was confirmed ($q < 0.05$), which is squared by a solid blue circle. (Color figure online)

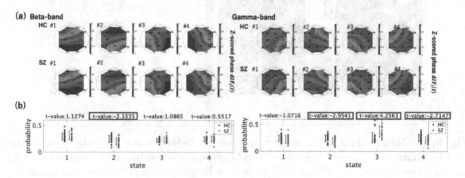

**Fig. 4.** (a) Topographs of the mean values of the $Z$-scored frequency $dIF_i(t)$ in the HC and SZ groups based on the duration for each state (#1, #2, #3, and #4) (cluster size $k = 4$, left and right parts exhibit the beta and gamma bands, respectively). In both the groups, region-specific leading phase, such as the right/left frontal and the right/left occipital leading phase, was confirmed. (b) Emergence probability of state #1, #2, #3, and #4. The significant large difference between the HC and SZ groups for the emergence probabilities was confirmed ($q < 0.05$), which is squared by a solid blue circle. (Color figure online)

for the frontal (occipital) leading phase of SZ significantly decreased (increased). Considering cluster size $k = 3$ (see Fig. 3), at both the beta and gamma bands, the states for the left frontal leading phase, the right frontal leading phase, and the occipital (especially, left occipital in gamma band) leading phase were confirmed. The emergence probabilities of the gamma band for the left occipital (right frontal) leading phase of SZ significantly decreased (increased). Considering cluster size $k = 4$ (see Fig. 4), at both the beta and gamma bands, the

## Beta-band

**(a)**                                                                        **(b)**

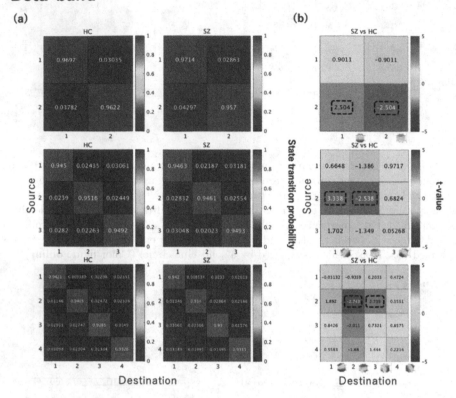

**Fig. 5.** (a) Mean values of the state transition probabilities for the beta band in the healthy control (HC) and schizophrenia (SZ) groups. (b) $t$-value of the state transition probabilities between the SZ and HC groups. A larger (smaller) $t$-value corresponds to a higher (lower) probability of SZ compared to that for HC. The parts surrounded by squares indicate the $t$-value satisfying the criteria of Benjamini-Hochberg false discovery correction. The probabilities of the SZ group were significantly smaller $q < 0.05$ than that of the HC group.

states for the left frontal leading phase, the right frontal leading phase, the left occipital leading phase and the right occipital leading phase were confirmed. In the beta band, the emergence probabilities for the right occipital leading phase of SZ significantly decreased. In the gamma band, the emergence probabilities for the left occipital leading phase of SZ significantly decreased; however, the emergence probabilities for the left frontal and right frontal leading phase of SZ significantly decreased.

The parts of (a) for Figs. 5 and 6 demonstrate the mean of the state transition probabilities of the HC and SZ groups in the beta and gamma bands, respectively. The parts of (b) for Figs. 5 and 6 demonstrate the $t$-values of the state transition probabilities between the HC and SZ groups. Considering the beta band (see

# Gamma-band

(a)                                                                              (b)

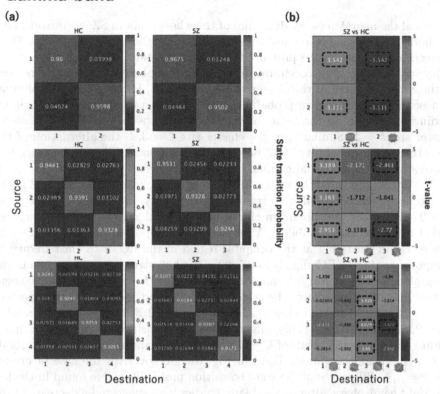

**Fig. 6.** (a) Mean values of the state transition probabilities for the gamma band in the healthy control (HC) and schizophrenia (SZ) groups. (b) $t$-value of the state transition probabilities between the SZ and HC groups. A larger (smaller) $t$-value corresponds to a higher (lower) probability of SZ compared to that for HC. The parts surrounded by squares indicate the $t$-value satisfying the criteria of the Benjamini-Hochberg false discovery correction. The probabilities of the SZ group were significantly smaller $q < 0.05$ than that of the HC group.

Fig. 5), the state transition probability of SZ becomes significantly larger from the occipital to the frontal leading phase state and smaller from the occipital to the occipital in $k = 2$ case, larger from the occipital to the left frontal leading phase state, and smaller from the occipital to the occipital in $k = 3$ case, larger from the right occipital to the left occipital leading phase state, and smaller from the right occipital to the right occipital in $k = 4$ case. Considering the gamma band (see Fig. 6), the state transition probability of SZ shows significantly larger transitions to the occipital leading phase state in $k = 2, 3, 4$ case; becomes smaller from the frontal to the frontal and occipital leading phase state in $k = 2$, smaller from the occipital and right frontal to the right frontal leading phase state in $k = 3$, and smaller from the left occipital to the left frontal leading phase state in $k = 4$.

## 4    Discussion and Conclusions

To reveal the neural network dynamics of the whole brain in SZ, we introduced a method for estimating IF micro-states distribution of the whole brain using EEG. Concerning the emergence probability of IF micro-state, the emergence probability for the state of the occipital (frontal) leading phase increased (decreased) in the SZ groups and all the cluster sizes ($k = 2, 3, 4$). Furthermore, the alternation of the state transition probability in SZ between the frontal and occipital leading phase state in the beta and gamma bands was observed in the case of cluster size $k = 2$. Similarly, in the cluster size $k = 3, 4$, the alternation of the state transition probability of SZ between the right and left hemispheric leading phase state in the beta and gamma bands was observed.

We further discussed the reason for the region specific alternation of SZ in the emergence probability and state transition probability. In the case of cluster size $k = 2$, differences were found in the frontal-occipital regions of the emergence and state transition probability. A previous study reported the enhancement of hub structure centering at the occipital region in the FC of the fast frequency components of the EEG signals in SZ [23]. The hub structure plays a role in the integration of the neural information process [24]; therefore, the enhancement of this hub structure affects the state occipital leading phase (see the result of $k = 4$ at the beta band in Fig. 4(b), the result of $k = 2, 3, 4$ at the gamma band in Figs. 2, 3, and 4(b)), especially, might induce the transition of the state occipital leading phase (see the result of $k = 2, 3$ at the beta band in Fig. 5 and the result of $k = 2, 3, 4$ at the gamma band in Fig. 6). Moreover, in the case of cluster size $k = 3, 4$, differences of the state transition probability were found in the left and right hemispheric alterations. Many studies have suggested a decrease in the connectivity of the corpus callosum in SZ [25]. The observed alternation of state transition between the left and right hemispheric is considered to be attributed to this abnormality of the corpus callosum in SZ.

Our study had certain limitations. Previous studies of hub location and node centrality reported reduced the average centrality across hubs, shorter path lengths, and significantly weaker centrality within frontal regions [24, 26]. The hub integrates information; however, it is currently only undirected, such as the PLI. The high transition probability to the occipital in the SZ IF micro-state transitions could be attributed to the centrality and hub structures. However, in order to correspond to the leading phase state transition in this study, the directed hub structure in the FC detected by the directed connectivity measures typified as directed PLI and transfer entropy should be considered. Moreover, although MEG is less clinically adaptable than EEG, applying IF-micro-state to MEG with high spatial resolution may lead more detailed characteristics than one for EEG. Additionally, this study used the k-means method as the most convetional clustering method for micro-state clustering; while, other methods such as Ward's method and furthest neighbor method may be used to capture the other aspects of state transitions. Thus, these points should be dealt with in future studies.

In conclusion, despite the limitations, our proposed IF micro-state based on the instantaneous frequency of the EEG approach succeeded in the detection of state transition alternations concerning SZ pathology. This approach could contribute in the elucidation of new aspects of dFC, which results in the establishment of a biomarker for SZ.

**Acknowledgment.** This study was supported by JSPS KAKENHI for a Grant-in-Aid for Scientific Research (C) (Grant No. 22K12183) (SN). This study was partially supported by the JST CREST (Grant No. JPMJCR17A4).

# References

1. Green, M.F., Horan, W.P., Lee, J.: Nonsocial and social cognition in schizophrenia: current evidence and future directions. World Psychiatry **18**(2), 146–161 (2019)
2. Correll, C.U., Schooler, N.R.: Negative symptoms in schizophrenia: a review and clinical guide for recognition, assessment, and treatment. Neuropsychiat. Dis. Treat. **16**, 519 (2020)
3. Friston, K.J.: The disconnection hypothesis. Schizoph. Res. **30**(2), 115–125 (1998)
4. Medaglia, J.D., Lynall, M.-E., Bassett, D.S.: Cognitive network neuroscience. J. Cogn. Neurosci.**27**(8), 1471–1491 (2015)
5. Cohen, J.R.: The behavioral and cognitive relevance of time-varying, dynamic changes in functional connectivity. NeuroImage **180**, 515–525 (2018)
6. Takahashi, T., Goto, T., Nobukawa, S., Tanaka, Y., Kikuchi, M., Higashima, M., Wada, Y.: Abnormal functional connectivity of high-frequency rhythms in drug-naïve schizophrenia. Clin. Neurophysiol. **129**(1), 222–231 (2018)
7. Sporns, O., Honey, C.J.: RolfKotter. Identification and classification of hubs in brain networks. PLoS ONE **2**(10), e1049 (2007)
8. Misic, B., Vakorin, V.A., Paus, T., McIntosh, A.R.: Functional embedding predicts the variability of neural activity. Front. Syst. Neurosci. **5**, 90 (2011)
9. Nobukawa, S., Kikuchi, M., Takahashi, T.: Changes in functional connectivity dynamics with aging: a dynamical phase synchronization approach. Neuroimage **188**, 357–368 (2019)
10. Takahash, T., et al.: Antipsychotics reverse abnormal EEG complexity in drug-naive schizophrenia: a multiscale entropy analysis. Neuroimage **51**(1), 173–182 (2010)
11. Fernández, A., Gómez, C., Hornero, R., José López-Ibor, J.: Complexity and schizophrenia. Progr. Neuro-Psychopharm. Biol. Psychiatr. **45**, 267–276 (2013)
12. Dong, D., Duan, M., Wang, Y., Zhang, X., Jia, X., Li, Y., Xin, F., Yao, D., Luo, C.: Reconfiguration of dynamic functional connectivity in sensory and perceptual system in schizophrenia. Cereb. Cortex **29**(8), 3577–3589 (2019)
13. Kang, J., Pae, C., Park, H.-J.: Graph-theoretical analysis for energy landscape reveals the organization of state transitions in the resting-state human cerebral cortex. PLoS ONE **14**(9), e0222161 (2019)
14. Allen, E.A., et al.: Tracking whole-brain connectivity dynamics in the resting state. Cerebral Cortex **24**(3), 663–676 (2014))
15. Tewarie, P., et al.: Tracking dynamic brain networks using high temporal resolution meg measures of functional connectivity. Neuroimage **200**, 38–50 (2019)
16. Hu, M.L., et al.: A review of the functional and anatomical default mode network iMn schizophrenia. Neurosci. Bull. **33**(1), 73–84 (2017)

17. Yin, W., et al.: The emergence of a functionally flexible brain during early infancy. Proc. Natl. Acad. Sci. **117**(38), 23904–23913 (2020)
18. Sporns, O., Betzel, R.F.: Modular brain networks. Ann. Rev. Psychol. **67**, 613–640 (2016)
19. Battiston, F., Cencetti, G., Iacopini, I., Latora, V., Lucas, M., Patania, A., Young, J.-G., Petri, G.: Networks beyond pairwise interactions: structure and dynamics. Phys. Rep. **874**, 1–92 (2020)
20. Nobukawa, S., Ikeda, T., Kikuchi, M., Takahashi, T.: Dynamical characteristics of state transition defined by neural activity of phase in alzheimer's disease. In: Mantoro, T., Lee, M., Ayu, M.A., Wong, K.W., Hidayanto, A.N. (eds.) ICONIP 2021. CCIS, vol. 1517, pp. 46–54. Springer, Cham (2021). https://doi.org/10.1007/978-3-030-92310-5_6
21. Khanna, A., Pascual-Leone, A., Michel, C.M., Farzan, F.: Microstates in resting-state EEG: current status and future directions: Neurosci. Biobeh. Rev. **49**, 105–113 (2015)
22. Strelets, V., et al.: Chronic schizophrenics with positive symptomatology have shortened EEG microstate durations. Clin. Neurophysiol. **14**, (11), 2043–2051 (2003)
23. Krukow, P., Jonak, K., Karpiński, R., Karakuła-Juchnowicz, H.: Abnormalities in hubs location and nodes centrality predict cognitive slowing and increased performance variability in first-episode schizophrenia patients. Sci. Rep. **9**(1), 1–13 (2019)
24. van den Heuvel, M.P., Mandl, R.C.W. Stam, C.J., Kahn, R.S., Hulshoff Pol, H.E.: Aberrant frontal and temporal complex network structure in schizophrenia: a graph theoretical analysis. J. Neurosci. **30**(47), 15915–15926 (2010)
25. Najjar, S., Pearlman, D.M.: Neuroinflammation and white matter pathology in schizophrenia: systematic review. Schizophrenia Res. **161**(1), 102–112 (2015)
26. Jalili, M., Knyazeva, M.G.: EEG-based functional networks in schizophrenia. Comput. Biol. Med. **41**(12), 1178–1186 (2011)

# Lagrange Programming Neural Networks for Sparse Portfolio Design

Hao Wang[1], Desmond Hui[2], and Chi-Sing Leung[3(✉)]

[1] College of Electronics and Information Engineering, Shenzhen University, Shenzhen, China
haowang@szu.edu.cn
[2] Department of Mechanical and Industrial Engineering, University of Toronto,
Toronto, Canada
desmond.hui@mail.utoronto.ca
[3] Department of Electrical Engineering, City University of Hong Kong, Hong Kong, China
eeleungc@cityu.edu.hk

**Abstract.** Although there were some works on analog neural networks for sparse portfolio design, the existing works do not allow us to control the number of the selected assets and to adjust the weighting between the risk and return. This paper proposes a Lagrange programming neural network (LPNN) model for sparse portfolio design, in which we can control the number of selected assets. Since the objective function of the sparse portfolio design contains a non-differentiable $\ell_1$-norm term, we cannot directly use the LPNN approach. Hence, we propose a new formulation based on an approximation of the $\ell_1$-norm. In the theoretical side, we prove that state of the proposed LPNN network globally converges to the nearly optimal solution of the sparse portfolio design. The effectiveness of the proposed LPNN approach is verified by the numerical experiments. Simulation results show that the proposed analog approach is superior to the comparison analog neural network models.

**Keywords:** Lagrange programming neural network · sparse portfolio optimization · financial data · analog neural network · analog optimization

## 1 Introduction

The analog neural network approach for optimization received a lot of attention in the last several decades. The pioneering works of analog models date back to the 1980s [1,2]. In [1], the Hopfield model was proposed for several optimization problems. In [3], the analog neural approach was demonstrated to be able to solve various quadratic optimization problems. Also, it is widely utilized in many optimization problems, such as feature selection [4] and sparse approximation [5]. Nevertheless, many existing models were developed for solving a set of particular problems.

The Lagrange programming neural network (LPNN) approach [5–7] is a general solver for various constrained optimization problems. Recently, the LPNN approach was adopted in many new areas, such as sparse approximation [5], robust target localization in multi-input multi-output and time-difference-of-arrival systems [6]. In these signal processing applications, the LPNN-based solver achieves comparable or better performance in comparison with numerical methods.

© The Author(s), under exclusive license to Springer Nature Switzerland AG 2023
M. Tanveer et al. (Eds.): ICONIP 2022, LNCS 13624, pp. 37–48, 2023.
https://doi.org/10.1007/978-3-031-30108-7_4

As a cornerstone of modern finance, portfolio optimization [8,9] is a key topic. It aims at selecting the assets to be invested and determining the percentages of investments on the selected assets. A classical approach for asset allocation is the mean-variance (MV) portfolio selection model [8,9]. Since a dense portfolio in the MV model creates some difficulties in asset management and high transaction costs, one of the research directions is to construct a sparse portfolio.

There were some works using continuous time neural networks [10,11] for sparse portfolio design, which are the constrained pseudoconvex optimization (CPO) [10] and the collaborative neurodynamic optimization (CNO) [11]. However, their formulations have some limitations. They do not allow users to control the number of the selected assets and to tune the weighting between risk and return. Hence it is interesting to explore new neural network models to overcome the limitations.

This paper introduces a neural model, based on the LPNN concept, for the sparse portfolio design. To achieve sparsity, there is an $\ell_1$-norm penalty term in the objective function. Since the $\ell_1$-norm term is nondifferentiable, we propose a new formulation of the LPNN framework based on an approximation of the $\ell_1$-norm. On the theoretical side, we show that the equilibrium point of the LPNN model is the optimal solution of the sparse MV model. In addition, we show that the state of the proposed LPNN network globally converges to the equilibrium point of the LPNN model. The effectiveness of the proposed LPNN approach is verified by numerical experiments on three data sets: Kenneth French 49 Industry (49Ind), 100 Fama French (100FF) and Standard & Poor's 500 (S&P 500). From the experimental results, there are no significant differences between our analog approach and the comparison digital method. Also, our method is superior to the two mentioned analog models, CPO and CNO.

The rest of this paper is arranged as follows. Section 2 presents the MV model and the Lagrange programming neural network. The proposed LPNN-based algorithm for portfolio optimization is given in Sect. 3. Section 4 includes the experimental results. Finally, conclusions are presented in Sect. 5.

## 2 Background

**Portfolio Optimization**
Consider $n$ risky assets and the daily return matrix $R \in \mathbb{R}^{D \times n}$, where each row vector in $R$ is the return vector in a particular day. Let $\mu \in \mathbb{R}^n$ and $C \in \mathbb{R}^{n \times n}$ be the mean return vectors and covariance matrix of return vectors, respectively. The classical MV model is given by:

$$\min_{w} \quad w^{\mathrm{T}} C w - \gamma \mu^{\mathrm{T}} w, \qquad \text{s.t.} \quad \mathbf{1}^{\mathrm{T}} w = 1, \tag{1}$$

where $w = [w_1, \ldots, w_n]^{\mathrm{T}} \in \mathbb{R}^n$ is the portfolio weight vector, $\gamma$ is the risk preference of an investor, and $\mathbf{1} \in \mathbb{R}^n$ is a vector whose elements are ones. Element $w_i$ represents the percentage of investment in the $i$-th assets. The values of $w_i$'s can negative. Note that "$w_i < 0$" means that short-selling is performed on the $i$th asset. The optimal solution of (1) is usually a dense vector. Also, the obtained weights are sensitive to the estimation errors and extreme samples in $R$. Some researchers suggest adding an $\ell_1$-norm penalty term, $\|x\|_1$, into the objective function. The modified MV model becomes the following non-smooth constrained optimization problem:

$$\min_{w} \quad w^{\mathrm{T}}Cw - \gamma\mu^{\mathrm{T}}w + \rho\|w\|_1, \qquad \text{s.t.} \quad \mathbf{1}^{\mathrm{T}}w = 1, \tag{2}$$

where $\rho > 0$ is the penalty parameter. In (2), the $\ell_1$-norm term is to compensate the effect of outliers and to limit the number of the selected assets. Although the constrained optimization problem, stated in (2), contains a non-differential term, we can solve it using the alternating direction method of multipliers (ADMM) concept [12].

**Lagrange Programming Neural Network**

In the analog neural approach, we solve an optimization problem by deriving a number of continuous time differential equations. We use a number of neurons to hold the values of the decision variables of the optimization problem. Those differential equations manages the state update of the neurons. After the neurons' state converges to an equilibrium state, we obtain the solution from the state of the neurons.

The LPNN approach aims at solving the following optimization problem:

$$\min_{x} \quad f(x), \qquad \text{s.t.} \quad \mathbf{h}(x) = \mathbf{0}, \tag{3}$$

where $x \in \mathbb{R}^n$ is the collection of decision variables, $f : \mathbb{R}^n \rightarrow \mathbb{R}$ is the objective function, and $\mathbf{h} : \mathbb{R}^n \rightarrow \mathbb{R}^m$ describes the $m$ equality constraints. When $f$ and $\mathbf{h}$ should be twice differentiable, we can define the Lagrangian function:

$$\mathcal{L}(x, \lambda) = f(x) + \lambda^{\mathrm{T}}\mathbf{h}(x), \tag{4}$$

where $\lambda \in \mathbb{R}^m$ is the Lagrange multiplier vector. In a LPNN, there are two classes of neurons, namely, variable neurons and Lagrange neurons. The variable neurons hold the decision variable vector $x$, while the Lagrange neurons hold multipliers vector $\lambda$. The dynamics of those neurons are defined by

$$\frac{dx}{dt} = -\nabla_x\mathcal{L}(x, \lambda), \text{ and } \frac{d\lambda}{dt} = \nabla_\lambda\mathcal{L}(x, \lambda). \tag{5}$$

## 3 $\ell_1$-LPNN for Sparse Portfolio Selection

### 3.1 Development of $\ell_1$-LPNN

Since the $\ell_1$-norm term in (2) is nondifferentiable, we cannot directly explore the LPNN concept for sparse portfolio selection. To handle the nondifferentiable issue, we utilize an approximation:

$$\|w\|_1 = \sum_{i=1}^{n} |w_i| \approx \sum_{i=1}^{n} \frac{\ln(\cosh(\alpha w_i))}{\alpha}, \tag{6}$$

where $\alpha$ should be a large positive number.

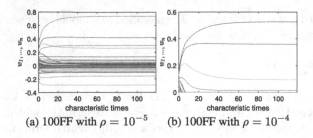

(a) 100FF with $\rho = 10^{-5}$        (b) 100FF with $\rho = 10^{-4}$

**Fig. 1.** Convergence examples of $\ell_1$-LPNN with $\gamma = 0.001$.

One may concern that the approximation of $\frac{\ln(\cosh(\alpha w_i))}{\alpha}$ is complicated and would be difficult to implement in an analog circuit. In fact, as shown in the rest of this section, with this approximation, we only need to implement the derivative of this approximation, i.e., $\tanh(\alpha w_i)$. Activation function $\tanh(\cdot)$ is a commonly used activation function in the neural network community.

With the approximation, the sparse portfolio optimization (2) becomes

$$\min_{\boldsymbol{w}} \quad \boldsymbol{w}^\mathrm{T} \mathbf{C} \boldsymbol{w} - \gamma \boldsymbol{\mu}^\mathrm{T} \boldsymbol{w} + \frac{\rho}{\alpha} \sum_{i=1}^{n} \ln(\cosh(\alpha w_i)), \qquad \text{s.t.} \quad \mathbf{1}^\mathrm{T} \boldsymbol{w} = 1. \qquad (7)$$

To explore the LPNN concept, we construct the Lagrangian of (7):

$$\mathcal{L}(\boldsymbol{w}, \lambda) = \boldsymbol{w}^\mathrm{T} \mathbf{C} \boldsymbol{w} - \gamma \boldsymbol{\mu}^\mathrm{T} \boldsymbol{w} + \lambda(\mathbf{1}^\mathrm{T} \boldsymbol{w} - 1) + \frac{\rho}{\alpha} \sum_{i=1}^{n} \ln(\cosh(\alpha w_i)), \qquad (8)$$

where $\lambda \in \mathbb{R}$ is the Lagrange multiplier. Applying (5) on (8), we can define the dynamics for (7) as follows:

$$\frac{d\boldsymbol{w}}{dt} = -\nabla_{\boldsymbol{w}} \mathcal{L}(\boldsymbol{w}, \lambda) = -(2\mathbf{C}\boldsymbol{w} - \gamma\boldsymbol{\mu} + \rho \tanh(\alpha\boldsymbol{w}) + \lambda\mathbf{1}), \qquad (9a)$$

$$\frac{d\lambda}{dt} = \nabla_{\lambda} \mathcal{L}(\boldsymbol{w}, \lambda) = \mathbf{1}^\mathrm{T} \boldsymbol{w} - 1. \qquad (9b)$$

Figure 1 shows the dynamics of the LPNN model on the 100FF dataset. The datasets are described in Sect. 4. It is observed that the network can settle down within 100 characteristic times.

### 3.2 Properties of the Sparse MV Problem and $\ell_1$-LPNN

Before we present the convergence properties, we first investigate how the equilibrium point relates to the optimal solution of the optimization solution. Since the sparse MV problem stated in (7) is convex, from basic optimization theories, we have the following lemma.

**Lemma 1.** *A point $\boldsymbol{w}^*$ is an optimal solution of the sparse MV problem, if and only if, there exists a $\lambda^*$ (Lagrange multiplier), such that*

$$2\mathbf{C}\boldsymbol{w}^* - \gamma\boldsymbol{\mu} + \rho \tanh(\alpha\boldsymbol{w}) + \lambda^*\mathbf{1} = \mathbf{1}, \qquad (10a)$$

$$\mathbf{1}^\mathrm{T} \boldsymbol{w}^* - 1 = 0. \qquad (10b)$$

In Lemma 1, (10) tells us the Karush Kuhn Tucker (KKT) conditions of the sparse MV problem. Since the problem is convex, the KKT conditions are sufficient and necessary. That means, if there exists $\{w^*, \lambda^*\}$ satisfying (10), then $w^*$ is the optimization solution of the sparse MV problem.

For the proposed $\ell_1$-LPNN model, at the equilibrium, we have

$$\frac{dw}{dt} = 0 \quad \text{and} \quad \frac{d\lambda}{dt} = 0. \tag{11}$$

Hence, we have the following lemma to describe the relationship between the optimal solution of the sparse MV problem and the equilibrium point of the $\ell_1$-LPNN.

**Lemma 2.** *A point $\{w^*, \lambda^*\}$ is an equilibrium point of the $\ell_1$-LPNN, if and only if, it is the optimal solution of the sparse MV problem.*

**Proof:** Let $\{w^*, \lambda^*\}$ be an equilibrium point of the $\ell_1$-LPNN. From (9) and (11), at this point, we have

$$2Cw^* - \gamma\mu + \rho\tanh(\alpha w) + \lambda^*\mathbf{1} = \mathbf{1}, \tag{12a}$$
$$\mathbf{1}^Tw^* - 1 = 0. \tag{12b}$$

Clearly, (12) is the same as the KKT conditions of the sparse MV problem, i.e., (12) is the same as (10). Hence, we conclude that $\{w^*, \lambda^*\}$ is an equilibrium point of the $\ell_1$-LPNN, if and only if, it is the optimal solution of the sparse MV problem. The proof is completed. ∎

## 3.3 Global Stability

The aforementioned subsection only tells us that an equilibrium point of our LPNN model is the optimal solution. But it does not tell us whether the state of the network converges to the optimal solution (equilibrium point) or not. Since the problem, stated in (6), is convex, the optimal solutions $\{x^*, \lambda^*\}$ exist. In addition, from Lemmas 1 and 2, any optimal solution corresponds to an equilibrium point. Hence, if we can prove that the neurons' state, according to (9), converges to an equilibrium, then the neurons' state converges to the optimal solution.

The rest of this section will prove that the neurons' state, according to (9), converges to an equilibrium. Mathematically, the convergence means that as $t \to \infty$, $w(t) \to w^*$ and $\lambda(t) \to \lambda^*$. Our proof has the following key points:

1. We define a scalar function $V(w, \lambda)$.
2. This function is lower bounded and is radially unbounded.
3. We then prove that the time derivative $\dot{V}(w, \lambda) = \frac{dV}{dt} = 0$ for $\{w, \lambda\} = \{x^*, \lambda^*\}$, and that $\dot{V}(w, \lambda) = \frac{dV}{dt} > 0$ for $\{w, \lambda\} \neq \{x^*, \lambda^*\}$.
4. From the well known Lyapunov theory, the state for any initial state $\{w(0), \lambda(0)\}$, as $t \to \infty$, $\{w(t), \lambda(t)\} \to \{w^*, \lambda^*\}$.

**Lyapunov function:**

For our LPNN model, we define the following Lyapunov function $V(w, \lambda)$:

$$V(w, \lambda) = \frac{1}{2}\|2Cw - \gamma\mu + \rho\tanh(\alpha w) + \lambda\mathbf{1}\|_2^2 + \frac{1}{2}\|\mathbf{1}^Tw - 1\|_2^2$$
$$+ \frac{1}{2}\|w - w^*\|_2^2 + \frac{1}{2}\|\lambda - \lambda^*\|_2^2. \tag{13}$$

Clearly, $V(w, \lambda) > 0$. Also, $\{w, \lambda\}$ is an equilibrium point, its corresponding $V$ is equal to zero. Thus $V(w, \lambda)$ is lower bounded. Besides, as $|w| \to \infty$ and $|\lambda| \to \infty$, $V(w, \lambda) \to \infty$. That implies that $V(w, \lambda)$ is radially unbounded.

**Time derivative of Lyapunov function:**

We separate the Lyapunov function $V(w, \lambda)$ into two parts:

$$V(w, \lambda) = V_1(w, \lambda) + V_2(w, \lambda), \tag{14a}$$

$$V_1(w, \lambda) = \frac{1}{2}\|2Cw - \gamma\mu + \rho\tanh(\alpha w) + \lambda\mathbf{1}\|_2^2 + \frac{1}{2}\|\mathbf{1}^Tw - 1\|_2^2, \tag{14b}$$

$$V_2(w, \lambda) = \frac{1}{2}\|w - w^*\|_2^2 + \frac{1}{2}\|\lambda - \lambda^*\|_2^2. \tag{14c}$$

The derivative of $V(w, \lambda)$ w.r.t. time is

$$\dot{V}(w, \lambda) = \dot{V}_1(w, \lambda) + \dot{V}_2(w, \lambda),$$
$$= \left[\frac{\partial V_1(w, \lambda)}{\partial w}\right]^T \frac{dw}{dt} + \left[\frac{\partial V_1(w, \lambda)}{\partial \lambda}\right]^T \frac{d\lambda}{dt}$$
$$+ \left[\frac{\partial V_2(w, \lambda)}{\partial w}\right]^T \frac{dw}{dt} + \left[\frac{\partial V_2(w, \lambda)}{\partial \lambda}\right]^T \frac{d\lambda}{dt}. \tag{15}$$

Let

$$g = 2Cw - \gamma\mu + \rho\tanh(\alpha w) + \lambda. \tag{16}$$

Thus, $\frac{dw}{dt} = -g$. The derivative $\dot{V}_1(w, \lambda)$ w.r.t. time is given by:

$$\dot{V}_1(w, \lambda) = -g^T(Ag + (\mathbf{1}^Tw - 1)\mathbf{1}) + (\mathbf{1}^Tw - 1)g^T\mathbf{1} = -g^TAg, \tag{17}$$

where $A = 2C + \alpha\rho B$ is positive definite, where $B$ is a diagonal matrix with diagonal elements equal to "$1 + \alpha\rho\tanh(\alpha w_i)$".

For $\dot{V}_2(w, \lambda)$, we have:

$$\dot{V}_2(w, \lambda) = -g^T(w - w^*) + (\mathbf{1}^Tw - 1)(\lambda - \lambda^*). \tag{18}$$

Based on (12), $\dot{V}_2(w, \lambda)$ can be rewritten as

$$\dot{V}_2(w, \lambda) = -(w - w^*)^T(2C(w - w^*) + \rho(\tanh(\alpha w) - \tanh(\alpha w^*)) + (\lambda - \lambda^*)\mathbf{1})$$
$$+ \mathbf{1}^T(w - w^*)(\lambda - \lambda^*)$$
$$= -2(w - w^*)^TC(w - w^*) - \rho(w - w^*)^T(\tanh(\alpha w) - \tanh(\alpha w^*)). \tag{19}$$

Now we consider three cases of $\{w, \lambda\}$.

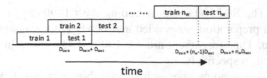

**Fig. 2.** Illustration of rolling windows.

- $\{w, \lambda\} = \{w^*, \lambda^*\}$:

  From (12) and (16), $g$ becomes a zero vector. Thus, from (17) and (19) $\dot{V}_1(w^*, \lambda^*) = \dot{V}_2(w^*, \lambda^*) = 0$. That is, $\dot{V}(w^*, \lambda^*) = 0$.

- $\{w = w^*, \lambda \neq \lambda^*\}$:

  In this case, $\{w, \lambda\}$ is not an equilibrium point. Since $w = w^*$, we have $\frac{d\lambda}{dt} = 0$ (from (12b)), and $\dot{V}_2(w^*, \lambda) = 0$. However, as $\lambda \neq \lambda^*$ and the point is not an equilibrium point, we have $g \neq 0$ and $\dot{V}_1(w^*, \lambda) < 0$ (from (17)), and $\dot{V}(w^*, \lambda^*) < 0$. Note that $A$ is positive definite.

- $\{w \neq w^*\}$:

  Clearly, for $w \neq w^*$, we have $g \neq 0$ and $\dot{V}_1(w, \lambda) < 0$. Since $\tanh(\cdot)$ is a strictly monotonic function, we get $-\rho(w - w^*)^T (\tanh(\alpha w) - \tanh(\alpha w^*)) < 0$. Hence, $\dot{V}_2(w, \lambda) < 0$. It can be seen that $\dot{V}(w, \lambda) = \dot{V}_1(w, \lambda) + \dot{V}_2(w, \lambda) < 0$.

From the above cases, we can conclude that $\dot{V}(w, \lambda) < 0$ for $\{w, \lambda\} \neq \{w^*, \lambda^*\}$ and $\dot{V}(w^*, \lambda^*) = 0$. In addition, $V(w, \lambda)$ is lower bounded and radially unbounded. Therefore, $V(w, \lambda)$ is a Lyapunov function for (9). The analog system has global asymptotic stability.

## 4    Experiments

This section has two objectives. **The first one** is to verify the effectiveness of our proposed LPNN method. Since we use an approximation in our LPNN method. The purpose of this section is to verify whether the performances of LPNN are similar to that of the digital numerical algorithm $\ell_1$-ADMM [8, 12].

**The second objective** is to compare the performance of our LPNN with some existing analog models. Although there are no other analog models for the sparse MV problem stated in (2), there are other analog models, CPO [10] and CNO [11], for sparse portfolio design. Hence, we compare the performance of LPNN with the CPO and CNO. Note that the formulations of the two comparison models do not allow us to control the number of the selected assets and to tune the weighting between risk and return.

### 4.1    Settling

We consider three real-world datasets: 49Ind, 100FF and S&P500, to verify our proposed model. In the 49Ind and 100FF datasets, there are 2788 trading days (3 Jan. 2005 to 29 Jan. 2016)[1]. The S&P500 contains trading days 1699 days (1 May 2009 to 29 Jan 2016)[2]. In data preparation, suspended and newly enlisted assets within the time period are excluded. Therefore, there are 49, 100 and 414 assets in the 49Ind, 100FF and S&P500 datasets, respectively.

We use the rolling window approach in [8], shown in Fig. 2. We use a training window with 500 days data to generate the portfolio and then use the following 100/200 days data to evaluate its out-of-sample performance. Here the test window size is called as "rebalancing period".

For the LPNN approach and $\ell_1$-ADMM, we consider three risk preferences: $\gamma = \{0.001, 0.005, 0.01\}$. The penalty parameter $\rho$ ranges from $10^{-6}$ to $10^{-4}$ for 49Ind and 100FF, and ranges from $10^{-7}$ to $10^{-4}$ for S&P500.

Two performance indicators are used for evaluation. One is the mean daily return (MDR) (mean daily return of test periods), denoted as $\mu$. For the $\tau$-th testing window, let $\boldsymbol{r}_\tau \in \mathbb{R}^n$ be the daily return vector over the testing period, where $[\boldsymbol{r}_\tau]_i$ is the daily return for holding the $i$-th assest for $l_{test}$ days. The MDR is defined as

$$\mu = \frac{1}{N_w} \sum_{\tau=1}^{N_w} \boldsymbol{y}_\tau^{\mathrm{T}} \boldsymbol{r}_\tau, \tag{20}$$

where $N_w$ is the number of testing periods.

Another one is the Sharpe ratio, denoted as $\mathcal{S}$. In finance management, a higher return usually results in a higher risk (variation of the returns). The Sharpe ratio is an indicator that balances the risk and return, given by

$$\mathcal{S} = \frac{\mu}{\sigma}, \text{ where } \sigma = \sqrt{\frac{1}{N_w - 1} \sum_{\tau=1}^{N_w} (\boldsymbol{y}_\tau^{\mathrm{T}} \boldsymbol{r}_\tau - \mu)^2}, \tag{21}$$

where $\sigma$ is the standard derivation of daily returns, i.e., the variation of returns.

In finance management, for two portfolios with the similar return, we should select the one with a higher Sharpe ratio. Similarly, for two portfolios with the similar Sharpe ratio, we should select the one with a higher return.

## 4.2 Sparsity

The sparse portfolio design is a kind of multi-objective optimization problems. A good portfolio should be with a small number of assets, a high return, and a high Sharpe ratio. This subsection studies the effect of the regularization $\rho$ on the number of the selected assets.

Since the objective formulation of our LPNN model is the same as the digital numerical algorithm $\ell_1$-ADMM [8], the properties of LPNN should be the same as those of the $\ell_1$-ADMM. A property of the sparse MV problem is that we can use the regularization parameter $\rho$ to control the number $K$ of the selected assets. When we increase

---

[1] http://mba.tuck.dartmouth.edu/pages/faculty/ken.french.

[2] https://finance.yahoo.com.

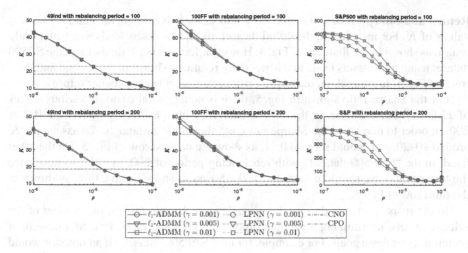

**Fig. 3.** Sparsity parameters versus the number of the selected assets.

$\rho$, the number $K$ of the selected assets becomes small. It should be noticed that the formulations of the CPO and the CNO models do not allow us to control the number of the selected assets.

Figure 3 shows the regularization parameter $\rho$ versus the number $K$ of the selected assets. Since we use the sliding window concept, the number of the selected assets in the resultant portfolios are not integers. It can be seen that as we increase $\rho$, the number $K$ of the selected assets gradually decreases. On the other hand, varying the risk parameter $\gamma$ and the rebalancing period does not affect the number $K$ of the selected assets. From the figure, there are no significant differences among the LPNN and $\ell_1$-ADMM. It is because these two algorithms come from the same formulation. Also, the figure confirms that the solutions obtained from our LPNN model are similar to those obtained from the digital $\ell_1$-ADMM algorithm.

Since the formulations of the CPO and the CNO models do not allow us to control the number of the selected assets, the numbers of the selected assets in the CPO and the CNO models are constants for a setting. For example, in the S&P500 dataset with the rebalancing period of 100 days, the numbers of selected assets in the CPO and the CNO models are 32.8 and 29.4, respectively.

### 4.3 Performance: Verification and Comparison

Figures 4 and 5 show the return and Sharpe ratio under various settings. From the figures, we can observe that in terms of return and Sharpe ratio, there are no significant differences between our proposed analog LPNN method and the $\ell_1$-ADMM method (a digital numerical method). Hence, we can conclude that our analog LPNN method is a good alternative for sparse portfolio optimization.

In the rest of this subsection, we discuss the influence of settings on the performance indicators (return and Sharpe ratio), and the comparison between our LPNN model and the two comparison analog models (CPO and CNO).

**Return and Sharpe Ratio.** For the daily return, there is no general trend on various values of $K$. For instance, in the 49Ind dataset, using less assets leads to a better daily return, as shown in the first row of Fig. 4. However, for the 100FF dataset and the SP500 dataset using more assets leads to a better daily return, as shown in the second and third rows of Fig. 4. In general, increasing the risk parameter $\gamma$ leads to better return.

For the Sharpe ratio shown in Fig. 5, there is no a general trend on various values of $K$ and $\gamma$. For example, in the "49Ind dataset" with rebalancing periods of 100 or 200, in order to maximize the Sharpe ratio, we should use around 10–20 assets, i.e. $K$ around 10–20, and should set $\gamma$ to 0.01, as shown in the first row of Fig. 5. On the other hand, in the "S&P500 dataset" with rebalancing period of 200, in order to maximize the Sharpe ratio, we should use around 100–150 assets and set $\gamma$ to 0.001, as shown in the third row of Fig. 5.

In the sparse portfolio design, there are three goals: minimizing the number of the selected assets, maximizing return and maximizing Sharpe ratio, Investors have their preferences on those goals. For example, for the "S&P500 dataset", if an investor would like to maximize the Sharpe ratio, they should use around 100–150 assets with the rebalancing period equal to 200 days, as shown in the third row of Fig. 5.

**Comparison with Other Analog Models.** Now we focus on the discussion on the comparison among the three analog models: our LPNN mode, CPO and CNO. In the "100FF dataset", shown in Figs. 4–5, our two proposed LPNN model is much better than the CNO and CPO models. For example, with the rebalancing period of 100 days, the return and Sharpe ratio of CNO are 0.0166 % (the second row of Fig. 4) and 0.1202 (the second row of Fig. 5), respectively. They are much lower than those of our LPNN model. For the CPO model, the return and Sharpe ratio are 0.0412 % (the second row of Fig. 4) and 0.3237 (the second row of Fig. 5), respectively. They are much lower than those of our models with $\gamma$ equal to 0.005 and 0.01.

In the "49InD dataset", the return and the Sparpe ratio of CPO are pretty low and we focus on the discuss on the comparison between CNO and our LPNN model. With the rebalancing period of 100 days, the return and the Sharpe ratio of CNO are 0.03576 % and 0.4378, respectively, as shown in the first row of Fig. 4 and the first row of Fig. 5. Also, the number of the selected assets from the CNO is around 18. Clearly, from the figures, our LPNN model has better performance with less assets. With the rebalancing period of 100 days, when we use 10 assets and set $\gamma$ to 0.01 in our LPNN model, the return and the Sharpe ratio of our model are 0.037 % and 0.5227, respectively, as shown in the first row of Fig. 4 and the first row of Fig. 5. These values are better than those of CNO.

In the "S&P500 data set", again, the return and the Sparpe ratio of CPO are pretty low. Hence we focus on the comparison between CNO and our LPNN model. For the rebalancing period of 100 days the return and the Sharpe ratio of CNO are equal to 0.0427 % and 0.61825, respectively, as shown in the first row of Fig. 4 and the first row of Fig. 5. Also, the number of the selected assets of CNO is around 30. Clearly, the performances of our model are better than those of CNO. For instance, for the rebalancing period of 100 days in our LPNN model, we can use 38 assets with $\gamma$ equal to 0.01. With this setting, the return and the Sharpe ratio of our LPNN model are equal

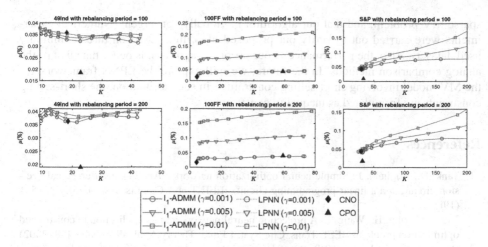

**Fig. 4.** Mean Return. The first row presents the results with rebalancing every 100 days, while the second row presents the results with re-balancing every 200 days.

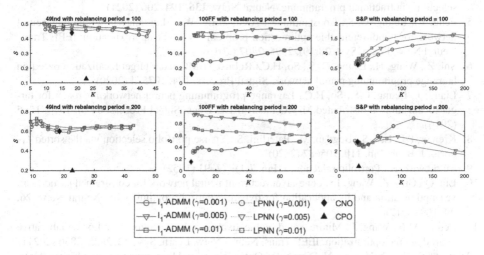

**Fig. 5.** Sharpe Ratio. The first row presents the results with rebalancing every 100 days, while the second row presents the results with re-balancing every 200 days.

to 0.0530 % and 0.8314, respectively. Of course, if investors would like to use more assets, such as around 120 assets, then the return and the Sharpe ratio of our LPNN are much better. They are equal to 0.1122 % and 1.6874, respectively.

## 5 Conclusion

This paper presented a LPNN-based approach for solving the sparse MV problem. We showed that the proposed analog algorithm is able to find out the optimal solution of the sparse MV problem, and that our algorithm is globally convergent. Numerical experiments were carried out to verify our proposed algorithm based on three commonly used datasets. Also, the performance of our proposed approach is better than that of the analog comparison methods. In the future, we will explore the LPNN framework for the MV model involving short-selling constraints. In such a situation, the short-selling constraints are formulated as inequality constraints.

# References

1. Tank, D., Hopfield, J.: Simple'neural'optimization networks: An a/d converter, signal decision circuit, and a linear programming circuit. IEEE Trans. Circuits Syst. **33**(5), 533–541 (1986)
2. Xia, Y., Leung, H., Wang, J.: A projection neural network and its application to constrained optimization problems. IEEE Trans. Circ Syst. I Fund. Theory Appl. **49**(4), 447–458 (2002)
3. Bouzerdoum, A., Pattison, T.R.: Neural network for quadratic optimization with bound constraints. IEEE Trans. Neural Netw. **4**(2), 293–304 (1993)
4. Wang, Y., Li, X., Wang, J.: A neurodynamic optimization approach to supervised feature selection via fractional programming. Neural Netw. **136**, 194–206 (2021)
5. Feng, R., Leung, C.-S., Constantinides, A.G., Zeng, W.J.: Lagrange programming neural network for nondifferentiable optimization problems in sparse approximation. IEEE Trans. Neural Netw. Learn. Syst. **28**(10), 2395–2407 (2016)
6. Shi, Z., Wang, H., Leung, C.-S., So, H.C.: Robust MIMO radar target localization based on lagrange programming neural network. Signal Process. **174**, 107574 (2020)
7. Liang, J., Leung, C.S., So, H.C.: Lagrange programming neural network approach for target localization in distributed MIMO radar. IEEE Trans. Signal Process. **64**(6), 1574–1585 (2016)
8. Kremer, P.J., Lee, S., Bogdan, M., Paterlini, S.: Sparse portfolio selection via the sorted $\ell_1$-norm. J. Bank. Fin. **110**, 105687 (2020)
9. Markowitz, H.: Portfolio selection. J. Fin. **7**(1), 77–91 (1952)
10. Liu, Q., Guo, Z., Wang, J.: A one-layer recurrent neural network for constrained pseudoconvex optimization and its application for dynamic portfolio optimization. Neural Netw. **26**, 99–109 (2012)
11. Leung, M.F., Wang, J.: Minimax and biobjective portfolio selection based on collaborative neurodynamic optimization. IEEE Trans. Neural Netw. Learn. Syst. **32**, 2825–2836 (2021)
12. Boyd, S., Parikh, N., Chu, N: Distributed Optimization and Statistical Learning Via the Alternating Direction Method of Multipliers. Now Publishers Inc (2011)

# An Adaptive Convolution Auto-encoder Based on Spiking Neurons

Chuanmeng Zhu[1], Jiaqiang Jiang[1], Runhao Jiang[2], and Rui Yan[1]($\boxtimes$)

[1] College of Computer Science, Zhejiang University of Technology, Hangzhou, China
{2112012040,ryan}@zjut.edu.cn
[2] College of Computer Science and Technology, Zhejiang University, Hangzhou, China

**Abstract.** Neural coding is one of the central questions in neuroscience for converting visual information into spike patterns. However, the existing encoding techniques require a preset time window and lack effective learning. In order to overcome these two problems, we design an adaptive convolutional auto-encoder based on spiking neurons in this paper. We first exploit the spike pixel mapping decoding approach to find the optimal value of the time window automatically. Next, we design a deep convolutional neural network to adapt the learning parameters by reconstruction errors to realize the spike encoding process. Then we can naturally get coding pre-training parameters for unifying the convolutional spike coding layer with back-end deep spiking neural networks (SNNs) for recognition tasks. Simulation results demonstrate that the proposed method can achieve better performance compared with other encoding methods.

**Keywords:** Neural coding · Convolutional auto-encoder · Encoding time window · Spike decoding · Reconstruction error · SNNs

## 1 Introduction

A central question in systems neuroscience about sensory systems: how neurons represent the input-output relationship between stimuli and their spikes. This problem is formulated as neural coding and consists of two essential parts: encoding and decoding. Spike neurons need to convert other forms of information into spike patterns, this information transformation process is spike coding [1]. There are two mainstream spike coding mechanisms: rate coding [2] and temporal coding [3]. Rate coding expresses information by the number of spikes at a selected time window, it is simple and noise-resistant, but the low efficiency makes it difficult to achieve a rapid response to external stimuli; the temporal coding encodes the precise release time of spikes, which can be achieved by using sparse spike sequences or even single spikes. However, this leads to the generated spike sequences being more sensitive to noise, and the model performance is poorer than the model using rate coding. The encoding time window is the first necessary timescale characterizing a neural coding [4]. It is defined as the

M. Tanveer et al. (Eds.): ICONIP 2022, LNCS 13624, pp. 49–61, 2023.
https://doi.org/10.1007/978-3-031-30108-7_5

window containing the particular response patterns. The encoding time window is not as long as possible; a long simulation time can lead to unbearable resource consumption when the network structure is deep. Existing coding methods do not have a reasonable and efficient way to preset the time window.

In general, after years of development, researchers have proposed various bio-inspired spike coding schemes for different modalities and problems. Still, there are more or fewer problems with both information expression efficiency and noise robustness. Especially the spike coding process often lacks an effective learning process, making it challenging to make full use of the multiple dimensions of the spike for collaborative information-bearing. The fragmentation of the back-end SNN learning algorithm module makes it difficult to tailor the learning to the task objectives. However, traditional coding methods do not fully utilize the temporal and spatial dimensions for coding due to requiring a preset time window and lacking effective learning. The encoding and decoding of spike neurons can be seen as a back-and-forth transformation of stimuli and spikes in different directions. In this paper, we propose an adaptive convolutional auto-encoder based on spiking neurons, which can optimize the convolutional spike encoding process using decoding.

Recently, a novel approximate Bayesian decoding technique [5] has been proposed, which uses non-linear deep neural networks (DNNs) to decode images from the spiking activity of populations of retinal ganglion cells (RGCs). The approach outperforms linear reconstruction techniques usually used to interpret neural responses to high-dimensional stimuli. Based on this work, a novel decoding framework based on DNNs, a spike-image decoder (SID) [6], has been presented for reconstructing visual scenes, including static images and dynamic videos, from experimentally recorded spikes of a population of RGCs. However, both of the above approaches only focus on the decoding phase, and the spike encoding phase did not get substantially changed. We believe that the encoding and decoding processes of spikes are complementary to each other, so we can use the reconstruction error between the decoded image and the original image for the purpose of optimizing the encoding process. First, we use the spike pixel mapping decoding approach to automatically determine the optimal value of the time window for spike coding. After that, we design a deep convolutional neural network to adapt the learning parameters by reconstruction errors to realize the spike coding process.

Unlike traditional artificial neural networks (ANNs) that use real values as valid information, SNNs use discrete spike sequences to characterize the data. To overcome the scalability issues with an increasing number of output classes in the fully-connected shallow SNNs, researchers in the brain-like field have proposed spiking convolutional neural networks (SCNNs) capable of extracting high-level features embedded in an image pattern and sharing learned features across different classes of ways [7]. Recent works [8,9] have achieved great progress in exploring how to train high-performance SCNNs. With the increase in simulation time and network depth, the complex spatio-temporal dynamics of SNNs also bring challenges to the learning and optimization process. Therefore, the learning algorithms need to consider the characteristics of the encoded spike signals entirely. This paper verifies the learning performance of SCNNs using optimized convolutional spike coding.

## 2  Methods

In this section, we design an adaptive convolutional auto-encoder (Fig. 1) based on spiking neurons. First, we introduce the Leaky Integrate-and-Fire (LIF) neuron [10] and describe the convolutional spike coding in detail. Two decoding methods are proposed to optimize convolutional spike coding, first using spike pixel mapping decoding to determine the value of the encoding time window, and then using deep convolutional decoding to determine the coding pre-training parameters.

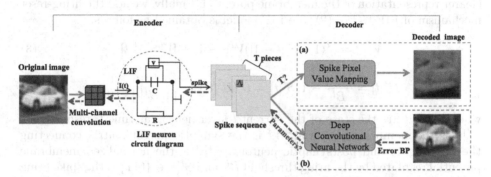

**Fig. 1.** An adaptive convolution auto-encoder based on spiking neurons. The encoding stage proposes a convolutional spike coding; The decoding stage: (a) spike pixel value mapping and (b) deep convolutional decoding.

### 2.1  Neuron Model

The LIF benefits from its computational efficiency and is the most widely used model in computational neuroscience. In its simplest form, the spiking neuron is modeled as a "leaky integrator" of its input current, defined as:

$$\tau_m \frac{d\nu}{dt} = -V(t) + RI(t), I(t) = \sum_j \omega_j \sum_{t_j^k \in T_\omega} K(t - t_j^k)H(t - t_j^k) \qquad (1)$$

where $V(t)$ is the neuron's membrane voltage at the moment $t$. $\tau_m = R \cdot C$ is the membrane time constant in this equation. $I(t)$ is the weighted sum of each postsynaptic current, and $W_j$ is the synaptic weight from the $j$th input neuron. $t_j^k$ is the arrival time of the $k$th presynaptic spike of the $j$th input neuron within the integration time window of $T_\omega$. $H(\cdot)$ is the Heaviside step function, and $K(\cdot)$ is the kernel function describing the time-decaying effect of synaptic currents. In Fig. 1, the LIF neuron model simulates the process of bioelectricity transmission by biological neurons through this circuit, and the black dot represents the capacitor. $I(t)$ passed by the synapse charges the circuit. $V(t)$ that has passed through the capacitor ($C$) is compared to the threshold ($v$) and an output spike is generated if $V(t) = v$ at moment $t_j^k$.

We convert the differential form of Eq. (1) into a discrete iterative form. The membrane potential equation is as follows:

$$V(t) = (1 - \frac{dt}{\tau_m})v(t-1) + \frac{dt}{\tau_m}I(t) = \lambda v(t-1) + \sum_j W_j s_j(t) \tag{2}$$

We simplify the $1 - \frac{dt}{\tau_m}$ term with an attenuation factor $\lambda$, expand $I(t)$ term as a weighted summation of the input spikes $\sum_j W_j s_j(t)$. $s_j$ denotes the spike firing state of the $j$th input neuron expressed as a binary (0 or 1), and the scaling effect of $\frac{dt}{\tau_m}$ is incorporated into the synaptic weight $W$ to obtain a clearer representation of the membrane potential. Finally, we add the firing-reset mechanism of LIF to Eq. (2), the LIF model is obtained as follows:

$$V^l[t] = \lambda^l(1 - S^l[t-1])V^l[t-1] + W^l S^{l-1}[t] \tag{3}$$

$$U^l[t] = \frac{V^l[t]}{B^l} - 1, S^l[t] = H\left(U^l[t]\right) = \begin{cases} 1 \ U^l[t] \geq 0 \\ 0 \ U^l[t] < 0 \end{cases} \tag{4}$$

where $l$ and $t$ are the state of the current $l$-layer neuron at time $t$, respectively, $V^l[t]$ is the membrane potential, $W^l$ is the synaptic weight matrix connecting the presynaptic and postsynaptic neurons, $U^l[t]$ is the normalized membrane potential regulated by the firing threshold $B^l$, and $S^l[t] \in \{0,1\}$ is the spike firing state controlled by $H(\cdot)$. The neuron emits spike $S^l[t] = 1$ when the normalized membrane potential is greater than zero, and the neuron emits no spike output $S^l[t] = 0$ when the normalized membrane potential is smaller than zero.

## 2.2  Convolutional Spike Coding

Fast, efficient, and accurate spike coding of perceptual information, modeled on how the brain processes perceptual information, has been critical research in SNNs. The best method is to automate the encoding of pixel values as spikes by simulating retinal neurons, eliminating the information loss caused by rate representation and shortening the encoding time window, thus reducing the simulation time and resource consumption.

The retina of living organisms consists of three layers of neurons: photoreceptors, bipolar cells, and ganglion cells (Fig. 2(a)). First, the photoreceptors capture signals from the visual scene. The signal is then sent to a bipolar cell that has a fixed receptive field and performs a linear filtering operation. Finally, signals processed by multiple bipolar cells are integrated into ganglion cells to generate a series of action potentials transmitted to various areas downstream of the brain via the optic nerve. All visual information about the environment is encoded in ganglion cells' spatiotemporal pattern of spikes output. Drawing on these biological retinal systems, and considering that the fundamental values of static images already preserve the color information associated with the optical signal, we propose a convolutional spike coding method (Fig. 2(b)). The standard multi-channel sliding convolution in deep learning is used to simulate the linear filtering operation of bipolar cells, LIF neurons are used to simulate the spontaneous, nonlinear spike response of ganglion cells to the filtered signal.

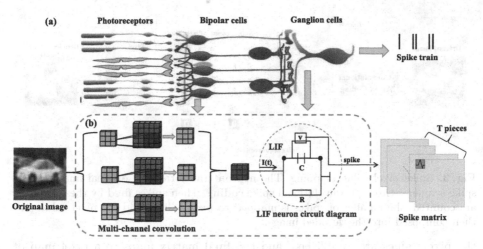

**Fig. 2.** Convolutional spike coding. (a) The retina of a living creature consists of three layers: photoreceptors, bipolar cells, and ganglion cells. (b) Schematic diagram of the convolutional spike encoding method.

In each time step, the convolutional kernel convolves the input channels of the original image. It sums up the convolution results of the kernels of the different channels. A single-channel feature map output is formed as the filtering result. The filtered output is considered the external input current $I$ of the LIF. With the help of the temporal integration property of the LIF neuron and the spike dynamics process (Eq. (3)), a spike sequence with temporal properties can be generated. Convolutional spike coding has significant advantages since it is quite suitable to unify the convolutional spike coding layer and the back-end deep SNN for co-training according to the task objectives.

## 2.3  Spike Pixel Value Mapping

Exploring the time window allows the spike coding stage to retain more valid information about the features. Ensuring invariance and robustness in the information representation is particularly important to improve the accuracy of information processing in spike coding and underlying spike learning algorithms. We compute information about all features using putative encoding windows whose length is varied parametrically and then characterize the range of encoding the window length as (10,20) to balance the adequate information of images and the computing consumption [4]. For convolutional spike coding, the spike pixel value mapping method (Fig. 3) can automatically find an optimal value of the time window by comparing the similarity between the decoded image and the original image when the time window changes.

Mapping decoding takes the spike matrix generated by the neuron of the convolutional spike coding as input. Each element of the matrix is the number of spikes fired by neurons within the specified time window. After pre-processing and normalization of the matrix, a mapping between the number of spikes and

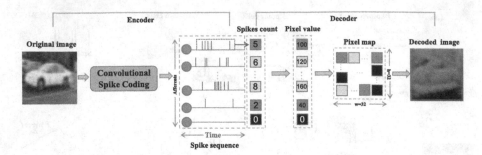

**Fig. 3.** Spike pixel value mapping. The neuron transforms the original image into a spike sequence through convolutional spike coding. Then spikes fired by these neurons are counted, the number of spikes is mapped to pixel values, and the pixel values are then composed into the decoded image.

the pixel values are established, and the final matrix maps to a pixel map of $n \times n$, where $n$ is the width and height of the image.

Then three typical reconstructed image metrics are used to evaluate image decoding results: mean squared error (MSE), describing the absolute difference of each pixel; the peak signal-to-noise ratio (PSNR), which represents the global quality; and the structural similarity index metric (SSIM), which captures detail or image distortion. The results are shown in Fig. 5. Thus, when the time window changes from 10 to 20, the optimal value of the time window for convolutional spike coding are 11 and 12 in processing the MNIST [11] and CIFAR10 [12] dataset via this method, respectively.

## 2.4 Deep Convolutional Decoding

After the value of the time window is determined, we introduce deep convolutional decoding (Fig. 4). Given an input image, it will generate spikes by convolutional spike coding, and then spikes are input to the deep convolutional decoding network for reconstruction operation. The reconstructed image is compared with the original image, and then the reconstruction error is back-propagated to optimize the network. Ultimately a clear and refined reconstructed image can be obtained. Notably, the convolutional spike coding learning process is optimized, and we can automatically get the coding pre-training parameters.

In this model (Fig. 4), a network extended from the densenet model [13] is used for the first time for the spike-to-image reconstruction tasks. In this study, the network is composed of $L + 1$ layers, where the nonlinear variation of each layer is noted as $C_i(\cdot)$, $C_i(\cdot)$ uses the combination of BN+ReLU+ Conv($3 \times 3$). Each network layer needs to implement a nonlinear transformation $C_i(\cdot)$ to ensure that the spike retains the maximum information flow in the intermediate layers. In the first $L$ layers of the network, we record the spike input of layer $i$ as $S_i$, it is written as $S_i = C_i([S_0, S_i, ..., S_{i-1}])$, where $[\cdot]$ stands for concatenation, *i.e.*, stitching together all the output spike features in $S_0$ and $S_{i-1}$ layers. The spike input of layer $i$ is related not only to the spike output of layer $i - 1$, but also to the spike outputs of all previous layers, then passes the spike features extracted by this layer to each subsequent layer. Here, five dense blocks are used

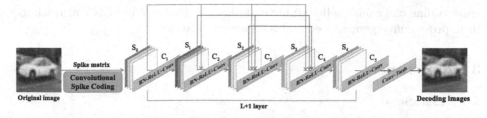

**Fig. 4.** Reconstruction model based on deep convolutional decoding.

**Table 1.** Network architecture for deep convolutional decoding.

| Layers | Output Size | Network architecture |
|---|---|---|
| Convolution layer | 128,32×32 | 3×3 conv,stride 1 |
| LIF | 128,32×32 | - |
| Dense block(0) | 32,32×32 | {1×1,stride 1}×128, {3×3,stride 1}×32 |
| Concatenate | 160,32×32 | - |
| Dense block(1) | 32,32×32 | {1×1,stride 1}×128, {3×3,stride 1}×32 |
| Concatenate | 192,32×32 | - |
| Dense block(2) | 32,32×32 | {1×1,stride 1}×128, {3×3,stride 1}×32 |
| Concatenate | 224,32×32 | - |
| Dense block(3) | 32,32×32 | {1×1,stride 1}×128, {3×3,stride 1}×32 |
| Concatenate | 256,32×32 | - |
| Dense block(4) | 32,32×32 | {1×1,stride 1}×128, {3×3,stride 1}×32 |
| Convolution layer | 3,32×32 | 3×3 conv,stride 1 |

to keep the size of the spike feature map of each block in the dense block the same for tandem operations. It consists of two convolutional layers, the first layer has a convolutional kernel with size of $1 \times 1$, stride of 1, and output channel of 128, the second layer has a convolutional kernel with size of $3 \times 3$, stride of 1, and output channel of 32.

Noteworthy, the part of the transition layer between the original dense blocks is omitted because there is no need to compress the extracted spike feature maps. In order to accommodate better spike decoding, the last layer of the network is the input without feature stacking, reducing the number of channels number so that the last layer can be output ($32 \times 32 \times 32$). Then all the extracted spike features are transformed into images by a convolution operation. The network structure is shown in Table 1.

This study explores three different loss functions to ensure that the errors can be back-propagated by comparing the reconstructed and original images. Here we use L1loss, MSEloss, and (L1+MSE) loss functions, as shown in Fig. 6, a detailed comparison of the reconstruction results of 30 example images inside the first training epoch processed with these three loss functions by MSE, PSNR and SSIM. The three typical values of the reconstructed image metrics show that we can obtain high-quality reconstruction results from the model. Convolutional spike coding can enable self-learning and play a crucial role in SCNN algorithms to improve convergence speed and recognition accuracy.

## 3   Experiment

### 3.1   DataSets

To evaluate our approach, the dataset we choose are the static datasets MNIST and CIFAR10. MNIST has ten classes (0–9) of handwritten digits, and a total of 70000 grayscale images with a pixel size of $28 \times 28$. 60000 images are set as training and 10000 images as the test set. CIFAR10 has ten classes of real objects (truck, car, plane, boat, frog, bird, dog, cat, deer, horse). The dataset has a total of 60000 RGB color images with a pixel size of $32 \times 32$. 50000 images are set as training and 10000 images as the test set.

### 3.2   Determination of the Time Window

In this experiment, the spike pixel mapping decoding method decodes the spike output from the convolutional spike coding into an image of $32 \times 32$. The time window is from 10 to 20, implying that each image is encoded and decoded 11 times.

**Table 2.** PSNR, SSIM and MSE values of sample images across different time windows.

| Metric | Time window | | | | | | | | | | |
|---|---|---|---|---|---|---|---|---|---|---|---|
| | 10 | 11 | 12 | 13 | 14 | 15 | 16 | 17 | 18 | 19 | 20 |
| PSNR | 8.2798 | 8.2876 | **8.3052** | 8.2781 | 8.1968 | 8.2322 | 8.2775 | 8.3027 | 8.1932 | 8.2674 | 8.1945 |
| SSIM | 0.0922 | 0.0943 | **0.0982** | 0.0961 | 0.0892 | 0.0932 | 0.0929 | 0.0932 | 0.0912 | 0.0903 | 0.0907 |
| MSE | 0.1485 | 0.1483 | **0.1477** | 0.1486 | 0.1514 | 0.1502 | 0.1486 | 0.1478 | 0.1515 | 0.1490 | 0.1515 |

Table 2 shows the results of comparing the 11 decoded images of the example test image with the original image by three typical reconstructed image metrics. For PSNR, a high value indicates good performance. And for MSE and SSIM,

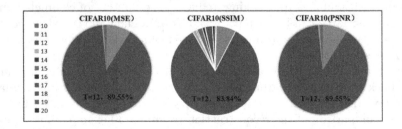

**Fig. 5.** Percentage of optimal value of the time $T$ found in the CIFAR10 dataset using MSE, PSNR and SSIM.

the values are expected to be close to 0 and 1, respectively. It can be seen that $T = 12$ (in bold font) is the optimal value of the encoding time window for the example image.

Then, this method automatically compares the 11 decoded images of each image in the CIFAR10 dataset with the original image using the three reconstructed image metrics and outputs the time $T$ corresponding to the optimal value of the MSE, PSNR, and SSIM among the 11 times, and then counts the proportion of each $T$ in the number of datasets. In CIFAR10, it can be seen that $T = 12$ is the optimal value of the time window with a percentage over 80% (blue), and in MNIST, $T = 11$ is the optimal value of the time window with a percentage over 99% using the same method. Due to the page length limit, we only show the time-window percentage plot for the CIFAR10 dataset in Fig. 5.

## 3.3  Determination of Coding Pre-training Parameters

In this task, three different methods of defining the reconstruction loss function are explored so that we can evaluate the performance of a deep convolutional decoding method by obtaining high-quality reconstruction results. The loss functions used here are L1loss, MSEloss and (L1+MSE) loss, Fig. 6(a) shows the reconstructed image results for 7 images with 30 images in the first training epoch.

It can be seen that the global and detailed content of the images is well reconstructed. Depending on the loss function used for training, the intermediate images may differ in color detail processing. However, the final reconstructed results are similar, with minor differences in the detailed textures. For these 30 images, all three loss functions give similar performances in terms of three metrics, where the MSEloss decoding performance is slightly higher. In this model,

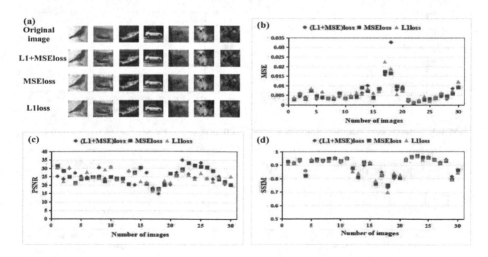

**Fig. 6.** (a) Reconstruction images from three different error functions; comparison of reconstruction errors between L1loss (gray triangles), MSEloss (orange square), and (L1+MSE) loss (blue rhombus) measured by MSE (b), PSNR (c), and SSIM (d). (Color figure online)

it can be seen that the reconstructed images obtained at different losses are highly similar to the original images, which means the model performance is high enough to pretrain the convolutional spike coding well, so we can automatically get optimal coding pre-training parameters.

### 3.4   Spike Convolutional Neural Networks (SCNNs)

SCNNs are mainly composed of a hierarchy of stacked convolutional layers for feature extraction followed by fully-connected layers for final classification [7]. We designed SCNNs of different depths to cope with different datasets. For the two-dimensional spatial structure of images in the visual classification task, we use SCNNs to directly encode and extract effective spatio-temporal features for recognition. To ensure sufficient perceptual field, SCNNs use convolutional kernels of $3 \times 3$ sizes (C3) and average pooling of $2 \times 2$ sizes (AP2). To adequately characterize the spatio-temporal features of the dataset, the network architecture of the MNIST dataset is 128C3-AP2-128C3-AP2-2048FC-10, and the network architecture of the CIFAR10 dataset is 128C3-256C3-AP2-512C3-AP2-1024C3-512C3-1024FC-512FC-10.

In the experiments with static data sets, the first convolutional spike layer of SCNN is changed to an encoding layer using convolutional spike coding by default. As shown in Table 3, for the simple MNIST dataset, the SCNN with optimized convolutional spike coding has a relatively higher test accuracy than the SCNN with rate coding and other visual recognition algorithms with coding. Still, for the relatively complex CIFAR10 dataset, our proposed coding method

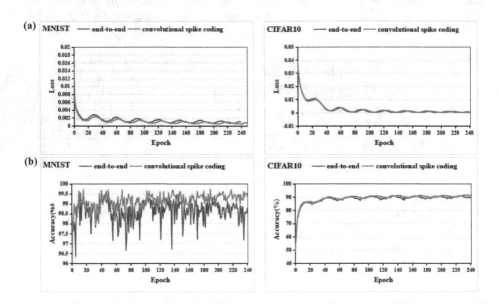

**Fig. 7.** End-to-end SCNN (blue) and SCNN with adaptive convolutional spike coding (orange) (a) training loss curve and (b) test accuracy curve when processing MNIST and CIFAR10 datasets. (Color figure online)

**Table 3.** Performance comparison between the proposed method and the state-of-the-art methods on different visual datasets.

| Method | Accuracy MNIST | Accuracy CIFAR10 |
|---|---|---|
| ANN2SNN + rate coding [14] | 99.44% | 90.85% |
| ANN2SNN + rate coding [15] | - | 91.55% |
| Spike-based BP + poisson coding [16] | 99.59% | 90.95% |
| HM2-BP + rate coding [17] | 99.49% | - |
| Whetstone + binary coding [18] | 99.53% | 84.67% |
| STCA + rate coding [19] | 98.60% | - |
| STBP + rate coding [20] | 98.89% | 90.53% |
| SCNN + rate coding | 99.21% | 85.48% |
| End to end SCNN | 99.35% | 90.97% |
| **SCNN + convolutional spike coding** | **99.70%** | **91.64%** |

significantly improves the test accuracy and performance. For example, the test accuracy is 6.16% higher than SCNN using rate coding. It is worth mentioning that compared with the better-performing end-to-end SCNN algorithm, which receives real-valued images directly and converts them into spikes without using a coding mechanism. Our coding using optimized convolutional spikes is comparable to end-to-end, and our training loss and test accuracy are slightly better than end-to-end on both datasets in 240 epochs (Fig. 7). Importantly, we demonstrate the advantages of our approach in terms of coding. This shows that effective spatio-temporal coding of static images is essential for the SNN learning process of complex recognition tasks. By optimizing the coding process from pixel intensities to spatio-temporal spike patterns and maximizing the retention of original image information, SNN can achieve a much better recognition performance.

## 4    Conclusion

This paper designs an adaptive convolutional auto-encoder based on spiking neurons. First, we propose a convolutional spike coding method and optimize the coding process using two decoding methods. We evaluate our work on both static image MNIST and CIFAR10 datasets. The optimal time window values for convolutional spike coding in processing the MNIST and CIFAR10 datasets are then determined to be 11 and 12, respectively, based on the spike-pixel mapping decoding method. Finally, the deep convolutional decoding approach is used to have suitable pre-training parameters for the spike coding stage in both static image datasets. Since the convolutional spike layer is a common module of the deep SNN, it is natural to combine the convolutional spike coding layer with the back-end deep SNN as a unified network. Experimental results show that SCNNs using optimized convolutional spike coding have superior performance over end-to-end SCNNs and algorithms using other codes across all datasets. In particular, the above two decoding methods can be used as a general approach to optimize the coding method. In conclusion, the SNNs algorithm using optimized convolutional spike coding can achieve better performance compared with other encoding methods.

**Acknowledgments.** This work was supported by the National Natural Science Foundation of China NSAF under Grant No. U2030204.

# References

1. Azarfar, A., Calcini, N., Huang, C., et al.: Neural coding: a single neuron's perspective. Neurosci. Biobehav. Rev. **94**, 238–247 (2018)
2. Van Rullen, R., Thorpe, S.J.: Rate coding versus temporal order coding: what the retinal ganglion cells tell the visual cortex. Neural Comput. **13**(6), 1255–1283 (2001)
3. VanRullen, R., Guyonneau, R., Thorpe, S.J.: Spike times make sense. Trends Neurosci. **28**(1), 1–4 (2005)
4. Panzeri, S., Brunel, N., Logothetis, N.K., et al.: Sensory neural codes using multiplexed temporal scales. Trends Neurosci. **33**(3), 111–120 (2010)
5. Parthasarathy, N., Batty, E., Falcon, W., et al.: Neural networks for efficient Bayesian decoding of natural images from retinal neurons. In: Advances in Neural Information Processing Systems, vol. 30 (2017)
6. Zhang, Y., Jia, S., Zheng, Y., et al.: Reconstruction of natural visual scenes from neural spikes with deep neural networks. Neural Netw. **125**, 19–30 (2020)
7. Pfeiffer, M., Pfeil, T.: Deep learning with spiking neurons: opportunities and challenges. Front. Neurosci. **12**, 774 (2018)
8. Zheng, H., Wu, Y., Deng, L., et al.: Going deeper with directly-trained larger spiking neural networks. arXiv preprint, arXiv:2011.05280 (2020)
9. Li, Y., Guo, Y., Zhang, S., et al.: Differentiable spike: rethinking gradient-descent for training spiking neural networks. In: Advances in Neural Information Processing Systems, vol. 34 (2021)
10. Orhan, E.: The leaky integrate-and-fire neuron model, no. 3, pp. 1–6 (2012)
11. LeCun, Y.: The MNIST database of handwritten digits (1998). http://yann.lecun.com/exdb/mnist/
12. Krizhevsky, A., Hinton, G.: Learning multiple layers of features from tiny images (2009)
13. Huang, G., Liu, Z., Van Der Maaten, L., et al.: Densely connected convolutional networks. In: Proceedings of the IEEE Conference on Computer Vision and Pattern Recognition, pp. 4700–4708 (2017)
14. Rueckauer, B., Lungu, I.A., Hu, Y.H., et al.: Conversion of continuous-valued deep networks to efficient event-driven networks for image classification. Front. Neurosci. **11**, 682 (2017)
15. Sengupta, A., Ye, Y., Wang, R., et al.: Going deeper in spiking neural networks: VGG and residual architectures. Front. Neurosci. **13**, 95 (2019)
16. Lee, C., Sarwar, S.S., Panda, P., et al.: Enabling spike-based backpropagation for training deep neural network architectures. Front. Neurosci. 119 (2020)
17. Jin, Y., Zhang, W., Li, P.: Hybrid macro/micro level backpropagation for training deep spiking neural networks. In: Advances in Neural Information Processing Systems, vol. 31 (2018)

18. Severa, W., Vineyard, C.M., Dellana, R., et al.: Training deep neural networks for binary communication with the whetstone method. Nat. Mach. Intell. **1**(2), 86–94 (2019)
19. Gu, P., Xiao, R., Pan, G., et al.: STCA: spatio-temporal credit assignment with delayed feedback in deep spiking neural networks. In: IJCAI, pp. 1366–1372 (2019)
20. Wu, Y.J., Deng, L., Li, G.Q., et al.: Direct training for spiking neural networks: faster, larger, better. In: Proceedings of the AAAI Conference on Artificial Intelligence, vol. 33, no. 01, pp. 1311–1318 (2019)

# Schizophrenia Detection Based on EEG Using Recurrent Auto-encoder Framework

Yihan Wu[1], Min Xia[1], Xiuzhu Wang[1], and Yangsong Zhang[1,2(✉)]

[1] School of Computer Science and Technology, Laboratory for Brain Science and Medical Artificial Intelligence, Southwest University of Science and Technology, Mianyang 621010, China
zhangysacademy@gmail.com

[2] MOE Key Lab for Neuroinformation, University of Electronic Science and Technology of China, Chengdu 610054, China

**Abstract.** Schizophrenia (SZ) is a serious mental disorder that could seriously affect the patient's quality of life. In recent years, detection of SZ based on deep learning (DL) using electroencephalogram (EEG) has received increasing attention. In this paper, we proposed an end-to-end recurrent auto-encoder (RAE) model to detect SZ. In the RAE model, the raw data was input into one auto-encoder block, and the reconstructed data were recurrently input into the same block. The extracted code by auto-encoder block was simultaneously served as an input of a classifier block to discriminate SZ patients from healthy controls (HC). Evaluated on the dataset containing 14 SZ patients and 14 HC subjects, and the proposed method achieved an average classification accuracy of 81.81% in subject-independent experiment scenario. This study demonstrated that the structure of RAE is able to capture the differential features between SZ patients and HC subjects.

**Keywords:** EEG · Schizophrenia detection · Auto-Encoder · Convolutional neural network

## 1 Introduction

Schizophrenia is a severe mental disorder. This disease affects approximately 24 million people in the world, reported by the World Health Organization [23]. One in 300, on average, people suffer from SZ, and this rate reaches up to one in 222 in adults [7]. However, the majority of patients with SZ have not received proper treatment. One of the most difficult issues is the absence of significant biological markers [11].

Benefiting from the advantages such as non-invasive, high temporal resolution, low cost, electroencephalography (EEG) has been widely used in the disease detection field [1,2,4,17]. With the development of machine learning, artificial features based on EEG signals have been rapidly employed in the field

© The Author(s), under exclusive license to Springer Nature Switzerland AG 2023
M. Tanveer et al. (Eds.): ICONIP 2022, LNCS 13624, pp. 62–73, 2023.
https://doi.org/10.1007/978-3-031-30108-7_6

of SZ detection. For example, Vázquez et al. [20] proposed a method using random forest to operate on the extracted connectivity metrics of generalized partial directed coherence (GPDC) and direct directed transfer function (dDTF) of 1-minute segments. They conduct subject-unaware partitioning and leave-$p$-subject-out experiments and obtain the area under the curve (AUC) of 0.99 and 0.87, respectively. Najafzadeh et al. [12] proposed a method based on the adaptive neuro fuzzy inference system (ANFIS). They tried to employ ANFIS, support vector machine (SVM), and artificial neural network (ANN) to detect the SZ using Shannon entropy, spectral entropy, approximate entropy, and the absolute value of the highest slope of auto-regressive coefficients and achieved accuracy of 99.92% in the subject-dependent experiment. Chandran et al. [13] introduced their method based on Long Short-Term Memory (LSTM). They calculated Katz fractal dimension, approximate entropy and the time-domain feature of variance as artificial feature, and fed them into the LSTM network to distinguish the SZ patients from HC subject. They obtained an accuracy of 99.0% in the subject-dependent experiment.

These methods utilized artificial features that are highly dependent on the prior knowledge of researchers. The outstanding high performance of deep learning makes end-to-end SZ detection possible. For instance, the CNN-LSTM model is proposed by Shoeibi et al. [19] They tried several combinations of 1D-CNN and LSTM to verify the best model. Their model achieved an accuracy of 99.25% in the subject-dependent experiment. Oh et al. [14] introduced a deep convolution neural network (CNN) to detect SZ. This model contains four convolution layers, five max-pooling layers and two fully connected layers. The experiments were conducted in both subject-dependent and subject-independent scenarios using 25 s segments. They achieved an accuracy of 98.07% and 81.26% respectively.

In most of the studies presented, the methods were evaluated in a subject-dependent scenario, which has a serious problem called data leakage. Due to the high correlation between continuous EEG signals, when the EEG signals collected in one subject were divided into several segments, and these segments were shuffled and partitioned simultaneously into training set and testing set. The training set and the testing set were inevitably intersecting. On the other hand, logically speaking, the subject-dependent method is unpractical, as it is unreasonable to detect SZ for subjects after knowing clearly whether they are patients or not.

Based on this consideration, we proposed a model named Recurrent Auto-Encoder (RAE), and evaluated its performance in a subject-independent scenario. It contains a recurrent auto-encoder to extract task-related features and a linear classifier to recognize the SZ and HC. We conducted experiments on a publicly accessed dataset containing 14 schizophrenia patients and 14 age-matched healthy control subjects. The results indicate that our RAE performed better than the current baseline methods.

This paper is organized as follows. Section 2 introduces the dataset and proposed model. Section 3 describes the experiment setting and result. In Sect. 4, the discussions and conclusions are present.

## 2   Materials and Methods

### 2.1   Dataset

In this study, we used a dataset collected by the Institute of Psychiatry and Neurology in Warsaw, Poland [16]. This dataset consists of EEG recording from 14 patients (7 males: $27.9 \pm 3.3$ years, 7 females: $28.3 \pm 4.1$ years) with SZ and 14 HC (7 males: $26.8 \pm 2.9$, 7 females: $28.7 \pm 3.4$ years). All the patients met International Classification of Diseases for paranoid schizophrenia (ICD-10, F20.0). The eyes-closed resting state EEG signals lasting for 15 min were collected with a sampling rate 250 Hz. The 19 electrodes were used, i.e., Fp1, Fp2, F7, F3, Fz, F4, F8, T3, C3, Cz, C4, T4, T5, P3, Pz, P4, T6, O1 and O2, which were placed according to the standard of international 10–20 system. More details could be found in the reference [16].

### 2.2   Pre-processing

To improve the signal-noise ratio, we first employed a bandpass filter with a frequency of 0.5–50 Hz. The data were then divided into segments of 5 s in length. The obtained segments should pass a threshold check to reduce the interference of electrooculography (EOG). We dropped the segment which peak value is out of range of $-100\,\mu V \sim 100\,\mu V$. Finally, the common reference and z-score normalization were applied to obtain the processed data.

### 2.3   Methods

The motivation of the proposed model is that: if the EEG data are recurrently processed by a encoder-decoder is beneficial to generate more discriminative embedding codes, the procedure is summarized as follows:

- Encode the data $D_1$ to obtain the embedding $Z_1$
- Decode the $Z_1$ to reconstruct $D_2$
- Process the $D_2$ as above did for several loops to obtain $D_n$ and $Z_n$.

On the assumption that the encoder and decoder are effective and stable enough, the embedding codes $Z_1$, ..., $Z_n$ should remain similar task-related property, although the waveform of $D_1$, ..., $D_n$ maybe not exactly the same. We termed the similarity as semantic invariance. On the other hand, if we optimize the encoder to improve the semantic invariance between $Z_1$, ..., $Z_n$, the optimization could be regarded as effective. In actual application, the true label can be defined as the task-related property. Improving the prediction accuracy of all embedding codes, especially $Z_2$, ... ,$Z_n$, can be regarded as improving semantic invariance. This is the key idea of this method.

Previous studies in the field of computer vision (CV) have proved that Auto-Encoder is a powerful frame of feature extraction and reconstruction [6]. Therefore, we leveraged the Auto-Encoder as the main architecture to design our model. EEGNet is a widely used baseline method in the field of EEG analysis [9]. It has stable performance and feature representation ability. We designed the encoder and the corresponding decoder modules using the similar operations in EEGNet.

The structure of RAE is shown in Fig. 1. It is consisted of a recurrent auto-encoder feature extractor and a fully-connected classifier. The fully-connected classifier is used to classify all embedding codes extracted by RAE. The semantic invariance is improved by optimizing the classification accuracy to improve the performance of encoder.

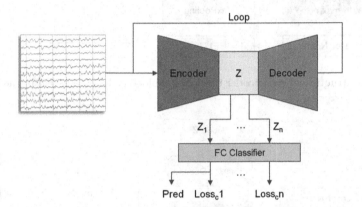

**Fig. 1.** The structure of RAE. $Z_n$ is the representation recurrently generated by the encoder in cycles for $n$ times.

**Backbone.** The Backbone of RAE structure is modifiable. In this work, the backbones of the encoder and decoder were comprised of the similar operation that used in the classical EEGNet model. To facilitate decoding, the sizes of all temporal convolution kernels were set to be odd so that the padding can be symmetric. For similar reasons, the average pooling after the second convolution layer was replaced by a max pooling layer. In addition, we used layer normalization in the model in order to reduce the interference of other samples in one mini-batch. The structure is shown in Fig. 2.

Decoder is the opposite procedure of encoder, which uses transposed convolution to realize deconvolution. In addition, layer normalization is applied in the end to keep each reconstructed sample separate from the others in one mini-batch. The structure of the decoder is shown in Fig. 3.

**Recurrent Auto-encoder.** First, the raw data $D_i \in \mathbb{R}^{C \times T}$ is input into the encoder block $En$ to generate the embedding code $Z_i$, which could be described as:

$$Z_i = En(D_i) \in \mathbb{R}^{N * C' * T'} \tag{1}$$

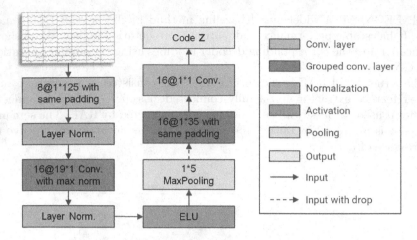

**Fig. 2.** The structure of the encoder block. This block is denoted as $En$ in the formula.

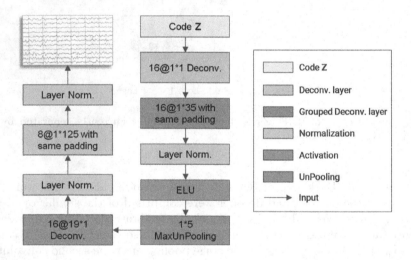

**Fig. 3.** The structure of the decoder block. This block is denoted as $De$ in the formula.

where $C'$ and $T'$ are the numbers of the electrode channels and time-dimension sampling point, which are equal to 1 and 250, respectively. $N$ denotes the number of the convolution kernels, which was set to 16 in this model.

The embedding code $Z_i$ is input into the decoder $De$ to reconstruct the data $D_{i+1} \in \mathbb{R}^{C \times T}$, which is illustrated in the following:

$$D_{i+1} = De(Z_i) \in \mathbb{R}^{C*T} \tag{2}$$

Then, the reconstructed data $D_{i+1} \in \mathbb{R}^{C \times T}$ was regarded as the input of block $En$ in the next cycle. After $n$ loop iterations, the model is able to generate embedding code $Z_1$, $Z_2$, ..., $Z_n$. For the task of SZ detection, the $n$ was set to 2 in the following experiments. All the embedding codes will be employed to calculate the loss and predict the class label as follows.

**Classifier and Loss.** The classifier structure is shown in the Fig. 4.

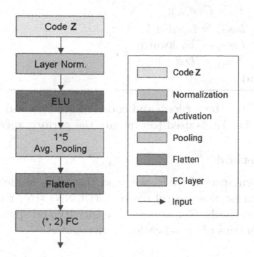

**Fig. 4.** The structure of the classifier block. This block is denoted as $Cls$ in the formula.

Embedding code $Z_i$ is input into the classifier $Cls$ to obtain the predicted label $\hat{y}_i$, which could be described as:

$$\hat{y}_i = \arg\max Cls(Z_i) \tag{3}$$

In this work, we employed the cross entropy loss between predicted labels and the corresponding true labels of the samples to optimize the model, which could be illustrated as:

$$L_i = loss\_fn(\hat{y}_i, y) \tag{4}$$

where $loss\_fn$ denotes cross entropy operator, and $y$ denotes true label.

**Model Training.** The parameters optimization of $En$, $De$ and $Cls$ blocks in each optimization loop were separate. The procedure is summarized as Algorithm 1:

---

**Algorithm 1:** Training procedure

**Input:** input parameters X, Y, nloops
**Output:** Classifier Cls($\cdot$)

1  $X_0 = X$;
2  **for** $i$ *in range(0, nloops)* **do**
3  |    $z_i = En(X_i)$;
4  |    $\hat{Y}_i = Cls(z_i)$;
5  |    $loss_i = \text{LossFn}(\hat{Y}, Y)$;
6  |    $loss_i \leftarrow$ backward;
7  |    $X_{i+1} = De(z_i)$;
8  **end**

---

In the testing stage, the for loop and code in lines 5, 6 and 7 are unessential. The output of $Cls$, i.e. $\hat{Y}_0$, is used to evaluate the performance of this model.

### 2.4  Baseline Methods

To verify the performance of RAE, we used three DL models, i.e., DeepConvNet [18], Deep Convolution neural network (DCNN) [15], EEGNet [9], as the compared baseline methods. The accuracy, sensitivity and specificity were served as the evaluation metrics of classification performance.

**DeepConvNet.** DeepConvNet is a deep convolution model proposed by Schirrmeister et al [18]. This model contains four convolution blocks. One temporal convolution filter and one max pooling layer are employed in each block. In particular, a spatial convolution layer is added additionally. Due to its robustness and high performance, DeepConvNet is widely used in the field of classification based on EEG [5,10].

**DCNN.** Deep Convolution neural network (DCNN) is a method specifically used for SZ classification proposed by Oh et al [15]. DCNN consists of five convolution layers, two max pooling layers, two average pooling layers, a global average pooling layer and a fully-connected layer. The convolution layers are able to extract features automatically, and the max pooling layers are able to capture the most significant feature extracted by the previous convolution layer. Finally, all features are used to classify the signal in the fully-connected layer. DCNN achieved 81.26% accuracy with the time window of length 25 s.

**EEGNet.** EEGNet is a compact convolutional neural network proposed by Lawhern et al [9]. They first introduced the use of depthwise convolution and separable convolution on the EEG data. EEGNet also applied several well-known ideas in the field of BCI, such as optimal spatial filtering and filter bank. Due to the compact structure and stable performance, EEGNet has been widely applied in EEG-based classification tasks, such as steady-state visual evoked potential [22], motor imagery [24], and emotion recognition [21], etc.

## 3 Experiments and Results

### 3.1 Model Implementation

To evaluate the performance of the methods, level-one-subject-out (LOSO) strategy was used. Specifically, the data from one subject was used as the testing data, those of the remaining subjects were adopted as the training data. This procedure repeated until all the subjects served as the testing subject once. Each method was run for 5 times, and then the average accuracy, sensitivity and specificity were calculated as evaluation indicators via equations (5) to (7), which are usually used in the disease detection field [3].

$$accuracy = \frac{TP + TN}{TP + FP + TN + FN} \tag{5}$$

$$sensitivity = \frac{TP}{TP + FN} \tag{6}$$

$$specificity = \frac{TN}{TN + FP} \tag{7}$$

where TP, TN, FP and FN denotes the total number of true positive, true negative, false positive and false negative examples, respectively.

For the RAE and all compared models, adaptive moment estimation (ADAM) optimizer was adopted as the optimization method [8], and the learning rate was set as 1e-4. The experiment was executed for 30 iterations, and the accuracy in the last epoch was employed to evaluate the performance of all the methods.

### 3.2 Results

The classification results of each subject obtained by all the methods are summarized in Table 1. For each subject, the average accuracy was calculated by averaging the accuracies of five times experiments, and the standard deviation was also calculated on these accuracies. We could find that the proposed RAE achieved better performance than all other methods, which yields the average accuracy of 81.81%. Besides, the results indicate that the RAE could yield more robust results with smaller standard deviations, such as those of subject No.4.

Since the intra-subject SEN and SPE have no significance owing to the unique label of data from each subject, we summarized the global confusion matrix on all the five experiments and calculated SEN and SPE across subjects. The results are shown in Table 2 and Fig. 5. ACC denotes the average accuracy across the subjects.

**Table 1.** The classification accuracies of each subject (mean±std.) for the RAE and all compared methods.

| Sub. | DeepConvNet | DCNN | EEGNet | RAE |
|------|-------------|------|--------|-----|
| 1 | 42.47 ± 52.76 | 95.21 ± 3.96 | 91.23 ± 6.75 | 97.95 ± 3.28 |
| 2 | 44.58 ± 50.13 | 15.52 ± 6.24 | 10.94 ± 11.53 | 21.67 ± 6.4 |
| 3 | 81.38 ± 41.65 | 100. ± 0. | 99.5 ± 0.52 | 99.12 ± 1.63 |
| 4 | 59.75 ± 54.27 | 53.74 ± 31.61 | 22.09 ± 41.6 | 94.6 ± 7.81 |
| 5 | 84.84 ± 31.55 | 95.05 ± 2.95 | 99.57 ± 0.7 | 99.78 ± 0.48 |
| 6 | 64.64 ± 36.98 | 24.9 ± 15.08 | 53.38 ± 15.03 | 61.99 ± 12.73 |
| 7 | 53.29 ± 47.85 | 49.76 ± 22.53 | 40.35 ± 14.29 | 46.59 ± 4.79 |
| 8 | 69.51 ± 45.01 | 82.68 ± 36.34 | 98.29 ± 1.69 | 88.05 ± 11.13 |
| 9 | 60. ± 54.77 | 80.25 ± 44.16 | 100. ± 0. | 99.88 ± 0.28 |
| 10 | 80. ± 44.72 | 84.07 ± 11.72 | 96.91 ± 2.83 | 97.28 ± 3.45 |
| 11 | 65.29 ± 48.36 | 77.53 ± 35.11 | 100. ± 0. | 99.88 ± 0.26 |
| 12 | 42.73 ± 47.2 | 33.29 ± 35.48 | 37.27 ± 12.18 | 24.6 ± 8.45 |
| 13 | 58.92 ± 46.28 | 98.44 ± 2.14 | 99.04 ± 0.91 | 98.56 ± 1.24 |
| 14 | 61.56 ± 52.71 | 47.19 ± 21.69 | 32.93 ± 19.6 | 62.16 ± 28.35 |
| 15 | 31.1 ± 43.1 | 53.05 ± 13.29 | 35.37 ± 13.45 | 36.46 ± 8.39 |
| 16 | 79.75 ± 44.58 | 72.03 ± 35.11 | 93.67 ± 4.54 | 86.33 ± 20.86 |
| 17 | 40. ± 54.77 | 78.77 ± 9.58 | 75.61 ± 7.51 | 68.16 ± 15.98 |
| 18 | 75.89 ± 43.2 | 74.25 ± 12.15 | 99.73 ± 0.61 | 97.53 ± 1.79 |
| 19 | 20. ± 44.72 | 90.26 ± 7.21 | 88.55 ± 9.04 | 98.95 ± 1.2 |
| 20 | 39.69 ± 39.57 | 39.43 ± 7.87 | 51.38 ± 9.2 | 76.48 ± 4.19 |
| 21 | 60. ± 54.77 | 96.58 ± 7.36 | 100. ± 0. | 99.79 ± 0.28 |
| 22 | 79.89 ± 44.36 | 92.28 ± 8.07 | 97.72 ± 1.41 | 98.8 ± 0.45 |
| 23 | 40. ± 54.77 | 94.29 ± 1.9 | 95.86 ± 3.79 | 92.02 ± 7.99 |
| 24 | 93.78 ± 13.91 | 99.27 ± 0.67 | 97.44 ± 2.74 | 98.17 ± 3.42 |
| 25 | 60. ± 54.77 | 30. ± 9.05 | 51.52 ± 12.97 | 55.22 ± 16.47 |
| 26 | 40. ± 53.22 | 93.93 ± 2.57 | 96.82 ± 1.43 | 92.94 ± 3.94 |
| 27 | 80. ± 44.72 | 99.64 ± 0.54 | 99.88 ± 0.27 | 99.28 ± 1.3 |
| 28 | 41.84 ± 53.23 | 98.37 ± 2.29 | 96.12 ± 6.76 | 98.37 ± 2.02 |
| **Mean** | 58.96±6.92 | 73.21±4.74 | 77.18±0.96 | **81.81±1.60** |

**Table 2.** Classification results of RAE and all compared methods. ACC, SEN and SPE denotes accuracy (mean±std), sensitivity and specificity, respectively.

| Methods | ACC(%) | SEN(%) | SPE(%) |
|---------|--------|--------|--------|
| DeepConvNet | 58.96±6.92 | 60.24 | 55.33 |
| DCNN | 73.21±4.74 | 71.91 | 75.18 |
| EEGNet | 77.18±0.96 | 74.58 | 79.36 |
| **RAE** | **81.81±1.60** | **80.30** | **83.37** |

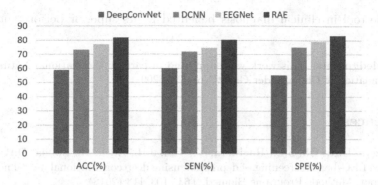

**Fig. 5.** Classification results of RAE and all compared methods.

## 4   Discussion and Conclusion

The results indicate that RAE is an effective method for SZ detection. It is worth mentioning that RAE could serve as a model framework, the detailed structure could be adjusted according to specific classification tasks. Namely, the backbone of the encoder can be adapted to the tasks, and the selection of backbone will greatly affect the performance of model. Besides, the number of loops ($n$) could be optimized according to the classification task. We conducted a series of experiments to obtain the best value of $n$, and each experiment was implemented five times. As shown in the Table 3, when $n$ was set to 2, the model obtained the best accuracy and relatively balanced sensitivity and specificity.

In the current study, only one dataset was used to evaluate the performance, more SZ datasets should be collected to verify the generalization of RAE. Besides, RAE is expected to be effective in detecting other mental diseases, such as major depressive disorder. We have conducted several preliminary experiments and will release the further results in the future studies.

**Table 3.** Result of experiments concerning $n$ selection.

| $n$ | ACC(%) | SEN(%) | SPE(%) |
|---|---|---|---|
| 1 | 77.74±2.03 | 75.55 | 79.13 |
| **2** | **81.81±1.60** | **80.30** | **83.37** |
| 3 | 79.61±1.36 | 78.46 | 80.74 |
| 4 | 78.26±2.72 | 76.08 | 82.13 |
| 5 | 77.67±1.48 | 75.68 | 79.00 |

In summary, we proposed a novel framework method for SZ detection with recurrent Auto-Encoder. This method achieved an average accuracy of 81.81%, sensitivity of 80.30%, and specificity of 83.37% in the LOSO experiments, which improved 4.62% than the best baseline method. The RAE is expected to be a feasible tool in clinical diagnosis benefited by its superior performance and stability.

**Acknowledgments.** This work was supported in part by the National Natural Science Foundation of China under Grant No. 62076209.

# References

1. Acharya, U.R., Oh, S.L., Hagiwara, Y., Tan, J.H., Adeli, H., Subha, D.P.: Automated EEG-based screening of depression using deep convolutional neural network. Comput. Methods Programs Biomed. **161**, 103–113 (2018)
2. Boostani, R., Sadatnezhad, K., Sabeti, M.: An efficient classifier to diagnose of schizophrenia based on the EEG signals. Expert Syst. Appl. **36**(3, Part 2), 6492–6499 (2009)
3. Ciprian, C., Masychev, K., Ravan, M., Manimaran, A., Deshmukh, A.: Diagnosing schizophrenia using effective connectivity of resting-state EEG data. Algorithms **14**(5), 139 (2021)
4. Durongbhan, P., et al.: A dementia classification framework using frequency and time-frequency features based on EEG signals. IEEE Trans. Neural Syst. Rehabil. Eng. **27**(5), 826–835 (2019)
5. Gao, Z., et al.: EEG-based spatio-temporal convolutional neural network for driver fatigue evaluation. IEEE Trans. Neural Netw. Learn. Syst. **30**(9), 2755–2763 (2019)
6. He, K., Chen, X., Xie, S., Li, Y., Dollár, P., Girshick, R.: Masked autoencoders are scalable vision learners. In: Proceedings of the IEEE/CVF Conference on Computer Vision and Pattern Recognition (CVPR), pp. 16000–16009 (2022)
7. Institute for Health Metrics and Evaluation: Institute of health metrics and evaluation (IHME). Global health data exchange (GHDx) (2019). https://ghdx.healthdata.org/gbd-results-tool?params=gbd-api-2019-permalink/27a7644e8ad28e739382d31e77589dd7
8. Kingma, D.P., Ba, J.: Adam: a method for stochastic optimization. arXiv preprint arXiv:1412.6980 (2014)
9. Lawhern, V.J., Solon, A.J., Waytowich, N.R., Gordon, S.M., Hung, C.P., Lance, B.J.: EEGNet: a compact convolutional neural network for EEG-based brain-computer interfaces. J. Neural Eng. **15**(5), 056013 (2018)
10. Li, Y., Zhang, X.R., Zhang, B., Lei, M.Y., Cui, W.G., Guo, Y.Z.: A channel-projection mixed-scale convolutional neural network for motor imagery EEG decoding. IEEE Trans. Neural Syst. Rehabil. Eng. **27**(6), 1170–1180 (2019)
11. Luo, Y., Tian, Q., Wang, C., Zhang, K., Wang, C., Zhang, J.: Biomarkers for prediction of schizophrenia: insights from resting-state EEG microstates. IEEE Access **8**, 213078–213093 (2020)
12. Najafzadeh, H., Esmaeili, M., Farhang, S., Sarbaz, Y., Rasta, S.H.: Automatic classification of schizophrenia patients using resting-state EEG signals. Phys. Eng. Sci. Med. **44**(3), 855–870 (2021). https://doi.org/10.1007/s13246-021-01038-7
13. Nikhil Chandran, A., Sreekumar, K., Subha, D.P.: EEG-based automated detection of schizophrenia using long short-term memory (LSTM) network. In: Patnaik, S., Yang, X.-S., Sethi, I.K. (eds.) Advances in Machine Learning and Computational Intelligence. AIS, pp. 229–236. Springer, Singapore (2021). https://doi.org/10.1007/978-981-15-5243-4_19

14. Oh, S.L., Vicnesh, J., Ciaccio, E.J., Yuvaraj, R., Acharya, U.R.: Deep convolutional neural network model for automated diagnosis of schizophrenia using EEG signals. Appl. Sci. **9**(14), 2870 (2019)
15. Oh, S.L., Vicnesh, J., Ciaccio, E.J., Yuvaraj, R., Acharya, U.R.: Deep convolutional neural network model for automated diagnosis of schizophrenia using EEG signals. Appl. Sci. **9**(14), 2870 (2019)
16. Olejarczyk, E., Jernajczyk, W.: Graph-based analysis of brain connectivity in schizophrenia. PLoS ONE **12**(11), e0188629 (2017)
17. Saeedi, A., Saeedi, M., Maghsoudi, A., Shalbaf, A.: Major depressive disorder diagnosis based on effective connectivity in EEG signals: a convolutional neural network and long short-term memory approach. Cogn. Neurodyn. **15**(2), 239–252 (2021)
18. Schirrmeister, R.T., et al.: Deep learning with convolutional neural networks for EEG decoding and visualization. Hum. Brain Mapp. **38**(11), 5391–5420 (2017)
19. Shoeibi, A., et al.: Automatic diagnosis of schizophrenia in EEG signals using CNN-LSTM models. Front. Neuroinform. 15 (2021)
20. Vázquez, M.A., Maghsoudi, A., Mariño, I.P.: An interpretable machine learning method for the detection of schizophrenia using EEG signals. Front. Syst. Neurosci. **15**, 652662 (2021)
21. Wang, Y., Huang, Z., McCane, B., Neo, P.: EmotioNet: a 3-D convolutional neural network for EEG-based emotion recognition. In: 2018 International Joint Conference on Neural Networks (IJCNN), pp. 1–7 (2018)
22. Waytowich, N., et al.: Compact convolutional neural networks for classification of asynchronous steady-state visual evoked potentials. J. Neural Eng. **15**(6), 066031 (2018)
23. World Health Organization: Schizophrenia (2022). https://www.who.int/news-room/fact-sheets/detail/schizophrenia
24. Wu, H., et al.: A parallel multiscale filter bank convolutional neural networks for motor imagery EEG classification. Front. Neurosci. **13**, 1275 (2019)

# Functional Roles of Amygdala and Orbitofrontal Cortex in Adaptive Behavior

Layla Chadaporn Antaket[1]([✉])(iD), Kazuki Hamada[1], and Yoshiki Kashimori[1,2]

[1] Department of Engineering Science, University of Electro-Communications, Chofu, Tokyo 182-8585, Japan
a1943002@edu.cc.uec.ac.jp, kashi@pc.uec.ac.jp
[2] Center of Brain and Engineering, Univ. of Electro-Communications, Chofu, Tokyo 182-8585, Japan
kashi@pc.uec.ac.jp

**Abstract.** It is important for survival of animals in nature to adapt their behavioral strategies to everchanging environment. To do so, they must evaluate their behaviors by reward. Experimental studies reported the involvement of orbitofrontal cortex (OFC) and basolateral amygdala (ABL) in value evaluation and outcome expectation. OFC and ABL play different functional roles: ABL is critical for acquiring cue-outcome association, while OFC is involved in generating cue-outcome expectation to guide adaptive behavior. However, the neural mechanism underlying these functional roles remains unclear. To address this issue, we develop a model of OFC/ABL circuit that accounts for theses functional roles. We also incorporated a reinforcement learning in the model. Using the model, we show that ABL learns the association between odor and taste information, depending on predictive values and reward prediction errors. The association in the ABL allows the OFC network to generate cue-outcome expectation for forthcoming food. In a reversal learning, the mechanisms similar to the first learning create the new association of odor and taste in ABL and the task-relevant cue-outcome expectation in OFC.

**Keywords:** Orbitofrontal cortex · Basolateral amygdala · Adaptive behavior

## 1 Introduction

To survive in nature, animals must adapt their behavioral strategies to everchanging environment. To do so, they must recognize the external world and evaluate their behaviors by reward, which also improve the ability of outcome prediction. However, it remains unclear how animals adapt their behaviors to the environmental changes.

Experimental studies on odor discrimination tasks in rats [1–3] and devaluation tasks in monkeys [4–6] have shown that orbitofrontal cortex (OFC) and basolateral amygdala (ABL), besides the brain reward system, play crucial roles

M. Tanveer et al. (Eds.): ICONIP 2022, LNCS 13624, pp. 74–85, 2023.
https://doi.org/10.1007/978-3-031-30108-7_7

in value evaluation and outcome prediction. OFC is directly interconnected with ABL [7]. OFC and ABL may work as an integrated system that brings associative learning to bear on decision making. Moreover, lesion studies of ABL and OFC in an odor discrimination task [1–3] and its reversal [8,9] demonstrated distinct roles of ABL and OFC: ABL is critical for acquiring cue-outcome association, while OFC is involved in generating cue-outcome expectation to guide adaptive behavior. However, it still remains unclear the neural mechanisms underlying the functions of these areas.

To address this issue, we develop a model of OFC/ABL circuit that accounts for the roles of OFC and ABL and the adaptive behaviors generated by the interaction between these areas [10]. We are concerned with an odor discrimination task. In the task, rats must discriminate between two odors to ingest forthcoming food. The rats choose go or no-go action on the basis of odor information, and can ingest a food after a delay period only for go action. We show that ABL learns the association between odor and taste information, depending on predictive values and reward prediction errors. The learning allows the OFC network to create cue-selective neurons, providing expectation. In a reversal learning, rats fail to perform the reversal task just after the switching to the reversal task. After several failures, the mechanism similar to the first learning allows the ABL network to learn the new association between odor and taste information, and the association allows the OFC network to generate cue-outcome expectation relevant to the reversal task. Our model offers the mechanisms of how ABL and OFC work to acquire adaptive behaviors in the odor discrimination task and its reversal.

## 2   Model

### 2.1   Odor Discrimination Task and Reversal Task

Rats were trained on a series of two-odor, go/no-go discriminations [1–3]. In the discrimination task, rats were presented with one of the two odors (odor1 and odor2) at an odor spot, as shown in Fig. 1a. They moved to a food spot after the odor presentation, and chose go or no-go action. The rats poked their noses to a food well in go action, and waited a forthcoming food during a delay period. One odor (odor1) signaled the availability of a rewarding sucrose solution, and other odor (odor2) signaled the delivery of an aversive quinine solution. The procedure of the task is shown in Fig. 1b. On the other hand, in no-go action, the rats did not poke their nose to the well, and were not available for food. After the training, the rats learned appetitive behavior for sucrose solution coupled with odor1 and aversive behavior for quinine solution associated with odor2.

After the learning of the odor discrimination task shown in Fig. 1b, the relationship between the odors and foods were reversed as shown in Fig. 1c [1–3]. In the reversal task [8,9], the odor1 signaled the availability of the aversive quinine, and the odor2 signaled the availability of the rewarding sucrose. In the early period of the reversal learning, the rats exhibited a panic behavior, but they were able to adapt to the reversal task as the training proceeded.

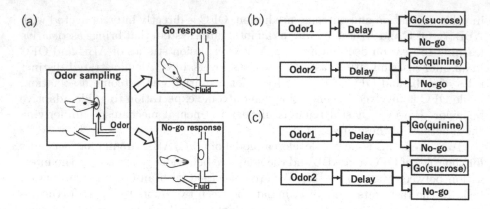

**Fig. 1.** Odor discrimination task. (a) Rats are presented with an odor at an odor spot, and then choose go and no-go action. The rats can ingest a food after a delay period in the go action. (b) Odor discrimination task. Odor1 is associated with sucrose, and odor2 is associated with quinine. (c) Reversal learning.

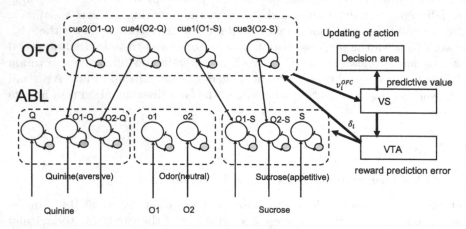

**Fig. 2.** The network model for an odor discrimination task. The model consists of the networks of ABL and OFC, the area of ventral striatum (VS) and ventral tegmental area (VTA), and a decision area. The white and gray circles indicate a main neuron and an inhibitory interneuron, respectively.

## 2.2   Overview of the Model

Figure 2 shows a model for the odor discrimination task by Schoenbaum et al. [1,2]. The model consists of the networks of ABL and OFC, the area of ventral striatum (VS) and ventral tegmental area (VTA), and a decision-making area. The ABL model contains odor- and taste-selective ABL neurons, receiving the odors and taste information, respectively. For an action of animals, the predictive value is updated in the VS area, and the reward prediction error is estimated in the VTA. The value is determined by the activity of OFC neurons, and the value-dependent activity of OFC neurons facilitates the association learning of odor and taste information in the ABL. The reward prediction error is utilized for Hebbian learning of ABL neurons. On the other hand, the OFC model contains cue-selective neurons, receiving inputs from a specific combination of odor- and taste-selective ABL neurons. The connections between ABL neurons and cue-selective ones are learned with Hebbian rule. The probability of action $a_i$ is determined by a function of the predictive values when cue-selective neurons are activated. Then the predictive value is updated by the action. Thus cue-selective neurons represent the causality of a cue (odor) and an outcome (taste). Until the cue-selective neurons are activated, rats make a random prediction of outcome (sucrose or quinine).

## 2.3   The Model of ABL

The model of ABL consists of three neuron units, neutral, appetitive, and aversive units, as shown in Fig. 3. The neutral unit consists of two groups of odor-selective neurons, odor1- and odor2-selective neurons. Each group consists of 10 neurons. The appetitive unit has three neuron groups, s-, o1-s, and o2-s neurons, responding to sucrose and the pair of sucrose and one of the two odors, respectively. The aversive unit also contains three neuron groups, q, o1-q, and o2-q neurons, similar to those in the appetitive unit. Moreover. each group has an interneuron, providing neurons within the group with an inhibition. A single neuron was modeled with leaky integrate-and-fire neuron (LIF) model. The membrane potential of the $(i, j)$th neuron, $V_{ij}^{ABL}$, is given by

$$\tau_{ABL} \frac{dV_{ij}^{ABL}}{dt} = -V_{ij}^{ABL} + \sum_{k,l} w_{ij,kl}^{ABL} S_{kl}^{ABL} + I_{ij}^{FF} + I_{OFC}(t), \tag{1}$$

where $w_{ij,kl}^{ABL}$ is the weight of the synaptic connection from the $(k, l)$th neuron to the $(i, j)$th one. $\tau_{ABL}$ is the time constant of $V_{ij}^{ABL}$. $I_{ij}^{FF}$ is the feedforward input from gustatory and olfactory sensory areas, and $I_{OFC}(t)$ is the input from the OFC network, depending on the value in the reward learning. An ABL neuron emits a spike when the membrane potential exceeds the firing threshold.

## 2.4   The Model of OFC

The model of OFC consists of four neuron groups, which are cue-selective neurons, cue1-, cue2-, cue3-, and cue4-selective neurons. Each group consists of 10 neurons. The cue1-selective neurons receive the inputs from o1-s neurons in the

ABL network. Other cue neurons also receive information of a pair of odor and taste, as shown in Fig. 2. A single neuron was modeled with the LIF model. The membrane potential of the $(i, j)$th neuron, $V_{ij}^{OFC}$, is given by

$$\tau_{OFC} \frac{dV_{ij}^{OFC}}{dt} = -V_{ij}^{OFC} + \sum_{kl} w_{ij,kl}^{OFC} S_{kl}^{OFC} + I_{ABL}^{X-Y}, \qquad (2)$$

where $w_{ij,kl}^{OFC}$ is the weight of the synaptic connection from the $(k, l)$th neuron to the $(i, j)$th one. $\tau_{OFC}$ is the time constant of $V_{ij}^{OFC}$. $I_{ABL}^{X-Y}$ ($X$ = O1, O2; $Y$ = s, q) are the inputs from $XY$-selective ABL neurons responding to an odor $X$ and a taste $Y$.

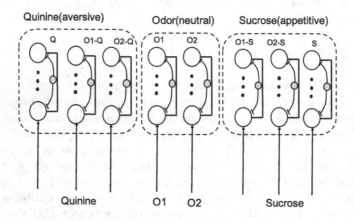

**Fig. 3.** The model of ABL. The model contains odor-selective (O1, O2), taste-selective (q, s), and odor-taste-selective (o1-s, o2-s, o1-q, o2-q) neurons. The white and gray circles indicate the neurons similar to Fig. 2

## 2.5   Reinforcement Learning

In the training of the odor discrimination task, the rats were trained to facilitate the sucrose ingestion coupled with odor1 and avoid quinine ingestion associated with odor2. Adaptive actions of the discrimination task were learned using a model of reinforcement learning, or Rescorla-Wagner algorithm [11]. For each odor stimulus, the rats choose a go or a no-go action and, then, receive sucrose or quinine only for go action. In the task, there are 4 events: positive go (PG), positive no-go (PNG), negative go (NG), and negative no-go (NNG), where "positive" means sucrose ingestion because sucrose is a preferred (sweet) food for rats and "negative" indicates quinine ingestion because quinine is an avoided (bitter) food. Other behaviors are to choose randomly a go or a no-go action. The random choice behaviors appear in the early period of learning until the odor-taste association is shaped in the ABL/OFC system. In our model, rats choose randomly a go and a no-go action until cue-selective neurons are activated. The behaviors are called random positive go (RPG), random negative go (RNG), random no-go (RNOG). The expected value for the action $a_i, v_i(t)$, is updated by

$$v_i(t+1) = v_i(t) + \rho \delta_i(t), \tag{3}$$

$$\delta_i(t) = R_i - v_i(t), \tag{4}$$

where $i =$ PG, PNG, NG, NNG for the activation of cue-selective neurons, and $i =$ RPG, RNG, and RNOG for the inactivation of cue-selective neurons. $\delta_i(t)$ is the reward prediction error. The expected values were set at $v_i(t) = 0$ for the 7 actions before the learning, and the actual rewards were set at $R_{PG} = 0.5$, $R_{NPG} = 0.1$, $R_{NG}(t) = -0.5$, and $R_{NNG} = -0.1$, $R_{RPG} = 0.5$, $R_{RNG} = -0.5$, and $R_{NNOG} = -0.1$. $\rho$ is the learning rate. Given the predictive values, the probability of the action $a_i$, under the odor $j$ presentation $O_j, p(a_i|O_j)$, is given by

$$p(a_i|O_j) = \frac{e^{\beta v_i}}{\sum_i e^{\beta v_i}} p(O_j), \tag{5}$$

where $\beta$ is the inverse temperature. The denominator was summed over $i =$ PG and PNG for odor1 and $i =$ NG and NNG for odor2. The probabilities of random behaviors were fixed at 0.5.

## 2.6   Learning of ABL and OFC Networks

The synaptic connections of ABL and OFC networks were learned in 100 trials of the odor discrimination task, consisting of 50 odor1-sucrose and 50 odor2-quinine trials. These trials were given randomly. The connections between ABL neurons were developed by a reward-dependent Hebbian learning [12]. The learning makes association between odors and tastes. The synaptic weight from the $(k, l)$th neuron to the $(i, j)$th one, $w_{ij,kl}^{ABL}$, is given by

$$\tau_w \frac{dw_{ij,kl}^{ABL}}{dt} = -w_{ij,kl}^{ABL} + \lambda_{ABL} S_{ij}^{ABL} S_{kl}^{ABL} |\delta_m(t)|, \tag{6}$$

where $\tau_w$ is the time constant of $w_{ij,kl}^{ABL}$, $\lambda_{ABL}$ is the learning rate, and $\delta_m(t)$ is the reward prediction error of the action $a_m$. The connections between cue-selective neurons and those between ABL and OFC neurons were learned with the learning rule similar to Eq. (6).

## 2.7   Two Trails After the Learning

After the learning, we investigated the neuronal responses of the ABL and OFC networks in 2 trails in the odor discrimination task. In the first trial, a trained rat was presented with odor1 at an odor spot, and the rat performed PG action and then received sucrose at a food well. In the second trial, the rat was presented with odor2, and performed NG action and then received quinine.

## 3   Result

### 3.1   Learning of Adaptive Behavior

Figure 4a shows the temporal courses of the predictive values for four actions during the learning of odor discrimination task. The values of no-go actions, PNG and NNG, were nearly zero. In the early learning period ($t < 4000\,\mathrm{ms}$), the predictive value of RPG was increased by sucrose ingestion, whereas that of RNG was reduced due to quinine ingestion. In the period, rats do not perceive the association between an odor and a taste, and exhibit random choice of go and no-go action. After the random actions ($t > 4000\,\mathrm{ms}$), the predictive value of PG was beginning to increase, and that of NG was beginning to decrease. Concurrently, the predictive values of RPG and RNG decayed. This indicates that the association of odors and tastes is generated in ABL and cue-selective neurons in OFC are activated. As the learning proceeds, the predictive values of PG and NG tend to the respective asymptotic values.

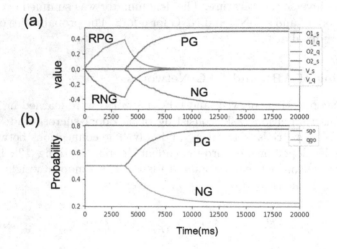

**Fig. 4.** Temporal variations of predictive values (a) and action probabilities (b) during the learning of odor discrimination task. Sucrose is available for positive go (PG) action, and quinine is ingested for negative go (NG) action. RPG and RNG indicate random choices of PG and NG action, respectively.

Figure 4b shows the action probabilities of PG and NG. In the early period ($t < 4000\,\mathrm{ms}$), the probabilities of any actions were not updated, and rats chose randomly go and no-go actions. After the random choice, the action probability of PG was increased, while that of NG was decreased. In the late period ($t > 15000\,\mathrm{ms}$), these probabilities converged to asymptotic values, indicating that rats facilitate the sucrose ingestion associated with odor1 and suppress the quinine ingestion coupled with odor2.

## 3.2  Formation of Odor-Taste Association in ABL and Expectation in OFC

Figure 5 shows the raster plots of ABL and OFC neurons during the learning of odor discrimination task, together with the time course of the action probabilities of PG and NG. ABL neurons are activated by an odor and a taste stimulus, and the connections between these neurons are learned by the input from OFC depending on the predictive values of the two actions. The learning elicits the activation of ABL neurons coding odor1 and sucrose or of those encoding odor2 and quinine, making the associations between these odor-taste pairs. On the other hand, OFC neurons were not activated in the early learning period, because OFC neurons do not receive directly odor inputs and the connections between ABL and OFC neurons are nearly zero in the period. As the learning proceeds, the connections between ABL and OFC neurons were developed with the changes of the predictive values of PG and NG, and OFC neurons encoding the specific pairs of odor and taste, or odor1-sucrose (O1-S) and odor2-quinine (O2-Q), were activated. The cue1-selective neurons encoding the coupling of odor1 and sucrose show a larger activation shown in Fig. 5, leading to the facilitation of sucrose ingestion. In contrast, the cue4-selective neurons encoding odor2 and quinine exhibited a sparse firing, causing the suppression of quinine ingestion.

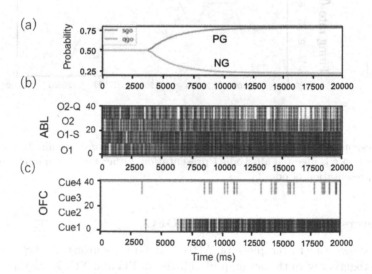

**Fig. 5.** Responses of ABL and OFC neurons during the learning of an odor discrimination task. (a) Temporal courses of the probabilities of PG and NG actions. (b), (c) Raster plots of ABL and OFC neurons during the learning.

## 3.3  Responses of OFC Networks for Two Task Trials

Figure 6 shows the raster plots of OFC neurons in the sequence of two odor discrimination trials. In one trial, a rat performed the PG and received sucrose after a delay period. In another trial, the rat performed the NG and received quinine. The odor1 stimulation evoked the firing of cue1-selective neuron, and it remained active during the delay period, working as expectation for forthcoming outcome (sucrose ingestion). The firing was enhanced during the delay period because sucrose is a preferable food for rats. On the other hand, the odor2 stimulation elicited the firing of cue4-selective neuron, sustained in a delay period followed by quinine ingestion. The sustained firing of cue4-selective neuron also works as expectation for the forthcoming ingestion (quinine ingestion), but it exhibited a sparse activity because of aversive availability of quinine.

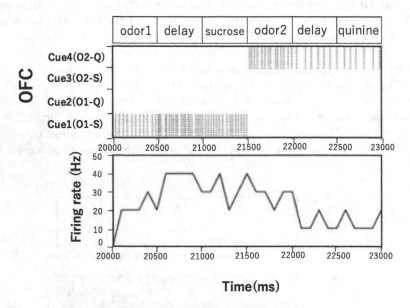

**Fig. 6.** Responses of OFC neurons for a sequence of two odor discrimination tasks. In one task, sucrose is ingested after odor1 presentation. Other task has quinine ingestion after the presentation of odor2.

## 3.4   Reversal Learning of ABL and OFC

Figure 7 shows the raster plots of ABL and OFC neurons during a reversal learning, together with the action probabilities of PG and NG. After the learning of the association of odor1-sucrose and odor2-quinine, the relationship of odors and tastes was rapidly reversed at $t = 25000$ ms. The new relationship is odor1-quinine and odor2-sucrose. In the reverse learning, the reinforcement learning reduced the probability of PG and increased that of NG. In the late period of the learning ($t > 40000$ ms), the two actions were adapted to the new relationship of odor and taste.

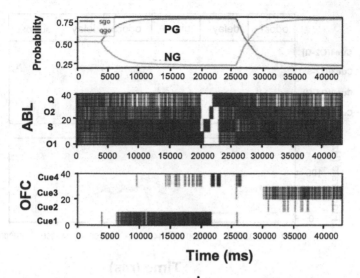

**Fig. 7.** Responses of ABL and OFC neurons in the first odor discrimination task and its reversal learning. The reversal learning was started at $t = 25000$ ms.

The ABL made the new associations in the similar way to the first learning. The connections between ABL neurons encoding odor1 and quinine and those encoding odor2 and sucrose were learned, while the connections involved in the first learning decayed. In association with the development of the new association in ABL network, OFC neurons were activated for the reversal relationship; the cue2- and cue3-selective neurons, encoding the pair of odor1 and quinine and that of odor2 and sucrose, respectively, were activated.

Figure 8 shows the raster plot and firing rate of OFC neurons in the sequence of two odor discrimination tasks after the reversal learning. The odor1 stimulus activated cue2-selective neurons, and their sparse activities were maintained during a delay period followed by quinine stimulation. Similarly, odor2 stimulus evoked the spiking of cue3-selective neurons and their sustained activities. The cue2-selective neurons exhibited sparse activity in the delay period, indicating the expectation for the aversive quinine ingestion, whereas the cue3-selective neurons caused the increased activity, reflecting the expectation of the preferred sucrose ingestion. The reversal learning shapes the cue-selective responses consistent with the reversal relationship between odor and taste information.

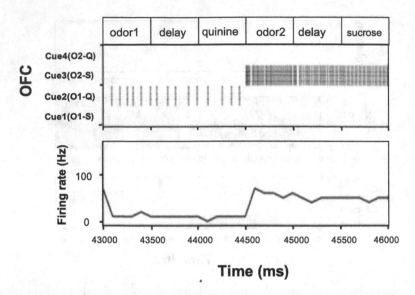

**Time (ms)**

**Fig. 8.** Responses of OFC neurons after the reversal learning. Odor1 elicit the sustained activity of cue2-selective neurons, leading to aversive ingestion of quinine. Similarly, odor2 elicits appetitive ingestion of sucrose. The relationship between odors and tastes are reversed for that in the first task.

## 4    Conclusion

We have presented the neural mechanisms by which ABL makes association between odors and tastes and OFC generates the expectation of cue-induced outcome. The learning of ABL and OFC is based on a reinforcement learning and Hebbian learning depending on reward prediction error. We have also shown that ABL and OFC modulate the representation of odors and tastes in a reversal leaning. The results provide insights into understanding the functional roles of ABL and OFC in adaptive behavior.

## References

1. Saddoris, M.P., Gallagher, M., Schoenbaum, G.: Rapid associative encoding in basolateral amygdala depends on connections with orbitofrontal cortex. Neuron **46**, 321–331 (2005)
2. Schoenbaum, G., Setlow, B., Saddoris, M.P., Gallagher, M.: Encoding predicted out-come and acquired value in orbitofrontal cortex during cue sampling depends upon input from basolateral amygdala. Neuron **39**, 855–867 (2003)
3. Schoenbaum, G., Roesch, M.: Orbitofrontal cortex, associative learning, and expectancies. Neuron **47**, 633–636 (2005)
4. Baxter, M.G., Parker, A., Lindner, C.C., Izquierdo, A.D., Murray, E.A.: Control of response selection by reinforce value requires interaction of amygdala and orbitofrontal cortex. J. Neurosci. **20**, 4311–4319 (2000)

5. Fluzat, E.C., Rhodes, S.E., Murray, E.A.: The role of orbitofrontal-amygdala interactions in updating action-outcome valuations in macaques. J. Neurosci. **37**, 2463–2470 (2017)
6. Rhodes, S.E., Murray, E.A.: Differentail effects of amygdala, orbital prefrontal cortex, and prelimbic cortex lesions on goal-directed behavior in rhesus macaques. J. Neuro-sci. **33**, 3380–3389 (2013)
7. Kolb, B.: Functions of the frontal cortex of the rat: a comparative review. Brain Res. Rev. **8**, 65–98 (1984)
8. Schoenbaum, G., Setlow, I.B., Nugent, S.L., Saddoris, M.P., Gallagher, M.: Lesions of orbitofrontal cortex and basolateral amygdala complex disrupt acquisition of odor-guided discriminations and reversals. Learn. Mem. **10**, 129–140 (2003)
9. Stalnaker, T.A., Franz, T.M., Singh, T., Schoenbaum, G.: Basolateral amygdala lesions abolish orbitofrontal-dependent reversal impairments. Neuron **54**(1), 51–58 (2007)
10. Takei, K., Fujita, K., Kashimori, Y.: A neural mechanism of cue-outcome expectancy generated by the interaction between orbitofrontal cortex and amygdala. Chem. Senses **45**, 15–26 (2020)
11. Rescorla, R.A., Wagner, A.R.: A theory of Pavlovian conditioning: variations in the effectiveness of reinforcement and nonreinforcement. In: Black, A.H., Prokasy, W.F. (eds.) Classical Conditioning II: Current Research and Theory, pp. 64–99. Appleton-Century-Crofts, New York (1972)
12. Roelfsema, P.R., Holtmaat, A.: Control of synaptic plasticity in deep cortical networks. Nat. Rev. Neurosci. **19**, 167–180 (2018)

# A Multiclass EEG Signal Classification Model Using Channel Interaction Maximization and Multivariate Empirical Mode Decomposition

Pankaj Kumar Jha, Anurag Tiwari, and Amrita Chaturvedi[✉][iD]

Department of Computer Science and Engineering, Indian Institute of
Technology (BHU), Varanasi, India
amrita.cse@iitbhu.ac.in

**Abstract.** Brain-Computer Interface (BCI) is an emerging technology
that facilitates a pathway between the human brain and external devices.
Electroencephalography (EEG) data are mainly employed in BCI systems
to reflect the underlying mechanism of different neural activities associated
with various limb motions or Motor Imagery (MI) activities. Multichannel EEG signal processing generally results in high-dimensional features,
which increases BCI's overall computational and temporal complexity. We
introduce a channel selection methodology using the mutual information-based three-way interaction scheme to reduce this burden due to many
channels. Our approach initializes a set of three candidate solutions for
a given MI classification task and subsequently determines a highly significant EEG channel set. It effectively balances relevance and redundancy
levels in the final channel subset during the selection and rejection of a
newly selected channel. The proposed scheme is evaluated on the BCI
Competition IV-2008 dataset with four MI classes (left hand, right hand,
tongue, and feet) and twenty-two channels. The performance of our scheme
is compared with three recently published state-of-the-art methods. The
proposed approach realized an average of 86.66% classification accuracy
using only nine channels on the data of nine participants. The comparative study shows that our approach realized better performance in terms
of higher classification accuracy and channel reduction rate than all three
baseline models. The results are promising for the online BCI paradigm
that requires low complexity while conducting multiple sessions of BCI
experiments for a larger group of participants.

**Keywords:** Brain-Computer Interface · Motor Imagery ·
Electroencephalography · High Dimensional Data · Channel
Reduction · Channel Interaction Maximization

## 1 Introduction

The EEG-based BCI system facilitates a communication framework between
the human brain and external intelligent devices by decoding the intrinsic cognitive patterns associated with different neural activities [29]. It takes in brain

M. Tanveer et al. (Eds.): ICONIP 2022, LNCS 13624, pp. 86–100, 2023.
https://doi.org/10.1007/978-3-031-30108-7_8

signals, analyses them, and translates them into computer-based control commands. These are further used to control machines such as smart home appliances, neuroprosthetics, and intelligent chairs without performing any muscular activity. A block diagram of a generalized BCI system is shown in Fig. 1. In BCI systems, Motor Imagery (MI) refers to a dynamic mental state in which a subject imagines a muscular activity without actually executing it. Several researchers defined the relationship between motor activities and corresponding brain states by classifying respective brain oscillations [16,27] into four groups (1) delta (<4 Hz), (2) theta (<4, 8> Hz), (3) alpha (<8.0, 13.0> Hz), and (4) beta waves ($\geq$13 Hz). It has been concluded that a complex EEG spectrum consisting of an upper range of alpha waves and a lower range of beta waves (<8, 30> Hz) represents spatial-temporal properties of MI-specific brain signals. In EEG signal processing, scalp sites are termed as channels or electrodes from which signals are recorded. Although the dense arrangement of electrodes reveals more information about cognitive activities, it increases the redundancy due to noise and results in high-dimensional data. Besides, the inclusion of a large number of channels increases the cost of the system. These factors further increase the effort involved in the BCI setup that reduces its practicality [20]. These limitations motivate researchers to adopt efficient schemes that select only relevant and non-redundant channels for developing a productive BCI system.

Mathematically, Optimal Channel Selection (OCS) is an NP-Complete (NPC) problem for which no efficient solution has been found within polynomial time. Several existing methods have been developed to filter significant channels by associating electrodes' location and respective MI activities, but their effectiveness is limited because of inter-subject variability. These methods often employ the neuro-physiological basis of the human brain to locate an initial set of candidate solutions before estimating the relevance of new ones in the optimal channel subset. The selected channels achieve better classification accuracy and minimize the computational cost involved in processing high-dimensional cognitive signals.

This study considers multichannel EEG signal processing for MI tasks a multidimensional classification problem. This problem has been addressed in earlier works by targeting muscular movements and cognitive task-related experiments. A detailed discussion on existing channel selection methods is presented in [5]. These methods can be grouped into three classes: (1) Filter, (2) Wrapper, and (3) Hybrid methods. Filter-based channel selection methods are fast, independent of the applied classification approach, and highly scalable. However, they suffer from poor classification accuracy because they ignore the relevance of newly selected channels to earlier selected channels. Hence their performance is limited by high redundancy associated with the selected channel subset. These methods often explore different information-theoretic concepts such as mutual information, correlation, entropy, and variance for channel selection. On the contrary, wrapper methods implement a classification algorithm iteratively to determine the effectiveness of the selected channels. Therefore, these approaches are relatively more accurate than filter methods. However, because of the involvement of the classifier, wrapper methods require high computation time for a dataset with many channels. In addition, they are prone to overfitting because they train machine learning models with different combinations of features extracted from selected channels. These methods employ a variety of metaheuristic algorithms in the channel selection process because of their ability to maintain a good balance between search space exploration and solution space exploitation. Finally, hybrid methods enjoy the benefits of both filter and wrapper methods in terms of effectiveness and overfitting issues. However, the effectiveness of hybrid methods depends on the compatibility of the participating methods; otherwise, their performance may deteriorate the classification accuracy of the BCI system.

The primary objective of existing optimization methods is to filter only task-specific channels that can effectively refer to the performed MI activity with minimum computational cost. Recently, various algorithms have been introduced to solve the channel selection problem. For example, Arvaneh et al. (2011) [4] introduced two variants of Sparse Common Spatial Patterns (SCSP). In the first variant, they selected the least number of channels within a constraint of classification accuracy, while in the second one, they determined the least number of channels without compromising the classification accuracy obtained by using all the channels. Both methods maximized variance between two MI classes by applying spatial filters. Yang et al. (2017) [30] computed the correlated channels by considering mutual information between Laplacian derivatives of power features extracted from the selected channels and the candidate channels. However, this method suffered from excessive redundancy because of ignoring the relevance of channels individually. Torres-Garcia et al. (2016) [26] developed a fuzzy system interface to obtain a Pareto front solution for classification accuracy maximization. In this problem, a bi-objective function with two criteria (error rate and the number of channels) was used to obtain a robust tradeoff between the number of channels and the classification accuracy. In recent work, the Multi-Objective Non-Sorting Genetic Algorithm (MO-NSGA) has been used for channel selection [23]. This method employed a hybrid signal feature set using Empirical Mode Decomposition (EMD) and Discrete Wavelet Transform (DWT) with the MO-NSGA algorithm and achieved 100% classification accuracy.

Wang et al. [29] reduced the irrelevant and redundant EEG channels using a threshold-based Normalized Mutual Information (NMI) measure. They constructed an NMI connection matrix to obtain the relationship between pairs of channels. Then, setting an appropriate threshold, optimal channels were selected for classification purposes. However, this method achieved better classification accuracy but ignored channels' relevance individually. Jiao et al. [17] developed an improved CSP variant to capture shared salient information across related spatial patterns using a multiscale optimization approach. They combined multi-view learning-based sparse optimization to jointly extract robust CSP features with the L2,1-norm regularization method. They achieved competitive results compared to original CSP and other state-of-the-art methods.

Considering the limitations of the above-discussed methods, we present an improved filter method by introducing a mutual-information-based three-way

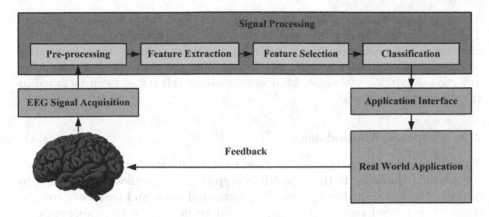

**Fig. 1.** Block diagram of a general BCI system

interaction scheme [6] to determine the optimal channel subset. Compared to conventional filter methods, our method follows the maximum relevance and minimum redundancy principle while selecting new channels. It computes an interaction score between randomly chosen and earlier selected electrodes to determine the significance of the newly inserted channels. In case of a positive score, the new channel is selected else rejected. Next, Multivariate Empirical Mode Decomposition (MEMD) [25] approach is applied to determine spatiotemporal features from the selected channels. Finally, classifier schemes were applied to discriminate four MI classes (left, right, tongue, feet) specific brain signals

The rest of the paper is organized as follows: Sect. 2 presents the material and methodology used in the proposed work. In Sect. 3, experimental results of classification accuracy and channel reduction rate are discussed. Finally, Sect. 4 concludes the research work with the future scope of the related domain.

## 2    Material and Methodology

This section presents a detailed description of the dataset and proposed channel selection approach. In Fig. 2, our methodology is shown in three sequential steps: (1) signal preprocessing, (2) channel selection & feature engineering, and (3) classification. A detailed description of all the steps is given in subsequent subsections.

### 2.1    Dataset Details

In our work, we use the BCI Competition IV- 2008 - II A dataset to validate our methodology. It comprises EEG signals collected from 9 healthy participants. This spectrum consists of 22 EEG channels and 3 EOG channels with the left mastoid as reference. It is a four-class MI task-based dataset where class 1 represents the left-hand movement, the right-hand gesture constitutes class 2, class

3 comprises the motion of both feet, and class 4 deals with the tongue activity. This dataset consists of individual training and validation EEG samples for all nine subjects to corroborate any classification scheme. Hence, there is no need to decompose given data samples into the training and the validation sets using any cross-validation technique. More details about this dataset can be found in the reference article [8].

## 2.2   Proposed Methodology

This study performs a sequence of steps to discriminate four MI classes using the selected channels. Initially, multiple preprocessing methods are applied to improve the signal quality by curtailing unwanted noise and frequency components. The refined signals were further used to determine the importance of respective channels in candidate solutions. A detailed discussion of applied steps is given below.

**Channel Setting.** As discussed above, the BCI dataset consists of 25 channels in which 22 channels refer to the EEG spectrum while the remaining 3 represent EOG waves. Here, only EEG signals are used to select the most optimal channel subset and for the classification of performed MI tasks. Therefore, EOG channels are directly eliminated and not considered in any data analysis step. In the next phase, oscillations of 22 EEG channels are used for cognitive pattern analysis.

**SNR Enhancement.** We make our data more precise by minimizing noise and outliers induced in the raw brain waves. This step helps to maximize Signal to Noise Ratio (SNR) of the EEG signals. Here, a 3rd order Savitzky-Golay filter [14] with an optimal window size of 1000 is used to optimize outliers' data points. This step is essential because it provides biased results and reduces classification accuracy. Next, the Fast Independent Component Analysis (FastICA) algorithm [21] eliminates noise and outliers from the optimized EEG signals. Since all the frequency components are not required to discriminate MI classes, we extract beta waves in $<12-30\,\mathrm{Hz}>$. Each step mentioned above is performed sequentially, and improvement was observed after each step. A pictorial representation of all three steps is shown in Fig. 3.

**Channel Selection.** The proposed channel selection approach utilizes a mutual information-based three-way channel interaction scheme to determine the relationship between newly selected channels, earlier selected ones, and three candidate channels. Our approach maximizes the global mutual information among all three categories of channels so that the selected channel maximizes the relevance and minimizes the redundancy score in the global channel subset. Our proposed channel selection methodology is motivated by the earlier proposed Feature Interaction Maximization (FIM) algorithm [6]. The original algorithm has been effectively used for large feature space optimization in data classification problems. Let $P(X)$ denote the Probability Density Function (PDF) of data sequence X, then the entropy of X can be defined as:

$$H(X) = -\sum_{i=1}^{n} p(x_i) \log(p(x_i)) \tag{1}$$

where $0 \leq H(X) \leq 1$. In the case of two variables X and Y, joint and conditional entropy is given as:

$$H(X|Y) = -\sum_{i=1}^{n}\sum_{j=1}^{m} p(x_i, y_i) \log(p(x_i|y_i)) \tag{2}$$

$$H(X,Y) = -\sum_{i=1}^{n}\sum_{j=1}^{m} p(x_i, y_i) \log(p(x_i, y_i)) \tag{3}$$

In information theory, joint and conditional entropy are related in the following manner:

$$H(X,Y) = H(X) + H(Y|X) \tag{4}$$

$$H(X,Y) = H(Y) + H(X|Y) \tag{5}$$

The mutual information and entropy can be correlated as

$$I(X;Y) = \sum_{i=1}^{N}\sum_{j=1}^{M} p(x_i, y_i) \log(\frac{p(x_i|y_i)}{p(x_i)p(y_i)}) \tag{6}$$

The value of MI in Eq. 6 is always positive. It is high if both variables are highly associated; MI is zero if both variables are independent. MI can be defined as a function of the entropies, as follows:

$$I(X;Y) = H(Y) - H(Y|X) \tag{7}$$

$$I(X;Y) = H(X) - H(X|Y) \tag{8}$$

$$I(X;Y) = H(X) + H(Y) - H(X,Y) \tag{9}$$

In the channel selection problem, the mutual information I (X; Y) represents the relation between channel X and Y. This relationship is also referred to as information gain; the channel with the highest mutual information is considered the most informative and given higher priority in the application. In our work, two information theory measures, namely (1) Conditional mutual information $I(X_j; Y/X_i)$ and (2) Three-way interaction information $((X_j; X_i; Y))$ are merged to determine the relevance of the selected channels. In both cases, the relation between a feature and target class is studied in the context of other features.

$$I(X_j; Y/X_i) = H(X_j; Y) - H(X_j/Y, X_i) \tag{10}$$

$$I(X_j; X_i; Y) = I(X_j, X_i; Y) - I(X_j; Y) - I(X_i; Y) \tag{11}$$

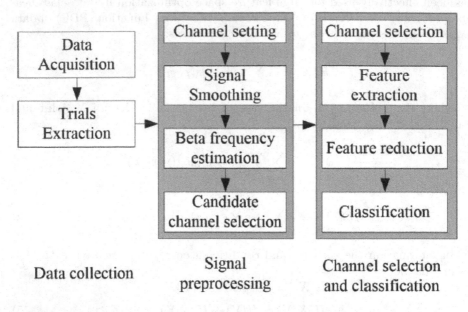

**Fig. 2.** Block diagram of proposed EEG channel selection methodology using CIM

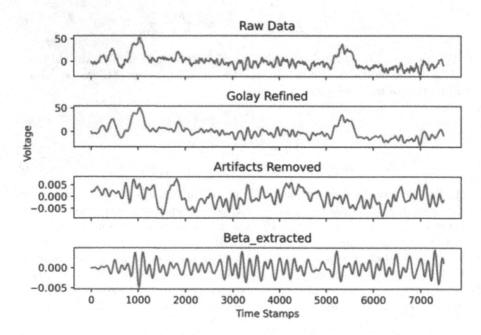

**Fig. 3.** Sequential representation of steps applied in SNR enhancement

Unlike MI, three-way interaction measures can be positive, negative, or zero. A positive score refers to the combined information associated with two channels that cannot be provided by each of them individually. It is negative when any two features compute the combined information. A zero score shows that both channels are independent and don't share any information. Suppose $X_i$ is a candidate solution and $X_s$ is a channel belonging to the subset $S$ (i.e., it has already been selected) and $C$ is a target class (attribute); channel interaction maximization (CIM) can be defined as

$$CIM = \arg\max(I(X_i; C) + \min_{X_s \in S}(I(X_i; X_s; C)))  \qquad (12)$$

where

$$I(x_i; X_x; C) = I(x_i; X_s; C) - I(X_i; C) - I(X_s; C)  \qquad (13)$$

In Eq. 12, the mutual information between features $X_i$ and $C$ computes the relationship between the candidate feature and the class attribute. The interaction information among $x_i X_s$, and $C$ is the redundancy term. The selected feature is the one that maximizes the objective function defined in Eq. 12. It has the maximum relevance to the class attribute and the minimum interaction with the selected features. The advantage of this criterion is its ability to select the features that have the highest discriminative power. The pseudocode of the three-way interaction maximization approach that we have employed in our work is given in Algorithm 1.

---

**Algorithm 1.** Channel Interaction Maximization

---

1: (Initialisation) Set U 'Initial set of 22 Channel'
2: S Contains Candidate Channels "c3,c4 and Cz". Set X contains 'Remaining 19 Channels'
3: (Mutual information with each Class label is calculated ) For every Channel $x_i \in X$, Calculate I(C;$x_i$)
4: **for** (Greedy selection) We will Repeat until all channels are selected,i.e, S = k **do**
5:     Calculation of the Mutual Information between elements) For all pairs of elements I($x_i; x_s$) with $x_i \in X$, $x_s$, calculate I($x_i; x_s$), only if it was not previously calculated.
6:     (For choosing Next Feature) feature $x_i$ is selected as the one that maximises the goal function: $I(C; xi) - \beta \sum_{x_s \in S} I(x_i; x_s)$. Set $X \leftarrow X x_i$; set $S \leftarrow S U x_i$
7: **end for**
8: Print the set S containing the chosen Channels in ranked order
9: Further we will take Top K channels from the set and perform the classification and compute accuracy.

---

**Feature Extraction.** The MI-specific signal parameters for the categorization of motor imagery activities are extracted using the channel chosen in the preceding section. Multivariate Empirical Mode Decomposition (MEMD) is a popular feature extraction used for dealing with nonlinear and non-stationary EEG data.

It's a multivariate variant of the conventional Empirical Mode Decomposition (EMD) technique in which numerous n-dimensional envelopes are generated by projecting EEG instances in n-variate spaces in every direction. The local mean is then calculated using these predictions. We chose MEMD over other common feature extraction methods because it employs cross-channel information to compute the behaviour of intrinsic EEG signals.

Let a multivariate EEG segment be represented by n-dimensional vectors $\{x(t)\}_{t=1}^{n} = \{x_1(t), x_2(t), \cdots, x_n(t)\}$ where $d^{\theta_k} = \{d_1^k, d_2^k, \cdots, d_n^k\}$ denotes a set of direction vectors along the directions given by angles $\theta^k = \{\theta_1^k, \theta_2^k, \cdots, \theta_n^k\}$ on an $(n-1)$ space. The steps used in MEMD computation are given below:

1. Select appropriate points for sampling on (n-1) planes.
2. Compute projection $\{p^{\theta_k}(t)\}_{t=1}^{T}$ along the direction $d^{\theta_k}$ of the input signal $\{x(t)\}_{t=1}^{n}$ for all $k$ resulting $\{p^{\theta_k}(t)\}_{k=1}^{K}$ to form a projection set.
3. Find the time series instants $t_1^{\theta_k}$ corresponding to the maxima $\{p^{\theta_k}(t)\}_{k=1}^{K}$.
4. Interpolate $\{t_1^{\theta_k}, x(t_1^{\theta_k})\}$ to obtain multivariate envelop curves $\{e^{\theta_k}(t)\}_{k=1}^{K}$.
5. Compute the mean function $m(t)$ by averaging all multivariate envelope curves, defined as follows:

$$m(t) = \frac{1}{n} \sum_{k=1}^{K} e^{\theta_k(t)} \tag{14}$$

6. Compute the detail using $c(t) = Y(t) - m(t)$. If the stopping criterion is satisfied, this detailed Intrinsic Mode Function (IMF) becomes Multivariate Intrinsic Mode Function (MIMF). Otherwise, $Y(t)$ is assigned to the remainder $c(t)$, and the process of identifying a new IMF is reiterated. The entire process is iteratively performed to compute all IMFs from the signal $Y(t)$.

Let $x_i \in R^k$ denote the EEG signal observation at a time instance where is the total number of optimal channels. The formal definition of the spatial covariance matrix is defined by $Cov = E\{(x_j - E\{x_j\})(x_j - E\{x_j\})^T\}$ where $E\{.\}$ denotes the expected value and is a superscript that represents the transposition of (.). In the BCI system development, each entry of the spatial covariance matrix is considered a feature of the respective observation.

**Feature Reduction.** In the last step, MEMD computes a large 3-D feature matrix because of multiple decomposition operations for each selected channel. Most of the produced features are redundant and static, and their role in MI-tasks discrimination is insignificant. Hence, it is essential to recognize and eliminate them before the classification process. Here, Principal Component Analysis (PCA) [1] removes the less significant features by fixing the 96.2% variance threshold on the originally produced MEMD feature set. In PCA, the feature matrix $F$ is used to compute the orthogonal matrix $W_{K \times M}$, which is further used to produce the transformed matrix $Y_{K \times N}$ using Eq. 15:

$$Y_{K \times N} = W_{K \times M} \times F_{M \times N} \tag{15}$$

where $M$ is the size of the feature set, $N$ is the number of instances, and $K \leq M$, $K \leq M$ represents the dimension of the output feature set where $K$ is selected based on the accumulation of the first few largest eigenvalues exceeding 96.2% of the total sum of all the eigenvalues computed from the feature matrix.

## 3  Results and Discussion

In our experiment, six classifiers are applied to find the compatibility of the proposed channel selection approach with different discrimination criteria. These classifiers are: (1) eXtreme Gradient Boosting (XgBoost) [1], (2) Random Boosting (RB) [7], (3) Light Gradient Boosting (LGB) [3], (4) Ensemble Learning Classifier (ELC) [11], (5) Support Vector Machine (SVM) [18], and (6) Spiking Neural Networks(SNNs) [31]. Except SNNs (a deep learning classification approach), all techniques are from the class of machine learning methods. In SNNs implementation, we executed the typical vanilla SNN architecture discussed in the article [31] and realized best set of hyper parameters as: (1) learning rate = 0.1, (2) exponential decay for the first moment estimates $(\beta_1) = 0.91$, (3) exponential decay rate for the second-moment estimates $(\beta_2) = 0.99$, and (4) $\epsilon$ (threshold to prevent any segmentation error during optimization) = $10^{-8}$. Moreover, the SNNs implementation was time consuming on the training data and hence not suitable for light weight BCI-based wearable device. Therefore, we

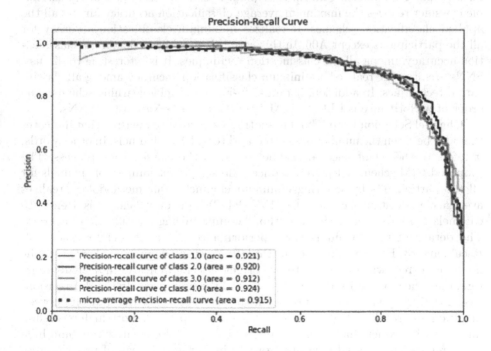

**Fig. 4.** Precision-Recall curve for all four MI classes

**Table 1.** Performance comparison of five classification techniques in the proposed channel selection approach.

| Subject | Selected channels | No. of channels | CSR | Classification Accuracy (CA) | | | | | |
|---|---|---|---|---|---|---|---|---|---|
| | | | | Xgboost | RB | LGB | ELC | SVM | SNNs |
| A01 | C3, C4, Cz, 5,3,9,14,21 | 8 | 0.37 | 71.98 | 87.52 | 77.42 | 93.36 | 87.41 | 71.66 |
| A02 | C3, C4, Cz, 2,4,9,13, 15,16,22 | 10 | 0.46 | 63.18 | 44.78 | 55.60 | 95.45 | 90.66 | 69.29 |
| A03 | C3, C4, Cz, 1,3,9,5, 11,16, | 11 | 0.50 | 70.33 | 80.30 | 65.12 | 94.21 | 82.54 | 67.58 |
| A04 | C3, C4, Cz, 1,3,9,5,11, 16, 14, 21 | 8 | 0.37 | 57.68 | 65.40 | 83.46 | 81.62 | 78.40 | 59.33 |
| A05 | C3, C4, Cz, 2,3,14,16,22 | 8 | 0.37 | 51.24 | 72.68 | 76.20 | 85.09 | 87.40 | 63.02 |
| A06 | C3, C4, Cz, 5,9,13,16,19,22 | 9 | 0.41 | 89.02 | 60.33 | 91.33 | 91.11 | 98.39 | 73.23 |
| A07 | C3, C4, Cz, 3,14,15,22 | 7 | 0.32 | 65.10 | 77.02 | 77.30 | 79.38 | 74.60 | 78.18 |
| A08 | C3, C4, Cz, 15,11,16,19,21 | 8 | 0.37 | 76.48 | 92.30 | 60.0 | 89.91 | 86.33 | 62.50 |
| A09 | C3, C4, Cz, 1,9,15,11,16,19,22 | 10 | 0.46 | 72.60 | 75.38 | 80.50 | 83.25 | 78.20 | 68.92 |
| Average | Highly Voted Channel: 16, 14, 22, 3, 19, 15, 5, 13, 11, 21, 9, 2, 1, 4 | 8.77 ≈ 9 | 0.40 | 68.62 | 72.85 | 74.10 | 88.15 | 84.87 | 68.19 |

mainly focused on machine learning classification techniques and found improved results compared to SNNs. Further, to obtain true nonlinear cognitive patterns, radial basis function is used as kernel in the SVM classifier. The proposed study employs five performance measures: [1] Classification Accuracy (CA), [2] Channel selection rate (CSR), [3] Precision, [4] Recall, and [5] F1-score in the demonstration of the results. Table 1 presents the subject-wise classification accuracy for all 9 participants using all six classifiers. It can be observed that the ensemble classifier realizes the maximum average classification accuracy among all the applied classification schemes. It realizes maximum classification accuracy for all the participants except A06. In this case, SVM achieves maximum classification accuracy among all the classification techniques. It is interesting to discuss, SNNs architecture realized a minimum classification accuracy among all classification techniques. In addition, it ranks after the ensemble learning scheme. The order of performance is ELC > SVM > LGB > RB > Xgboost > SNNs.

Channel Selection Rate (CSR) is another performance measure that indicates the ratio between the number of selected and total EEG channels. In other words, it refers to the set of selected channels used in the classification process. The proposed CIM scheme effectively reduces the significant number of channels for all the participants by selecting minimum channels. Our methodology realizes average classification accuracy of 88.15% (ELC) using only 9 channels. Hence, 13 channels are effectively reduced without compromising classification accuracy. The details of the remaining three performance measures: (1) Precision, (2) Recall, and (3) F1-score, are given in Table 2. Precision determines 'how much the model is correct when it claims to be correct.' Recall indicates 'how many more right ones the model missed when it presented the right ones.' In Fig. 4, Precision-Recall (PR) curve is plotted for different thresholds to show the tradeoff between precision and recall. A high area under the curve represents both high recall and high precision, where high precision relates to a low false-positive rate, and high recall relates to a low false-negative rate. The harmonic mean of precision and

**Table 2.** Class wise details of precision, recall, and F1-score

| Metrics | Class 1 | Class 2 | Class 3 | Class 4 |
|---|---|---|---|---|
| Precision | 0.8506 | 0.8722 | 0.8715 | 0.8737 |
| Recall | 0.8951 | 0.8608 | 0.8482 | 0.8612 |
| F1-score | 0.8734 | 0.8665 | 0.8597 | 0.8674 |

**Table 3.** Performance comparison between our proposed channel selection scheme and three state-of-the-art methods. CA and P refer to classification accuracy and the number of selected features, respectively.

| Subject | IBGSA | | GSO | | RSS-SFSM | | Proposed Methodology | |
|---|---|---|---|---|---|---|---|---|
| | CA | P | CA | P | CA | P | CA | P |
| A01 | 71.30 | 7 | 68.31 | 11 | 73.91 | 9 | 93.36 | 8 |
| A02 | 66.04 | 10 | 56.91 | 13 | 70.08 | 6 | 95.45 | 10 |
| A03 | 81.54 | 8 | 79.68 | 13 | 85.02 | 5 | 94.21 | 11 |
| A04 | 81.91 | 9 | 69.43 | 14 | 77.29 | 12 | 81.62 | 8 |
| A05 | 76.66 | 13 | 73.33 | 9 | 80.00 | 11 | 85.09 | 8 |
| A06 | 66.66 | 13 | 83.52 | 9 | 79.52 | 6 | 91.11 | 9 |
| A07 | 73.57 | 11 | 66.33 | 14 | 82.45 | 9 | 79.38 | 7 |
| A08 | 79.32 | 7 | 63.10 | 5 | 83.11 | 8 | 89.91 | 8 |
| A09 | 87.52 | 12 | 91.33 | 7 | 89.28 | 13 | 83.25 | 10 |
| Average | 76.05 | 10 | 72.43 | 11 | 79.85 | 8.77 | 88.15 | 8.77 ≈ 9 |

recall is the F1-score. Here, we obtain maximum precision for class 4 while the best recall is achieved for class 1.

We compare our results with three state-of-the-art methods: (1) Glow Swarm optimization, (2) Improved Binary Gravitational Search Algorithm (IBGSA), and (3) Robust and Subject-Specific Sequential Forward Search Method. In the first method, Gonzalez et al. 2014 [13] introduced Glow Swarm Optimization (GSO) for channel reduction with Common Spatial Pattern (CSP) features. Finally, extracted features were classified using the Naïve Bayes classifier with better results than the conventional K-Nearest Neighbor (KN) and channel-optimized KNN approach. In the second method, the SNR of EEG signals is correlated with the channel optimization process, and the Improved Binary Gravitation Search Algorithm (IBGSA) is applied for EEG channel reduction [12]. They extracted statistical and temporal features from central beta frequency after channel reduction and achieved 80% classification accuracy at the maximum on BCI Competition 2008: 2a dataset. In [2], a Robust and Subject-Specific Sequential Forward Search Method (RSS-SFSM) is proposed for optimal channel selection. The proposed algorithm's main limitation was its inadequate validation of 100 iterations, affecting the classification accuracy when used in real-time BCI systems.

The comparative results between the proposed approach and the three base-line channel selection approaches mentioned above are shown in Table 3. It can be observed that our method achieves maximum classification accuracy compared to IBGSA except for two participants (A04 and A09). However, in both cases, our method selects fewer channels than IBGSA. In the second comparison, the proposed CIM method achieves higher classification accuracy than the GSO algorithm for all participants except A09. In addition, GSO selects a fewer number of channels compared to the proposed method for two participants, A08 and A09. Compared to RSS-SFSM, our method realizes inferior classification accuracy for A07 and A09. Interestingly, both methods have an equal number of average channels for all 9 participants.

## 4   Conclusion and Future Scope

This study develops a novel channel selection algorithm using a mutual information-based three-way interaction scheme for multichannel BCI systems. In this method, we minimize the redundancy level by reducing mutual information among newly chosen, early selected channels, and target attributes. Initially, we used a set of three channels (C3, C4, Cz) as a candidate solution to determine the relevance of the new channel. This procedure provides a sequence of all 22 channels based on their high relevance and low redundancy level in known solution. A MEMD feature extraction approach was applied to compute spatial-temporal properties of selected channels. Finally, five classifiers were used to find the suitable discrimination criteria for associated cognitive patterns. The classification results conclude that our method realizes superior classification accuracy than three state-of-the-art methods (GSO, IBGSA, RSS-SFSM), using fewer channels for most participants. In the future, some advanced methods such as channel map association using graph theory [10], information-guided search strategy [24], and clustering-oriented metaheuristics with Markov blanket [15] can be used to group the most significant EEG channels. Deep learning algorithms such as similarity-based Graph Neural Networks (Sim-GNNs) [19] and multi-input Deep Neural Networks (DNNs) [22] can also be used to determine correlated channel sets.

## References

1. Abdi, H., Williams, L.J.: Principal component analysis. Wiley Interdiscip. Rev. Comput. Stat. **2**(4), 433–459 (2010)
2. Aydemir, O., Ergün, E.: A robust and subject-specific sequential forward search method for effective channel selection in brain computer interfaces. J. Neurosci. Methods **313**, 60–67 (2019)
3. Alzamzami, F., Hoda, M., El Saddik, A.: Light gradient boosting machine for general sentiment classification on short texts: a comparative evaluation. IEEE Access **8**, 101840–101858 (2020)

4. Arvaneh, M., Guan, C., Ang, K.K., Quek, C.: Optimizing the channel selection and classification accuracy in EEG-based BCI. IEEE Trans. Biomed. Eng. **58**(6), 1865–1873 (2011)
5. Baig, M.Z., Aslam, N., Shum, H.P.: Filtering techniques for channel selection in motor imagery EEG applications: a survey. Artif. Intell. Rev. **53**(2), 1207–1232 (2020)
6. Bennasar, M., Setchi, R., Hicks, Y.: Feature interaction maximisation. Pattern Recognit. Lett. **34**(14), 1630–1635 (2013)
7. Biau, G., Scornet, E.: A random forest guided tour. TEST **25**(2), 197–227 (2016). https://doi.org/10.1007/s11749-016-0481-7
8. Brunner, C., Leeb, R., Müller-Putz, G., Schlögl, A., Pfurtscheller, G.: BCI Competition 2008-Graz data set A. Institute for Knowledge Discovery (Laboratory of Brain-Computer Interfaces). Graz University of Technology, vol. 16, pp. 1–6 (2008)
9. Chen, T., Guestrin, C.: Xgboost: a scalable tree boosting system. In: Proceedings of the 22nd ACM SIGKDD International Conference on Knowledge Discovery and Data Mining, pp. 785–794 (2016)
10. Das, A.K., Goswami, S., Chakrabarti, A., Chakraborty, B.: A new hybrid feature selection approach using feature association map for supervised and unsupervised classification. Expert Syst. Appl. **88**, 81–94 (2017)
11. Dietterich, T.G.: Ensemble learning. In: The Handbook of Brain Theory and Neural Networks, vol. 2, no. 1, pp. 110–125 (2002)
12. Ghaemi, A., Rashedi, E., Pourrahimi, A.M., Kamandar, M., Rahdari, F.: Automatic channel selection in EEG signals for classification of left or right hand movement in Brain Computer Interfaces using improved binary gravitation search algorithm. Biomed. Signal Process. Control **33**, 109–118 (2017)
13. Gonzalez, A., Nambu, I., Hokari, H., Wada, Y.: EEG channel selection using particle swarm optimization for the classification of auditory event-related potentials. Sci. World J. **2014** (2014)
14. Gorry, P.A.: General least-squares smoothing and differentiation by the convolution (Savitzky-Golay) method. Anal. Chem. **62**(6), 570–573 (1990)
15. Fu, S., Desmarais, M.C.: Markov blanket based feature selection: a review of past decade. In: Proceedings of the World Congress on Engineering, vol. 1, pp. 321–328. Newswood Ltd., Hong Kong, China (2010)
16. Imperatori, C., et al.: Coping food craving with neurofeedback. Evaluation of the usefulness of alpha/theta training in a non-clinical sample. Int. J. Psychophysiol. **112**, 89–97 (2017)
17. Jiao, Y., et al.: Sparse group representation model for motor imagery EEG classification. IEEE J. Biomed. Health Inform. **23**(2), 631–641 (2018)
18. Joachims, T.: Svmlight: support vector machine. SVM-Light Support Vector Machine http://svmlight. joachims. org/, University of Dortmund **19**(4), 25 (1999)
19. Li, Y., Guo, Z., Zhang, H., Li, M., Ji, G.: Decoupled pose and similarity based graph neural network for video person re-identification. IEEE Signal Process. Lett. **29**, 264–268 (2021)
20. Handiru, V.S., Prasad, V.A.: Optimized bi-objective EEG channel selection and cross-subject generalization with brain-computer interfaces. IEEE Trans. Hum.-Mach. Syst. **46**(6), 777–786 (2016)
21. Maino, D., et al.: All-sky astrophysical component separation with fast independent component analysis (FASTICA). Mon. Notices Royal Astron. Soc. **334**(1), 53–68 (2002)

22. Miikkulainen, R., et al.: Evolving deep neural networks. In: Artificial Intelligence in the Age of Neural Networks and Brain Computing, pp. 293–312. Academic Press (2019)

23. Moctezuma, L.A., Molinas, M.: EEG channel-selection method for epileptic-seizure classification based on multi-objective optimization. Front. Neurosci. **14**, 593 (2020)

24. Nakariyakul, S.: High-dimensional hybrid feature selection using interaction information-guided search. Knowl.-Based Syst. **145**, 59–66 (2018)

25. Rehman, N., Mandic, D.P.: Multivariate empirical mode decomposition. Proc. R. Soc. A Math. Phys. Eng. Sci. **466**(2117), 1291–1302 (2010)

26. Torres-García, A.A., Reyes-García, C.A., Villaseñor-Pineda, L., García-Aguilar, G.: Implementing a fuzzy inference system in a multi-objective EEG channel selection model for imagined speech classification. Expert Syst. Appl. **59**, 1–12 (2016)

27. Vimala, V., Ramar, K., Ettappan, M.: An intelligent sleep apnea classification system based on EEG signals. J. Med. Syst. **43**(2), 1–9 (2019)

28. Wang, Y.K., Chen, S.A., Lin, C.T.: An EEG-based brain-computer interface for dual task driving detection. Neurocomputing **129**, 85–93 (2014)

29. Wang, Z.M., Hu, S.Y., Song, H.: Channel selection method for EEG emotion recognition using normalized mutual information. IEEE Access **7**, 143303–143311 (2019)

30. Yang, Y., Chevallier, S., Wiart, J., Bloch, I.: Subject-specific time-frequency selection for multiclass motor imagery-based BCIs using few Laplacian EEG channels. Biomed. Signal Process. Control **38**, 302–311 (2017)

31. Zenke, F., Ganguli, S.: Superspike: supervised learning in multilayer spiking neural networks. Neural Comput. **30**(6), 1514–1541 (2018)

# MVNet: Memory Assistance and Vocal Reinforcement Network for Speech Enhancement

Jianrong Wang[1], Xiaomin Li[1], Xuewei Li[1], Mei Yu[1], Qiang Fang[2], and Li Liu[3(✉)]

[1] College of Intelligence and Computing, Tianjin University, Tianjin, China
[2] Institute of Linguistics, Chinese Academy of Social Sciences, Beijing, China
[3] Shenzhen Research Institute of Big Data, The Chinese University of Hong Kong, Shenzhen, China
liuli@cuhk.edu.cn

**Abstract.** Speech enhancement improves speech quality and promotes the performance of various downstream tasks. However, most current speech enhancement work was mainly devoted to improving the performance of downstream automatic speech recognition (ASR), only a relatively small amount of work focused on the automatic speaker verification (ASV) task. In this work, we propose a MVNet consisted of a memory assistance module which improves the performance of downstream ASR and a vocal reinforcement module to boosts the performance of ASV. In addition, we design a new loss function to improve speaker vocal similarity. Experimental results on the Libri2mix dataset show that our method outperforms baseline methods in several metrics, including speech quality, intelligibility, and speaker vocal similarity.

**Keywords:** Speech enhancement · Complex network · Speaker similarity · Memory assistance · Vocal reinforcement

## 1 Introduction

The interference of additive noise with speech can seriously reduce the perceptual quality and intelligibility of speech, which increases the difficulty and complexity of speech-related recognition tasks [17]. In some scenarios, the security of algorithms for tasks such as speech recognition and speaker verification can be seriously threatened by noise interference [2]. Speech enhancement (SE) is an important speech processing task dedicated to improving the perceptual quality as well as the intelligibility of the disturbed speech and to restore the performance of downstream tasks.

A good SE algorithm should obtain the output speech that is closer to the clean speech. And the output speech often has better speech quality and intelligibility than the input speech. In recent years, deep learning methods [9,15,27] were widely applied to SE tasks and achieved good results. Deep learning based methods can be classified into time domain and frequency domain depending on

© The Author(s), under exclusive license to Springer Nature Switzerland AG 2023
M. Tanveer et al. (Eds.): ICONIP 2022, LNCS 13624, pp. 101–112, 2023.
https://doi.org/10.1007/978-3-031-30108-7_9

how the input speech is processed. The common practice of time domain methods [3,21,31] is to map the time domain waveform of noisy speech directly to the time domain waveform of clean speech, through the learned mapping relationship. The frequency domain approach [9,11] obtains a mask by inputting the noisy speech spectral features into the network. Then the clean speech is obtained by multiplying the mask and the noisy speech.

Most of the previous work focused on improving speech quality as a training goal, and the current mainstream metrics are also based on speech quality. Several studies proposed to train SE models directly with speech quality metrics (PESQ and STOI), including quality-net [6], MetricGAN-u [7] and hifi-gan [14]. These methods achieved a significant improvement in speech quality. However, ASR and ASV pay different attention to speech features. ASR pays more attention to the intelligibility of speech, while ASV pays more attention to speaker vocal similarity. The optimization focus of the two is not consistent. Speech with higher speech quality can have more outstanding performance in the downstream ASR task, while less outstanding in ASV.

Current methods greatly improve speech quality (PESQ and STOI), ignoring the importance of vocal information. However, inconsistent vocals will lead to inconsistencies between speakers and increased distortion, which in turn affects the performance of downstream ASVs. We call this the vocal distortion problem. PFPL [10] started to demonstrate the importance of phonetic information. Their work demonstrates that adding the necessary speech information can guarantee speech details as well as speech quality. This provides us with ideas to alleviate the vocal distortion problem.

In this work, to adapt to both ASR and ASV at the same time, and to achieve the improvement of speech quality and vocal consistency, we propose a MVNet consisted of a memory assistance module and a vocal feature reinforcement module. Vocal reinforcement module is to extract the vocal information. We consider it important for vocal distortion problem. Memory assistance module is to improve the enhanced performance of the complex network. It reduces the loss from forgetting valid information in long sequences by the network while enhancing the gain from focusing on important information. Besides, we design a similarity joint loss that aims to alleviate vocal distortion problem. The experiments verify that our method can alleviate the vocal distortion problem while further improving the speech quality.

## 2    Related Work

### 2.1    Complex Structure of CRN

The traditional Convolutional Recurrent Neural Network (CRN) [25] is symmetric. It uses an encoder-decoder architecture in the time-domain, usually with an LSTM layer in the middle to model the temporal dependencies. The encoder-decoder block consists of convolution and deconvolution layers, batch normalization and activation functions.

To improve the performance of convolution in the complex domain, Tan *et al.* [26] proposed an one-encoder two-decoders convolution method. Unlike previous CRN that only targets amplitude mapping in the real domain, this network structure is also capable of modeling phase mapping in the complex domain. Compared with the traditional enhancement model, this structure can enhance the amplitude and phase of the speech at the same time, and the enhanced speech no longer needs to reuse the phase of the noisy speech. However, this one-encoder two-decoders structure actually divided the input into two channels, the real part and the imaginary part, and processed them as real numbers, which did not strictly follow the operation rules of complex numbers.

The above approaches did not directly utilize the prior knowledge of the magnitude and phase correlations of complex arithmetic. Hu *et al.* provided a complex domain convolution model DCCRN [11], which used a complex encoder-decoder combined with a complex LSTM to enhance speech. This network provided the ability to simulate complex multiplication, further enhancing the network's ability to capture the correlation between magnitude and phase. DCCRN has been shown to be effective, we take this as our baseline model.

## 2.2 Speech Feature Information

With the research in the signal processing, researchers developed different speech features according to the characteristics of different tasks. Speech feature extraction methods such as MFCC [28] and i-vector [8] were often used in various speech signal processing tasks such as speech recognition [18], speaker recognition [30], and phoneme detection. These feature representations focus on different speech information. A suitable feature representation can strongly promote the performance of a specific task.

Hsieh *et al.* [10] proposed a perceptual loss (PFPL) for SE task. They pointed out that phonetic feature information is the key to optimizing human perceptual. PFPL first proposed the idea of adding phonetic feature information to the original speech. This self-supervised SE method is based on DCCRN and wav2vec [1]. Their experimental results showed effectiveness of phonetic information. Thus we take PFPL as another baseline model.

## 3   Method

In this work, we propose a MVNet as shown in Fig. 1. In general, we extract the speaker vocal features through vocal reinforcement module, and fuse it with the noisy speech spectrum. A complex mask is then estimated by the memory assistance speech enhancement module and multiplied by the noisy spectrum to obtain the enhanced speech. Besides, we use the proposed similarity joint loss to alleviate vocal distortion problem.

Our method is based on DCCRN which excels in speech quality. We propose the memory assistance module to further improve speech quality and make the model pay more attention to the vocal features. To improve the vocal similarity of speech, we propose the vocal reinforcement module and the similarity joint loss.

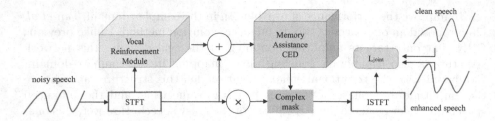

**Fig. 1.** The overall structure of MVNet.

## 3.1  Memory Assistance

In order to make the model further improve the speech quality, and at the same time make it have the ability to focus on the vocal features. We propose the memory assistance module under the DCCRN framework, as shown in Fig. 2.

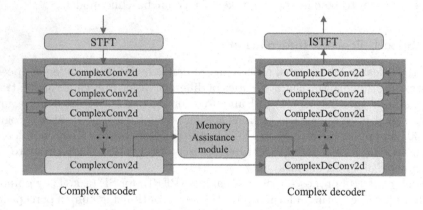

**Fig. 2.** The memory assistance CED structure.

We use 6 complex convolution blocks and symmetric 6 deconvolution blocks to implement the construction of the encoder-decoder with the number of channels set to {32, 64, 128, 256, 256, 256}, where each complex convolution block contains complex Conv2d, complex batch normalization and real-valued PReLU. Each complex Conv2d contains a real conv2d and an imaginary conv2d as in DCCRN [11].

**Memory Assistance Module.** The overall framework of the DCCRN model is based on CED, and the speech enhancement is mainly realized by the LSTM with causal modeling ability. The LSTM network controls the memory state of information in the long-term transmission process through gates, retains important information and forgets the information that the network considers unimportant. It plays the role in information filtering. Vocal information is a very detailed speech feature that can only be noticed from a global perspective. The core logic of the attention mechanism is the global attention, which can capture the vocal features. But only from a global perspective will ignore some local characteristics of speech. Thus, we combine it with the LSTM to form the memory assistance module, which focuses on the both global features and local details of speech.

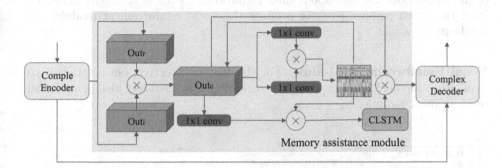

**Fig. 3.** Memory assistance speech enhancement module. $Out_r$ and $Out_i$ is the real and the imaginary part of the encoder output, respectively. $Out_c$ is the fused complex feature map. CLSTM is the complex LSTM layer.

Placing the attention before LSTM can amplify the memory ability, further improving the memory ability of LSTM for global vocal characteristics. If it is placed in the back, LSTM will forgot this information. At this time, the global characteristics of this information will be destroyed, and the global attention will not be able to pay attention to this information. Placing the attention in the back aggravates the forgetting ability. Thus, the final vocal reinforcement is as Fig. 3.

We utilize the features on the crisscross path to achieve global attention through two loops while controlling the memory consumption. Module collects contextual information in both horizontal and vertical directions to enhance the expressiveness of feature maps. As shown in Fig. 3, the noisy speech is passed through the complex encoder to obtain the feature maps of both the real and imaginary parts. The feature maps is fused and sent into three one dimension convolutional layers (Conv1ds). The horizontal and vertical attention map is obtained from the first two Conv1ds and then passed back to the input to obtain the global attention map. The global attention map and the output of the third Conv1d are concatenated and fed into the complex LSTM. The output is concatenated with the output of the encoder and fed to the decoder for further processing.

## 3.2    Vocal Reinforcement

Vocal feature is an important factor affecting the distortion degree of the final enhanced speech. When the vocal features of the enhanced speech and the clean speech are quite different, the speech sounds lack of uniform speaker characteristics, and it does not sound like the original speaker. That causes vocal distortion problem.

The problem of missing vocal characteristics in speech enhancement can be considered from two perspectives. One is that the model does not have the ability to discover such characteristics, and the other is that the optimization direction of the model does not care about this. To improve the vocal similarity of speech from these two perspectives, we propose the vocal reinforcement module and similarity joint loss.

**Vocal Reinforcement Module.** For the first perspective, our solution is to explicitly add vocal features to the network, which is the direct idea of our vocal reinforcement module.

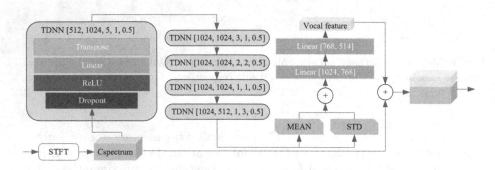

**Fig. 4.** Vocal reinforcement module.

The way of combining MFCC with TDNN [13] is a common way to obtain speaker representation in ASV, which has a strong expressive ability for vocal features. But MFCC is a compact speech representation. Since it uses mel filter to ignore the dynamics and distribution of speech energy, it still loses some speech details in essence [5]. It is a coarse-grained speech representation. Therefore, we adopt STFT (short-time Fourier transform) to obtain the spectral representation of speech and preserve the temporal information of speech. Combined with TDNN, a fine-grained vocal feature extraction method suitable for speech enhancement is realized. ASV directly uses the speech representation obtained by MFCC and TDNN to do the recognition task. But speech enhancement requires more than just vocal information. We therefore combine the obtained representation as an auxiliary feature with the spectrum obtained by STFT as the input to CED. This forms our vocal reinforcement module.

The proposed vocal reinforcement module is shown in Fig. 4. The spectrum obtained by STFT is sent to 5 TDNNs connected in sequence, the first four output 1024 channels, the last collapses the channels to 512, and the average and standard deviation of the last TDNN's output are calculated and connected to the original output, and then go through two linear layers in turn to get the vector about the vocal features. We fuse the feature vector and the original input into Memory Assistance CED.

**Similarity Joint Loss.** As mentioned above, to improve the vocal consistency, there are two perspectives. Designing a new loss function is from the second, changing the direction of model optimization.

The complete information of the speech signal is jointly represented by the amplitude and the phase, and the phase contains more detailed information of the speech. In the previous speech enhancement models, SI-SNR [19] was mostly used as the loss function. Although SI-SNR takes into account the vector direction of speech, the calculation process still depends on the signal amplitude. The cosine similarity has a stronger constraint on the consistency of the vector direction. In order to make the model pay more attention to the vector direction, we introduce the cosine similarity to our loss function. The ability of the model to improve the vocal consistency is enhanced by strengthening the constraint of the loss function on the consistency of the vector direction.

The proposed similarity joint loss is to make some improvements on the basis of the loss function of SI-SNR. We take the additive inverse of SI-SNR in our $\mathcal{L}_{\text{SI-SNR}}$, so that larger calculated results indicate less ideal separation. $\mathcal{L}_{\text{SI-SNR}}$ is defined as:

$$\begin{cases} s_{\text{target}} = \frac{\langle s, \hat{s} \rangle s}{\|s\|^2} \\ e_{\text{noise}} = \hat{s} - s_{\text{target}} \\ \mathcal{L}_{\text{SI-SNR}} = 10 \log_{10} \frac{\|s_{\text{target}}\|^2}{\|e_{\text{noise}}\|^2} \end{cases} \tag{1}$$

where $< \cdot, \cdot >$ represents the dot product of two vectors, $\| \cdot \|^2$ is the euclidean norm (L2 norm), $s$ is the clean speech, and $\hat{s}$ means the enhanced speech. SI-SNR is commonly used in papers.

In addition to $\mathcal{L}_{\text{SI-SNR}}$, we propose to use cosine similarity to improve the speaker vocal consistency. The similarity loss is defined as:

$$\mathcal{L}_{\text{smi}} = \alpha \log_{10}(1 - \cos_{\text{smi}}(\hat{s}, s) + \delta), \tag{2}$$

where the hyperparameter $\alpha$ is the scaling factor, which we set to 100.

The value range of the cosine similarity function $\cos_{\text{smi}}(\cdot, \cdot)$ is [-1, 1]. We take the $-\cos_{\text{smi}}$, so that the higher the calculation result, the more dissimilar the two speeches are. We add a constant number 1 to fix the range in [0,2]. $\delta$ is an extremely small number used to avoid zero values. We smooth the change of the curve through a logarithmic function, so that $\mathcal{L}_{\text{smi}}$ has a consistent change trend with $\mathcal{L}_{\text{SI-SNR}}$. Finally, we combine these two functions to propose our similarity joint loss $\mathcal{L}_{\text{joint}}$, defined as:

$$\mathcal{L}_{\text{joint}} = \mathcal{L}_{\text{SI-SNR}} + \mathcal{L}_{\text{smi}}. \tag{3}$$

### 3.3  Training Target

Our training target is to obtain a complex ratio mask (CRM) [33] to estimate clean speech. We adopt the method of signal estimation, that is, the noisy signal and the estimated mask are directly multiplied to obtain the enhanced signal. We improve the performance of the model by minimizing the $\mathcal{L}_{\text{joint}}$ between the enhanced speech and the clean speech.

## 4  Experiment and Result

### 4.1  Experimental Setup

**Dataset.** In our experiments, we use the Librispeech [20] as the clean data, which has 1252 speakers, each speaking for about 25 min, for a total of 478 h of speech duration. The noise data in the experiment comes from the noise dataset WHAM! [32], which consists of real ambient noises.

We mix the Librispeech and WHAM! datasets in the same way as the LibriMix [4], resulting in a training set with 921 speakers for a total of 364 h, and a validation set and a test set with 40 speakers for a total of 5.4 h. The dataset SNR we get from the mix is between −15 bB and 5 dB.

**Evaluation Metrics.** The evaluation of our experiments is based on several general metrics of speech quality, including Perceptual Evaluation of Speech Quality (PESQ) [23], Short-Time Objective Intelligibility (STOI) [24], the predicted Mean Opinion Score of signal distortion (CSIG) [12], background noise distortion (CBAK) [12], overall quality (COVL) [12], the scale-invariant signal-to-noise ratio improvement (SI-SNRi) [29], segmental SNR (segSNR) [22] and SIMI (a measure of vocal similarity). SIMI is the proposed new metric to measure the degree of vocal distortion, which is calculated by the speaker recognition algorithm provided by Deep-speaker [16]. The higher the score, the higher the probability that the speaker will be judged to be the same in the ASV task.

**Baseline.** We sample waveforms at 16 kHz, and set the window length and number of hops to 25 ms and 6.25 ms, respectively. The FFT length is 512. We use Adam optimizer. The initial learning rate is set to 0.001, and when the validation loss increases, the learning rate decreases by 0.5. We train for 200 epochs and record the top PESQ ranked model parameters as our best model for related experiments. For fairness, we run the official codes of baseline models (PFPL and DCCRN) with the same training configuration as ours for comparison.

## 4.2    Ablation Study

**Ablation Study for Memory Assistance.** We propose the memory assistance module to further improve speech quality and make the model pay more attention to the vocal features.

**Table 1.** Ablation study for memory assistance. $Ours_{ma}$ and $Ours_{bma}$ represents the result of placing the memory assistance module before the LSTM layer and after the LSTM, respectively. DCCRN represents the results obtained without memory assistance.

| Metric | noisy | $Ours_{bma}$ | DCCRN | $Ours_{ma}$ |
|--------|-------|--------------|-------|-------------|
| PESQ   | 1.18  | 2.52         | 2.65  | **2.70**    |
| STOI   | 0.54  | 0.81         | 0.84  | **0.87**    |
| SIMI   | 0.36  | 0.39         | 0.43  | **0.46**    |

From Table 1, it can be seen that memory assistance module has the best results when placed before LSTM. Memory assistance module can amplify the importance of effective information before the LSTM forgets some information, so that LSTM continues to amplify those effective information. When the memory assistance module is placed behind LSTM, the degree of forgetting of the LSTM will be aggravated, thereby reducing the performance of the model.

Memory assistance module outperforms DCCRN in PESQ, STOI and SIMI. This indicates its ability to further improve speech quality while empowering the model to focus on vocal features.

**Ablation Study for Vocal Reinforcement.** To improve the vocal similarity of speech, we propose the vocal reinforcement module and the similarity joint loss. We compare with PFPL(with phonetic information) and DCCRN(without any speech information). Results are shown in Table 2, our results are significantly better than the above methods.

**Table 2.** Ablation study for vocal reinforcement. $Ours_{vr}$ is the model with only the vocal reinforcement. $Ours_{MVL}$ is the MVNet with both memory assistance and vocal reinforcement.

| Metric   | noisy | DCCRN | PFPL | $Ours_{vr}$ | $Ours_{MVL}$ |
|----------|-------|-------|------|-------------|--------------|
| PESQ     | 1.18  | 2.65  | 2.71 | **2.90**    | 2.88         |
| STOI     | 0.54  | 0.84  | 0.78 | **0.91**    | **0.91**     |
| SIMI     | 0.36  | 0.43  | 0.51 | **0.52**    | **0.52**     |
| SegSNR   | 2.07  | 6.49  | 5.58 | 5.88        | **6.55**     |
| CSIG     | 2.02  | 2.10  | 2.37 | **2.47**    | 2.44         |
| SI-SNR$_i$ | -   | 6.98  | 6.27 | 9.88        | **9.97**     |

PFPL explicitly added the additional information (phonetic information) to the DCCRN. This additional information contained more speech details, which made the PESQ, SIMI and CSIG of PFPL higher than those of DCCRN. However, PFPL did not modify the original CED structure of DCCRN, and its model did not have the ability to adapt to this fine-grained information. Thus, its STOI, SegSNR and SI-SNR$_i$ were degraded. Compared with DCCRN and PFPL, both Ours$_{vr}$ and Ours$_{MVL}$ have better SIMI and CSIG scores, which prove that the vocal reinforcement can improve the vocal consistency. Since Ours$_{vr}$ lacks the memory assistance module in Ours$_{MVL}$, the local details (SegSNR) and overall performance (SI-SNR$_i$) of Ours$_{vr}$ are worse than Ours$_{MVL}$.

### 4.3    Comprehensive Evaluation

We comprehensively evaluate the performance of our method on various metrics, as shown in Table 3. Our model outperforms the baseline models in all metrics and lower distortion can be guaranteed while maintaining higher speech quality.

**Table 3.** Comprehensive evaluation.

|  | PESQ | STOI | CSIG | CBAK | CVOL | SIMI | SegSNR | SI-SNR$_i$ |
|---|---|---|---|---|---|---|---|---|
| noisy | 1.18 | 0.54 | 2.02 | 1.99 | 1.75 | 0.36 | 2.07 | - |
| DCCRN | 2.65 | 0.84 | 2.10 | 2.11 | 1.91 | 0.43 | 6.49 | 9.04 |
| PFPL | 2.71 | 0.78 | 2.37 | 2.45 | 2.12 | 0.51 | 5.58 | 8.34 |
| Ours$_{ma}$ | 2.70 | 0.87 | 2.26 | 2.34 | 2.08 | 0.46 | 6.37 | 10.61 |
| Ours$_{vr}$ | **2.90** | **0.91** | **2.47** | **2.60** | **2.61** | **0.52** | 5.88 | 11.94 |
| Ours$_{MVL}$ | 2.88 | **0.91** | 2.44 | 2.59 | 2.27 | **0.52** | **6.55** | **12.04** |

## 5    Conclusions

In this work, we propose the MVNet consisted of a memory assistance module and a vocal reinforcement module. Memory assistance module is proposed to further improve the speech quality while making the model focus more on vocal features. Vocal reinforcement module explicitly introduces vocal features to improve the speaker vocal similarity. Besides, we design a similarity joint loss, which aims to improve the speaker vocal consistency. Experiments verify that the MVNet can further improve speech quality while maintaining the increase in speaker similarity and the decrease in speech distortion, respectively. In the future, we will explore real-time adaptive speech feature extraction methods.

## References

1. Baevski, A., Zhou, Y., Mohamed, A., Auli, M.: wav2vec 2.0: a framework for self-supervised learning of speech representations. In: Advances in Neural Information Processing Systems, vol. 33, pp. 12449–12460 (2020)

2. Chen, G., et al.: Who is real bob? Adversarial attacks on speaker recognition systems. In: 2021 IEEE Symposium on Security and Privacy (SP), pp. 694–711. IEEE (2021)
3. Chen, J., Mao, Q., Liu, D.: Dual-path transformer network: direct context-aware modeling for end-to-end monaural speech separation. arXiv preprint arXiv:2007.13975 (2020)
4. Cosentino, J., Pariente, M., Cornell, S., Deleforge, A., Vincent, E.: Librimix: an open-source dataset for generalizable speech separation. arXiv preprint arXiv:2005.11262 (2020)
5. Faraji, F., Attabi, Y., Champagne, B., Zhu, W.P.: On the use of audio fingerprinting features for speech enhancement with generative adversarial network. In: 2020 IEEE Workshop on Signal Processing Systems (SiPS), pp. 1–6 (2020). https://doi.org/10.1109/SiPS50750.2020.9195238
6. Fu, S.W., Liao, C.F., Tsao, Y.: Learning with learned loss function: speech enhancement with quality-net to improve perceptual evaluation of speech quality. IEEE Signal Process. Lett. **27**, 26–30 (2019)
7. Fu, S.W., Yu, C., Hung, K.H., Ravanelli, M., Tsao, Y.: Metricgan-u: unsupervised speech enhancement/dereverberation based only on noisy/reverberated speech. In: ICASSP 2022–2022 IEEE International Conference on Acoustics, Speech and Signal Processing (ICASSP), pp. 7412–7416. IEEE (2022)
8. Garcia-Romero, D., Espy-Wilson, C.Y.: Analysis of i-vector length normalization in speaker recognition systems. In: Twelfth Annual Conference of the International Speech Communication Association (2011)
9. Hao, X., Su, X., Horaud, R., Li, X.: Fullsubnet: a full-band and sub-band fusion model for real-time single-channel speech enhancement. In: ICASSP 2021–2021 IEEE International Conference on Acoustics, Speech and Signal Processing (ICASSP), pp. 6633–6637. IEEE (2021)
10. Hsieh, T.A., Yu, C., Fu, S.W., Lu, X., Tsao, Y.: Improving perceptual quality by phone-fortified perceptual loss using wasserstein distance for speech enhancement. arXiv preprint arXiv:2010.15174 (2020)
11. Hu, Y., et al.: DCCRN: deep complex convolution recurrent network for phase-aware speech enhancement. arXiv preprint arXiv:2008.00264 (2020)
12. Hu, Y., Loizou, P.C.: Evaluation of objective quality measures for speech enhancement. IEEE Trans. Audio Speech Lang. Process. **16**(1), 229–238 (2007)
13. Ji, W., Chee, K.C.: Prediction of hourly solar radiation using a novel hybrid model of ARMA and TDNN. Sol. Energy **85**(5), 808–817 (2011)
14. Kong, J., Kim, J., Bae, J.: HiFi-GAN: generative adversarial networks for efficient and high fidelity speech synthesis. Adv. Neural. Inf. Process. Syst. **33**, 17022–17033 (2020)
15. Li, A., Liu, W., Zheng, C., Fan, C., Li, X.: Two heads are better than one: a two-stage complex spectral mapping approach for monaural speech enhancement. IEEE/ACM Trans. Audio Speech Lang. Process. **29**, 1829–1843 (2021). https://doi.org/10.1109/TASLP.2021.3079813
16. Li, C., et al.: Deep speaker: an end-to-end neural speaker embedding system. arXiv preprint arXiv:1705.02304 (2017)
17. Liu, L., Feng, G., Beautemps, D., Zhang, X.P.: Re-synchronization using the hand preceding model for multi-modal fusion in automatic continuous cued speech recognition. IEEE Trans. Multimedia **23**, 292–305 (2020)
18. Liu, L., Hueber, T., Feng, G., Beautemps, D.: Visual recognition of continuous cued speech using a tandem CNN-HMM approach. In: Interspeech, pp. 2643–2647 (2018)

19. Luo, Y., Mesgarani, N.: Conv-TasNet: surpassing ideal time-frequency magnitude masking for speech separation. IEEE/ACM Trans. Audio Speech Lang. Process. **27**(8), 1256–1266 (2019)

20. Panayotov, V., Chen, G., Povey, D., Khudanpur, S.: Librispeech: an ASR corpus based on public domain audio books. In: 2015 IEEE International Conference on Acoustics, Speech and Signal Processing (ICASSP), pp. 5206–5210. IEEE (2015)

21. Pandey, A., Wang, D.: Densely connected neural network with dilated convolutions for real-time speech enhancement in the time domain. In: ICASSP 2020–2020 IEEE International Conference on Acoustics, Speech and Signal Processing (ICASSP), pp. 6629–6633. IEEE (2020)

22. Quackenbush, S.R.: Objective measures of speech quality (subjective) (1986)

23. Rix, A.W., Beerends, J.G., Hollier, M.P., Hekstra, A.P.: Perceptual evaluation of speech quality (PESQ)-a new method for speech quality assessment of telephone networks and codecs. In: 2001 IEEE International Conference on Acoustics, Speech, and Signal Processing. Proceedings (Cat. No. 01CH37221), vol. 2, pp. 749–752. IEEE (2001)

24. Taal, C.H., Hendriks, R.C., Heusdens, R., Jensen, J.: An algorithm for intelligibility prediction of time-frequency weighted noisy speech. IEEE Trans. Audio Speech Lang. Process. **19**(7), 2125–2136 (2011)

25. Tan, K., Wang, D.: A convolutional recurrent neural network for real-time speech enhancement. In: Interspeech, vol. 2018, pp. 3229–3233 (2018)

26. Tan, K., Wang, D.: Complex spectral mapping with a convolutional recurrent network for monaural speech enhancement. In: ICASSP 2019 2019 IEEE International Conference on Acoustics, Speech and Signal Processing (ICASSP), pp. 6865–6869. IEEE (2019)

27. Tan, K., Wang, D.: Towards model compression for deep learning based speech enhancement. IEEE/ACM Trans. Audio Speech Lang. Process. **29**, 1785–1794 (2021). https://doi.org/10.1109/TASLP.2021.3082282

28. Tiwari, V.: MFCC and its applications in speaker recognition. Int. J. Emerg. Technol. **1**(1), 19–22 (2010)

29. Vincent, E., Gribonval, R., Févotte, C.: Performance measurement in blind audio source separation. IEEE Trans. Audio Speech Lang. Process. **14**(4), 1462–1469 (2006)

30. Wang, J., et al.: Three-dimensional lip motion network for text-independent speaker recognition. In: 2020 25th International Conference on Pattern Recognition (ICPR), pp. 3380–3387. IEEE (2021)

31. Wang, K., He, B., Zhu, W.P.: TSTNN: two-stage transformer based neural network for speech enhancement in the time domain. In: ICASSP 2021–2021 IEEE International Conference on Acoustics, Speech and Signal Processing (ICASSP), pp. 7098–7102. IEEE (2021)

32. Wichern, G., et al.: Wham!: extending speech separation to noisy environments. arXiv preprint arXiv:1907.01160 (2019)

33. Williamson, D.S., Wang, Y., Wang, D.: Complex ratio masking for monaural speech separation. IEEE/ACM Trans. Audio Speech Lang. Process. **24**(3), 483–492 (2015)

# Learning Associative Reasoning Towards Systematicity Using Modular Networks

Jun-Hyun Bae[1], Taewon Park[1], and Minho Lee[1,2]([⊠])

[1] Kyungpook National University, 80, Daehak-ro, Buk-go, Daegu, Republic of Korea
{junhyun.bae,ptw4570,mholee}@knu.ac.kr
[2] NEOALI, 80, Daehak-ro, Buk-go, Daegu, Republic of Korea

**Abstract.** Learning associative reasoning is necessary to implement human-level artificial intelligence even when a model faces unfamiliar associations of learned components. However, conventional memory augmented neural networks (MANNs) have shown degraded performance on systematically different data since they lack consideration of systematic generalization. In this work, we propose a novel architecture for MANNs which explicitly aims to learn recomposable representations with a modular structure of RNNs. Our method binds learned representations with a Tensor Product Representation (TPR) to manifest their associations and stores the associations into TPR-based external memory. In addition, to demonstrate the effectiveness of our approach, we introduce a new benchmark for evaluating systematic generalization performance on associative reasoning, which contains systematically different combinations of words between training and test data. From the experimental results, our method shows superior test accuracy on systematically different data compared to other models. Furthermore, we validate the models using TPR by analyzing whether the learned representations have symbolic properties.

**Keywords:** Associative reasoning · Memory augmented neural networks · Systematic generalization

## 1 Introduction

Humans can constantly imagine new things and infer the expected outcomes of what they do. One of the reasons for its promising ability is that the human brain can relate to multiple distinct experiences by memorizing past events and recalling appropriate knowledge. This capability, called associative reasoning, allows humans to cope with unfamiliar situations which they have not previously experienced. Modern deep learning approaches for memory augmented neural networks (MANNs) show glittering advances in associative reasoning [1,7,10, 12,14,19]. However, unlike humans, conventional MANNs still fail to generalize associations when there are systematic differences between training and test data [16,17].

---

J.-H. Bae and T. Park—Equal Contribution.

The main reason for the failure of systematic generalization with deep neural networks is that representations are learned only to achieve minimum training risk without considering systematical differences in the future. Therefore, current approaches relying on learned representations cannot generalize to unseen combinations of learned components [9]. To achieve systematic generalization, we hypothesize that it is essential to (a) learn representations for components as well as (b) provide representations for their systematical combinations.

In this work, we aim to achieve those two goals for systematic generalization in associative reasoning of MANNs. We propose a modular encoder network to learn symbolic representations, which are reusable pieces that can be combined to represent unusual associations of known entities, and a Tensor Product Representation (TPR) [18] based external memory, which can bind such learned representations. The TPR is an embedding method for symbolic structures which binds symbolic representations with the tensor product. Using the TPR-based external memory, the associations can be systematically stored in the memory and systematically recalled. Since the TPR method assumes pre-defined symbolic representations for a given symbolic structure, we focus on learning to extract recomposable representations from natural language to strengthen TPR-based memory. We adopt recent approaches using competitive learning of the modular structure of RNNs to specialize each RNN module to have its own independent mechanisms [6]. These modules then encode given input according to their own dynamics, with only the modules relevant to the input mainly participating. As a result, the input is encoded into several pieces of specialized representations, and their collection represents the original input. This symbolic characteristic of representations provides validity for using the TPR for associative memory.

The main contribution of our work is providing an effectual method to achieve systematic generalization for associative reasoning. To validate systematic generalization ability of our method, we design a synthetic task called *Systematic Associative Recall* (SAR). In the experiments, our approach shows improved performance than not only existing MANN-based models [3,7,8,12] but also TPR-based MANNs such as *fast weight memory* (FWM) [16]. We found that the modular structure can effectively learn symbolic representations for the TPR memory than their work with a conventional RNN encoder.

## 2   Related Work

Deep neural networks require sufficient capacity to understand the context existing in long sequential data. MANNs expand the capacity of neural networks by adding an external memory. The objective of these networks is to store given sequential data in the memory, retrieve meaningful information from memory, and utilize that to solve problems.

Content-based addressable memory networks, such as Differentiable neural computer (DNC) [7] and its variations [2,5,14,15], exploited content-based addressing for memory writing and reading operations and have shown remarkable strength in basic reasoning tasks [21]. Also, many researchers adopted state-of-the-art techniques of deep neural networks to enhance the reasoning ability of

MANNs [1,10,14]. Meta-learned neural memory (MNM) [12] interprets the memory as a rapidly adaptable function, and Transformer-XL (TXL) [3] introduces a recurrence mechanism to grant the memory ability to transformer architecture [20]. Those approaches have shown significant progress in the reinforcement learning task and language modeling task. Despite their achievements, few studies investigated the performance degradation of MANNs in environments where there is a systematic difference between training and test data.

Recently, some researchers have researched the systematic generalization of MANNs using the TPR method. TPR-RNN [17] firstly introduces a TPR-based memory approach to learn combinatorial representations, and FWM [16] provides a more general method for TPR-RNN to expand on longer sequences. We also follow their approach using TPR as external memory; however, unlike their work that relies solely on TPR constraints to learn symbolic representations, our method efficiently learns such representations using modular networks with competitive learning. In effect, our method provides a more valid approach for using TPR, and it also achieves better generalization performance on systematically different data distribution.

## 3   Proposed Method

We focus on learning symbolic representations for external memory representations to achieve systematic generalization. Our method utilizes a modular structure of RNN to encode symbolic representations and a Tensor Product Representation (TPR) to bind them. Each RNN module encodes an input into symbolic representations based on its own dynamics, and memory representations are obtained by binding them with the TPR method.

### 3.1   Tensor Product Representation

TPR provides an embedding of symbolic structures to represent systematicity [4] by using the tensor product of component representations, so-called roles and fillers. For example, to represent associations of a pair of objects in a set {*John*, *Apple*, *Three*}, the filler *Apple* may be the value for the role of *second element*. In another perspective, the filler *Apple* and *Three* may be the values for the role *John*. If roles and fillers are specified, and their distributed representations are given, the role/filler relations can be described by binding them with a tensor product. From the TPR literature, multiple role/filler relations are superposed to represent an entire symbolic structure. Formally, a distributed representation for an input that consists of $N$ role/filler relations $\{\mathbf{r}_k/\mathbf{f}_k\}_{k=1}^N$ is expressed as a superposition of tensor products of roles and fillers:

$$T = \sum_{k=1}^{N} \mathbf{r}_k \otimes \mathbf{f}_k \qquad (1)$$

where $T$ is distributed representation for the input, $\otimes$ is the tensor product operator, $\mathbf{r}_k/\mathbf{f}_k$ are role/filler representations for $k$-th role/filler relation.

In order to guarantee the correctness of binding information, the TPR method provides unbinding operator, which extracts filler for a particular role from TPR. Let $\mathbf{u}_i$ be an unbinding vector for role $\mathbf{r}_i$. For a simple case where the role representations are orthonormal and $\mathbf{u}_i = \mathbf{r}_i$, it is sufficient to unbind filler from a particular role $\mathbf{r}_i$ by using the inner product

$$\mathbf{u}_i \cdot T = \mathbf{u}_i \cdot \sum_{k=1}^{N} \mathbf{r}_k \otimes \mathbf{f}_k = \mathbf{f}_i \tag{2}$$

where $\cdot$ is the inner product operator.

## 3.2   TPR-Based External Memory

In our work, the external memory for MANNs is formulated as the superposition of TPR from each time step. We utilize one of the recent TPR-based memory approach called *fast weight memory* [16] as a baseline model. At time step $t$, the encoder encodes the input $x_t$ into a hidden representation $h_t = \text{Encoder}(x_t, h_{t-1})$. We then extract a role representation $\mathbf{r}_t$ and a filler representation $\mathbf{f}_t$ from the hidden state $h_t$ as follows:

$$\mathbf{r}_t = \tanh(W_{\mathbf{r}} h_t) \tag{3}$$
$$\mathbf{f}_t = \tanh(W_{\mathbf{f}} h_t) \tag{4}$$
$$\beta = \sigma(W_\beta h_t) \tag{5}$$

where $\sigma(\cdot)$ is sigmoid function and $\beta$ is scalar for memory write strength. For using $\beta$, it is a common approach for conventional MANN models, replacing the previous values with a mixture of previous and current values. The representation of association between role and filler is embedded in TPR as $\mathbf{r}_t \otimes (\beta \mathbf{f}_t - (1 - \beta) \mathbf{f}_{t-1})$, and it is written on the previous memory $\mathbf{M}_{t-1}$ as follows:

$$\mathbf{M}_t = \mathbf{M}_{t-1} + \mathbf{r}_t \otimes (\beta \mathbf{f}_t - (1 - \beta) \mathbf{f}_{t-1}). \tag{6}$$

Since information is continuously added to memory, we normalize it to avoid divergence.

To read previous information from memory, unbinding vector $\mathbf{u}_t$ for time step $t$ is extracted from the hidden state $h_t$ when the model requires to use memory. The final read output $\mathbf{o}_t$ is obtained by inner producting $\mathbf{u}_t$ and memory state $\mathbf{M}_t$.

$$\mathbf{u}_t = \tanh(W_{\mathbf{u}} h_t) \tag{7}$$
$$\mathbf{o}_t = W_{\mathbf{o}}(\mathbf{u}_t \cdot \mathbf{M}_t) \tag{8}$$

To increase the capacity of memory, one can extract two role vectors $\mathbf{r}_t^1$ and $\mathbf{r}_t^2$, and use outer product of them to derive $\mathbf{r}_t = \mathbf{r}_t^1 \otimes \mathbf{r}_t^2$. In effect, each role representation learns to manifest its own role, such as entity or action [17]. In this case, unbinding vector for memory read also should be formulated with two unbinding vectors. For complex reasoning such as multi-hop reasoning problems, it is possible to read memory multiple times before the linear layer $W_o$.

### 3.3 Recurrent Encoder Modules

Since the original work of TPR assumes pre-defined roles/fillers for a given symbolic structured data and their representations, extracting them and learning their representations are essential for TPR to obtain proper representations of symbolic structures. To this end, we utilize an efficacious approach to learn symbolic representations using modular encoder RNNs with competitive learning, specializing each module to a specific mechanism [6,13]. The intuition behind this modular encoder concept is that each activated module participates in encoding input so that the hidden state becomes a combination of specialized encoding mechanisms. Specifically, we exploit Recurrent Independent Mechanisms (RIMs) [6] for our encoder network. For recurrent encoder modules with RIMs, each module is determined whether to be activated based on its relevance to the current input. The activated modules then encode the input with their encoding mechanisms, exploiting outputs from non-activated modules. The final hidden state of the encoder is derived by concatenating the hidden state from each module. During the training process, only the activated module can be updated and this competitive learning leads modules to have their own independent mechanisms, as demonstrated by recent works [6,11,13].

## 4    Experiments

We evaluate our contributions in two different levels, a synthetic level and a realistic level. First, we design a new synthetic task called *Systematic Associative Recall* (SAR) to analyze the systematic generalization of MANNs. The purpose of this level is to show the limitations of existing MANNs and the improvement of our proposed methods, directly. Next, we validate our proposed method on long sequential question answering task, comparing to other models.

### 4.1    Systematic Associative Recall Task

The SAR is an associative reasoning task designed to measure the ability to memorize combinatorial associations when there is a systematic difference between training and test data. In this task, the combinatorial associations are formed with multiple objects by binding them. Concretely, we consider three object sets such as human name $S_h$, fruit name $S_f$, and number name $S_n$, and the data are constructed by concatenating the word embedding vectors sampled from each object set. The main goal of this task is to reason the association between human name objects and other objects when the relationship has not been exposed to the model. For evaluating the systematic generalization, we divide $S_h$ into three

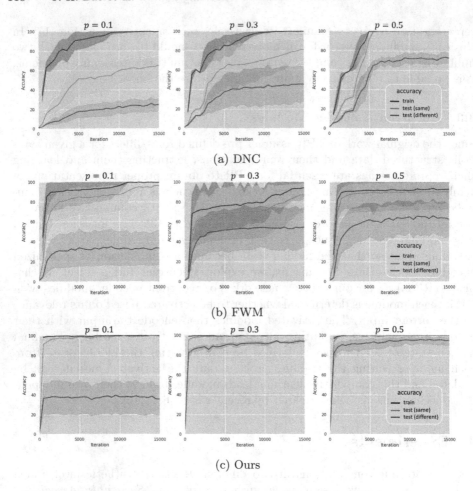

(a) DNC

(b) FWM

(c) Ours

**Fig. 1.** Training and test accuracy on the SAR task over $15k$ iterations.

subsets, $S_h^1$, $S_h^2$, and $S_h^3$, and each subset is given to infer objects from specific sets associated with its human objects: $S_h^1 \rightarrow S_n$, $S_h^2 \rightarrow S_f$, and $S_h^3 \rightarrow S_f \cup S_n$ where $A \rightarrow B$ indicates that the model should infer B associated with A. The set $S_h^3$ provides more diverse combinatorial associations to the model, and this can affect the generalization performance of the model.

In the test, the model is required to infer the associations of two cases: (a) associations similar to the training data: $S_h^1 \rightarrow S_n$ and $S_h^2 \rightarrow S_f$ and (b) systematically different from the training: $S_h^1 \rightarrow S_f$ and $S_h^2 \rightarrow S_n$. Therefore, in order to solve case (b), the model should learn representations for objects included in combinatorial data, and generalize to associations between objects. Additionally, we consider the effect for the degree of systematic difference between training and test data. For fixed value of $|S_h|$, we adjust the proportion of $|S_h^3|$ with a

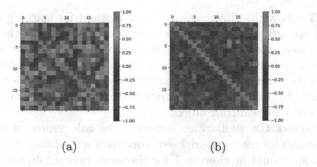

(a)                                    (b)

**Fig. 2.** The similarity matrices between role vector $r_t$ and unbinding vector $u_t$ for different human objects on (a) FWM and (b) our method. The x-axis and the y-axis indicate $r_t$ and $u_t$ for sequence, respectively.

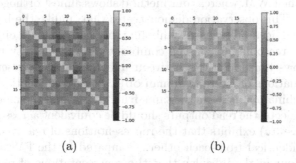

(a)                                    (b)

**Fig. 3.** The similarity matrices between read vectors $o_t$ for same fruit objects on (a) FWM and (b) our method. The x-axis and the y-axis indicate $o_t$ for various time steps.

value $p = \frac{|S_h^3|}{|S_h|}$. Since lower value of $p$ means larger systematic difference, the SAR task becomes more challenging as the value of $p$ decreases.

**Results.** We evaluate DNC [7], FWM [16], and our method on the SAR task to verify their systematic generalization ability. We consider three experimental settings with $p = 0.1, 0.3$, and $0.5$, and the results are shown in Fig. 1. DNC and FWM show performance degradation for all settings with large gaps between case (a) (plotted as test (same) in the figure) and case (b) (plotted as test (different)), whereas our model successfully achieves systematic generalization for $p = 0.3$ and $0.5$. Even if FWM considers systematicity by using the TPR method, it fails on the SAR task since FWM does not explicitly learn symbolic representations.

**Analysis.** To show the effectiveness of our method for learning symbolic representations, we analyze learned representations generated by the baseline and our model in two aspects. We investigate the representational analogy between (i) role vectors and unbinding vectors, and (ii) unbound vectors for the identical target object. In this analysis, we consider the association $S_h^1 \rightarrow S_f$ corresponding to the case (b), but the combinatorial data are consisting of multiple human objects and one identical fruit object.

Figure 2 shows the similarities between the role representations and the unbinding vectors for combinatorial sequence data with different human objects. The similarities should be close to 1 for the same human objects and 0 for different human objects to unbind appropriate filler representations from TPR memory; therefore, the role vectors and unbinding vectors should be orthogonal. From the analysis, role representations and unbinding vectors are not perfectly orthogonal for the FWM, whereas our method shows almost orthogonal results as shown in Fig. 2. This observation demonstrates that, unlike the baseline model, our method could accurately learn symbolic representations for objects and perform associative reasoning for unseen combinatorial data in a systematic way.

Figure 3 shows the similarities between unbound filler representations from different combinations of multiple human objects and one identical fruit object. Since every combination includes the same fruit object, and the model is required to recall fruit objects, the read outputs should be equivalent across multiple combinations. Figure 3(a) exhibits that the representations of read vectors from the FWM are not identical from each other. Compared to the FWM, the similarity patterns in our method display that the representations of read vectors are nearly identical for every association, as shown in Fig. 3(b). These results also demonstrate that our method performs associative reasoning in systematic way.

**Table 1.** The average test accuracy for 3 runs of different MANN models on the catbAbI task. * indicates experimental results from our trial. Note that our experiments of FWM cannot reach official results from [16].

|  | LSTM | TXL | MNM | FWM | FWM* | **Ours*** |
|---|---|---|---|---|---|---|
| Test accuracy | 80.88% | 87.66% | 88.97% | **96.75%** | 94.94% | **96.63%** |

## 4.2  Concatenated-bAbI

The concatenated-bAbI (catbAbI) [16] is a more challenging question-answering task than the bAbI task [21] with an infinite sequence of stories from the bAbI. Note that the catbAbI task does not explicitly target systematic generalization. Hence, our intention of this experiment is to verify the performance of models on general associative reasoning problems.

In the evaluated results on catbAbI in Table 1, FWM and our approach show the highest performance, while conventional methods show low test accuracy. Also, our model still achieves comparable results on long sequential text understanding problems. Especially in our experiment, our model shows improved test results than FWM by using a module-based encoder.

# 5   Conclusion

In this paper, we focused on learning representations of symbols to generalize their associations for associative reasoning. We proposed a novel MANN model with a modular encoder network and TPR-based external memory and a new task called *Systematic Associative Recall* (SAR) to analyze the systematic generalization of MANNs. Our proposed method has shown that it can learn the symbolic representations for objects individually and perform associative reasoning in a systematic way. Furthermore, our proposed method achieved state-of-the-art results on the SAR and a large-scale task.

**Acknowledgment.** This work was partly conducted by Center for Applied Research in Artificial Intelligence (CARAI) grant funded by Defense Acquisition Program Administration (DAPA) and Agency for Defense Development (ADD) (UD190031RD) (50%). It was also supported by the MSIT (Ministry of Science and ICT), Korea, under the ITRC (Information Technology Research Center) support program (IITP-2022-2020-0-01808) supervised by the IITP (Institute of Information & Communications Technology Planning & Evaluation) (50%).

# References

1. Banino, A., et al.: Memo: a deep network for flexible combination of episodic memories. In: International Conference on Learning Representations (2020)
2. Csordás, R., Schmidhuber, J.: Improving differentiable neural computers through memory masking, de-allocation, and link distribution sharpness control. In: International Conference on Learning Representations (2019)
3. Dai, Z., Yang, Z., Yang, Y., Carbonell, J., Le, Q.V., Salakhutdinov, R.: Transformer-xl: attentive language models beyond a fixed-length context. arXiv preprint arXiv:1901.02860 (2019)
4. Fodor, J.A., Pylyshyn, Z.W.: Connectionism and cognitive architecture: a critical analysis. Cognition **28**(1–2), 3–71 (1988)
5. Franke, J., Niehues, J., Waibel, A.: Robust and scalable differentiable neural computer for question answering. In: Proceedings of the Workshop on Machine Reading for Question Answering, pp. 47–59 (2018)
6. Goyal, A., et al.: Recurrent independent mechanisms. arXiv preprint arXiv:1909.10893 (2019)
7. Graves, A., et al.: Hybrid computing using a neural network with dynamic external memory. Nature **538**(7626), 471 (2016)
8. Hochreiter, S., Schmidhuber, J.: Long short-term memory. Neural Comput. **9**(8), 1735–1780 (1997)

9. Hupkes, D., Dankers, V., Mul, M., Bruni, E.: Compositionality decomposed: how do neural networks generalise? J. Artif. Intell. Res. **67**, 757–795 (2020)
10. Le, H., Tran, T., Venkatesh, S.: Self-attentive associative memory. In: International Conference on Machine Learning, pp. 5682–5691. PMLR (2020)
11. Madan, K., Ke, N.R., Goyal, A., Schölkopf, B., Bengio, Y.: Fast and slow learning of recurrent independent mechanisms. arXiv preprint arXiv:2105.08710 (2021)
12. Munkhdalai, T., Sordoni, A., Wang, T., Trischler, A.: Metalearned neural memory. In: Advances in Neural Information Processing Systems, vol. 32 (2019)
13. Parascandolo, G., Kilbertus, N., Rojas-Carulla, M., Schölkopf, B.: Learning independent causal mechanisms. In: International Conference on Machine Learning, pp. 4036–4044. PMLR (2018)
14. Park, T., Choi, I., Lee, M.: Distributed associative memory network with memory refreshing loss. Neural Netw. **144**, 33–48 (2021)
15. Rae, J.W., et al.: Scaling memory-augmented neural networks with sparse reads and writes. In: Proceedings of the 30th International Conference on Neural Information Processing Systems, pp. 3628–3636 (2016)
16. Schlag, I., Munkhdalai, T., Schmidhuber, J.: Learning associative inference using fast weight memory. arXiv preprint arXiv:2011.07831 (2020)
17. Schlag, I., Schmidhuber, J.: Learning to reason with third order tensor products. In: Advances in Neural Information Processing Systems, vol. 31 (2018)
18. Smolensky, P.: Tensor product variable binding and the representation of symbolic structures in connectionist systems. Artif. Intell. **46**(1–2), 159–216 (1990)
19. Sukhbaatar, S., Weston, J., Fergus, R., et al.: End-to-end memory networks. In: Advances in Neural Information Processing Systems, pp. 2440–2448 (2015)
20. Vaswani, A., et al.: Attention is all you need. In: Advances in Neural Information Processing Systems, vol. 30 (2017)
21. Weston, J., et al.: Towards AI-complete question answering: a set of prerequisite toy tasks. arXiv e-prints pp. arXiv-1502 (2015)

# Identifying Dominant Emotion in Positive and Negative Groups of Navarasa Using Functional Brain Connectivity Patterns

Pankaj Pandey[1(✉)], Richa Tripathi[3], Gayatri Nerpagar[2], and Krishna Prasad Miyapuram[1,2]

[1] Computer Science and Engineering, IIT Gandhinagar, Palaj, India
pankaj.p@iitgn.ac.in
[2] Centre for Cognitive and Brain Sciences, IIT Gandhinagar, Palaj, India
[3] Center for Advanced Systems Understanding (CASUS),
Helmholtz-Zentrum Dresden-Rossendorf, Görlitz, Germany

**Abstract.** Natyashastra is an ancient wisdom of Indian performing arts, written by Bharat Muni, and remarkable for its Rasa theory. Rasa means nectar, the sentiments or emotion felt by the spectator while watching a performance. There are nine Rasas (nine emotions), famously known as Navarasa. Eminent scholars have contemplated Rasa as a superposition of several emotions with the dominance of a particular emotion. Empirical brain research may provide evidence to understand the previous theoretical work. Hence, we carry out this research to understand the most dominant Rasa in positive and negative emotion Rasa groups. By dominance, we mean how well the Rasa of the positive emotion group was distinguished from negative Rasa and vice versa. Our analysis is based on EEG data collected on participants while watching movie clips based on these Rasas with capturing time-varying activity using three functional connectivity metrics. Network properties are extracted from networks and utilized to feed as features for Random Forest classifier. We obtained maximum accuracy (greater than 90%) in five pairs between negative and positive emotions. We find the two most dominant Rasas are Sringaram (Love) and Bibhatsam (Disgust), representing positive and negative emotions, respectively. We observe that weaker connections in delta and gamma bands with the lowest network feature values significantly aid in classifying emotions. The strongest connections of delta and gamma connections involve inter-hemispheric and intra-hemispheric engagement patterns respectively, which suggest global and local information processing while watching emotional clips. Beta waves generate strong connections across regions, which suggest inline findings with previous works on beta for the western classification of emotions.

**Keywords:** Rasa · Emotion · Machine Learning · EEG · Brain Networks

© The Author(s), under exclusive license to Springer Nature Switzerland AG 2023
M. Tanveer et al. (Eds.): ICONIP 2022, LNCS 13624, pp. 123–135, 2023.
https://doi.org/10.1007/978-3-031-30108-7_11

# 1   Introduction

Natyashastra is an ancient Indian treatise on performing arts written by Bharat Muni (Saint). It has an influence on Indian dance, music, and literature. Indian classical dance and music are based on Natyashastra [3]. It contains a total of 36 chapters, subjected to drama composition, acting, dance movements, body movements, construction of the stage, music, etc. Chap. 6 of Natyashastra talks about Rasa or sentiments. Rasa is the audience's experience when they are observing a performance [13]. The performer's primary goal is to make them experience what they are performing, for example, if the performer is crying the audience should feel sad. What a performer performs is bhava and what an audience feel is a Rasa. Rasa and Bhava are interrelated and no Rasa without bhava and no Bhava without Rasa. Rasa is what a spectator experience when he/she is observing a stimulus (which is performance/act according to Natyashastra), and Bhava is what is presented by the stimulus. The bhava expressed in the performance gets translated into the emotional experience of the spectator. In Fig. 1, we have shown the eight Rasas and their division into positive and negative groups. The last Rasa Santam (peace) is not considered for this study.

| Navrasas | Dominant State |
|---|---|
| Hasyam (Comic) | Hasyabhava (Mirth) |
| Sringaram (Erotic) | Rati (Love) |
| Adbhutam (Marvellous) | Vismay (Astonishment) |
| Veeram (Heroic) | Utsah (Energy) |
| Raudram (Rage) | Krodha (Anger) |
| Bhayanakam (Terror) | Bhay (Fear) |
| Bibhatsam (Odious) | Jugupsha (Disgust) |
| Karunayam (Pathos) | Shoka (Sorrow) |

**Fig. 1.** [Left] Facial Expressions depict different Rasas. [Right] In the table, we provide closest English translation and the corresponding dominant emotional state (or Sthayi Bhava).

Electroencephalography (EEG) is the most widely used neuroimaging technique due to its temporal precision. The multi-variate time series of EEG can be considered to extract the time-varying brain regional interaction by using brain connectivity metrics against primary five brain waves. In this study, we discuss the emotion attached to each Rasa. Hence EEG data was collected while watching movie clips narrating these Rasas pertaining to a particular emotion. In recent years, several studies discussed the classification of emotions by EEG-based functional connectivity patterns [5,7,10]. Lee and Hsieh computed functional connectivity including correlation, coherence and phase synchronizaion and observed the connectivity patterns differentiating emotional states: neutral, positive, or negative [5]. Liu et al., examined the subject-independent discriminative connection using phase lag index (PLI) to identify the positive, neutral and negative emotions and obtained maximum accuracy of 87.03% in the beta band, along with role of frontal and temporal lobes in emotion-relation activities [9]. They also discussed the network property such as global efficiency was more distinguishable to positive compared to neutral emotion. Zhang and colleagues reported the differences between positive and negative emotion using connectivity network and highlighted the role of prefrontal region in emotional processing interactions with other regions [19].

We compute functional networks using phase lag index (PLI), weighted phase lag index (wPLI), and corrected imaginary part of Phase Locking Value (ciPLV) for five brain waves and further exploited the network topology by extracting fourteen network features. These features are used for classification between positive and negative Rasas. We use Rasa as a representation of emotion, so we interchangeably use Rasa or emotion in this study as shown in Fig. 1. Based on the dimensional model of emotion classification by Russell and Barrett [14], we hypothesize that the Rasas can be divided into two categories based on the valance: positive and negative. We considered positive valance Rasa to be Hasyam, Shringaram, Veeram, and Adbhutam, and negative valance Rasa to be Raudram, Bhayanakam, Bibhatsam, and Karunayam.

The objectives of this study are a) Identify the dominant emotion in positive and negative sets of Rasas using classification b) Role of functional networks and frequency bands in identifying the significantly distinguishable pairs c) Interpreting the outcome of classifiers using network metrics.

Dominant refers to a Rasa which is mostly distinguishable to other group across functional metrics and bands based on the significance of classification output. This work is novel in identifying the differences using three functional networks against five brain waves and is an extension of recent work [12] on Rasa by introducing positive and negative emotion rasa groups.

## 2    Data Description

### 2.1    Stimuli Selection and Participants

Nine Bollywood movies were selected as a stimulus for each Rasa, based on the highest rating given by a group of people for each Rasa independently as shown in Table 1. Bharat muni originally defined eight Rasas, the ninth Rasa shantam (peace) was later added and we have excluded for this study and focus on positive and negative sets of Rasa. The duration of the movie clip was based on the time that the content of the movie needed to evoke that particular Rasa, hence the time was different for different Rasa. The language used in movie clips was Hindi. The release date of movies varied from 1980 to recent that is four decades. 20 participants (mean age: 26 years, 16 males, 4 females), right-handed students of IIT Gandhinagar participated in the study. Before conducting the study, the informed concern was provided to all the participants. All participants were proficient in the Hindi language. The instruction about the study was given prior to the study. The participants were asked to watch nine selected Bollywood movie clips; a fixation cross appeared on a screen for 10 s following each movie clip. The order of movies was randomized for each participant. The Institute Ethical Committee (IEC) of the Indian Institute of Technology, Gandhinagar approved this study.

**Table 1.** Movie clips used in EEG data collection

| Emotion | Rasa Genre | Film Name | Year | Start Time | End Time |
|---------|-----------|-----------|------|-----------|----------|
| Positive | Hasyam | 3 Idiots | 2009 | 59m 55s | 1h 2m 28s |
| Positive | Sringaram | Umrao Jaan | 1981 | 43m 08s | 43m 50s |
| Positive | Adbhutam | Mr. India | 1987 | 1h 1m 40s | 1h 3m 28s |
| Positive | Veeram | Lagaan: Once Upon a Time in India | 2001 | 2h 10m 57s | 2h 13m |
| Negative | Raudram | Ghajini | 2008 | 2h 38m 43s | 2h 40m 52s |
| Negative | Bhayanakam | Bhoot | 2003 | 1h 2m 57s | 1h 4m 31s |
| Negative | Bibhatsam | Rakhta Charitra | 2010 | 43m 55s | 45m 7s |
| Negative | Karunayam | Kal Ho Naa Ho | 2003 | 2h 47m 41s | 2h 50m 18s |

## 2.2    EEG Data Recording and Preprocessing

A High-density Geodesic system of 128 channels was used for this acquisition with a sampling rate 250 Hz. The experiment was designed and run on E-prime TM and recording was captured using Net-Station TM. The preprocessing was performed using the Matlab EEGLAB package. High-frequency signals 60 Hz were filtered to avoid noise effects. Movements and eye-blinks artifacts were removed using artifact subspace reconstruction. Following this, the data was chunked respective to each Rasa across subjects and used for further analysis.

## 3    Methodology

To identify functional signatures of Rasas, we obtain functional brain networks using standard functional connectivity metrics. As the functional activity of the brain is encoded into neural oscillation frequencies, we obtain these signatures in different brain frequency bands by frequency decomposition of the metrics into five frequency bands: delta (1 Hz–4 Hz), theta (4 Hz–7 Hz), alpha (8–13 Hz), beta (13 Hz–30 Hz), and gamma (30 Hz–45 Hz). The nodes in these functional brain networks are the EEG electrodes, and the edges representing functional co-dependence are obtained using three measures- Phase Lag Index (PLI), weighted Phase Lag Index (wPLI) and corrected imaginary part of Phase Locking Value (ciPLV). We use three connectivity metrics to identify robust Rasa divisions into positive and negative categories, and also to investigate which metric capture the Rasa divisions more accurately than others.

### 3.1    Functional Connectivity

The definitions of the three metrics defining correlations between two timeseries EEG signals ($x$ and $y$) and hence the edge between the corresponding nodes in the brain network are explained here. For cross and auto-power spectral densities of these signals depicted as $P_{xy}$ and ($P_{xx}$ and $P_{yy}$), the definitions are,

– **PLI:** The PLI value between $x$ and $y$ is:

$$PLI = |E[sign(I(P_{xy}))]|, \qquad (1)$$

where is $E$ is the average over epochs, $sign$ denotes positive or negative sign of the quantity over which it is applied, $I$ is the imaginary part of the complex number over which it is applied, and $|.|$ denotes absolute value of the quantity.

– **wPLI:** The wPLI is defined as

$$wPLI = \frac{|E[I(P_{xy})]|}{E[|I(P_{xy})|]}, \qquad (2)$$

where symbol meanings are same as in $PLI$ definition.

– **ciPLV:** The ciPLV is defined as:

$$ciPLV = \frac{|E[I(P_{xy}/|P_{xy}|)]|}{\sqrt{1 - |E[R(P_{xy}/|P_{xy}|)]|^2}}, \qquad (3)$$

where $R$ denotes the real part of the complex number over which it is applied.

## 3.2   Thresholding of Functional Networks

Functional networks mostly preserve weak and erroneous connections, which may conceal the topology of crucial connections [15]. For all the networks constructed based on above defined network metrics, we perform a thresholding of the connections to retain the important edges and discard spurious ones. Thresholding is commonly used to remove a percentage of the weakest links to retain a usable sparse network. We applied the thresholding process as implemented in the paper [1]: the network should be 97% connected, and the average degree should be greater than $2 * log(n)$, while maintaining the highest threshold value for edge weights, where n is the number of nodes.

## 3.3   Network Metrics as Features

To characterize the network differences of one Rasa from the other, we calculated the 14 network properties of the brain networks constructed using the three functional connectivity measures. The network metrics are listed along with their definitions in Table 2.

## 3.4   Machine Learning and Evaluation

Random forests were trained using [number of subjects $\times$ 14] features. Due to the complexity of the EEG setup and data collection, most EEG studies include 15–20 participants. Hence, some techniques are proposed to determine the significance of Machine Learning performance estimates with small sample size. Based on the recent article [16], it is essential to use rigorous analysis methods rather than relying on K-fold Cross-Validation alone. We therefore used permutation tests with 500 rounds of five-fold cross validation to produce robust and unbiased estimates regardless of the sample size. In this paper, we reported the accuracy based on the significance of classification using the permutation test [11].

**Table 2.** The table give definitions of the fourteen network metrics used for network analysis in the present work. The numerical computation of the metric values were carried out using NetworkX module of python [4]

| Network Measure | Symbol | Definition |
|---|---|---|
| Average Degree | AD | The node's degree is the number of its direct neighbours. Average degree is the mean over the degrees of the nodes in the network |
| Maximum Degree | MD | It is the maximum degree existing in the network |
| Average Edge Weight | AEW | Edge weight is the numerical value of the metric governing the connection between the network nodes. Average edge weight is the mean over all edge weights of the network edges |
| Maximum Edge Weight | MEW | It is the maximum edge weight existing in the network |
| Network Density | D | It is the ratio of number of existing edges in the network to the total number of potential connections in the network |
| Average Clustering Coefficient | ACC | The average clustering coefficient is the fraction of closed triplets to the total number of all open and closed triplets present in the network |
| Average Local Efficiency | ALE | Local Efficiency is defined for a particular node, as the inverse of the average shortest path connecting all its neighbours. ALE is average over all these values |
| Global Efficiency | GE | It is defined as the inverse of the average characteristic path length between all pairs of nodes existing in the network |
| Number of Communities | NC | It is the number of clearly identifiable modules, such that the connectivity between the nodes within a module is higher than across module connectivity |
| Modularity | M | Vaued between 0 and 1, modularity is the value depicting how nice is the division of nodes into modules or communities |
| Transitivity | T | Transitivity is the ratio of three times the number of triangles of nodes to the number of connected triples of nodes in the network |
| Average Degree Centrality | ADC | The degree centrality of a node is a fraction of the number of links a node has to all the possible links it can have |
| Average Node Betweenness Centrality | NBC | The betweenness of a node is the measure of how frequently the node lies in the shortest paths in the network. NBC is the average of this quantity over all nodes |
| Average Edge Betweenness Centrailty | EBC | The betweenness of an edge is the measure of how frequently the edge lies in the shortest paths in the network. EBC is the average of this quantity over all edges in the network |

# 4    Results

## 4.1    Identifying the Dominant Emotion in Positive and Negative Sets of Rasas Using Classification

We first examined the discriminating pair of Rasas with maximum accuracy. The maximum discriminating pair for positive emotions are shown in Table 3. We observed that Bibhatsam (Disgust) formed the maximum discrimination with three positive Rasas, including Hasyam (Comic), Adbhutam (Astonishment), and Veeram (Heroic). Sringaram (Love) and Karunayam (Sorrow) showed maximum accuracy of 97.5%. The minimum accuracy was 80% in Adbhutam. The network generated in PLI showed the maximum discrimination, including delta, theta, and gamma bands. The bottom of Table 3 for negative pair shows that Sringaram formed the maximum discriminating pair with every negative emotion. All the accuracies were above 90%. The most crucial finding was that PLI preserved the critical network information for discrimination and seven maximum pairs out of eight pairs of emotions shown in Table 3.

Table 3. [Top] Maximum classification accuracy obtained in each positive emotion. [Bottom] Maximum classification accuracy obtained in each negative emotion

| Positive | Hasyam | Sringaram | Adbhutam | Veeram |
|---|---|---|---|---|
| Negative | Bibhatsam | Karunayam | Bibhatsam | Bibhatsam |
| Method | PLI | PLI | PLI | PLI |
| Band | delta | theta | gamma | delta |
| Accuracy | 0.9 ±0.12 | 0.975 ±0.05 | 0.8 ±0.1 | 0.875 ±0.0 |
| Precision | 0.883 ±0.15 | 0.96 ±0.08 | 0.79 ±0.13 | 0.92 ±0.1 |
| Recall | 0.95 ±0.1 | 1.0 ±0.0 | 0.85 ±0.12 | 0.85 ±0.12 |
| F1Score | 0.91 ±0.11 | 0.978 ±0.04 | 0.81 ±0.09 | 0.87 ±0.02 |
| pvalue | 0.002 | 0.002 | 0.006 | 0.004 |

| Negative | Raudram | Bhayanakam | Bibhatsam | Karunayam |
|---|---|---|---|---|
| Positive | Sringaram | Sringaram | Sringaram | Sringaram |
| Method | PLI | ciPLV | PLI | PLI |
| Band | gamma | theta | gamma | theta |
| Accuracy | 0.95 ±0.06 | 0.95 ±0.06 | 0.95 ±0.06 | 0.975 ±0.05 |
| Precision | 0.96 ±0.08 | 0.96 ±0.08 | 0.92 ±0.1 | 0.96 ±0.08 |
| Recall | 0.95 ±0.1 | 0.95 ±0.1 | 1.0 ±0.0 | 1.0 ±0.0 |
| F1Score | 0.949 ±0.06 | 0.949 ±0.06 | 0.956 ±0.05 | 0.978 ±0.04 |
| pvalue | 0.002 | 0.002 | 0.002 | 0.002 |

We reduced the higher dimensional features to lower-dimensional features to visualize the differentiation between Rasas. In Fig. 2, we plotted the pair for Adbhutam and Veeram. We can observe the significant distinction between pairs of emotions (Rasas), and each data point represents a participant. Adbhutam and Veeram achieved 80% and 87.5% accuracy, which can be inferred from t-SNE plots showing 3–4 data points in opposite direction. Therefore, the network features captured a significant distinction. Similarly, we observed separation between Raudram and Sringaram in Fig. 2. Lower-dimensional visualization is a way to understand the distribution present in high-dimensional space. Therefore, in our presented figures, we observed a distinction between the features of the two emotions. Moreover, the classification algorithm also plays a part in learning the linear and non-linear boundaries.

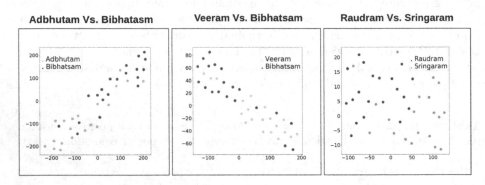

**Fig. 2.** t-SNE (t-distributed Stochastic Neighbor Embedding) visualization of features in two-dimension for Adbhutam (Astonishment), Veeram (Heroic), Bibhatasam (Disgust), Sringaram (Love) and Raudram (Anger). Each data point represents a participant. The maximum number of data points of each Rasa are clustered in the opposite direction to other Rasa, reflecting differences in features between Rasas.

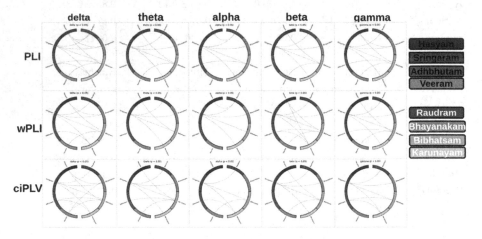

**Fig. 3.** Each connection indicates a significant (p < 0.05) distinguishable pair obtained in each functional network against frequency bands.

## 4.2  Role of Functional Networks and Frequency Bands for Identifying the Significantly Distinguishable Pairs

In Fig. 3, we illustrated the significant (p < 0.05) distinguishable pairs of emotions for every functional network against each frequency band. There is significant discrimination between every pair of Rasas using delta features of PLI and ciPLV. The most striking observation is that the low-frequency oscillations delta provided the maximum discrimination, followed by beta and gamma, as mentioned in Table 4. Sringaram and Bibhatsam showed the maximum discrimination in the delta band across all functional networks. The alpha band showed the minimum number of distinguishable pairs. We also derived a scale based on the maximum number of distinguishable pairs observed across functional networks against bands. In Fig. 4, we observed that Sringaram and Bibhtastam formed the maximum number of distinguishable pairs.

**Table 4.** Number of discriminating pairs (p < 0.05) observed in functional networks against frequency bands [Highlighted digit shows the maximum across bands]

| FC | delta | theta | alpha | beta | gamma |
|---|---|---|---|---|---|
| PLI | **10** | 6 | 6 | 6 | 8 |
| wPLI | **8** | 5 | 5 | 9 | 7 |
| ciPLV | **10** | 7 | 5 | 6 | 6 |
| Total | **28** | 18 | 16 | 21 | 21 |

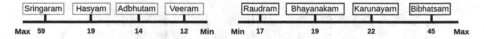

**Fig. 4.** A scale showing the maximum number of discriminating significant pairs (p < 0.05) observed in each Rasa after summing the pairs of three functional networks including all bands.

**Table 5.** Network features of PLI networks after averaging of bands and subjects for Rasas

| Positive | Average Degree | Density | Clustering Coeff | Global Effc | No. of communities |
|---|---|---|---|---|---|
| Hasyam | 21.713 | 0.171 | 0.34368 | 0.518 | 3.930 |
| Sringaram | **15.194** | **0.120** | **0.2395** | **0.477** | **3.860** |
| adbhutam | 21.414 | 0.169 | 0.33191 | 0.515 | 4.070 |
| veeram | 20.948 | 0.165 | 0.33582 | 0.515 | 3.990 |
| Negative | | | | | |
| Raudram | 22.780 | 0.179 | 0.34747 | 0.524 | 4.040 |
| Bhayanakam | 19.013 | 0.150 | 0.30981 | 0.503 | 4.010 |
| Bibhatsam | **17.858** | **0.141** | **0.2902** | **0.492** | 3.990 |
| karunayam | 22.059 | 0.174 | 0.34754 | 0.524 | **3.920** |

### 4.3    Interpreting Outcome of Classifier Using Network Properties

In Fig. 5, we have plotted the 5% strongest network connections of PLI. We observed a significant pattern of cross-connection across the regions of the brain hemisphere in the delta. In contrast, cross-connections were significantly reduced in the gamma band. In Fig. 5, equal number of edges are shown, however more sparse connectivity is observed in Sringaram and Bibhatsam, whereas others have strongest connections with near by nodes. And this is also inferred by Table 5, Sringaram and Bibhatsam had the minimum number of network features. Therefore classifier results were clearly observed in the scale mentioned in Fig. 4. Stronger connections are observed in parietal and occipitial regions in alpha and beta bands.

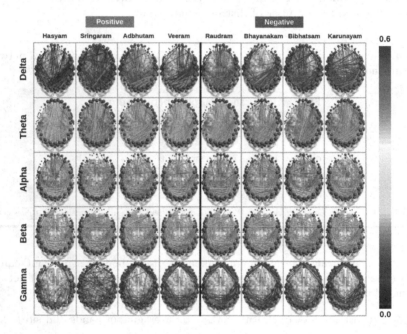

**Fig. 5.** Functional networks (PLI) of averaged across 20 subjects are displayed for 5% of the strongest connections. Node size denotes the degree and edge width, and color represents the strength.

**Table 6.** Average Degree of PLI is shown against each Rasa and Band

| Positive | delta | theta | alpha | beta | gamma |
|---|---|---|---|---|---|
| Hasyam | 16.206 | 33.788 | 25.766 | 19.884 | 12.919 |
| Sringaram | **12.991** | **20.235** | **18.318** | **14.103** | **10.325** |
| Adbhutam | 19.512 | 28.218 | 26.453 | 21.114 | 11.771 |
| Veeram | 17.620 | 30.484 | 23.491 | 21.327 | 11.816 |
| Negative | | | | | |
| Raudram | 20.519 | 30.527 | 28.178 | 21.664 | 13.012 |
| Bhayanakam | 17.961 | 23.945 | 23.307 | 18.311 | **11.541** |
| Bibhatsam | **15.264** | **22.196** | **21.991** | **17.349** | 12.491 |
| Karunayam | 21.412 | 25.484 | 26.602 | 23.519 | 13.280 |

## 5 Discussion

Weaker connections are observed in delta and gamma bands as compared to theta, alpha, and beta waves. Sringaram on delta waves have spread connections across the brain, which seem to generate the maximum classification, as shown in Table 5, 6 and Fig. 5. The cross-connections across hemisphere in delta may suggest the global information processing across brain, whereas gamma waves are primarily involved in processing local computation, higher brain and cognitive functions [2]. Therefore, the most striking visible difference observed of strong inter- and intra-hemispheric engagement in delta and gamma bands, respectively. Both rhythms have different dynamical mechanisms to perform. Buzsáki et al. discuss the magnitude of gamma oscillation modulated by slower rhythms and coupling between active patches of cortical circuits may take place via cross-frequency coupling [2]. Delta band has been recently documented to preserve functional connectivity patterns for discriminating the emotions engaging parietal and occipital sites [17]. Higher frequency bands (gamma and beta) has been reported widely to be significant for the classification of different emotions [6,8,18].

## 6 Conclusion

We have presented findings using network properties of three functional networks between positive and negative emotion groups of Rasa. The most crucial part of our research is the interpretability of the obtained outcome of classifiers. We derive a scale that defines the differences between two emotional categories. Sringaram (Love) and Bibhatsam (Disgust) form the maximum distinguishable pairs. Delta and gamma brain waves generate the weaker connections, whereas theta, alpha and beta show strongest connections. These findings are crucial in their novelty and an initial attempt to bring the ancient wisdom of performing arts into modern brain research. We have provided a limited interpretation in this work, but there seems to be ample opportunity to understand the theta, alpha, and beta roles in Rasas for future work.

**Acknowledgements.** We acknowledge SERB and PlayPower Labs for their support of PMRF to Pankaj Pandey and FICCI for facilitating PMRF. This work was partially funded by the Center of Advanced Systems Understanding (CASUS), which is financed by Germany's Federal Ministry of Education and Research (BMBF) and by the Saxon Ministry for Science, Culture and Tourism (SMWK) with tax funds on the basis of the budget approved by the Saxon State Parliament.

# References

1. Bassett, D.S., Meyer-Lindenberg, A., Achard, S., Duke, T., Bullmore, E.: Adaptive reconfiguration of fractal small-world human brain functional networks. Proc. Natl. Acad. Sci. **103**(51), 19518–19523 (2006)
2. Buzsáki, G., Wang, X.J.: Mechanisms of gamma oscillations. Annu. Rev. Neurosci. **35**, 203 (2012)
3. Chakravorty, P.: Hegemony, dance and nation: the construction of the classical dance in India. South Asia J. South Asian Stud. **21**(2), 107–120 (1998)
4. Hagberg, A.A., Schult, D.A., Swart, P.J.: Exploring network structure, dynamics, and function using networkx. In: Varoquaux, G., Vaught, T., Millman, J. (eds.) Proceedings of the 7th Python in Science Conference, Pasadena, CA USA, pp. 11–15 (2008)
5. Lee, Y.Y., Hsieh, S.: Classifying different emotional states by means of EEG-based functional connectivity patterns. PLoS ONE **9**(4), e95415 (2014)
6. Li, M., Lu, B.L.: Emotion classification based on gamma-band EEG. In: 2009 Annual International Conference of the IEEE Engineering in Medicine and Biology Society, pp. 1223–1226. IEEE (2009)
7. Li, P., et al.: EEG based emotion recognition by combining functional connectivity network and local activations. IEEE Trans. Biomed. Eng. **66**(10), 2869–2881 (2019)
8. Lin, Y.P., et al.: EEG-based emotion recognition in music listening. IEEE Trans. Biomed. Eng. **57**(7), 1798–1806 (2010)
9. Liu, X., et al.: Emotion recognition and dynamic functional connectivity analysis based on EEG. IEEE Access **7**, 143293–143302 (2019)
10. Martini, N., et al.: The dynamics of EEG gamma responses to unpleasant visual stimuli: from local activity to functional connectivity. Neuroimage **60**(2), 922–932 (2012)
11. Ojala, M., Garriga, G.C.: Permutation tests for studying classifier performance. J. Mach. Learn. Res. **11**(6) (2010)
12. Pandey, P., Tripathi, R., Miyapuram, K.P.: Classifying oscillatory brain activity associated with Indian rasas using network metrics. Brain Inform. **9**(1), 1–20 (2022)
13. Pathloth, V.: Rasa prakaranam - the aesthetics of sentiments and their interpretation in kuchipudi dance. IJCRT **8**, 1–23 (2020)
14. Russell, J.A.: A circumplex model of affect. J. Pers. Soc. Psychol. **39**(6), 1161 (1980)
15. Sun, S., et al.: Graph theory analysis of functional connectivity in major depression disorder with high-density resting state EEG data. IEEE Trans. Neural Syst. Rehabil. Eng. **27**(3), 429–439 (2019)

16. Vabalas, A., Gowen, E., Poliakoff, E., Casson, A.J.: Machine learning algorithm validation with a limited sample size. PLoS ONE **14**(11), e0224365 (2019)
17. Wu, X., Zheng, W.L., Lu, B.L.: Identifying functional brain connectivity patterns for EEG-based emotion recognition. In: 2019 9th International IEEE/EMBS Conference on Neural Engineering (NER), pp. 235–238. IEEE (2019)
18. Yang, K., Tong, L., Shu, J., Zhuang, N., Yan, B., Zeng, Y.: High gamma band EEG closely related to emotion: evidence from functional network. Front. Hum. Neurosci. **14**, 89 (2020)
19. Zhang, J., Zhao, S., Huang, W., Hu, S.: Brain effective connectivity analysis from EEG for positive and negative emotion. In: Liu, D., Xie, S., Li, Y., Zhao, D., El-Alfy, E.S. (eds.) Neural Information Processing, pp. 851–857. Springer, Cham (2017). https://doi.org/10.1007/978-3-319-70093-9_90

# A Cerebellum-Inspired Model-Free Kinematic Control Method with RCM Constraint

Xin Wang[1], Peng Yu[1], Mingzhi Mao[2], and Ning Tan[1]([✉])

[1] School of Computer Science and Engineering, Sun Yat-sen University,
Guangzhou, China
tann5@mail.sysu.edu.cn
[2] School of Software Engineering, Sun Yat-sen University, Zhuhai, China

**Abstract.** Minimally invasive surgery requires to reduce trauma to patients during the movement of the inserted robot end-effector. The cerebellum is able to control limbs in many scenarios, with high precision and robustness. This article designs a cerebellum-inspired model-free scheme for the tracking control of redundant robot manipulators with remote center of motion (RCM) constraint. The scheme is formed by coupling liquid state machines (LSM) and zeroing neural network (ZNN). The ZNN is able to generate approximate joint angle commands as teaching signals without a perfect robot model to train the cerebellum model based on LSM. The output of the LSM is used as the control commands for current moment, which includes managing the constraint on RCM. Finally, demonstration simulations and experiments are conducted to verify the efficacy of the proposed control strategy.

**Keywords:** Liquid State Machines · Remote center of motion ·
Redundant manipulator · Zeroing Neural Network · Spiking Neural
Network

## 1 Introduction

With the development of artificial intelligence in the past few years, redundant robots are increasingly used in various fields, especially in medical surgery [4]. Minimally invasive surgery requires a small incision in the patient's abdominal wall to allow for the insertion of surgical instruments. The small incision will have a restraint on the inserted robot end-effector, often called the remote center of motion (RCM) constraint [13]. This constraint requires the robot end-effector to perform its task while surrounding the incision as much as possible to avoid

This work is partially supported by the Key-Area Research and Development Program of Guangzhou (202007030004), the National Natural Science Foundation of China (62173352), the Guangdong Basic and Applied Basic Research Foundation (2021A1515012314), and the Open Project of Shenzhen Institute of Artificial Intelligence and Robotics for Society (AC01202005006).

M. Tanveer et al. (Eds.): ICONIP 2022, LNCS 13624, pp. 136–147, 2023.
https://doi.org/10.1007/978-3-031-30108-7_12

enlarging it or causing other damage to the patient. The surgical instruments are usually operated by a human assistant during the procedure. This requires the human assistant to have proficient experience in operating the robot end-effector to avoid secondary damage to the patient. In order to reduce the mental and physical burden of the surgeon, some automatic control strategies have been applied in medical surgery, such as visual servo automatic control strategies [3,19]. Although a mechanical implementation is a safer alternative, it requires complex structures and more calibration procedures. Therefore, it is more flexible and convenient to implement tasks such as visual servo and trajectory tracking through control algorithms. But how to maintain the RCM constraint during the surgical operations becomes a challenging problem.

In recent years, many approaches have taken the RCM constraint into account in control strategies [1,8,12]. Aghakhani et al. [1] designed a variable associated with the end-effector insertion depth to obtain an augmented RCM Jacobian matrix. The final Jacobian matrix was obtained by splicing the task Jacobian matrix and the augmented RCM Jacobian matrix. Compared to [1], Sadeghian et al. [12] introduced a Jacobian matrix with minimal dimension to reduce the computational complexity. Li et al. [8] proposed a recurrent neural network with the RCM constraint. However, in practical application scenarios, there may be cases where some of the kinematic parameters are unavailable or the calibrated parameters do not match the actual parameters, making these model-based control methods less feasible. In recent years, model-free based control methods are also developing rapidly, such as [17,18]. However, these model-free control methods did not involve RCM constraint. The development of artificial intelligence and the fact that humans have powerful learning and precise motion capabilities motivate us to search for a model-free cerebellum-inspired scheme.

The human brain can be viewed as a complex, non-linear and parallel cognitive machine. Artificial neural networks (ANNs) are the hardware or software implementations inspired by the development of brain science in the last hundred years. In [10], ANNs are divided into three stages, the first one is the perceptrons consisting of binary gates, where each boolean function can be implemented by a suitable network structure. The second stage is defined by more complex neuronal models using continuous activation functions. Based on the continuous behavior of neurons, a back propagation training method was proposed. However, the second generation neural networks are biologically inaccurate and do not simulate well the operating mechanisms of real biological cerebellum. Some works based on cerebellum-like models have borrowed the structure of the motor nervous system [18], however they did not use spiking neurons and still did not fully exploit the potential of cerebellum-like models.

The third generation of ANNs is known as spiking neural networks (SNNs) with spiking neurons (e.g. integrate and fire neurons) as the basic unit. SNNs have rich neurodynamic properties, numerous coding mechanisms and event-driven low energy consumption properties that have attracted increased research interest [11]. Maass [9] stated that compared to the previous two generations of neural networks, spiking neural networks use a smaller number of neurons and therefore possess more computational power. Even in the worst case, SNNs possess the same computational power. Compared to the second generation of neural networks, SNNs are still in the early stages of rapid development, and due to the discontinuity of spiking events, the back propagation algorithm is not directly applicable to SNNs, and there is no uniform or clear training method. Inspired by the structure of the mammalian central nervous system, Maass [10] proposed a reservoir computational system based on spiking neurons, called a liquid state machine (LSM) (which is used in this paper).

Humans are capable of accurately performing a variety of complex tasks despite their complex skeletal structures and non-deterministic sensory delay feedback. Inspired by the biology of cerebellar motor control, in this work we will develop a spiking neural networks-based control strategy for redundant robot arm trajectory tracking tasks under RCM constraint. The main contributions of this work are summarized as follows.

- A model-free cerebellum-inspired control scheme, which based on liquid state machine, is designed to solve the tracking control problem of redundant manipulators with surgical tooltip, considering RCM constraint.
- Compared with some RCM constrained robot arm control schemes, the proposed control scheme does not require modeling of the robot arm, which is more robust and portable.
- Compared to general control strategies base on liquid state machine, the proposed control scheme does not require pre-training and reduces the time overhead.

## 2   Preliminaries

In this section, we discuss the concepts and formulas of the liquid state machines. In addition, the RCM constraint and its kinematics control problem of redundant manipulators investigated in this paper are also described.

### 2.1   Liquid State Machine

Liquid states machine is a reservoir computing system where many input signals are injected in a nonlinear dynamic system using spiking neurons and then spiking sequences are collected and calculated in an external layer called *readout* [10]. The nonlinear dynamic system has the ability to represent the current and past inputs, which is capable of solving time-series data in [5,20]. LSM usually consists of three layers: input layer, liquid layer and readout layer. Liquid layer generally is composed of Leaky Integrate-and-Fire (LIF) neurons [6] connected in a recurrent way. Readout layer can handle the high-dimensional space of the liquid layer using a linear classifier such as a least squares linear regression [14]. As illustrated in Fig. 1, the input layer propagates the sensorimotor information towards liquid layer at each simulation time step (2 ms). There is no uniform way how to encode sensor data into spiking events, and common methods include population code and probabilistic spike sampling code. In this paper, the input layer is divided into $K$ mossy fiber neuron groups, corresponding to $K$ input signals. Each input neuron group has $L$ unique mossy fiber (MF)

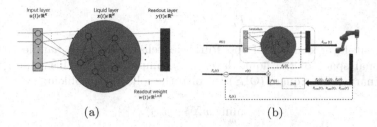

(a)                                    (b)

**Fig. 1.** (a) The diagram of the cerebellum model based on the LSM. (b) The schematic diagram of the proposed cerebellum scheme for the tracking control of a redundant manipulator with RCM constrain.

neurons representing each variable using the simplified population code. Only $K$ MF neurons in the input layer are activated to represent the input signals at each simulation time step.

Liquid layer consists of the granular cells, which are modelled as Leaky Integrate-and-Fire (LIF) neurons, that act as a reservoir and are responsible for remembering past and present coded sensorimotor information from the input layer. There are several spiking neuron models proposed, and [7] compares the biological plausibility and computational efficiency between models. The LIF neuron with exponential synaptic dynamics is used in this work, which can be defined by a set of equations as follows:

$$\frac{dv}{dt} = \frac{i_e(t) + i_i(t) + i_{offset} + i_{noise}}{c_m} + \frac{v_{rest} - v}{\tau_m} \tag{1}$$

$$\frac{di_e}{dt} = -\frac{i_e}{\tau_{syn_e}} \tag{2}$$

$$\frac{di_i}{dt} = -\frac{i_i}{\tau_{syn_i}} \tag{3}$$

where $v$ is the membrane potential, $v_{rest}$ is the resting membrane potential, $\tau_m$ is the membrane time constant, $c_m$ is the membrane capacitance, $i_{offset}$ and $i_{noise}$ can be seen as coming from different noise sources, $\tau_{syn_e}$ is the decay time of the excitatory synaptic current while $\tau_{syn_i}$ is the decay time of the inhibitory one. When the membrane potential of a LIF neuron is greater than the membrane threshold($V_{threshold} = 15mV$), the neuron sends a spike.

Liquid layer is created using $N$ LIF neurons forming a 3D structure where 80% excitatory(E) neurons and 20% inhibitory(I) neurons. The liquid layer is divided into $K$ subgroups corresponding to the number of mossy fiber neurons in the input layer, and mossy fiber neurons are connected in a fully connected manner to the excitatory neurons in the corresponding liquid subgroups. The probability of connection generation between neurons depends on the type and distance of neurons, as shown in (4) where the value of $\lambda$ is equal to 1.2 and $C$ depends on the type of neurons: 0.3(EE), 0.2(EI), 0.4(IE) and 0.1(II).

$$P_{i,j} = Ce^{-\frac{D(i,j)^2}{\lambda^2}}. \tag{4}$$

Readout layer consists of the Purkinje cells, which can handle the high-dimensional space of the liquid layer in a simple way. In this paper, we use the Ordinary Least Square (OLS) to solve the following problem:

$$\min_{w} \|wX - y_l\|_2^2. \tag{5}$$

The matrix $X$ is usually composed of a sequence of the liquid spikes, $y_l$ is the variable values to be learned and $w$ is the readout weight. After the input signal $u(t) \in \mathbb{R}^K$ is fed to input layer, the spiking sequence $x(t) \in \mathbb{R}^N$ at this current moment is obtained and the final output $y(t) \in \mathbb{R}^L$ of the readout layer is model as a weighted sum of the liquid layer:

$$y(t) = w(t)x(t). \tag{6}$$

## 2.2   Manipulator Kinematics Model

For a redundant manipulator with $m$ joints, the forward kinematics model is formulated as follows:

$$f(\theta(t)) = P_a(t) \tag{7}$$

where $\theta(t) \in \mathbb{R}^m$ is the joint angle of the manipulator at the time $t$, $P_a(t) \in \mathbb{R}^d$ is the Cartesian space position and $f(\cdot) : \mathbb{R}^m \implies \mathbb{R}^d$ represents the nonlinear mapping function of the manipulator. For the highly nonlinear of (7), the forward kinematic model is obtained at the velocity level by taking the derivative of (7) as below:

$$J(t)\dot{\theta}(t) = \dot{P}_a(t) \tag{8}$$

where $J \in \mathbb{R}^{d \times m}$ is the Jacobian matrix of the manipulator. The error of the trajectory tracking task is defined as $e_t(t) = P_a(t) - P_d(t)$, where $P_d(t) \in \mathbb{R}^d$ is the desired end-effector position in the Cartesian space.

## 2.3   Remote Center of Motion Constraint

During the surgery procedure, the surgical tooltip of the robot needs to pass through the initial insertion point ($P_{trocar} \in \mathbb{R}^d$) without moving violently with the movement of the robot arm, as shown in Fig. 2(b).

For an m-DoF robot manipulator with a surgical tooltip, the mapping from its joint space to the Cartesian coordinate of the robot arm end-effector $P_m \in \mathbb{R}^d$ and the surgical tooltip end-effector $P_{m+1} \in \mathbb{R}^d$ can be described by the following function:

$$\begin{aligned} P_m &= f_m(\theta), \\ P_{m+1} &= f_{m+1}(\theta). \end{aligned} \tag{9}$$

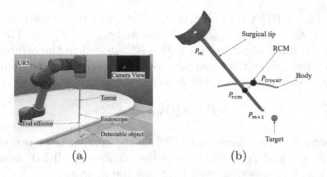

**Fig. 2.** Robotic endoscope system. (a) UR5 robot coupled with a flexible endoscope. (b) Remote center of motion.

In order to satisfy the RCM constraint, $P_{trocar}$ should be as close to the straight line $L_1$ between $P_m$ and $P_{m+1}$. The RCM point $P_{rcm}$ is defined as the projection of the $P_{trocar}$ on the straight line $L_1$ as below:

$$
\begin{aligned}
k &= -((x_m - x_{trocar})(x_{m+1} - x_m) \\
&\quad + (y_m - y_{trocar})(y_{m+1} - y_m) + (z_m - z_{trocar})(z_{m+1} - z_m)), \\
x_{rcm} &= \frac{k(x_{m+1} - x_m)}{(x_{m+1} - x_m)^2 + (y_{m+1} - y_m)^2 + (z_{m+1} - z_m)^2} + x_m, \\
y_{rcm} &= \frac{k(y_{m+1} - y_m)}{(x_{m+1} - x_m)^2 + (y_{m+1} - y_m)^2 + (z_{m+1} - z_m)^2} + y_m, \\
z_{rcm} &= \frac{k(z_{m+1} - z_m)}{(x_{m+1} - x_m)^2 + (y_{m+1} - y_m)^2 + (z_{m+1} - z_m)^2} + z_m
\end{aligned}
\tag{10}
$$

where $P_{rcm} := [x_{rcm}, y_{rcm}, z_{rcm}]^T$.

Then the error of RCM can be represented as follows:

$$
e_r = P_{rcm} - P_{trocar}.
\tag{11}
$$

The $P_{rcm}$ should satisfy the following equation:

$$
J_r(t)\dot{\theta}(t) = \dot{P}_{rcm}(t)
\tag{12}
$$

where $J_r \in \mathbb{R}^{d \times m}$ is the Jacobian matrix corresponding to the RCM point.

## 3    Control Scheme Design

### 3.1    Cerebellum Model

For a redundant manipulator with $m$ joints, a cerebellum-like spiking model based on LSM is created as a controller and computes the joint velocity at this moment, as illustrated in Fig. 1(b). The parameters about (1), (2) and (3) are chosen as: $i_{noise} \sim U(13.5, 14.5)(nA)$, $i_{noise} \sim \mathcal{N}(0, 0.01)(nA)$, $c_m = 30(nF)$, $v_{rest} = 0(mV)$, $\tau_m = 30(ms)$, $\tau_{syn_e} = 3(ms)$ and $\tau_{syn_i} = 6(ms)$. The joint angle $\theta(t) \in \mathbb{R}^m$ as the input signal $u(t) = \theta(t) \in \mathbb{R}^m$ are fed to the cerebellum model and the number of mossy fiber neuron groups $K$ is equal to the number of joint $m$. Next, the cerebellum model will generate the output signal

$$y(t) = x(t)w(t) \in \mathbb{R}^m \tag{13}$$

as joint velocity $\dot{\theta}_{out}(t)$, where $x(t) \in \mathbb{R}^N$ is the spiking sequence of the liquid layer at moment $t$ and $N = 100 \times m$ is the number of all LIF neurons in the liquid layer, which means that each joint has 100 neurons.

However, many previous control methods based on liquid state machines [2, 10] required pre-training the readout weight parameters $w$, which requires to model the robot arm and collect certain data in advance. Since the training of the readout weight is based on supervised learning algorithms, in the following, we will combine zeroing neuron network (ZNN) [21] to present a model-free, pre-training-free and on-line learning approach.

## 3.2   Tracking Control Scheme

According to (8) and (12), we can obtain the two desired joint velocities for the trajectory tracking task and the RCM constraint task, respectively, as below:

$$\dot{\theta}_t(t) = J^\dagger(t)\dot{e}_t(t),$$
$$\dot{\theta}_r(t) = J_r{}^\dagger(t)\dot{e}_r(t) \tag{14}$$

with $(\cdot)^\dagger$ being the pseudo-inverse operation where $\dot{e}_t(t)$ is the derivative of $e_t(t)$ and $\dot{e}_r(t)$ is the derivative of $e_r(t)$. How to handle these two desired joint velocities appropriately is a key point, since the two task affect each other in most cases. Inspired by [15], the following joint velocity solution is designed by:

$$\dot{\theta}_d(t) = \dot{\theta}_t(t) + \tilde{J}^\dagger(t)(\dot{e}_r(t) - J_r(t)\dot{\theta}_t(t)) \tag{15}$$

where

$$\tilde{J} = J_r(I - J^\dagger J) \tag{16}$$

with $I$ being an identity matrix. However, matrices $J$ and $J_r$ require accurate prior modeling of the robotic arm model, and the Jacobian matrix $J_r$ also involve the specific insertion point$P_{trocar}$. Therefore, ZNN is introduced to estimate the Jacobian matrices $\hat{J}$ and $\hat{J}_r$. A vector-valued error function is defined to measure the error when the estimated Jacobian matrix $\hat{J}$ used in (8):

$$\varepsilon(t) = \dot{P}_a(t) - \hat{J}\dot{\theta}(t) \in \mathbb{R}^m. \tag{17}$$

The following equation is defined according to znn design idea:

$$\dot{\varepsilon}(t) = -\phi\varepsilon(t) \tag{18}$$

(a)              (b)              (c)              (d)

**Fig. 3.** Some snapshots of the UR5 manipulator for tracking the circular trajectory

where $\phi$ is a positive parameter for the convergence adjustment and $\dot{\varepsilon}(t)$ is the derivative of $\varepsilon(t)$ with respect to time. Substituting (17) into (18) yields

$$\dot{\hat{J}}(t) = (\ddot{P}_a(t) - \hat{J}(t)\ddot{\theta}(t) + \phi(\dot{P}_a(t) - \hat{J}(t)\dot{\theta}(t)))\dot{\theta}^\dagger(t) \tag{19}$$

where $\dot{\hat{J}}(t)$ represents the derivative of $\hat{J}(t)$ with respect to $t$, $\ddot{P}_a(t)$ is the derivative of $\dot{P}_a(t)$, $\ddot{\theta}(t)$ is the derivative of $\dot{\theta}(t)$ and $\dot{\theta}^\dagger(t)$ represents the pseudo-inverse of the $\dot{\theta}(t)$. Correspondingly, we can similarly obtain the equation for $\hat{J}_r$

$$\dot{\hat{J}}_r(t) = (\ddot{P}_{rcm}(t) - \hat{J}_r(t)\ddot{\theta}(t) + \nu(\dot{P}_{rcm}(t) - \hat{J}_r(t)\dot{\theta}(t)))\dot{\theta}^\dagger(t) \tag{20}$$

where $\dot{\hat{J}}_r(t)$ represents the derivative of $\hat{J}_r(t)$ with respect to $t$, $\ddot{P}_{rcm}(t)$ is the derivative of $\dot{P}_{rcm}(t)$ and $\nu \in \mathbb{R}^+$ is another design parameter of ZNN. Finally, we can obtain the training signal $\dot{\hat{\theta}}_d(t)$ used to calculate the readout weight $w$ of the LSM

$$\begin{aligned} \dot{\hat{\theta}}_d(t) &= \dot{\hat{\theta}}_t(t) + \tilde{\hat{J}}^\dagger(t)(\dot{e}_r(t) - \hat{J}_r(t)\dot{\hat{\theta}}_t(t)), \\ \dot{\hat{\theta}}_t(t) &= \hat{J}^\dagger(t)\dot{e}_t(t), \\ \tilde{\hat{J}} &= \hat{J}_r(I - \hat{J}^\dagger\hat{J}). \end{aligned} \tag{21}$$

In order to achieve real-time training of the readout weight $w(t+1)$, we take the spiking sequences of the liquid layer at the previous $g$ moments to form $X_g \in \mathbb{R}^{g \times N}$ and the corresponding training signals calculated by (21) to form $Y_k = [\dot{\hat{\theta}}_d(t-g+1), ..., \dot{\hat{\theta}}_d(t-1), \dot{\hat{\theta}}_d(t)]^\mathrm{T} \in \mathbb{R}^{g \times L}$, and the readout weight $w(t+1)$ is calculated according to (4).

To summarize the above proposed control scheme, the joint angle as input signal is input into the liquid state machine and the corresponding joint velocities are obtained after a cerebellum-like simulation time. The proposed ZNN combined with RCM constraint is used to get the training signal to iteratively update the parameters of the readout layer of the liquid state machine.

## 4    Simulations and Experiments

### 4.1    Simulations

The liquid state machine is implemented using the Brian2 neuron simulator [16]. To validate the effectiveness of the proposed control scheme, simulations are performed in the virtual robot experimentation platform (V-REP).

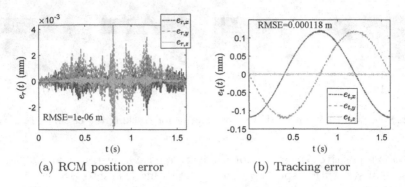

(a) RCM position error          (b) Tracking error

**Fig. 4.** Results of the circular trajectory tracking of the endoscope with RCM constraint

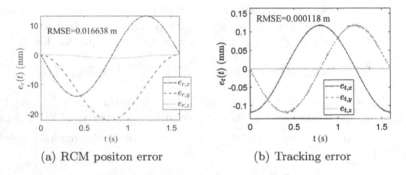

(a) RCM positon error          (b) Tracking error

**Fig. 5.** Results of the circular trajectory tracking of the endoscope without RCM constraint

**Simulation Setting.** As shown in Fig. 2(b), simulations are conducted on a UR5 manipulator with a endoscope in V-REP. The tracking control task is to control the end-effector of the endoscope to move along a circular trajectory. The time step for this tack is set to 2 ms as the LSM simulation time step. Since the sixth joint of UR5 is a rotating joint, it does not affect the tracking task and RCM constraint, so it is set to a constant value (0 rad). The initial joint angles of the robotic arm are set as $\theta(0) = [0, \pi/6, 2\pi/9, \pi/9, -\pi/2]^T$ rad. The ZNN design parameters of (19) and (20) are set as $\phi = \nu = 0.01$. The relevant parameters of the structure of LSM also need to be specified. For the UR5 robot arm with six joints, the number of mossy fiber neuron groups is set as $K = 5$, the number of LIF neurons is set as $N = 500$ and the number of the readout layer is $L = 5$, which expressed as first five joint speeds as well as the number of moments to train the readout weight $g = 3$.

**Simulation Results.** Some snapshots of the UR5 manipulator for tracking the circular trajectory are depicted in Fig. 3, where surgical tooltip passed through the trocar point. The specific results of the tracking control task is illustrated in

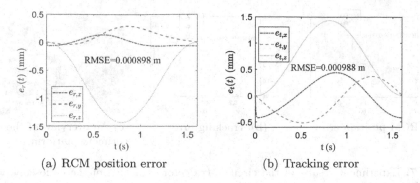

(a) RCM position error                    (b) Tracking error

**Fig. 6.** Results of the circular trajectory tracking synthesized by the accelerated RNN model

Fig. 4. As shown in Fig. 4(b), the proposed control method has achieved a root mean squared error(RMSE) of $1.18 \times 10^{-4}$ m in the trajectory tracking task. Figure 4(a) shows that the RCM point is at a small distance from the starting point during the motion with the RMSE being $1.0 \times 10^{-6}$ m. Meanwhile, in order to verify the effectiveness of the RCM constraint proposed in (21), we change the training signal of the liquid state machine to $\dot{\hat{\theta}}_d(t) = \dot{\hat{\theta}}_t(t)$ to obtain the results shown in Fig. 5. The results of the simulation without considering RCM constraint show that the RMSE of trajectory tracking is $1.18 \times 10^{-4}$ m as shown in Fig. 5(b), which is similar to the case with RCM constraint. However, as shown in Fig. 5(a), the RCM error ($1.66 \times 10^{-2}$ m) is much larger than that of the case considering RCM constraint. It is undesirable in surgical applications. These simulation results have verified the effectiveness of the proposed method in tracking control task and complying RCM constraint.

**Comparison.** To further reveal the advantage of the proposed cerebellum-inspired control scheme, we adopt a model-based approach [8] based on accelerated RNN model as comparison. The parameters of the accelerated RNN model is $\rho = \zeta = \xi = 1$ and $p = 0.5$. As shown in Fig. 6(b), the accelerated RNN model has achieved an RMSE of $9.88 \times 10^{-4}$ m in the tracking task. However, the RMSE of the RCM constraint is $8.98 \times 10^{-4}$ m, which is larger than that of the proposed cerebellum-inspired control scheme ($1.0 \times 10^{-6}$ m). This illustrates the advantage of the proposed method since the smaller the RCM error, the smaller the possibility of secondary injury to the surgical wound of patients.

## 4.2   Experiment

Physical experiment based on the KINOVA JACO Gen3 manipulator with a endoscope is conducted to verify the efficacy of the proposed control scheme as shown in Fig. 7(c). The initial joint angles of the Jaco[3] are set as $\theta(0) = [8.56, 2.91, 171.84, 267.71, 2.38, 274.63]^{\mathrm{T}}$ degree. The ZNN design parameters of

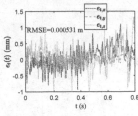

(a) RCM position error

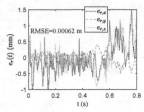

(b) Tracking error

(c) An overview of the experiment platform.

**Fig. 7.** Experiment results of the circular trajectory tracking of the endoscope with RCM constraint

(19) and (20) are set as $\phi = \nu = 0.0005$. Experimental results can be found in Fig. 7. As shown in Fig. 7(a), the manipulator has achieved the trajectory tracking task successfully with an RMSE of $5.31 \times 10^{-4}$ m. It can be seen that the RCM error is maintained in a small range with the RMSE being $6.2 \times 10^{-4}$ m as depicted in Fig. 7(b). In summary, the experiment results have also validated the efficacy of the proposed control method for robot-assisted minimally invasive surgery.

## 5   Conclusion

In this paper, we have presented a cerebellum-inspired network based on LSM. Then, based on the cerebellum-inspired network, a model-free tracking control scheme has been proposed for robotic manipulators in minimally invasive surgery. The control scheme is capable of dealing with two vital problems simultaneously, namely trajectory tracking and RCM constraint. Simulations and experiments have been designed to verify the efficacy of the proposed cerebellum-inspired scheme in the tracking control task with RCM constraint. The results have shown that the robot can achieve the trajectory tracking task successfully as well as complying the RCM constraint by means of the proposed scheme. The advantage of the proposed method has also been revealed by comparison against existing method.

## References

1. Aghakhani, N., Geravand, M., Shahriari, N., Vendittelli, M., Oriolo, G.: Task control with remote center of motion constraint for minimally invasive robotic surgery. In: IEEE International Conference on Robotics and Automation, pp. 5807–5812 (2013)
2. de Azambuja, R., Klein, F.B., Adams, S.V., Stoelen, M.F., Cangelosi, A.: Short-term plasticity in a liquid state machine biomimetic robot arm controller. In: International Joint Conference on Neural Networks (IJCNN), pp. 3399–3408 (2017)

3. Bihlmaier, A.: Learning Dynamic Spatial Relations: The Case of A Knowledge-Based Endoscopic Camera Guidance Robot. Springer, Cham (2016). https://doi.org/10.1007/978-3-658-14914-7
4. Bouteraa, Y., Abdallah, I.B., Ghommam, J.: Task-space region-reaching control for medical robot manipulator. Comput. Electr. Eng. **67**, 629–645 (2018)
5. Burgsteiner, H., Kröll, M., Leopold, A., Steinbauer, G.: Movement prediction from real-world images using a liquid state machine. Appl. Intell. **26**(2), 99–109 (2007)
6. Gerstner, W., Kistler, W.M.: Spiking neuron models: single neurons, populations, plasticity. In: Spiking Neuron Models: Single Neurons, Populations, Plasticity. Cambridge University Press, Cambridge (2002)
7. Izhikevich, E.: Which model to use for cortical spiking neurons? IEEE Trans. Neural Networks **15**(5), 1063–1070 (2004)
8. Li, W., Chiu, P.W.Y., Li, Z.: An accelerated finite-time convergent neural network for visual servoing of a flexible surgical endoscope with physical and RCM constraints. IEEE Trans. Neural Netw. Learn. Syst. **31**(12), 5272–5284 (2020)
9. Maass, W.: Networks of spiking neurons: the third generation of neural network models. Neural Netw. **10**(9), 1659–1671 (1997)
10. Maass, W., Natschläger, T., Markram, H.: Real-time computing without stable states: a new framework for neural computation based on perturbations. Neural Comput. **14**(11), 2531–2560 (2002)
11. Roy, K., Jaiswal, A., Panda, P.: Towards spike-based machine intelligence with neuromorphic computing. Nature **575**(7784), 607–617 (2019)
12. Sadeghian, H., Zokaei, F., Hadian Jazi, S.: Constrained kinematic control in minimally invasive robotic surgery subject to remote center of motion constraint. J. Intell. Robot. Syst. **95**(3), 901–913 (2019)
13. Sandoval, J., Su, H., Vieyres, P., Poisson, G., Ferrigno, G., Momi, E.D.: Collaborative framework for robot-assisted minimally invasive surgery using a 7-DoF anthropomorphic robot. Robot. Auton. Syst. **106**, 95–106 (2018)
14. Schürmann, F., Meier, K., Schemmel, J.: Edge of chaos computation in mixed-mode vlsi - "a hard liquid", pp. 1201–1208 (2004)
15. Siciliano, B., Slotine, J.J.: A general framework for managing multiple tasks in highly redundant robotic systems. In: Fifth International Conference on Advanced Robotics Robots in Unstructured Environments, vol. 2, pp. 1211–1216 (1991)
16. Stimberg, M., Brette, R., Goodman, D.: Brian 2: an intuitive and efficient neural simulator (2019)
17. Tan, N., Yu, P.: Robust model-free control for redundant robotic manipulators based on zeroing neural networks activated by nonlinear functions. Neurocomputing **438**, 44–54 (2021)
18. Tan, N., Yu, P., Ni, F.: A cerebellum-inspired network model and learning approaches for solving kinematic tracking control of redundant manipulators. IEEE Trans. Cogn. Dev. Syst. 1–12 (2022)
19. Taniguchi, K., Nishikawa, A., Sekimoto, M., Kobayashi, T., Miyazaki, F.: Classification, design and evaluation of endoscope robots. Robot Surg. **1**, 172 (2010)
20. Zhang, Y., Li, P., Jin, Y., Choe, Y.: A digital liquid state machine with biologically inspired learning and its application to speech Recognition. IEEE Trans. Neural Netw. Learn. Syst. **26**(11), 2635–2649 (2015)
21. Zhang, Y., Ma, W., Yi, C.: The link between newton iteration for matrix inversion and zhang neural network (ZNN). In: 2008 IEEE International Conference on Industrial Technology, pp. 1–6 (2008)

# Instrumental Conditioning with Neuromodulated Plasticity on SpiNNaker

Pavan Kumar Enuganti[1], Basabdatta Sen Bhattacharya[1,2]([✉]) [iD],
Andrew Gait[2] [iD], Andrew Rowley[2] [iD], Christian Brenninkmeijer[2],
Donal K. Fellows[2], and Stephen B. Furber[2]

[1] Birla Institute of Technology and Science Pilani, Goa Campus, Sancoale, India
{p20180053,basabdattab}@goa.bits-pilani.ac.in
[2] University of Manchester, Manchester, UK
{andrew.gait,andrew.rowley,christian.brenninkmeijer,donal.k.fellows,
steve.furber}@manchester.ac.uk
https://www.binnlabs-goa.in/,
http://apt.cs.manchester.ac.uk/projects/SpiNNaker/

**Abstract.** We present a work-in-progress on implementing reinforcement learning by instrumental conditioning on SpiNNaker. Animals learn to behave by exploring the changing environment around them such that, over a period of time, their behaviour gives a good outcome (reward) i.e. a perception of 'satisfaction'. While inspired by animal learning, reinforcement learning adopts a goal-directed strategy of maximising rewards in a dynamic environment. Instrumental conditioning is a strategy to strengthen the association between an action and the environmental state when the state-action pair is rewarded i.e. the reward is instrumental in forming the association. However, in the real world, the delivery of a reward is often delayed in time, known as the distal reward problem. Using the concept of eligibility traces and spike-time dependant plasticity (STDP), Izhikevich (2007) simulated both classical and instrumental conditioning in a spiking neural network with Dopamine (DA)-modulated STDP. The current implementation of DA-modulated plasticity on SpiNNaker using trace-based STDP is reported by Mikaitas et al. (2018), who demonstrated classical conditioning with a similar experimental set up as Izhikevich. Our results show that using delayed DA-modulation of STDP on SpiNNaker, we can condition a neural population to maximise its reward over a period of time by firing at a higher rate than another competing population. Ongoing work is looking into a dynamic conditioning scenario where different actions can be selected within the same run as is the case in real world scenarios.

BSB and EPK are supported by Science and Engineering Research Board, Department of Science and Technology, Govt. of India, Grant no. CRG/2019/003534. AG, AR, CB, DKF and SBF are supported by European Union's Horizon 2020 Framework Programme for Research and Innovation under the Specific Grant Agreement No. 945539 (Human Brain Project SGA3).

M. Tanveer et al. (Eds.): ICONIP 2022, LNCS 13624, pp. 148–159, 2023.
https://doi.org/10.1007/978-3-031-30108-7_13

**Keywords:** Instrumental conditioning · Neuromodulated plasticity · Balanced random network · Dopamine-modulated STDP · SpiNNaker · Delayed reward

# 1 Introduction

Behavioural learning in animals evolves by exploring and interacting with their immediate environment in a closed-loop manner. While exploring several possible actions in an unknown environmental state (S), a 'good' outcome (R) after taking a certain action (A) strengthens an association between the state S and action A, such that the animal is likely to repeat the 'behaviour', i.e. selecting A when repeat encountering S. Conditioning experiments are used by psychologists and neuroscientists to understand how animals learn to behave in unknown environments. One such paradigm is instrumental conditioning, where animals are trained by 'reinforcing', their good behaviour by delivering a preferred food or beverage. Over a period of time, by trial-and-error, the animals learn to optimise their behaviour such as to obtain the maximum reinforcement, commonly called reward (R). This biological strategy to learn to adapt to, and navigate in, new environments forms the inspiration for the field of Reinforcement Learning (RL), where the goal has been to build algorithms for machines that can navigate the environment to optimise their rewards in the long term [19]. The notion of training a robot by instrumental conditioning was investigated in [21], which in turn was inspired by the works of B.F. Skinner (1963) on 'operant' (a nomenclature that is conceptually similar to instrumental) conditioning. The authors in [21] coined the term 'Skinnerbot' for training a robot using strategies adopted during instrumental conditioning of animals by human trainers. We have been working on building brain-inspired frameworks for action-selection on SpiNNaker [18], a neuromorphic hardware that has potential for low power robotic applications [7]. However, our previous application on SpiNNaker was 'hardwired' (static) to associate with a stimulus [18]. The brain is known to learn adaptively by forming (discarding) new (unused) connections between its neurons, a phenomenon that is termed as 'synaptic plasticity' [12]. In this work, we present a work-in-progress on implementing instrumental conditioning on SpiNNaker by parameterising a balanced random network with conductance-based Izhikevich's neuron models as its compute nodes, and neuromodulated plasticity as implemented in the toolchain sPyNNaker [17].

The environment around us is noisy; by the time an animal receives R for A while in some $S_1 \in S$, the environment will have changed to $S_2 \in S$. In behavioural literature, this is known as the distal (delayed) reward problem [11], whereby the brain needs to work out the causality between R, A and $S_1$, even if R was received while in $S_2$. This led to the proposition that neurotransmitters in the brain leave 'traces' corresponding to the pair $(S_1,A)$ that facilitate the assigning of credit to this state-action pair for R. In RL, the distal reward problem is referred to as the credit assignment problem, and can be addressed by using the concept of 'eligibility traces' [1,22]. Thus, every time there is an

R caused by the state-action pair $(S_1, A)$, the synaptic trace of this activity is likely to be higher than all other state-action pairs, thus making it eligible for assigning the credit, albeit with a degree of uncertainty that is implicit in a noisy environment. The neurotransmitter Dopamine (DA) in the brain is now known to cause 'satisfaction' after a good outcome (R) corresponding to $(S_1, A)$ by 'modulating' (controlling) synaptic targets in specific areas of the brain. Such a mechanism is referred to as DA-modulated (also, neuromodulated) synaptic plasticity. To the best of our knowledge, Izhikevich (2007) [11] for the first time addressed the distal reward problem in a DA-modulated spiking neural network (SNN), demonstrating RL using both classical (a reward prediction conditioning strategy, also called Pavlovian) and instrumental conditioning.

Following the experimental set up in [11], classical conditioning using DA-modulated plasticity was first demonstrated in [13]. While classical conditioning is suited for reward prediction, it is not straightforward to use it for action-selection. Our interest is in implementing the action-selection mechanism on SpiNNaker using DA-modulated plasticity for instrumental conditioning. Furthermore, we have been using an implementation of conductance-based Izhikevich (IZK) neuron models on SpiNNaker [5,18]. Besides computing efficiently, IZK neurons can be used to simulate a wide range of spike patterns as observed in the brain; also, the parameter space corresponding to the rich repertoire of spike dynamics are well known owing to several previous works. However, all our previous applications with the IZK neurons were with static synapses. The work presented here is the first implementation of plastic synapses on an IZK neuron-based network on SpiNNaker. Our experimental set up for demonstrating instrumental conditioning in a balanced random spiking neural network is adapted from [11]. Two sub-populations of neurons as a part of a larger population, represent two competing 'actions' (A and B). We (i.e. the human trainers) decide on a preferred 'behaviour' (or 'policy' in RL nomenclature), say, selecting A over B; we provide reinforcement R only if A is selected, and at a delayed time, i.e. after 'evaluating' the network behaviour at regular intervals during the simulation; If the network doesn't follow the specified behaviour, no R is delivered. Preliminary results from our work show instrumental conditioning in the network—over a period of time, the network learns to 'behave' such as to maximise the rewards by selecting the preferred action. Furthermore, the network remembers this selected action even if we stop rewarding after a period of training. In the next phase of this work, we aim to simulate the exact instrumental conditioning demonstrations in [11] on SpiNNaker, where policies can be changed dynamically in the network. Our short term goal is to map these strategies onto an existing brain-inspired architecture on SpiNNaker [18] using DA-modulated plasticity as implemented on sPyNNaker.

In Sect. 2 we present the relevant background to this work. In Sect. 3, we present the simulation methodologies and the results from our work-in-progress. In Sect. 4, we summarise and critique our work and outline future directions.

## 2   Background

For readability, below we explain briefly the brain inspired theory of spike-time dependant plasticity (STDP) that forms the substrate for neuromodulated plasticity in our spiking neural network. Our narrative is in the context of reinforcement learning by instrumental conditioning, and existing implementations on SpiNNaker.

### 2.1   STDP and Synaptic Trace

STDP is a widely used technique to incorporate adaptability, i.e. plasticity, in spiking neural networks, and is supported by physiological evidence [2]. The underlying theory is similar to Hebbian correlation-based learning; the connection strength between two neurons that fire closer in time are modified based on the temporal order of their firing. If a pre-synaptic neuron fires before a post-synaptic neuron, and if the time-interval between their two spikes is within an 'eligible' time-window, then the synaptic strength between the two neurons is increased, thereby strengthening their association, a phenomenon that is called long term potentiation (LTP). Unlike the Hebbian learning though, in the converse case of the post- firing before pre-synaptic neuron, the implicit assumption is that there is a lack of association between the two neurons, thereby decreasing the synaptic strength between them; this is called long term depression (LTD). The change in synaptic strength, commonly referred to as 'synaptic weight', due to STDP is an exponentially decreasing function of the temporal order as well as the distance between the pre- and the post-synaptic spikes [14]. A biologically plausible implementation of STDP is by means of 'trace' variables that simulate the gradual decay of neurotransmitter concentration in the synaptic cleft, and is defined below [12]:

$$\frac{ds_i}{dt} = -\frac{s_i}{\tau_+} + \sum_{t_i^f} \delta(t - t_i^f) \tag{1}$$

$$\frac{ds_j}{dt} = -\frac{s_j}{\tau_-} + \sum_{t_j^f} \delta(t - t_j^f) \tag{2}$$

$$\Delta w_{ij}^-(t_i^f) = F_-(w_{ij})s_j(t_i^f) \tag{3}$$

$$\Delta w_{ij}^+(t_j^f) = F_+(w_{ij})s_i(t_j^f) \tag{4}$$

where $s_{i/j}$ are pre- and post-synaptic trace variables; $\tau_{+/-}$ are time constants corresponding to post-synaptic decay of neurotransmitter concentration in the synaptic cleft; $t_i^f$ $t_j^f$ are pre- and post-synaptic spike times respectively; the Dirac delta function ($\delta$) denote a spike; $\Delta w_{ij}^-$ and $\Delta w_{ij}^+$ are LTD and LTP respectively; $F_\pm$ define the amplitudes of the weight changes and are a function of the current weights.

Neuromodulators in the brain facilitate STDP-based learning by 'reshaping' the eligibility windows of LTP and LTD during which the weight changes are effective [4,15]. In the following section, we describe how the neuromodulation of STDP by DA addresses the credit assignment problem of RL and facilitates instrumental conditioning in the brain.

## 2.2  Dopamine and Neuromodulated Plasticity

A major issue in real world applications of RL is that of delayed reinforcement, for example when an animal decides on 'fight or flight' in response to an environmental situation, the 'perception' (good or bad) of the outcome, is delayed in time. However, several other incidents may have happened in the meanwhile, and yet, the animal learns to associate the outcome to the specific response under a given circumstance. The theory of eligibility trace is one way to address the problem of correctly assigning credit in spite of temporal delay between action and reinforcement [20]. According to this theory [1]:

> "Whenever a neuron fires, those synapses that were active during the summation of potentials leading to the discharge become eligible to undergo changes in their transmission effectiveness. If the discharge is followed by further depolarization, then the eligible excitatory synapses become more excitatory. If the discharge is followed by hyperpolarization, then eligible inhibitory synapses become more inhibitory. In this way a neuron will become more likely to fire in a situation in which firing is followed by further depolarization and less likely to fire in a situation in which firing leads to hyperpolarization."

It is now known that delayed reinforcement in the brain is facilitated by DA, and the timing of dopamine release is thought to be crucial in behavioural instrumental conditioning [22]. In the context of STDP, the concentration of DA is reported to rise steeply within tens of milliseconds, followed by a rapid decay. These concepts were used in [11] to demonstrate both classical and instrumental conditioning using the following equations:

$$\dot{c} = -c/\tau_c + STDP(\tau)\delta(t - t_{pre/\ post}) \tag{5}$$

$$\dot{g} = c \cdot d \tag{6}$$

$$\dot{d} = -d/\tau_d + DA(t) \tag{7}$$

where $c$ is the eligibility trace variable; $g$ is the synaptic weight; $d$ is the extracellular DA concentration variable; $\tau_d$ is the time constant for DA uptake after the synapse; $DA(t)$ simulates the DA concentration in the extracellular space. Interested readers may refer to [11] for further details. Below we explain the implementations on SpiNNaker.

## 2.3  STDP and DA-modulation on SpiNNaker

SpiNNaker [9] is a neuromorphic computer made up from chips containing up to 18 ARM-968 200 MHz CPUs with 64 KB of local data memory and 32 KB of local instruction memory for each core, coupled with 128 MB of SDRAM shared between the cores. Boards of 48 such chips then couple together the chips with a unique network architecture designed to multicast small messages, akin to spikes in a neural network, to multiple targets simultaneously, replicating the highly connected networks in the brain. The largest SpiNNaker system to date has been built in Manchester from 1200 boards, making a single machine with more than 1 million cores. Neural network simulation software for SpiNNaker, called sPyNNaker [17], uses the PyNN [6] neural network language with some extensions.

The implementation of STDP on SpiNNaker is trace-based as indicated in Eqs. (1)–(4) and was introduced for the first time by [8]. The current implementation on SpiNNaker is as in [12], which is an improvement in terms of algorithmic complexity over the first implementation. All applications on SpiNNaker using STDP has thus far used Leaky-integrate-and-fire (LIF) neuron models. This is the first formal work where we have implemented STDP in a IZK neuron based network on SpiNNaker.

The DA-modulated STDP by Mikaitis et al [13] was a significant addition to sPyNNaker. The equations implementing eligibility trace based DA reinforcement are similar to Eqs. (5)–(7), after [11]. However, because of the event-driven nature of processing on SpiNNaker, the equations needed adaptation. The specific algebraic derivations and algorithmic implementations are discussed in detail in the aforementioned work [13]. The implementation is designed with its modularity in mind, whereby neuromodulation can simply be added to an existing STDP network with little modification. At the time of writing this paper, the neuromodulation implementation was made available in the development version of sPyNNaker and testing of this has been added to the daily integration testing of the software to ensure it continues to work after future changes and in future releases; the experiments in this work were performed using the "master" branches on GitHub.

# 3   Methodology and Results

The balanced random network (BRN) proposed by Brunel [3] is widely used to study the dynamics in a closed loop network of excitatory-inhibitory spiking neural populations with recurrent connections. The BRN is implemented on SpiNNaker with LIF neurons. Here, we implement the BRN parameterised for conductance-based IZK neurons. We test the model output dynamics to be within the Asynchronous-Irregular (A-I) regime using two attributes viz. irregularity and synchrony (see Sect. 3.2). Next, we add neuromodulated plasticity in specific network pathways of the BRN to demonstrate the effects of positive (reward) and negative (punishment) reinforcements. We then parameterise the model to demonstrate instrumental conditioning. The simulation design, experimental set up, and results are detailed below.

## 3.1  Simulation Methods

The current implementation for solving the neuron model differential equations on SpiNNaker follows the methods indicated in [10]. The simulation time-step for IZK neurons is 0.1 ms, which executes on SpiNNaker in 1 ms wall-clock time. The synaptic conductance parameters $g_x$ simulate the synaptic 'weights' corresponding to a synaptic projection $x$. To record the progression of plastic weights, the simulation is stopped at every 1000 ms interval within the same run i.e. without resetting the machine. The peri-stimulus time histogram (PSTH) is computed with time bin $\Delta t = 1000$ ms; the firing rate ($\nu_{\Delta t}$) for each time bin is computed thus: $\nu_{\Delta t} = \frac{1}{N.\Delta t} \sum_{t=t'}^{t'+\Delta t} \sum_{n=0}^{N-1} \delta_n(t - t^f)$, where $N$ is the total number of neurons in the population, $t$ denotes time, and $t^f$ is the times in the spike train. To take into account the non-stationary behaviour of the output due to the noisy input, we run 10 trials, each with randomly generated seed for the Poisson input, and average the PSTH across all trials.

The plastic projection parameters defined in Eqs. (1)–(4) are thus:$\{\tau_+ : 1\,ms, \tau_- : 2\,ms, F_+ : 1, F_- : 1\}$. Plastic weight is bound between 0 and $2\,\mu S$. Neuromodulation parameters in Eqs. (5)–(7) are thus:$\{\tau_c : 100\,ms, \tau_d : 5\,ms\}$. The initial concentration of DA is set as $0.05\,\mu M$.

## 3.2  The Balanced Random Network Parameterised for IZK Neurons

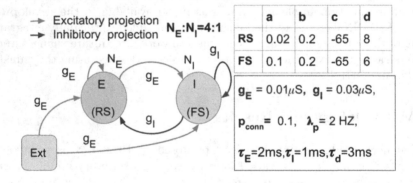

**Fig. 1.** The balanced random network computed with conductance-based Izhikevich's neurons parameterised (a,b,c,d) in the Regular (RS) and Fast (FS) spiking modes respectively for Excitatory (E) and Inhibitory (I) populations. Each synaptic projection is defined by its weight, which are the conductance values ($g_{E/I}$), probability ($p_{conn}$) and delay ($t_d$). The proportion of neurons in the E ($N_E$) and I ($N_I$) populations is in the ration $4:1$. The number of neurons in the Poisson spike train input ($Ext$) to the network is $N_E + N_I$, firing at $\lambda_p = 2$ Hz.

The BRN parameterised for IZK neurons (BRN-IZK) to operate in the A-I regime is shown in Fig. 1. Recently, we have used the conductance-based IZK

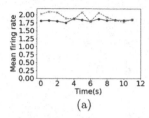 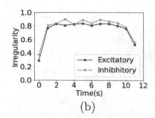 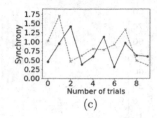

         (a)                          (b)                       (c)

**Fig. 2.** (a) Peri-stimulus time histogram and (b) irregularity for the balanced random network computed over 10 trials, where each trial of T=12 s duration is divided into 12 bins of 1s each and run with random seed for 2 Hz Poisson input. (c) Synchrony computed for 10 trials.

neuron models to propose a reduced-scale cortical network [5] that was set in the A-I regime based on upper- and lower-bounds for two attributes, viz. irregularity and synchrony, as specified in [16]; we use the same bounds in this study. The PSTH in Fig. 2(a) shows a uniform distribution for both the FS and RS neurons with a mean firing rate of $\approx 2$ Hz, similar to the input rate.

For any neuron $(n)$ in a population, the irregularity $(\iota_{\Delta t})$ of its spike train output over a time-bin width of $\Delta t$ is measured as the coefficient of variation of the inter-spike-intervals (isi), which is defined as a ratio of the standard deviation $(\sigma^n_{\Delta t})$ to the mean $(\mu^n_{\Delta t})$ of the isi:$\iota_{\Delta t} = \frac{1}{N}\sum_{n=0}^{N}\frac{\sigma^n_{\Delta t}}{\mu^n_{\Delta t}}$. Bounds for A-I regime is defined as $0.7 < \iota < 1.2$ [5], after [16]. The irregularity histogram for our BRN-IZK is shown in Fig. 2(b). Being a function of isi, the plot is sensitive to the transient effects of the simulation start and end times across all the 10 trials. In the stable region, the measure in our network is an uniform distribution and within the bounds defining the A-I regime.

Synchrony $(\kappa)$ is measured as a dispersion in the spike count histogram $(\eta)$ for each trial run $j$ thus:$\kappa_j = \frac{\sigma^2_{\eta_j}}{\mu_{\eta_j}}$, the numerator(denominator) specifying the variance(mean) of $\eta_j$. The synchrony measured over 10 trial runs is shown in Fig. 2(c), and is within the specified bound $\kappa < 8$ specified for the A-I regime.

### 3.3 Neuromodulated Plasticity in the BRN

This is the first time that the neuromodulated plasticity is being implemented on the BRN presented in Sect. 3.2 on SpiNNaker. Only the recurrent projection of the E population is made plastic and modulated by reward and punishment; all other projections remain static. Readers may note that there is no action-selection in this example; the intention is to observe the effects of DA-modulated STDP in a tractable manner. Towards this, reward (DA) is applied at the $2^{nd}$, $3^{rd}$ and $4^{th}$ second for a duration of 1, 10 and 100 ms respectively; punishment is applied for similar progressively increasing duration respectively at the $6^{th}$, $7^{th}$ and $8^{th}$ second. The total simulation duration is $T = 12.5$ s. Such fixed duration reward and punishment is as during the initial demonstration of the neuromodulated plasticity implementation on SpiNNaker using a basic pre-post population

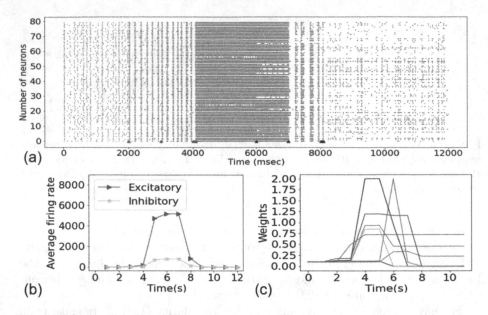

**Fig. 3.** (a) Spike raster of the E population of the BRN in Fig. 1; the green and red triangles indicate the times of reward and punishment respectively. (b) The PSTH of both E and I populations with 1000 ms time-bin and averaged across 10 trials, demonstrating increase (decrease) in firing rate corresponding to DA-modulated reward (punishment). (c) Weight progression for a few of the plastic projections in the network showing increase and decrease corresponding to reward and punishment respectively, and clipped at $2\,\mu S$.

set up computed with LIF neurons [13]. In this work, we have parameterised the BRN to demonstrate similar behaviour as in the initial demonstrative framework. The results are shown in Fig. 3.

The spike raster of the E Population of the BRN and its PSTH in Figs. 3(a) and (b) respectively show the network responding to the rewards (increase in firing rate) and punishments (decrease in firing rate). The corresponding progression of the plastic weights recorded at 1000 ms time bins is shown in Fig. 3 (c); for readability, we have shown only a few projections selected randomly from the full list generated during a single trial run. Due to the unsupervised nature of the STDP algorithm, the weight progressions are not deterministic, i.e. all weights are not guaranteed to increase, as can be seen in Fig. 3 (c); also, the responses to reward and punishment by individual projections are not synchronised in time and amplitude.

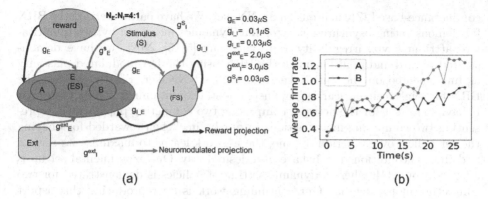

**Fig. 4.** (a) The instrumental conditioning set up and parameter values. (b) PSTH of the A and B populations corresponding to reward for firing rate $\nu_A > \nu_B$.

### 3.4 Instrumental Conditioning in the BRN

The set up for instrumental conditioning experiment with the BRN is shown in Fig. 4 and is adapted from [11]. Two populations A and B with number of neurons $N_A = N_B = \frac{1}{10}(N_E + N_I)$ is randomly selected from E, and their respective firing rates $(\nu_{A/B})$ observed for action-selection, where each of these populations is assumed to correspond to a motor task. A periodic stimulus (S) is provided as input to both E and I populations, for a duration of 50 ms with isi of 20; the S to E projection is made plastic, and modulated by DA. An 'action' preference, i.e. a behavioural policy is set for the network, say $\nu_A > \nu_B$. If this condition is satisfied at any observation instant $\Delta t_1$ then all of the E population is rewarded at a delay of $\Delta t_1 + 10$ ms, i.e. delayed reward. The reward is delivered by a neural population with same number of neurons as the E population, and is activated by a current pulse of amplitude 10 nA and width 4 ms. The network is then allowed to run freely for another 200 ms before repeating the cycle of applying S and evaluating the network behaviour. Note that, if the reward criteria is not met, we do not punish the network, which is similar to the experimental methods in [11]. We observe that over a period of time, the network learns to maximise the reward, and the preferred 'action' is selected, indicated by a higher PSTH in Fig. 4(b). At this point, even if the delivery of reward is stopped, the circuit 'remembers' the associations and continues to prefer the 'action' A. We have also tested the network for the converse situation where $\nu_A < \nu_B$; not shown here for brevity.

## 4 Conclusions

We have presented a work-in-progress on implementing instrumental conditioning in a balanced random network (BRN) on SpiNNaker using conductance-based Izhikevich's (IZK) neuron models. This is the first implementation of BRN (a popular spiking neural network proposed by N. Brunel [3]) with conductance-based IZK neurons on SpiNNaker. We have parameterised the BRN for demonstrating asynchronous-irregular dynamics measured by two quantitative attributes viz. irregularity and synchrony [5,16]. Next, we have demonstrated reward and punishment in the BRN using the DA-modulated plasticity as implemented on the SpiNNaker toolchain sPyNNaker. Finally, we test the BRN for reinforcement learning by instrumental conditioning. A policy (desired behaviour) is set for the network comprising two competing populations, representing two competing motor actions in our brain. When rewarded for obeying the set policy, over a period of time, the network learns to maximise its reward and thus, is conditioned to behave in a desired way. Our experimental set up is adapted from [11], where a dynamic setting of policies is demonstrated for two competing motor actions. Our continuing work is on reproducing this aspect on the SpiNNaker implementation presented here. In the short term, our objective will be to implement DA-modulated instrumental conditioning in an existing brain-inspired model on SpiNNaker [18], where indeed action-selection is demonstrated, but there is no plasticity in the network, unlike in the brain, and the DA-modulation is simulated by direct scaling of excitatory and inhibitory synapses. We believe our developing work will contribute to research in reinforcement learning by instrumental conditioning, a promising direction for brain-inspired robotics.

# References

1. Barto, A.G., Sutton, R.S., Brouwer, P.S.: Associative search network: a reinforcement learning associative memory. Biol. Cybern. **40**(3), 201–211 (1981)
2. Bi, G., Poo, M.: Synaptic modifications in cultured hippocampal neurons: dependence on spike timing, synaptic strength, and postsynaptic cell type. J. Neurosci. **18**(24), 10464–10472 (1998)
3. Brunel, N.: Dynamics of sparsely connected networks of excitatory and inhibitory spiking neurons. J. Comput. Neurosci. **8**(3), 183–208 (2000)
4. Brzosko, Z., Mierau, S.B., Paulsen, O.: Neuromodulation of spike-timing-dependent plasticity: Past, present, and future. Neuron **103**(4), 563–581 (2019)
5. Chiplunkar, C., et al.: A reduced-scale cortical network with Izhikevich's neurons on spinnaker. In: 2021 International Joint Conference on Neural Networks (IJCNN), pp. 1–8. IEEE (2021)
6. Davison, A.P., et al.: PyNN: a common interface for neuronal network simulators. Front. Neuroinform. **2**, 11 (2009)
7. Denk, C., Llobet-Blandino, F., Galluppi, F., Plana, L.A., Furber, S., Conradt, J.: Real-time interface board for closed-loop robotic tasks on the SpiNNaker neural computing system. In: Mladenov, V., Koprinkova-Hristova, P., Palm, G., Villa, A.E.P., Appollini, B., Kasabov, N. (eds.) ICANN 2013. LNCS, vol. 8131, pp. 467–474. Springer, Heidelberg (2013). https://doi.org/10.1007/978-3-642-40728-4_59
8. Diehl, P.U., Cook, M.: Efficient implementation of STDP rules on spinnaker neuromorphic hardware. In: 2014 International Joint Conference on Neural Networks (IJCNN), pp. 4288–4295. IEEE (2014)
9. Furber, S.B., Galluppi, F., Temple, S., Plana, L.A.: The spinnaker project. Proc. IEEE **102**(5), 652–665 (2014)

10. Hopkins, M., Furber, S.: Accuracy and efficiency in fixed-point neural ODE solvers. Neural Comput. **27**(10), 2148–2182 (2015)
11. Izhikevich, E.M.: Solving the distal reward problem through linkage of STDP and dopamine signaling. Cereb. Cortex **17**(10), 2443–2452 (2007)
12. Knight, J.C.: Plasticity in large-scale neuromorphic models of the neocortex. PhD Thesis, The University of Manchester, UK (2016)
13. Mikaitis, M., Garcia, G.P., Knight, J., Furber, S.: Neuromodulated synaptic plasticity on the spinnaker neuromorphic system. Front. Neurosci. **30**(30), 10127–10134 (2018)
14. Morrison, A., Aertsen, A., Diesmann, M.: Spike-timing-dependent plasticity in balanced random networks. Neural Comput. **19**(6), 1437–1467 (2007)
15. Pedrosa, V., Clopath, C.: The role of neuromodulators in cortical plasticity. A computational perspective. Front. Synaptic Neurosci. **8**(38), 1–9 (2017)
16. Potjans, T.C., Diesmann, M.: The cell-type specific cortical microcircuit: relating structure and activity in a full-scale spiking network model. Cereb. Cortex **24**(3), 785–806 (2014)
17. Rhodes, O., et al.: sPyNNaker: a software package for running PyNN simulations on spinnaker. Front. Neurosci. **12**, 816 (2018)
18. Sen Bhattacharya, B., et al.: Building a spiking neural network model of the basal ganglia on spinnaker. IEEE Trans. Cognitive Dev. Syst. **10**(3), 823–836 (2018)
19. Sutton, R.S., Barto, A.G.: Reinforcement Learning: An Introduction. MIT press, Cambridge (2018)
20. Sutton, R.S.: Temporal credit assignment in reinforcement learning. PhD Thesis, University of Massachusetts Amherst, USA (1984)
21. Touretzky, D.S., Saksida, L.M.: Operant conditioning in skinnerbots. Adapt. Behav. **5**(3–4), 219–247 (1997)
22. Wickens, J., Kötter, R.: Cellular models of reinforcement. In: James C. Houk, Joel L. Davis, D.G.B. (ed.) Models of Information Processing in the Basal Ganglia. The MIT Press, Cambridge, November 1994

# A Phenomenological Deep Oscillatory Neural Network Model to Capture the Whole Brain Dynamics in Terms of BOLD Signal

Anirban Bandyopadhyay[1], Sayan Ghosh[1], Dipayan Biswas[1],
Raju Bapi Surampudi[2], and V. Srinivasa Chakravarthy[1(✉)]

[1] Computational Neuroscience Lab, Department - Biotechnology,
Indian Institute of Technology Madras, Chennai, India
schakra@ee.iitm.ac.in
[2] Brain,Cognition, and Computation Lab, International Institute of Information
Technology Hyderabad, Hyderabad, India

**Abstract.** A large-scale model of brain dynamics, as it is manifested in functional neuroimaging data, is presented in this study. The model is built around a general trainable network of Hopf oscillators, the dynamics of which are described in the complex domain. It was shown earlier that when a pair of Hopf oscillators are coupled by power coupling with a complex coupling strength, it is possible to stabilize the normal phase difference at a value related to the angle of the complex coupling strength. In the present model, the magnitudes of the complex coupling weights are set using the Structural Connectivity information obtained from Diffusion Tensor Imaging (DTI). The complex-valued outputs of the oscillator network are transformed by a complex-valued feedforward network with a single hidden layer. The entire model is trained in 2 stages: in the $1^{st}$ stage, the intrinsic frequencies of the oscillators in the oscillator network are trained, whereas in the $2^{nd}$ stage, the weights of the feedforward network are trained using the complex backpropagation algorithm. The Functional Connectivity Matrix (FCM) obtained from the network's output is compared with empirical Functional Connectivity Matrix, a comparison that resulted in a correlation of 0.99 averaged over 5 subjects.

**Keywords:** BOLD Signal · Functional Connectivity · Hopf Oscillator

## 1 Introduction

Recent advancements in neuroimaging techniques have opened new opportunities in basic and clinical neuroscience, and have inspired a large body of computational modeling literature. The BOLD (Blood-Oxygen-Level Dependent) signal

Supported by DBT,Govt. of India, Centre of Computational System and Dynamics, IIT Madras.

measured by fMRI (functional Magnetic Resonance Imaging) is widely used to understand the nature of neural activity underlying hemodynamic changes in the brain under normal and pathological conditions like Ischemic Cerebral Stroke, Traumatic Brain injury (TBI) etc [1]. On the other hand, MRI also has been used to understand the structural connectivity between different brain regions with the help of Diffusion Tensor Imaging (DTI). The current study intends to develop a phenomenological model of the whole brain to understand the BOLD signal along with the functional connectivity associated with it.

Several computational and mathematical models have been developed over the last decade to understand the relation between the functional activity of the brain, and the BOLD signal [2]. They can be categorized into three main types—1) single neuron—based models (often integrated with the neurovascular coupling), 2) neural mass models, and 3) abstract models like the non-linear oscillator models [2,3]. Our current model falls under the last category of an abstract network of non-linear oscillators, wherein each brain region is modeled by a single Hopf oscillator. The network architecture which consists of a layer of Hopf oscillators with lateral connections, followed by a feedforward neural network with a single hidden layer of sigmoidal neurons,is used to simulate the BOLD signal recorded from the whole brain. The network is trained in two stages: in the $1^{st}$ stage, the intrinsic frequencies of the oscillators are trained, and in the $2^{nd}$ stage, the feedforward network is trained by supervised learning. The lateral connections among the oscillators are set using the structural connectivity information obtained from DTI.

The current paper is divided into five sections—first, the introduction section; in the second section, the model development section that presents equations for model dynamics and learning; the third section provides the simulation results, the fourth presents the discussion, and the last one outlines the conclusions and future goals.

## 2  Mathematical Model

### 2.1  Database Used

These days fMRI data is widely available for research in public repositories. However, we are using the processed data from the paper by Morellec et al. [4]. [The dataset can be found here - https://figshare.com/articles/dataset/Paris HCP brain_connectivity data/3749595 This repository has data from 40 unrelated participants, collected over approximately 55 min long sessions, taken with a repetition time of 0.72 s. In this study, we only take the 1st session data of 1196 time points spread out over 15 min. In this study, we only take the data from the first five participants consisting of 160 Region of Interest (ROIs). More information about the parcellation and the ATLAS used can be found in the original source [4].

### 2.2  The Basic Model

The proposed network architecture for modeling fMRI signals consists of two components: 1) an oscillatory layer and 2) a feedforward network. The oscillatory layer consists of Hopf oscillators connected in an all-to-all fashion. The dynamics of a single oscillator is described in complex domain and therefore the coupling coefficients are complex numbers. The oscillators are coupled by a special form of coupling called "power coupling" described earlier [5]. The second component, the feedforward network, consists of a complex-valued multilayer perceptron with a single hidden layer. The outputs of the oscillator layer are presented as inputs to the feedforward network. The output of the feedforward network approximates the fMRI data on which the network is trained. The dynamics of the oscillator layer are described below. A typical hopf oscillator can be given by this—

$$\dot{Z} = Z(\mu + i\omega - |Z|^2) \tag{1}$$

and, in polar form (r,$\phi$),

$$\dot{r} = \mu r - r^3; \dot{\phi} = \omega; \tag{2}$$

Now, after considering the coupling ($W_{ij}$) and the external signal ($D(t)$), the original Hopf oscillator equation as shown in the Eq. 1 for single oscillator turns out to be like this—[5]

$$\dot{Z}_i = Z_i(\mu + i\omega_i - |Z_i|^2) + \sum_{j=1, j\neq i}^{N} A_{ij} e^{i\frac{\theta_{ij}}{\omega_j}} Z_j^{\frac{\omega_i}{\omega_j}} + \epsilon e(t) \tag{3}$$

The second component of the network architecture, the feedforward network, is used in two forms, depending on the stage of learning: in the 1st stage learning, it is a single linear stage, whereas in the 2nd stage of learning it is a two stage network with a hidden layer, as shown below Fig. 1. Note that e(t) above is defined in Eq. 8 below.

## 2.3   $1^{st}$ Stage of Learning

In this stage of learning, we train the intrinsic frequencies, $\omega_i$, of the oscillator layer using Eq. 4 below. The lateral connections, which are complex numbers, $W_{ij}$, are partly trained and partly set using experimental data. While the magnitude of the lateral connections is set using structural connectivity information from DTI, the angle of the lateral connections are trained using a Hebb-like learning rule (shown in the Eq. 7) that is applicable to complex-valued weights. The feedforward network in this stage of learning simply consists of a linear stage, whose weights $\alpha_i$, are trained using Eq. 6. Note that $\alpha_i$ are trained by supervised learning, with the objective of minimizing the squared error between the network's output and the desired fMRI signal that it is trying to approximate. The network on the whole performs a Fourier like decomposition of the desired signal, with the $\alpha_i$ playing the role analogous to the Fourier coefficients. The equations governing training are given below [5].

$$\dot{\omega}_i = \beta_w e(t) \sin \phi_i \tag{4}$$

$$W_{ij} = A_{ij} e^{i\theta_{ij}/\omega_{ij}} \tag{5}$$

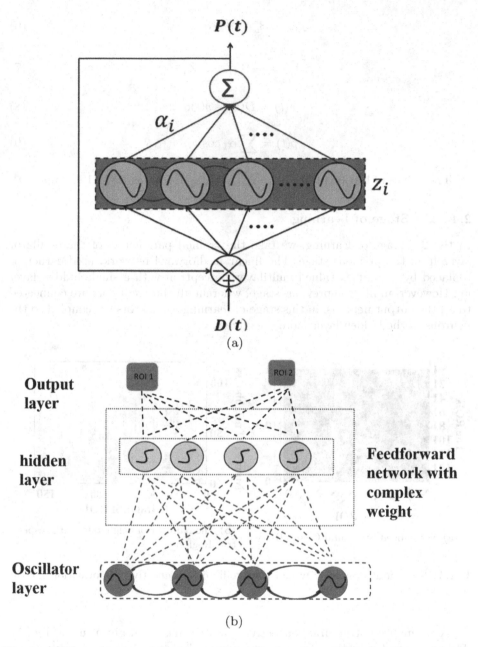

**Fig. 1.** (a)The network architecture used in the 1st stage of learning. It consists of the oscillator layer and a linear stage connecting the oscillators with the output layer. (b) shows the network architecture for second stage of learning involving one hidden layer consisting of 30 hidden neuron. In such feedforward network network has complex valued weights.

$$\dot{\alpha}_i = \beta_\alpha e(t) r_i \cos \phi_i \tag{6}$$

$$\tau_w \dot{W}_{ij} = -W_{ij} + Z_i Z_j^{*\frac{\omega_i}{\omega_j}} \tag{7}$$

$$e(t) = D(t) - p(t); \tag{8}$$

$$p(t) = \sum_{i=1}^{N} \alpha_i \cos \phi_i; \tag{9}$$

The values of learning rates like—$\beta_\alpha$, $\beta_\omega$ , $\tau_w$ are set as $10^{-4}$, $10^{-4}$, and $10^4$.

## 2.4    2$^{nd}$ Stage of Learning

In the 2$^{nd}$ stage of learning, we take the trained parameters of the oscillator layer from the previous stage. The linear feedforward network, used earlier, is replaced by a complex-valued multilayer perceptron with a single hidden layer [5]. However, unlike the previous stage, wherein all the oscillators are connected to all the output neurons, in this stage of learning, oscillators are coupled to the neurons in the hidden layer more selectively.

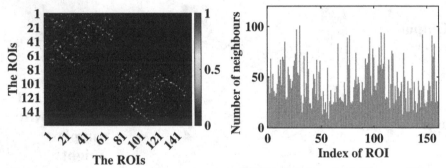

(a) Normalized Structural Connectivity

(b) The Number of neighbours with respective Regions

**Fig. 2.** Structural Connectivity and the oscillator's connectivity information for the model

A normalized structural connectivity matrix has been given in the Fig. 2a. The number of neighbours according to the index of oscillators is given in the Fig. 2b. Note the structural connectivity network is not a fully connected network; it is a sparse one since every brain region or ROI is not connected to every other ROI in the brain. Therefore, each ROI is associated with a single (say, $i^{th}$) oscillator, a single output neuron, and a hidden layer of size K, mediating between the two. The oscillator corresponding to a given ROI projects to all the K hidden neurons associated with it. In addition, all the oscillators to which the $i^{th}$ oscillator is connected, also project the same set of K hidden neurons corresponding to the $i^{th}$ ROI as shown in Fig. 3.

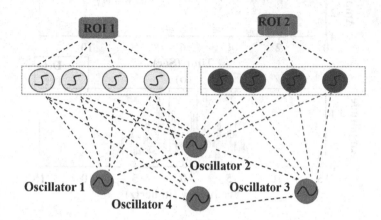

**Fig. 3.** A more detailed depiction of the network architecture that Fig. 1. The figure shows the separate hidden layer and an oscillator are associated with each ROI. See text for details.

The feedforward network in this stage is a complex-valued multilayer perceptron with a single hidden layer. It is trained by complex-valued backpropagation algorithm, with the aim of minimized squared output error for each ROI. The back propagation of the network follows gradient descent rule, where the real part and the complex part of the weights are updated individually [6].

## 3   Results

In this section, we briefly describe the simulated results from the first stage and second stage of learning. The learning rate of the back-propagation is set at a constant rate of 0.05 for both the weight stages (input to hidden layer and hidden layer to output). The core idea of such graph like structure is that, the dynamics of the network of trained oscillators, after the transformation by the feedforward network can approximate the desired ROI activity. All the simulations are done on the MATLAB 2021$b$ platform. The differential equations for the $1^{st}$ stage of learning are solved with the forward Euler's rule.

In the 1st phase of learning, the adaptive nature of the Hopf oscillator is leveraged following the governing equations described in the Subsect. 2.3. The time-series estimation and the frequency domain analysis are given below in the Fig. 4.

Now we can focus on the $2^{nd}$ stage of learning results. As we have discussed in the Subsect. 2.3 above, each oscillator positioned according to the structural

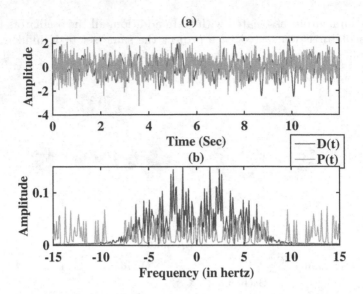

**Fig. 4.** Simulated result after 1st phase of learning. Figure 4. (a) shows the time domain analysis , and 4. (b) denotes the Fourier transform (magnitude plot). $D(t)$ is a empirical signal, and $P(t)$ is for simulated one

connectivity matrix is associated with its neighbours to participate in a feed-forward single hidden layered network. The index of the oscillator with their number of neighbours is shown in the Fig. 2b. Following the learning rule set up in the Subsect. 2.4 the model is able to regenerate the BOLD signal signal consisting of all the frequency components with great accuracy as shown in Figs. 5a, and 5b. Here we use K = 30 hidden units for each ROI's BOLD signal generation. However, the correlation coefficient does not vary even if we decrease the size of the hidden layer up to K = 20 hidden units as shown in the Fig. 5d. However there is a change in Root Mean Square Error (RMSE) observed when decreasing the number of hidden nodes as shown in the Fig. 5c. Root Mean Square Error (RMSE) value of each ROI is captured and as an example, two ROI's simulated and empirical BOLD signal comparison given in the Figs. 5a, and 5b along with RMSE value with respect to number of epochs. It can be seen that the model can regenerate the ROI signals with high accuracy. All the 160 ROI channel simulations are not shown here; only a few representative signals are shown. Totally 60,000 epochs are run for each ROI. How the RMSE and the correlation value varies with the number of epochs are shown in the Figs. 5e, and 5f. It discloses that the RMSE gradually decreases when the number of epochs are being increased.

One of the benchmarks for analyzing such a model is to check the functional connectivity of the ROIs, with Pearson's correlation coefficient. A correlation coefficient matrix is estimated and compared with the empirical BOLD signal's correlation based connectivity matrix or the functional connectivity matrix. The

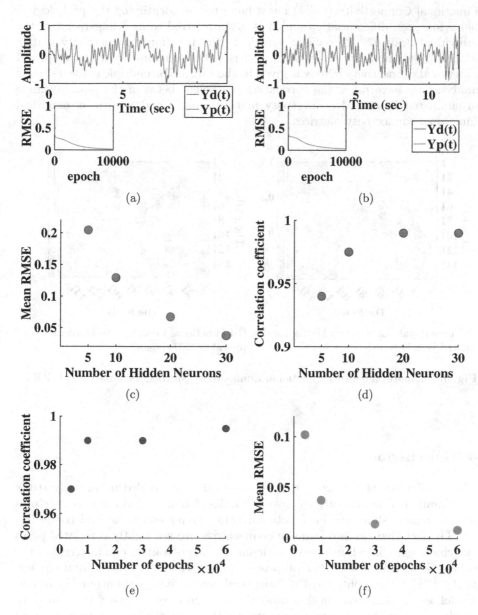

**Fig. 5.** (a), (b) represents the comparison between simulated and empirical time—series signal for ROI1, and ROI2 respectively. $Y_d$- empirical signal, and $Y_p$ simulated signal. (c), and (d) show how the correlation coefficient, and mean RMSE varies with number of hidden nodes when number of epochs is fixed at 10000; (e) and (f) shows how the number of epochs can affect the correlation coefficient, and mean RMSE, when the number of hidden nodes is fixed, $K = 30$. Note that the result is shown only for the first participant.

Functional Connectivity (FCM) has a huge role in identifying the pathological behaviour, sex differentiation and it is often referred to as a fingerprint of the individual brain [7]. This model simulates the functional connectivity matrix with correlation coefficient 0.99. Comparison between simulated and empirical functional connectivity matrix is given in the Fig. 6. For multiple subjects based analysis, we have taken the correlation coefficient between the grand average of simulated functional connectivity matrices and grand average of empirical functional connectivity matrices.

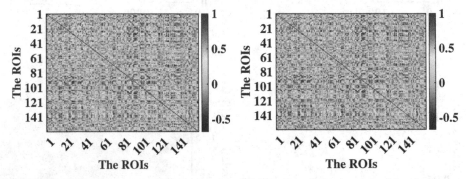

(a) Functional Connectivity Matrix for simulated BOLD signals

(b) Functional Connectivity Matrix for empirical BOLD signals

**Fig. 6.** Comparison between Functional Connectivity Matrices (simulated and BOLD)

## 4    Discussion

The significance of resting state BOLD signal was revealed in an early study by Biswal et al [8]. Since then several studies demonstrated the existence of a unique relationship between functional connectivity and structural connectivity. The fact that the two forms of connectivity are not tightly correlated poses the challenge, clinical and computational, of understanding one in terms of the other. Computational studies proposed models wherein the structural connectivity will be the input, and the functional connectivity the output [9]. In this model, we develop a system that takes the structural connectivity, and also accurately produces the functional connectivity. However, the current study achieves comparable results with existing computational models in terms of accuracy and the Pearson's correlation coefficient, which is often used as a benchmark. A comparison table is given below which compares our results with several recent studies 1. Note that, other models are simulated on different datasets of varying sizes.

The proposed model has several positive features. Use of Hopf oscillator permits adaptation of an explicit frequency parameter, $\omega_i$, which does not exist

**Table 1.** Discussion between Different Models with the current model [10–13]

| Model | Hopf-oscillator Model | Structural Connectivity | Correlation Coeffcient | Model description |
|---|---|---|---|---|
| Model 1[10] | √ | √ | 0.75 | Structural Connectivity Dependent Hopf oscillator model with non-linear coupling system |
| Model 2 [11]) | × | √ | 0.80 | Multiple kernel learning model with modified Wilson-Cowan based neuron activation |
| Model 3 [12] | √ | √ | 0.82 | Hopf oscillator based model and parameters are optimized with Monte Carlo simulation |
| Model 4 [13] | √ | √ | 0.82 | Hopf scillator based model with detailed description about impact of parameters and lesions |
| Current Model | √ | √ | 0.99 | Fourier like decomposition, and retrieved with oscillatory neural network model |

in other low-dimensional neuron models like Wilson-Cowan, FitzHugh-Nagumo, Morris-Lecar etc. The complex-valued coupling weights among oscillators encode time delays in the form of the phase angles. Even the learning rule for the coupling weights has a simple Hebbian form, without the need for use of complicated optimization methods. The step-size of the model during solving the differential equation is down sampled to 0.01 s. The results are also with the same time reference.

Another important question that will be asked is whether this model can able to simulate the full 55 min data, and whether it is applicable for any BOLD signal data-set. Figure 7a shows that the whole time series data can be regenerated for ROI 1 for participant 1; and Fig. 7b shows that the model is equally efficient in case of another data set (known as the Paris dataset) having only 200 data points. Among 21 unrelated healthy participants, the first participant's first indexed ROI is simulated [4]. However, in this paper, we computed the functional connectivity matrix with respect to one participant; it will be not it will not be applicable at group level or multiple participants' analysis. Pursuing the methodology described by Deco et al. [10], five simulations for first five participants taken from HCP dataset were done individually [Fig. 8], and compute the average simulated functional connectivity with average empirical functional connectivity with correlation coefficient value of 0.99(10000 epochs and 30 hidden neurons have been employed for simulation).

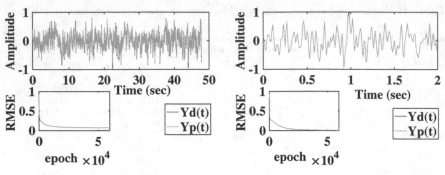

(a) ROI 1 from HCP dataset.    (b) ROI 1 simulation from Paris dataset.

**Fig. 7.** Validity of the Model

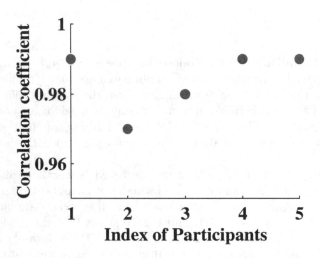

**Fig. 8.** Pearson Correlation coefficient for five participants .

## 5    Conclusion and Future Goal

The aim of the current work is to devise a trainable model of brain dynamics that can account for resting state BOLD responses. The novelty of the current model lies in the aforementioned two stages of the learning process-the Fourier like decomposition to attribute a certain frequency to each oscillator; a novel power coupling strategy; and a distinguished backpropagation algorithm in the second stage. This model also eliminates the rigorous parameter space identification for approximating the BOLD signal used in earlier modelling works. One of the criticisms of our work is that it requires so many hidden nodes to perform; our next goal will be reducing the hidden nodes, and incorporating some of the biological mechanisms like conduction delay, small- worldness of brain network, global connection strength in our model.

# References

1. Franck, A., et al.: Assessment of cerebrovascular dysfunction after traumatic brain injury with fMRI and fNIRS. NeuroImage: Clin. **25**, 102086 (2020)
2. Griffeth, V.E.M., Buxton, R.B.: A theoretical framework for estimating cerebral oxygen metabolism changes using the calibrated-BOLD method: modeling the effects of blood volume distribution, hematocrit, oxygen extraction fraction, and tissue signal properties on the BOLD signal. Neuroimage **58**(1), 198–212 (2011)
3. Cakan, C., Jajcay, N., Obermayer, K.: neurolib: a simulation framework for whole-brain neural mass modeling. Cogn. Comput. 1–21 (2021)
4. Marrelec, G., Messé, A., Giron, A., Rudrauf, D.: Functional connectivity's degenerate view of brain computation. PLoS Comput. Biol. **12**(10), e1005031 (2016)
5. Biswas, D., Pallikkulath, S., Chakravarthy, V.S.: A complex-valued oscillatory neural network for storage and retrieval of multidimensional aperiodic signals. Front. Comput. Neurosci. **15**, 38 (2021)
6. Georgiou, G.M., Koutsougeras, C.: Complex domain backpropagation. IEEE Trans. Circuits Syst. II: Analog Digital Sig. Process. **39**(5), 330–334 (1992)
7. Menon, S.S., Krishnamurthy, K.: A comparison of static and dynamic functional connectivities for identifying subjects and biological sex using intrinsic individual brain connectivity. Sci. Rep. **9**(1), 1–11 (2019)
8. Biswal, B., Zerrin Yetkin, F., Haughton, V.M., Hyde, J.S.: Functional connectivity in the motor cortex of resting human brain using echo-planar MRI. Magn. Resonan. Med. **34**(4), 537–541 (1995)
9. Pathak, A., Roy, D., Banerjee, A.: Whole-brain network models: from physics to bedside. Front. Comput. Neurosci. **16** (2022)
10. Deco, G., Kringelbach, M.L., Jirsa, V.K., et al.: The dynamics of resting fluctuations in the brain: metastability and its dynamical cortical core. Sci. Rep. **7**, 3095 (2017)
11. Surampudi, S.G., Naik, S., Surampudi, R.B., Jirsa, V.K., Sharma, A., Roy, D.: Multiple kernel learning model for relating structural and functional connectivity in the brain. Sci. Rep. **8**(1), 1–14 (2018)
12. Iravani, B., Arshamian, A., Fransson, P., Kaboodvand, N.: Whole-brain modelling of resting state fMRI differentiates ADHD subtypes and facilitates stratified neuro-stimulation therapy. Neuroimage **231**, 117844 (2021)
13. Hahn, G., et al.: Signature of consciousness in brain-wide synchronization patterns of monkey and human fMRI signals. Neuroimage **226**, 117470 (2021)

# Explainable Causal Analysis of Mental Health on Social Media Data

Chandni Saxena[1(✉)], Muskan Garg[2], and Gunjan Ansari[3]

[1] The Chinese University of Hong Kong, Hong Kong SAR, China
csaxena@cse.cuhk.edu.hk
[2] University of Florida, Gainesville, Florida, USA
muskangarg@ufl.edu
[3] JSS Academy of Technical Education, Noida, India
gunjanansari@jssaten.ac.in

**Abstract.** With recent developments in *Social Computing*, *Natural Language Processing* and *Clinical Psychology*, the social NLP research community addresses the challenge of automation in mental illness on social media. A recent extension to the problem of multi-class classification of mental health issues is to identify the cause behind the user's intention. However, multi-class causal categorization for mental health issues on social media has a major challenge of wrong prediction due to the overlapping problem of causal explanations. There are two possible mitigation techniques to solve this problem: (i) Inconsistency among causal explanations/ inappropriate human-annotated inferences in the dataset, (ii) in-depth analysis of arguments and stances in self-reported text using discourse analysis. In this research work, we hypothesise that if there exists the inconsistency among F1 scores of different classes, there must be inconsistency among corresponding causal explanations as well. In this task, we fine tune the classifiers and find explanations for multi-class causal categorization of mental illness on social media with LIME and Integrated Gradient (IG) methods. We test our methods with CAMS dataset and validate with annotated interpretations. A key contribution of this research work is to find the reason behind inconsistency in accuracy of multi-class causal categorization. The effectiveness of our methods is evident with the results obtained having category-wise average scores of 81.29% and 0.906 using cosine similarity and word mover's distance, respectively.

**Keywords:** causal analysis · explainability · mental health · text categorization

## 1 Introduction

People express their thoughts more conveniently on social media than during in-person (often analytical) sessions with experts. As per the National Institute of Mental Health report of 2020[1], 52.9 million adults in the USA suffer from mental illness. "The Health at a Glance Europe 2020" report[2] noted that the COVID-19 pandemic and the subsequent economic crisis caused a growing burden on the mental well-being of the citizens,

---

[1] https://www.nami.org/mhstats.
[2] https://health.ec.europa.eu/system/files/2020-12/2020_healthatglance_rep_en_0.pdf.

M. Tanveer et al. (Eds.): ICONIP 2022, LNCS 13624, pp. 172–183, 2023.
https://doi.org/10.1007/978-3-031-30108-7_15

with evidence of higher rates of stress, anxiety and depression. Previous studies support social media's powerful role in measuring the public's social well-being [8]. To this end, we obtain Reddit social media posts demonstrating mental health issues for mental health analysis.

In this research work, we narrow down the problem of *mental health analysis* to *the identification of reasons behind users' intent in their social media posts*. The sequence to sequence (Seq2Seq) models are applied to solve the problem of causal categorization over CAMS dataset[3]. The ground-truth of CAMS dataset contains two-fold annotations (i) *causal category* and (ii) *interpretations*. The textual segments of *interpretation* support decision making for identifying causal categories. However, there exists a major challenge of responsibility and explainability for multi-class causal analysis while applying fine-tuned Seq2Seq models. In this context, we find explanations for inconsistency among resulting accuracy of different classes/ categories. Another key contribution is to find distance among *inferences* and *explanations* to obtain semantic similarity over distributional word representation: (i) *cosine similarity* and (ii) *word mover distance*.

**Definition 1: Inferences -** The inferences are set of interpreted textual segments by trained human-annotators which appears as ground-truth information in CAMS dataset.

**Definition 2: Explanations -** The results obtained as the set of top-keywords using explainable AI approaches for multi-class causal categorization of Reddit posts is termed as explanations.

We further discuss a potential instance to define this problem of explainable causal analysis in this section. Consider a given sample $A$ where a user $U$ post $A$: "*Five years now and still no job. I am done with my life.*" The user $U$ is upset about his financial problems/ career due to *unemployment*. We consider this text as the user-generated social media data which demonstrates mental health issues. The intent of a user is '*to end life*' and a key challenge is to find the reason behind this intent. This cause-and-effect relationship aids the causal categorization. The category for sample $A$ is identified as '*Jobs and careers*' because the reason is associated with unemployment. There are five causal categories in annotated CAMS dataset, namely, *(i) bias or abuse, (ii) jobs and careers, (iii) medication, (iv) relationships, and (v) alienation*.

In this research work, we use the CAMS dataset for explanations on multi-class causal categorization. We have made three major contributions in this work. First, we fine-tune deep learning models for multi-class causal categorization. Second, we obtain explainable text for causal categorization using *Local Interpretable Model-Agnostic Explanations (LIME)* and IG. Third, two semantic similarity measures: *cosine similarity* and *word mover distance* assist the validation of resulting explainable snippets with annotated inferences. Our experimental results explains the inconsistency among accuracy of different classes and validates the consistency of inferences made by model and human annotators, thereby defining the need of discourses and pragmatics for this problem of causal analysis. All code[4] used are publicly available.

---

[3] https://github.com/drmuskangarg/CAMS.

[4] https://github.com/CMOONCS/CausalExplanationMHA.git.

## 2  Background

Our task is defined as a domain-specific problem to find *reasons behind the intent of a user on social media*. After extensive literature surveys, we observe minimal work on this problem. A domain-specific dataset is available for public use to examine the inferences (reasons) and causal categories (multi-class classification) task for mental health data as CAMS dataset [4]. The existing solution of a task of causal analysis is given as the use of machine learning and neural models for multi-class categorization of causal categories. The resulting values of f-measure vary for different classes and raise a new research question: *To what extent causal categorization is responsible?* We choose to resolve this problem by finding and validating the explainable texts.

To find the explanations for causal categorization, we explore existing explainable AI methods for natural language processing [6]. Some well-established surveys and tutorials categorize explainable approaches into *local vs global, post hoc vs self explaining* and *model agnostic vs model specific* [3]. We choose to observe local explanations with given input features for post-hoc interpretability methods which require less information. To this end, we identify two explainability approaches which are suitable for this study: (i) LIME and (ii) IG.

LIME samples nearby observations and uses model estimates to fit the logistic regression [7]. The parameters of logistic regression represent the importance measure and larger the parameters, greater effect will have on the output. The IG is an attempt to assign an attribution value to each input feature which measures the extent to which an input contributes to the final prediction [12]. A recent study is carried out to set a benchmark over three representative NLP tasks (sentiment analysis, textual similarity and reading comprehension) for interpretability of both neural models and saliency methods [14] thereby emphasizing the need of LIME and IG for downstream NLP tasks.

The explainable methods give output in the form of important words/ text segments which serve as the most important input features. As we have available human annotated inferences for causal categorization in the form of text, we use these inferences as ground truth information (text-reference) and resulting explanations (RE) (text-observation). Thus, we use two semantic similarity measures to evaluate the performance of explainable methods for causal categorization- Cosine similarity and Word Mover's distance (WMD). Cosine similarity [9] calculates similarity between two words, sentences, paragraph, piece of text etc. and evolves from the squared Euclidean distance measure which is used to measure how similar the documents are irrespective of their size. Word Mover's Distance (WMD) outperforms Bag-of-words and TF-IDF in terms of document classification error rates [5].

## 3    Framework

In this section, we give a brief overview of the proposed framework for Fig. 1 which represents the workflow for explainable causal analysis of mental health on social media data. We bifurcate our framework into three phases:

– Causal Categorization: The use of neural models for causal categorization of Reddit posts depicting mental illness.
– Explanations: Finding explanations in the form of text-observations and obtaining top-keywords.
– Evaluations: Validate the resulting text-observations by comparing them with the human annotations available in the CAMS dataset.

Consider a given set of self-reported short-text documents as $D$ where $D = d_1, d_2, ..., d_n$. In *Phase 1: causal categorization*, we segregate $D$ into training, validation and test set and give training set as an input and we fine-tune the multi-class classifier build model for our task. The model prediction are given as an input to *Phase 2: Finding explanations* along with Reddit posts to obtain explanations. We further obtain these resulting explanations and human-annotated inferences present in the CAMS dataset for *Phase 3: Evaluations* to test and validate the resulting explanations. Furthermore, we discuss three phases of our proposed framework in this section.

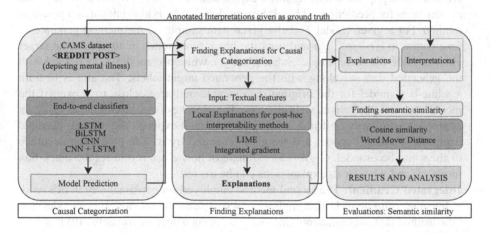

**Fig. 1.** Overview of the proposed framework for explainable causal analysis of mental health on social media data. The framework is divided into three phases - Phase 1: Causal categorization, Phase 2: Finding explanations, Phase 3- Semantic similarity.

### 3.1   Phase 1: Causal Categorization

To solve the problem of causal categorization, we employ four learning based multi-class classifiers [1]. We exploit following deep learning models and fine-tuned them for prediction:

– **LSTM.** Long-Short Term Memory (LSTM:) is a popular advanced Recurrent neural network architecture for modeling sequential data which allows the information to persist and is trained by taking the sequence of the embedding feature vector.
– **BiLSTM:** A Bidirectional LSTM trains two hidden layers on the input sequence. The additional layer reverses the direction of information flow which means that the input sequence flows backward in an additional LSTM layer.
– **CNN:** The CNN model efficiently extracts higher level features of the text using convolutional layers and max-pooling layers.
– **CNN-LSTM:** A Hybrid CNN-LSTM Model uses CNN layers for feature extraction on input text combined with LSTMs to support sequence prediction.

### 3.2   Phase 2: Finding Explanations

We obtain local explanations by using following two post-hoc interpretability models:

– **LIME:** It is a popular model-agnostic explainable method [7] which provides local explanations for predictions of black-box models. LIME is also known as a post-hoc method. For a given model $\dot{F}$ and a given data sample $\alpha$, the method generates a fake dataset $\alpha 1, \alpha 2, \alpha 3 .. \alpha n$ and uses the black box model, $\dot{F}$ to obtain the target class or value for each sample. Subsequently, a white box model, $\bar{G}$ is trained with the generated data set along with the generated target labels. The aim is to train a white-box model for the original data sample and areas close to it even if the model does not perform as well globally. The closeness can be estimated using an appropriate similarity or distance metric. LIME then explains the original example using the white-box model and weights generated by it. The prediction accuracy of the white-box model, $\bar{G}$ gives an estimate of how close it mimics the black-box model, $\dot{F}$ and whether its explanations can be trusted.
– **Integrated Gradient:** The second method employed for explainability in this work is Integrated Gradients [12], a gradient-based explanation method. It is a model specific method that uses gradients (for example, using a deep neural network) to assess the importance of a feature on the model's output. It employs the knowledge associated with the internal model for calculating the gradients of the model's layers. It computes an attribution score corresponding to each feature by considering the integral of the gradients calculated along a straight path from a baseline instance $u'$ to the input instance $u$.

### 3.3   Phase 3: Evaluations with Semantic Similarity

The *human-annotated inferences* in CAMS dataset, which represents the causal explanation in the post, is validated by a senior clinical psychologist and it serves as a ground truth for our predicted explanations. There are two types of similarity measures for identifying document similarity (i) syntactic similarity and (ii) semantic similarity. We omit exact string matching algorithms due to varying number of words in each chunk and inconsistency among length of inferences and resulting keywords. The semantic similarity among texts validates the effectiveness of classifiers. We employ two most widely used semantic similarity measures:

- **Cosine similarity:** It is a widely used metric in information retrieval which models text as vector of terms [9]. The similarity of two input sentences (documents) can be derived by calculating cosine values of term vectors for the given input using the following equation. The similarity between two vectors of given input documents ($Doc_1 Doc_2$) can be defined as:

$$Sim(Doc_1, Doc_2) = \frac{Doc_1 \cdot Doc_2}{||Doc_1|| \, ||Doc_2||} = \frac{\sum_{i=1}^{n} A_i \cdot B_i}{\sqrt{\sum_{i=1}^{n} A_i^2} \sqrt{\sum_{i=1}^{n} B_i^2}} \qquad (1)$$

where $A_i$ and $B_i$ represent the components of vectors $Doc_1$ and $Doc_2$, respectively.
- **Word Mover's Distance(WMD):** It is a novel distance metric [5] that is used to measure the dissimilarity between two text documents. The method is different from the conventional models that work on syntactic similarity rather than semantic similarity. The method employs word embedding like Glove and Word2Vec to learn semantically meaningful representations of sentences. It computes distance between two documents A and B as the minimum cumulative distance that the embedded words of document A need to travel to reach the embedded words of document B. WMD is computed using the cost-matrix having $x_i$ and $x_j$ be embedding of word i and j. The cost matrix $CM \in \mathbb{R}^m \times \mathbb{R}^m$ is the distance of embeddings, such that $CM_{ij} = ||x_i - x_j||^2$ as referred to in Eq. 2. The distance between two documents $Doc_1$ and $Doc_2$ is the optimum value of the following problem:

$$P \in \mathbb{R}^{m \times m} \sum_{ij} CM_{ij} P_{ij} \qquad (2)$$

such that $P_{ij} \geq 0$ Intuitively, $P_{ij}$ represents the amount of word $i$ that is transported to word $j$. WMD is defined as the minimum total distance to convert one document to another document.

## 4    Experiments and Evaluation

This section covers the dataset description, experimental setup, results and performance evaluation of the proposed study.

### 4.1    CAMS Dataset

CAMS dataset consists of 5051 instances (1896 from SDCNL dataset and (ii) 3155 Reddit posts which are available with subreddit r/depression using Python Reddit API Wrapper (PRAW)[5]) to categorize the *direct causes* of mental disorders through mentions by users in their posts. Annotation is carried out manually by annotators who

---
[5] https://praw.readthedocs.io/en/stable/.

are proficient in the language. They work independently for each post and follow the given guidelines. Each annotator takes one hour to annotate about $15 - 25$ Reddit posts. The annotated files are verified by a clinical psychologist and a rehabilitation counselor. Furthermore, the validation of three annotated files is carried out by Fliess' Kappa inter-observer agreement study. The trained annotators have 61.28% agreement for annotations of CAMS dataset. Despite the increased subjectivity of the task, the trained annotators *substantially agree* with their judgements.

## 4.2   Experimental Setup

Considering a CAMS dataset, we divide it into training, validation and testing set consisting of 1699, 117 and 370 instances, respectively. After preprocessing of the given documents $D$ (Reddit posts), we employ four deep learning methods to predict the causal category, namely, LSTM, BiLSTM, CNN and CNN-LSTM. At the initial layer of the neural network, we use GloVe, a distributional word embedding with dimension vectors of 100. The GloVe embedding extracts semantics by using information available in neighbouring spaces. For experimental study, we consider a batch size of 128, trained on 20 epochs, with 265 maximum length of tokens. We use Adam optimizer for all models with one or more dropout layers and optimal learning rate. We fine-tune CNN-LSTM model with a learning rate of $0.0005$ and set a learning rate of $1.46 * 10^{-3}$ for all other classifiers.

## 4.3   Experimental Results

We perform experiments over the given dataset and obtain results as shown in Fig. 2. To illustrate the effectiveness of our models, we give explanations for self-reported text of each causal-category. The given input is a *self-reported text* of the CAMS dataset. The human-annotations are two-fold: (i) human-annotated interpretations (inferences) and (ii) causal category. We further perform explainable causal categorization to compare and contrast the *inferences* with resulting top-keywords (*explanations*). We observe minimal connection among words for Cause 0: No reasons followed by Cause 3: medications. However, the other causal categories seem to have high similarity among inferences and explanations.

**Error Analysis:**   Other than the examples given in Fig. 2, the medical terms mentioned in inferences and explanations may vary. For instance, prescriptions like *propranalol*, name of diseases, heart problems, specific type of cancer and other antidepressants. This variation induces mismatch in semantic similarity among inferences and explanations for class 3.

## 4.4   Performance Evaluation

We use performance evaluation measures of the confusion matrix to evaluate the results of multi-class categorization. We analyze the results for each causal category and find overall accuracy of the model. Furthermore, we use two evaluation metrics of finding semantic similarity to evaluate explanations obtained by LIME and IG.

| Self-reported text | Inferences | Causal category | Explanations |
|---|---|---|---|
| Five years now and still no job. I am done with my life | Five years now, no job | Cause 2: Jobs and Career | no job, done, life |
| Bad things happen to me, and worst of all, they're inevitable. I think I'm better off dead than alive. | bad things, inevitable, worst | Cause 5: Alienation | Bad, worst, happen, alive |
| I reached my limit, nothing helps anymore, I tried everything I could get my hands on. My family doesn't even believe in depression. I'm just a big fat dumb loser, but it doesn't matter anymore, because tomorrow I'll put an end to all of this. | family, big fat dumb loser, reached limit, nothing helps | Cause 4: Relationships | dumb, loser, fat, depression, big, tomorrow, helps |
| I've been lying here for hours just doing nothing. I can't go out today because I'm having a chronic illness flare-up. But, I can't seem to find anything to interest me at home, either. | chronic illness flare up, hobbies seems dull | Cause 3: Medications | illness, chronic, lying, hours, find |
| I can't believe how fucking dumb I am. I fucking ruined  one of the best things that happened to me. I want to die . I want to die. I need to kill myself soon. Please god don't let me wake up tomorrow | ruined, dumb | Cause 1: Bias or Abuse | dumb, fucking, ruined, best |
| I've been crying almost every night for almost 2 months now. At first it hurt my eyes from crying so hard but not anymore, now I just feel sick when I cry, I just want a week where I don't feel like crying | crying, feel sick | Cause 0: No reason | crying, almost, hurt, every, night |

**Fig. 2.** Experimental results for explainable causal categorization for six different categories.

**Causal Categorization:** We categorize the text into one of the six categories as mentioned in experimental results section and present the resulting values for multi-class classifiers in Table 1. We observe the inconsistency in results for among different classes but consistency in variation among classes for different classifiers. To this end, we observe lowest F1 scores for causal category 1: *Bias or Abuse*. The demonstration indicates errors among predictions for *Alienation/ Relationship* as they overlap with *Bias or Abuse*. The complex interactions illustrated the perceivable overlap between *Bias or Abuse* and *Relationship* in the following example:

> *My friends are ignoring me and I am feeling bad about it. I have lost all my friends and don't want to live anymore.*

The given example is associated with *biasing* and *friendship*, in a case where someone feels ostracized by their friends. The emphasis on *friends* tips the balance in favor of the class *Relationship*. However, the major challenge is to train the model in such a way that it understands the inferences and then chooses the most emphasized *causal category* using optimization techniques. We view this challenge as an open research direction.

There are two possible mitigation techniques to solve this problem: (i) Inconsistency among causal explanations/ inappropriate human-annotated inferences in the dataset, (ii) in-depth analysis of arguments and instances in self-reported text using discourse

**Table 1.** Performance evaluation of multi-class classifiers for causal categorization of mental illness on social media data where F1:C0, F1: C1, F1:C2, F1:C3, F1:C4 and F1:C5 defines F1-score for 6 categories: cause 0: 'No reason', cause 1: 'Bias or abuse', cause 2: 'jobs and careers', cause 3: 'medication', cause 4: 'relationships', and cause 5: 'alienation', respectively.

| Classifier | F1:C0 | F1:C1 | F1:C2 | F1:C3 | F1:C4 | F1:C5 | Accuracy |
|---|---|---|---|---|---|---|---|
| LSTM | 0.55 | 0.30 | 0.36 | 0.45 | 0.55 | 0.25 | 0.4514 |
| BiLSTM | 0.59 | 0.25 | 0.53 | 0.44 | 0.58 | 0.43 | 0.5054 |
| CNN | 0.57 | 0.26 | 0.53 | 0.54 | 0.58 | 0.35 | 0.4919 |
| CNN-LSTM | 0.57 | 0.17 | 0.38 | 0.46 | 0.48 | 0.52 | 0.4784 |

analysis. In this research work, we hypothesise that if there exists the inconsistency among F1 scores of different classes, there exists an inconsistency among corresponding causal explanations as well. We find causal explanations and validate the results with human-annotated inferences. To this end, we choose to handle the first mitigation approach, thereby, enlisting new frontiers.

**Explainability:** In this section, we present evaluation of resulting top-keywords using LIME and IG methods. We use *word mover distance* and *cosine similarity* over distributional word representations of both inferences and resulting keywords. As observed in Table 2, the explanations of *Class 0: 'No reason'* have maximum distance from human-annotated inferences for all methods. The reason is well-justified with the fact that Reddit posts having no reason behind intent of a user may or may not choose random words from the entire text. These random words does not describe any reason and thus, are the most far away from human-generated inferences. Low values for all other classes signifies the presence of patterns among explanations for other classes. We find *class 2: jobs and careers*, and *class 4: relationships* as the semantically most similar explanations achieved by deep learning methods.

**Table 2.** Values obtained for semantic similarity among resulting top-keywords and human-annotated inferences using *Word Mover Distance*: More distance indicates less similarity among two different texts.

| Method used | Class0 | Class1 | Class2 | Class3 | Class4 | Class5 |
|---|---|---|---|---|---|---|
| LSTM+LIME | 1.029 | 0.854 | 0.857 | 0.896 | **0.838** | 0.889 |
| LSTM+IG | 1.097 | 0.890 | 0.870 | 0.926 | **0.867** | 0.906 |
| BiLSTM+LIME | 1.029 | 0.880 | 0.865 | 0.886 | **0.852** | 0.876 |
| BiLSTM+IG | 1.117 | 0.900 | 0.898 | 0.919 | **0.870** | 0.908 |
| CNN+LIME | 1.042 | 0.820 | 0.831 | **0.817** | 0.823 | 0.843 |
| CNN+IG | 1.123 | 0.907 | 0.882 | 0.912 | **0.880** | 0.913 |
| CNN-LSTM+LIME | 1.018 | 0.843 | **0.831** | 0.848 | 0.851 | 0.863 |
| CNN-LSTM +IG | 1.117 | 0.913 | **0.869** | 0.918 | 0.874 | 0.890 |

We further analyse the results for cosine similarity as shown in Table 3. We give input as a string, tokenize the text, use GloVe word embeddings to obtain word vectors, and find the mean of word vectors (obtained for each token). Experimental results demonstrate *class 2: jobs and careers*, and *class 4: relationships* as the most similar explanations to the human-annotated inferences. *Class 3: Medication* , being associated with medical terms are expected to be semantically least similar as we would need domain-specific distributional word representation for evaluation in this category. Thus, class 3 and class 0 are illustrating low scores as compared to other classes.

**Table 3.** Values obtained for semantic similarity among resulting top-keywords and human-annotated inferences using *Cosine Similarity*: The distance lies between 0 and 1

| Method used | Class0 | Class1 | Class2 | Class3 | Class4 | Class5 |
|---|---|---|---|---|---|---|
| LSTM+LIME | 0.787 | 0.825 | **0.889** | 0.751 | 0.881 | 0.854 |
| LSTM+IG | 0.723 | 0.779 | **0.870** | 0.701 | 0.869 | 0.813 |
| BiLSTM+LIME | 0.784 | 0.821 | **0.881** | 0.751 | 0.867 | 0.857 |
| BiLSTM+IG | 0.716 | 0.773 | **0.866** | 0.709 | 0.865 | 0.814 |
| CNN+LIME | 0.776 | 0.835 | **0.898** | 0.822 | 0.894 | 0.861 |
| CNN+IG | 0.729 | 0.765 | **0.863** | 0.689 | **0.863** | 0.818 |
| CNN-LSTM+LIME | 0.781 | 0.831 | **0.878** | 0.811 | 0.868 | 0.852 |
| CNN-LSTM+IG | 0.728 | 0.789 | 0.851 | 0.690 | **0.870** | 0.815 |

## 4.5 Ethical Considerations

NLP researchers are responsible for transparency about computational research with sensitive data accessed during model design and deployment. We understand the significance of ethical issues while dealing with a delicate subject of mental health analysis. We use the publicly available dataset and do not plan to disclose any sensitive information about the stakeholders (social media users) thereby preserving the privacy of a user [2].

We use publicly available pre-trained base models for our demonstration to avoid any ethical conflicts. We assure that we adhere to all ethical guidelines to solve this task. Development of fair AI technologies in mental healthcare supports unbiased clinical decision-making [13]. Our research work is fair and there is no intentional bias as we consider explainable causal categories for mental health on CAMS dataset.

## 5   Conclusion and Future Scope

We find the explanations for causal categorization of mental health in social media posts by using LIME and IG methods, followed by performance evaluation by using human-annotated inferences in CAMS dataset. We conclude our work with three key takeaways: (i) less variations among resulting values of all classes for causal explanations as compare to F1 scores in causal categorization validates the human-annotated inter-pretations for causal categorization; (ii) the results for *Class 0: No reason* and *Class 3: Medication* are least explainable due to randomization and the need of domain-specific analysis, respectively; (iii) the performance evaluation of explanations obtained using explainable NLP is possible with semantic similarity methods if human-annotated inter-pretations are predefined.

One of the path-breaking work is performed for causal explanation on social media which is obtained in the form of text [10]. The authors mentioned the complexity of this problem and made an attempt to resolve this issue by using discourses. However, the experiments were performed over a limited amount of Facebook data (*often referred as Causal Explanation Analysis (CEA) dataset*) to classify the texts containing causal explanations and thereby extracting causal explanations. Furthermore, the causal expla-nation detection takes place on CEA dataset by capturing the salient semantics of dis-courses contained in their keywords with a bottom graph-based word-level salient net-work [15]. In this context, we choose to propose domain-specific discourse relation embeddings [11] as a potential future research direction of causal analysis.

# References

1. Chen, B., Huang, Q., Chen, Y., Cheng, L., Chen, R.: Deep neural networks for multi-class sentiment classification. In: 2018 IEEE 20th International Conference on High Performance Computing and Communications; IEEE 16th International Conference on Smart City; IEEE 4th International Conference on Data Science and Systems (HPCC/SmartCity/DSS), pp. 854–859. IEEE (2018)
2. Conway, M., et al.: Ethical issues in using twitter for public health surveillance and research: developing a taxonomy of ethical concepts from the research literature. J. Med. Internet Res. **16**(12), e3617 (2014)
3. Danilevsky, M., Qian, K., Aharonov, R., Katsis, Y., Kawas, B., Sen, P.: A survey of the state of explainable AI for natural language processing. arXiv preprint: arXiv:2010.00711 (2020)
4. Garg, M., et al.: CAMS: an annotated corpus for causal analysis of mental health issues in social media posts. In: Language Resources and Evaluation Conference (2022)
5. Kusner, M., Sun, Y., Kolkin, N., Weinberger, K.: From word embeddings to document dis-tances. In: International Conference on Machine Learning, pp. 957–966. PMLR (2015)
6. Madsen, A., Reddy, S., Chandar, S.: Post-hoc interpretability for neural NLP: a survey. arXiv preprint: arXiv:2108.04840 (2021)
7. Ribeiro, M.T., Singh, S., Guestrin, C.: "why should i trust you?" Explaining the predictions of any classifier. In: Proceedings of the 22nd ACM SIGKDD International Conference on Knowledge Discovery and Data Mining, pp. 1135–1144 (2016)
8. Robinson, P., Turk, D., Jilka, S., Cella, M.: Measuring attitudes towards mental health using social media: investigating stigma and Trivialisation. Soc. Psychiatry Psychiatr. Epidemiol. **54**(1), 51–58 (2019). https://doi.org/10.1007/s00127-018-1571-5
9. Salton, G., Buckley, C.: Term-weighting approaches in automatic text retrieval. Inf. Process. Manage. **24**(5), 513–523 (1988)
10. Son, Y., Bayas, N., Schwartz, H.A.: Causal explanation analysis on social media. arXiv preprint: arXiv:1809.01202 (2018)
11. Son, Y., Schwartz, H.A.: Discourse relation embeddings: representing the relations between discourse segments in social media. arXiv preprint: arXiv:2105.01306 (2021)

12. Sundararajan, M., Taly, A., Yan, Q.: Axiomatic attribution for deep networks. In: International Conference on Machine Learning, pp. 3319–3328. PMLR (2017)
13. Uban, A.S., Chulvi, B., Rosso, P.: On the Explainability of automatic predictions of mental disorders from social media data. In: Métais, E., Meziane, F., Horacek, H., Kapetanios, E. (eds.) NLDB 2021. LNCS, vol. 12801, pp. 301–314. Springer, Cham (2021). https://doi.org/10.1007/978-3-030-80599-9_27
14. Wang, L., et al.: A fine-grained interpretability evaluation benchmark for neural NLP. arXiv preprint: arXiv:2205.11097 (2022)
15. Zuo, X., Chen, Y., Liu, K., Zhao, J.: Towards causal explanation detection with pyramid salient-aware network. In: Sun, M., Li, S., Zhang, Y., Liu, Y., He, S., Rao, G. (eds.) CCL 2020. LNCS (LNAI), vol. 12522, pp. 113–128. Springer, Cham (2020). https://doi.org/10.1007/978-3-030-63031-7_9

# Brain-Inspired Attention Model for Object Counting

Abhijeet Sinha⬤, Sweta Kumari⬤, and V. Srinivasa Chakravarthy(✉)⬤

Computational Neuroscience (CNS) Lab, Department of Biotechnology,
Indian Institute of Technology Madras, Chennai, India
sinha@alumni.iitm.ac.in, bt17d019@smail.iitm.ac.in, schakra@ee.iitm.ac.in

**Abstract.** We develop a sequential Q-learning model using a recurrent neural network to count objects in images using attentional search. The proposed model, which is based on visual attention, scans images by making a sequence of attentional jumps or saccades. By integrating the information gathered by the sequence of saccades, the model counts the number of targets in the image. The model consists primarily of two modules: the Classification Network and the Saccade Network. Whereas the Classification network predicts the number of target objects in the image, the Saccade network predicts the next saccadic jump. When the probability of the best predicted class crosses a threshold, the model halts making saccades and outputs its class prediction. Correct prediction results in positive reward, which is used to train the model by Q-learning. We achieve an accuracy of 92.1% in object counting. Simulations show that there is a direct relation between the number of glimpses required and the number of objects present to achieve a high accuracy in object counting.

**Keywords:** Attention · Object Counting · Q-learning

## 1 Introduction

The ability to count multiple objects in an image has several applications including counting cells in a micrograph [19], monitoring wildlife [16], traffic surveillance [13], inventory management and tracking objects through surveillance cameras [22]. A significant amount of work in the field of computer vision has been done for counting and locating objects in images. Segui et al. [21] developed a convolutional neural network and used its features to generate a confidence map which they use for detecting object occurrences. Inbar et al. [5] developed a novel weakly-supervised convolutional neural network training for finding and counting iterative objects that can be applied to single-frame scenarios. In the aforementioned approaches, counting is treated as a classification problem. The classification approach is helpful when the number of occurrences of the object is sufficiently large. The network takes two inputs; the input image and the object of interest marked in the image. Using supervised learning it learns to mark

M. Tanveer et al. (Eds.): ICONIP 2022, LNCS 13624, pp. 184–193, 2023.
https://doi.org/10.1007/978-3-031-30108-7_16

multiple occurrences of the objects. Hui et al. [11] developed a convolutional neural network called LaoNet for one-shot object counting. In this network, the counting model should consider only one instance and count objects of the new category. The network takes two inputs: the input image and image of the object of interest. From this input, the network uses supervised learning to create a density map of the objects of interest. In the case of the LaoNet, the approach to training is to minimize the mean squared error between the predicted density map and the ground truth density map. Jack et al. [12] developed a sequential learning model using reinforcement learning and recurrent neural network architecture to count and recognize objects in images. They did not use the complete image like the previous models but they used the attentional approach in which, analogous to the human visual system, an attentional window scans the image by making attentional jumps to count the target objects in the image. An instance of an attentional approach to counting is the recurrent attention model (RAM) model [14] in which the image is analyzed using a sequence of glimpses. Research on approaches to object counting using bio-inspired or attentional architectures is rather scarce. Their work introduced a method for image classification that uses a sequence of glimpses on different regions of the image to predict the class.

In this paper, we propose a brain-inspired attention model for object counting. The proposed attention model does not scan the image at once but it learns to find important locations in the image and extracts salient parts from those locations of the image. The architecture of the attention model is inspired by the anatomy of the vision processing network of the human visual system (HVS).

## 2    Method

We consider the attention problem as a sequential decision process, where the agent interacts with a visual environment and makes sequential decisions. This approach is realized in the form of an attention search-based deep Q neural network [14] applied to object counting. To perform object counting, we create a dataset containing images that consist of patches as objects with a count ranging from 1 to 10. The images are of size $64 \times 64$. The patches in the image are placed at random pixel locations (Fig. -1). To generate the image containing the patches, we used a bivariate gaussian function (Eq. -1). The size of the cells is determined by the value of the bi-variate gaussian function. A set of 14400 images were generated in each of the 10 count classes. In the dataset, 50% (7200) images in the training set, 25% (3600) images in the validation set, and 25% (3600) images in the testing set were considered for the experiment.

$$A_{ij} = min(255, \sum_{k=1}^{c} 255(exp(-(x_k - i)^2 - (y_k - j)^2)/2\sigma^2)) \tag{1}$$

where, $c$ is the count-class, $A$ is the matrix representation of the image, $A_{ij}$ represents the matrix's elements with $i^{th}$ row and $j^{th}$ column, and $(y_k, x_k)$ represent the $c$ randomly chosen row and column, respectively i.e. the center of a patch. The variance $\sigma$ determines how concentrated the values are around $(y_k, x_k)$, therefore $\sigma$ controls the size of the patches or objects in the image (Fig. 1).

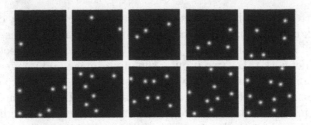

**Fig. 1.** Sample of images generated in dataset for each class.

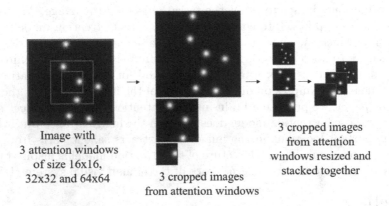

Image with
3 attention windows
of size 16x16,
32x32 and 64x64

3 cropped images
from attention windows

3 cropped images
from attention
windows resized and
stacked together

**Fig. 2.** Attention glimpse generation from image.

The attention model takes two inputs: the cropped-out attention glimpse from the image and the heatmap representation of the attentional window in the entire background image (Fig. 2). To generate an attention glimpse, three concentric attention windows of size $16 \times 16$, $32 \times 32$, and $64 \times 64$ respectively, centered on the currently attended point in the image, are cropped out of the original image. They are resized into one common size ($16 \times 16$), and arranged as a stack (Fig.-2). On the other side, to generate the attentional heatmap representation of the currently attended location, pixel values of the $16 \times 16$ window at the center location $l$ in a $64 \times 64$ array are assigned to 1, otherwise 0 (Fig. 3). The architecture of the complete system consists of three CNNs. The first CNN is called the Classifier Network, the second is the Eye Position network and the third network is the Saccade Network. The Classifier Network and the Saccade Network take the attentional glimpse as the input and the Eye Position Network takes the attentional heatmap representation of the currently attended point as the input. The inputs are processed in parallel by passing through the three constituent pathways of the proposed architecture.

To solve the counting problem, by taking a sequence of glimpses from the image, the network must necessarily hold a memory of the past glimpses. In the

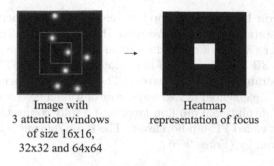

Image with
3 attention windows
of size 16x16,
32x32 and 64x64

Heatmap
representation of focus

**Fig. 3.** Attentional Heatmap representation.

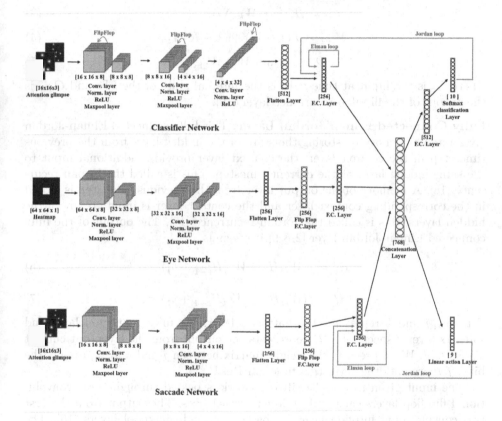

**Fig. 4.** Details of the attention model architecture for counting problem.

proposed network, this memory is held by two mechanisms: 1) use of Flip-flop
neurons in the hidden layers, 2) use of Elman and Jordan feedback recurrence
layers. These two features are now described.

**Flip Flop Neuron Layer:** A flip-flop is an electronic component used in digital circuits to store state information. We used JK flip-flop [10] neurons in place of LSTM neurons for reasons of more efficient memory utilization [10]. Holla et al. have shown that SR and Toggle flip-flops outperform the LSTM by using half of the training parameters comparatively. The JK flip-flop neuron layer in the current model is implemented in two ways: convolutional layer with JK flip-flops, which is called Convolutional flip-flop layer and a fully connected layer with JK flip-flops which is called FC flip-flop layer. The input/output relationship of the JK flip-flop is given by (Eqns. 2–5).

$$J = \sigma(W_j.X_t) \tag{2}$$

$$K = \sigma(W_k.X_t) \tag{3}$$

$$V_{t+1} = J.(1 - V_t) + (1 - K).V_t \tag{4}$$

$$O_{out} = \tanh(W_{out}.V_{t+1}) \tag{5}$$

where $X_t$ is the input at time $t$; $V_t$ is the internal state of the FF; and $O_{out}$ is the output of the flip flop recurrence layer [10].

**Fully Connected Elman Jordan Layer:** This fully connected Elman-Jordan layer retains memory by storing the state of the hidden layer from the previous time step in the context layer; the context layer provides additional input to the same hidden layer at the current timestep. This is called the Elman recurrence [18]. A memory of the output layer from the previous time step is stored in the corresponding context layer and the context layer is input to any of the hidden layers; this is called the Jordan recurrence [18]. The output of the fully connected Elman-Jordan layer [2,8,9] is given as

$$H_t^j = f(W_{ij}H_t^i + W_{jj}H_{t-1}^j + b_j) \tag{6}$$

$$H_t^j = f(W_{ij}H_t^i + W_{kj}H_{t-1}^k + b_j) \tag{7}$$

Where $i, j$ and $k$ represents the flattened layer, the fully connected layer and softmax layer respectively. $H_t^j$ represents the output from the layer $j$ at current time step $t$. $W_{ij}$ represents the weight matrix between $j$ and $i$. $b_j$ represents the bias. $f$ represents the activation function ReLU.

The input given to the Classifier network is passed through three convolutional flip-flop layers each with different kernel sizes. The output from the first two convolutional flip-flop layers is passed through maxpool layers [15]. The output of the last convolution layer is then flattened and passed through a fully connected Elman-Jordan layer. The Eye Position Network takes the heatmap representation as an input. The input is passed through two convolutional layers [17], and two maxpool layers. The output of the last maxpool layer is then flattened and is passed through one fully connected flip-flop layer and one fully connected layer. The Saccade Network is responsible for learning the previous actions of the glimpse locations and giving an optimal direction to the glimpse movement in the image. The Saccade Network takes the attention as an input.

The input is passed through two convolutional layers and two maxpool layers. The output of the last convolutional layer is flattened and passed through one fully connected flip-flop layer and one fully connected Elman-Jordan layer.

Output from the fully connected Elman-Jordan layer of the classifier network, the fully connected layer of the eye network, and the fully connected Elman-Jordan layer of the saccade network is concatenated into a single layer. The concatenation layer is split into two separate parallel paths: one is to predict the class and the other is to predict the next location of the attentional glimpse. In one pathway, the concatenation layer first passes through one fully connected layer then through one softmax layer [3] to give the output of classification probabilities. When the probability of a class is greater than the threshold value $\lambda$, the softmax classification layer makes a class prediction and receives a reward if the prediction is true. The reward scheme is described by the equation:

$$R_{t+1} = \begin{cases} 1, & \arg\max_{i \in n}(p_i) == \arg\max_{i \in n}(t_i) \\ & \max(p) > \lambda \\ 0, & \text{otherwise} \end{cases}$$

Where $p_i$ is the predicted probability for count class $i$ and $t_i$ is the actual probability for the count class $i$. $\lambda$ is the threshold value equal to 0.51.

In another pathway, the concatenation layer directly passes through one linear layer to give the output of action probabilities (Fig.-4). The agent makes a decision about the attentional window location for the next timestep. For the next location, the agent has 9 options to choose from. i.e. right, left, up, down, top left, top right, bottom left, bottom right, and nowhere. The glimpse moves a certain fixed amount of distance called "jump length" equal to 8 from the previous location $l$ in the given direction of movement.

**Training and Testing.** Cross entropy loss [20] and mean square error of temporal difference [1] was calculated from classification probabilities and action probabilities respectively. Both losses were added and backpropagated using the Adam optimizer [6]. First, we trained the network with as many as 128 glimpses for each image. During the testing phase, however, we limit the number of glimpses to $N$, where $N$ is equal to $1, 2, 4, 8, 16$ respectively, the model makes the decision to count class prediction if the maximum value of the predicted class probabilities crosses a threshold of 0.6. In doing so we could analyze the model's intermediate performance in estimating the count of objects after only a few glimpses.

## 3    Results

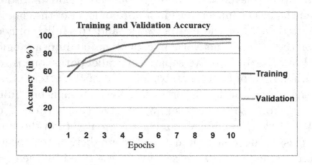

**Fig. 5.** Demonstration of the efficiency of the model with training and validation data.

The initial location of the attentional window was chosen to be random. During training, the agent takes actions (saccadic jumps) for 128 timesteps, while at each timestep deciding for the next window locations determined by the Saccade Network. After tuning the hyperparameters, the learning rate is chosen to be 0.001 with a decay factor of 0.5. All the weights in the network were regularized with a beta value of 0.1. The batch size for training, validation and testing was 512. The value of $\gamma$ for Q-learning [23] was set to 0.4. The network was trained up to 10 epochs. After training the attention model, using three concentric windows for 10 epochs, our attention model obtained a testing accuracy of 92.10%.

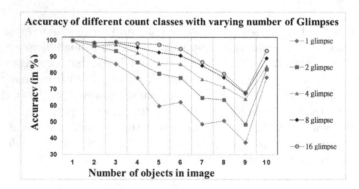

**Fig. 6.** Demonstration of the efficiency of the glimpse location decision mechanism.

The attention model successfully learns to focus on the most salient areas of the image (Fig.-5). During the testing phase, we limited the number of glimpses to N. The network has shown images which have fewer objects that were correctly classified even when the model was restricted to fewer glimpses (Fig.-6).

Images which had more objects required relatively more glimpses for correct classification. There exists a close relationship between the number of glimpses required and the number of objects in the image. The model has successfully learned how to choose correct glimpse locations that correspond to objects in the image.

## 4 Conclusion and Future Work

The proposed attention-based model, loosely fashioned after the what and where pathways of the primate visual system, could be trained successfully to count up to 10 visual items at a reasonable accuracy. It clearly learned how to count objects by identifying the relevant locations to focus upon, rather than processing the complete image. We have seen the effectiveness of the attention mechanism; the success of this model is due to the correct glimpse behavior. Once trained, it can process images more rapidly since it will process less data because it only considers relevant areas to focus upon. A future objective in this area is to build an attention recurrent model that not only aims to maximize counting accuracy but also minimize the time taken (or the number of glimpses) to perform counting. In this regard, it would be interesting to seek inspiration from counting strategies adopted by humans, and saccade patterns exhibited by the human visual system engaged in the counting task [12]. Another application of the proposed model is in change detection ( [4,7]), where the attention can move in the image and make the decision that which part of the image is changed.

**Acknowledgement.** We acknowledge the support from Pavan Holla and Vigneswaran for the implementation of flip-flop neurons. Sweta Kumari acknowledges the financial support of the Ministry of Human Resource Development for a graduate assistantship.

## References

1. Cai, Q., Yang, Z., Lee, J.D., Wang, Z.: Neural temporal-difference learning converges to global optima. In: Advances in Neural Information Processing Systems **32** (2019)
2. Chopra, D., Kumari, S., Chakravarthy, V.S.: Modelling working memory using deep convolutional Elman and Jordan neural networks. In: Journal of Computational Neuroscience, vol. 49, pp. S49–S50. Springer Van Godewijckstraat 30, 3311 GZ Dordrecht, Netherlands (2021)
3. Gao, Y., Liu, W., Lombardi, F.: Design and implementation of an approximate softmax layer for deep neural networks. In: 2020 IEEE International Symposium on Circuits and Systems (ISCAS), pp. 1–5. IEEE (2020)
4. Goyette, N., Jodoin, P.M., Porikli, F., Konrad, J., Ishwar, P.: Changedetection. net: a new change detection benchmark dataset. In: 2012 IEEE Computer Society Conference on Computer Vision and Pattern Recognition Workshops, pp. 1–8. IEEE (2012)
5. Huberman-Spiegelglas, I., Fattal, R.: Single image object counting and localizing using active-learning. In: Proceedings of the IEEE/CVF Winter Conference on Applications of Computer Vision, pp. 1310–1319 (2022)

6. Jais, I.K.M., Ismail, A.R., Nisa, S.Q.: Adam optimization algorithm for wide and deep neural network. Knowl. Eng. Data Sci. **2**(1), 41–46 (2019)

7. Keshvari, S., Van den Berg, R., Ma, W.J.: No evidence for an item limit in change detection (open access). Technical report, BAYLOR COLL OF MEDICINE HOUSTON TX HOUSTON United States (2013)

8. Kumari, S., Aravindakshan, S., Jain, U., Srinivasa Chakravarthy, V.: Convolutional Elman Jordan neural network for reconstruction and classification using attention window. In: Sharma, M.K., Dhaka, V.S., Perumal, T., Dey, N., Tavares, J.M.R.S. (eds.) Innovations in Computational Intelligence and Computer Vision. AISC, vol. 1189, pp. 173–181. Springer, Singapore (2021). https://doi.org/10.1007/978-981-15-6067-5_20

9. Kumari, S., Aravindakshan, S., Srinivasa Chakravarthy, V.: Elman and Jordan recurrence in convolutional neural networks using attention window. In: Gupta, D., Khanna, A., Bhattacharyya, S., Hassanien, A.E., Anand, S., Jaiswal, A. (eds.) International Conference on Innovative Computing and Communications. AISC, vol. 1165, pp. 983–993. Springer, Singapore (2021). https://doi.org/10.1007/978-981-15-5113-0_83

10. Kumari, S., Vigneswaran, C., Chakravarthy, V.S.: The flip-flop neuron-a memory efficient alternative for solving challenging sequence processing and decision making problems. BioRxiv (2021)

11. Lin, H., Hong, X., Wang, Y.: Object counting: you only need to look at one. arXiv preprint: arXiv:2112.05993 (2021)

12. Lindsey, J., Jiang, S.: Visual attention models of object counting. Glimpse, vol. 90, pp. 100

13. Meshram, S.A., Lande, R.S.: Traffic surveillance by using image processing. In: 2018 International Conference on Research in Intelligent and Computing in Engineering (RICE), pp. 1–3. IEEE (2018)

14. Mnih, V., Heess, N., Graves, A., et al.: Recurrent models of visual attention. In: Advances in Neural Information Processing Systems, vol. 27 (2014)

15. Nagi, J., et al.: Max-pooling convolutional neural networks for vision-based hand gesture recognition. In: 2011 IEEE International Conference on Signal and Image Processing Applications (ICSIPA), pp. 342–347. IEEE (2011)

16. Nguyen, H., et al.: Animal recognition and identification with deep convolutional neural networks for automated wildlife monitoring. In: 2017 IEEE International Conference on data Science and Advanced Analytics (DSAA), pp. 40–49. IEEE (2017)

17. O'Shea, K., Nash, R.: An introduction to convolutional neural networks. arXiv preprint: arXiv:1511.08458 (2015)

18. Pham, D.T., Karaboga, D., Duc Truong Pham and Dervis Karaboga: Training Elman and Jordan networks for system identification using genetic algorithms. Artif. Intell. Eng. **13**(2), 107–117 (1999)

19. Riccio, D., Brancati, N., Frucci, M., Gragnaniello, D.: A new unsupervised approach for segmenting and counting cells in high-throughput microscopy image sets. IEEE J. Biomed. Health Inform. **23**(1), 437–448 (2018)

20. Ruby, U., Yendapalli, V.: Binary cross entropy with deep learning technique for image classification. Int. J. Adv. Trends Comput. Sci. Eng. **9**(10), 2020

21. Seguí, S., Pujol, O., Vitria, J.: Learning to count with deep object features. In: Proceedings of the IEEE Conference on Computer Vision and Pattern Recognition Workshops, pp. 90–96 (2015)
22. Verma, N.K., Sharma, T., Rajurkar, S.D., Salour, A.: Object identification for inventory management using convolutional neural network. In:2016 IEEE Applied Imagery Pattern Recognition Workshop (AIPR), pp. 1–6. IEEE (2016)
23. Watkins, C.J.C.H., Dayan, P.: Q-learning. Mach. Learn. **8**(3), 279–292 (1992)

# Human Centered Computing

# Emotion Detection in Unfix-Length-Context Conversation

Xiaochen Zhang[1]([✉]) and Daniel Tang[2]

[1] Australian National University, Canberra, Australia
xiaochen.zhang@anu.edu.au
[2] University of Luxembourg, Esch-sur-Alzette, Luxembourg

**Abstract.** Emotion Detection in conversation is playing more and more important role in dialogue system. Existing approaches to Emotion Detection in Conversation (EDC) use a fixed context window to recognize speakers' emotion, which may lead to either scantiness of key context or interference of redundant context. In response, we explore the benefits of variable-length context and propose a more effective approach to EDC. In our approach, we leverage different context windows when predicting the emotion of different utterances. New modules are included to realize variable-length context: 1) two speaker-aware units, which explicitly model inner- and inter-speaker dependencies to form distilled conversational context and 2) a top-k normalization layer, which determines the most proper context windows from the conversational context to predict emotion. Experiments and ablation study show that our approach outperforms several strong baselines on three public datasets.

**Keywords:** Conversation · Emotion Detection · Transformer

## 1 Introduction

Emotion Detection in Conversation (EDC) is the task of predicting the speaker's emotion in conversation according to the previous context and current utterance. Great technical breakthroughs of EDC promote the development of applications in an army of domains, such as healthcare, political elections, consumer products and financial services [11,15,18]. Figure 1 shows an example of EDC. Existing approaches [4,6] consider a fixed context window (i.e., the number of preceding utterances), which may suffer from two issues: (1) semantic missing due to a small window; or (2) redundancy problem in big context text, making it difficult to choose the right context in the task CHQA. Therefore, knowing the current speaker is Harry is beneficial to choosing the right context window since one of the preceding utterances explicitly mentions Harry, which indicates that it may contain information relevant to the current utterance. That is, speaker dependencies are the key indicators to determine the right context window.speaker dependencies are both critical to conversation understanding [5], where speaker dependencies can be further categorized into inner- and inter-speaker dependencies [8]. Firstly, we model the above dependencies by an attention-based utterance encoder and two speaker-aware units to generate conversational context representation,

M. Tanveer et al. (Eds.): ICONIP 2022, LNCS 13624, pp. 197–206, 2023.
https://doi.org/10.1007/978-3-031-30108-7_17

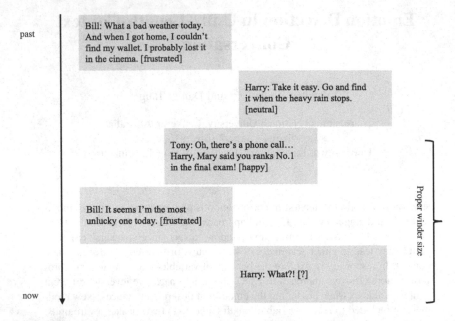

**Fig. 1.** A multi-party EDC example. The ideal context window to Harry's emotion would include exactly two preceding utterances, among which Tony provides evidence for Harry being happy. Utterances ahead of Tony are redundant since they are irrelevant to the current turn of conversation.

where inner- and inter-speaker dependencies are explicitly modeled to help detect the ideal context windows. Next, a top-k normalization layer generates top-k best context windows and their probability weights based on the dimension-reduced context representation. Lastly, we predict the emotion of current utterance by softly leveraging the top-k best context windows. Experiments show that our approach achieves competitive performance on three public conversational datasets: 66.35% F1 on IEMOCAP [2]; 61.22% F1 on DailyDialog [10]; and 38.93% F1 on EmoryNLP [20]. Extensive ablation study demonstrate the contribution of each component in our approach as well as the necessity of using variable-length context.

We summary our contributions as threefold:

- For the first time, we alleviate the context scantiness and context redundancy problems in EDC by varying the length of context..
- We propose a new approach that considers different context windows for different instances to conduct emotion prediction, where 1) speaker dependency is explicitly modeled by new speaker-aware units to help the detection of ideal context windows and 2) a new top-k normalization layer that generates top-k best context windows as well as their weights.
- We achieve competitive results on three public EDC datasets and conduct elaborate ablation study to verify on the effectiveness of our approach.

## 2    Related Work

Recent EDC studies are based on Deep Learning, which can be further categorized into three main kinds: RNN-based, GCN-based and Transformer-based models. RNN-based models have been well explored in the last few years. Poria et al. (2017) [16] first modeled the conversational context of EDC using Recurrent Neural Networks (RNNs) [14]. Hazarika et al. (2018) [8] took speaker information into account and Hazarika et al. (2018) [7] first modeled Inter-speaker dependencies. Majumder et al. (2019) [13] kept track of speakers' states and their method could be extended to multi-party conversations. Lu et al. (2020) [12] proposed an RNN-based iterative emotion interaction network to explicitly model the emotion interaction between utterances. Ghosal et al. (2019) [6] and Sheng et al. (2020) [19] adopted relational Graph Convolutional Networks (GCN) to model EDC, where the whole conversation was considered as a directed graph and they employed graph convolutional operation to capture the dependencies between vertices (utterances). However, converting conversations to graphs loses temporal attributes of original conversation. Owing to the excellent representation power of transformers [3], some researchers adapted them to EDC and got favorable results [9]. Recently, Ghosal et al. (2020) [4] incorporated commonsense knowledge extracted from pretrained commonsense transformers COMET [1] into RNNs and obtained favorable results on four public EDC datasets. However, none of the above models regarded the context scantiness or the context redundancy problem as us.

## 3    Our Method

### 3.1    Problem Formulation

A conversation consists of n temporally ordered utterances $\{x_1, \ldots, x_n\}$ and their speakers $\{s_1, \ldots, s_n\}$. $x_i$ is the $i$-$th$ word in the sequence. At time step $t$, the goal of EDC is to identify the most-likely categorical emotion label $\hat{y}_t$ for speaker $s_t$ given the current and preceding utterances as well as their speakers: $\hat{y}_t = argmaxp(y_t|x_{1:t}, s_{1:t})$, where $1:t$ means set of the former $t$ elements.

### 3.2    Model

As depicted in Fig. 2, our approach consists of the following modules: (1) an utterance encoder that encodes sequential dependencies among utterances; (2) two speaker-aware units that explicitly encodes inner-and inter-speaker dependencies to help detect the ideal context windows; (3) a multi-layer perception and a top-k normalization layer that generate distribution over different context windows, from which we determine top-k best context windows and their corresponding weights; and (4) a prediction module that generates emotion distribution from the top-k best context windows with different probability weights. Utterance Encoder The input of utterance encoder is a sequence of tokens with speaker information. At time step t, we generate the input sequence by prepending speaker information (i.e. the name of speaker) to each utterance and then concatenating utterances up to time step t into a single sequence of tokens. The name

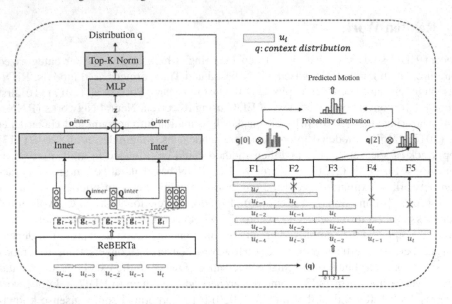

**Fig. 2.** Overall architecture of our approach.

of speaker and the utterance are separated by special [SEP] token. The input sequence is fed into the base version of RoBERTa [12] to encode the sequential dependencies among utterances and generate contextual representation for each utterance:

$$u_i = s_i \oplus [SEP] \oplus x_i,$$
$$[g_1, \ldots, g_t] = RoBERTa(\oplus_{i=1}^{t} u_i) \tag{1}$$

where $g_i$ represent the contextual representation for utterance at $i$, which is the RoBERTa output corresponding to the first token of $u_i$ With a context window considering up to M previous time steps, the encoder outputs a sequences of vectors $[g_{t-M}, \ldots, g_{t-1}, g_t]$, where $g_i \in \mathcal{R}^d$.

Speaker-Aware Units: Our approach incorporates speaker dependencies to guide the detection of ideal context windows. Concretely, we propose two speaker-aware units to explicitly capture inner-speaker and inter-speaker dependencies. The two units have the same attention-based structure, but they do not share parameters. We first divide utterance contextual representations $[g_{t-M}, \ldots, g_{t-1}]$, into two subsets Ginner and Ginter depending on whether their corresponding speakers are the same as the current one. Each speaker-aware unit then takes the corresponding subset G and gt as input, and applies multi-head attention with layer normalization to incorporate speaker dependencies:

$$o = LayerNorm(c + g_t),$$
$$c = Concat((head_1, \ldots, head_h), \Phi_1), \tag{2}$$
$$head_i = Attention((g_t, G, G)^T(\Phi_2, \Phi_3, \Phi_4)), \Phi \in \mathcal{R}$$

where $\Phi$s are the parameters of different layer in our model. Finally, we concatenate $o^{inter}$ and $o^{inner}$ into the vector $z$ as $z = [o^{inter}; o^{inner}] \in \mathcal{R}^{2d}$.

Context Window Distribution: Using the distilled vector z, we generate a probability distribution over context windows ranging from 0 to M. This is done via: (1) a multi-layer perceptron (MLP) which maps the distilled vector to scores of context windows, and (2) a top-k normalization layer which generates distribution over context windows.

Specifically, we first feed the distilled vector z into a two-layer MLP to get scores of context windows s:

$$h = ReLU(z, \Phi_5) \in \mathcal{R}_h^d,$$
$$s = MLP(h; \Phi_6) \in \mathcal{R}^{M+1} \tag{3}$$

Emotion Prediction from top-K best Context Windows Instead of using the context window with the highest probability to predict emotion, we use $q = softmax(s+m)$ as soft labels and leverage all top-K context windows in prediction. As shown in Fig. 2, our prediction module contains M + 1 context fields from 0 to M, where field i corresponds to the use of context window i. The input of each field, with a [CLS] at its front, is encoded by a field-specific contextual encoder, which has the same architecture of our utterance encoder. We use a field-specific linear classifier to the encoder output for [CLS], $g_{[CLS]}^i \in \mathcal{R}^d$, to compute the emotion label distribution $p^i$ given context window i:

$$p^i = softmax(g_{[CLS]}^i; \Phi_7) \in \mathcal{R}^c. \tag{4}$$

The final emotion label distribution $\hat{p}$ combines top-K context window distribution and emotion label distributions given different context windows:

$$\hat{p} = \Sigma_{i \in top-K} 1[i] p^i \in \mathcal{R}^c. \tag{5}$$

### 3.3   Training

We optimize cross-entropy loss $\mathcal{L}$ for each mini-batch **B** of conversations:

$$\mathcal{L} = \Sigma_{i=1}^{|B|} \Sigma_{j=1}^{|B_i|} - log\hat{p}^{ij}[y_{ij}], \tag{6}$$

## 4   Experiment Design

### 4.1   Dataset

We evaluate our approach on four publicly available datasets, IEMOCAP [2], DailyDialog [10], MELD [17] and EmoryNLP [20]. They differ in the number of interlocutors, conversation scenes, and the emotion labels. As shown in Fig. 3, the average conversation lengths of the four datasets differ a lot, with the maximum 49.23 for IEMOCAP and minimum 7.85 for DailyDialog. Moreover, the datasets hold varied data capacity and average utterance lengths. Following existing approaches, models for the datasets are independently trained and evaluated. For preprocessing, we follow Zhong et al. (2019) [21] to lowercase and tokenize the utterances in the datasets using Spacy.

## 4.2 Baselines

To demonstrate the effectiveness of our approach, we compare it with several strong baselines as follows:

- DialogueRNN, an RNN-based ERC model that keeps track of the states of context, emotion, speakers and listeners by several separate GRUs.
- DialogueGCN, a GCN-based ERC model, where they adopt relational graph neural networks to model different types of relations between utterances in the conversation according to their temporal order and speakers.
- KET , a transformer-based model, which leverages external knowledge from emotion lexicon NRC VAD and knowledge base ConceptNet to enhance the word embeddings. They adopt a hierarchical attention-based strategy to capture the contextual information.
- RoBERTa-BASE, the base version of RoBERTa. The inputs are concatenated utterances and the representation of the first subword from the last layer is fed to a simple linear emotion classifier. If the input length exceeds the limitation of RoBERTa, we discard the remote utterances at utterance level.
- COSMIC, a strong ERC model which extracts relational commonsense features from COMET and utilizes several GRUs to incorporate the features to help emotion classification.

# 5   Experimental Result

## 5.1   Main Results

| Approach | IEMOCAP | DailyDialog | MELD | EmoryNLP |
|---|---|---|---|---|
| DialogueRNN [15] | 62.75 | 50.65 | 57.03 | 31.70 |
| DialogueGCN [5] | 64.18 | - | 58.10 | - |
| RoBERTa-BASE* [12] | 62.46 | 58.41 | 63.22 | 35.44 |
| KET [27] | 59.56 | 53.37 | 58.18 | 34.39 |
| COSMIC [3] | 65.28 | 58.48 | **65.21** | 38.11 |
| COSMIC without CSK | 63.05 | 56.16 | 64.28 | 37.10 |
| Ours* | **66.35**±0.21 | **61.22**±0.16 | 64.42±0.18 | **38.93**±0.23 |

**Fig. 3.** Main results. The best F1 scores are highlighted in bold. - signifies the unreported results. CSK is the abbreviation of commonsense knowledge. means the results obtained by our implementation

The main results are reported in Fig. 3. Our approach achieves the best performance on IEMOCAP, DailyDialog and EmoryNLP datasets, surpassing COSMIC by 1.07%, 2.74% and 0.82% F1 scores respectively. We owe the better performance of our approach over COSMIC to the consideration of variable-length context. Moreover, unlike COSMIC, our approach does not rely on external knowledge. For MELD, the result of our approach is also competitive, outperforming all the baselines except COSMIC. We show that the slightly better performance of COSMIC is due to the use of common-sense knowledge (CSK). Our approach performs better than COSMIC without CSK. This indicates that external knowledge could be benefitial to the prediction of short utterances.

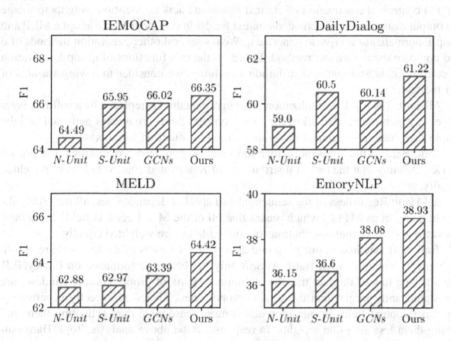

**Fig. 4.** Ablation for the speaker-aware units on the test sets of four datasets.

**Ablation Study.** In order to expose the contribution of different components in our approach, we conduct ablation experiments on the main components: the speaker-aware units and the generation method of context window distribution. Speaker-Aware Units We compare the speaker-aware units with following modeling methods of speaker dependencies: N-Unit: N-Unit shares the same structure with the inner- (inter-) speak-eraware unit. Different from the speaker-aware units, the keys and values of its inputs are all the previous utterance representations regardless of their innerand inter-speaker relationships. N-unit is non-speaker-aware. S-Unit: S-Unit concatenates one-hot vectors, which indicates the speaker of each utterance, to the utterance representations and conducts the same operation as N-Unit. GCNs: Method from [5], where multiple graph convolution layers captures the speaker dependencies. Nodes are utterances and edge weights are obtained by a similarity based attention module. We add a max pooling layer and a linear layer after it to get the vector z. The inputs of GCNs are the outputs of our utterance encoder. Fig 4 shows the comparison results. We attribute the superior performance of our method over S-Unit to the explicitly modeling of inner- and inter-speaker dependencies. S-Unit surpasses N-Unit, indicating that speaker information is indispensable in the context modeling of EDC. Moreover, our speaker-aware units gain over the best of the other three methods by 0.33% and 0.72% F1 scores on dyadic datasets (IEMOCAP and DailyDialog), less than those on multiparty datasets (MELD and EmoryNLP), 1.03% and 0.85%. We attribute this to more complex speaker dependencies in multi-party conversations than dyadic conversations. Our method is better at capturing speaker dependencies when more speakers participated in the conversation. Generation method of Context Window Distribution Context window distribution q (see Eq. 11) controls the activation of context fields and acts as attention weights to merge the output distributions of activated context fields. In our method, we adopt a MLP and a top-k normalization layer to generate q. We try several other generation methods of q and compare them with our method. Based on the two functions of q, top-k activation of context fields and output distribution weighting, we consider following variants of our method:

All-Soft: The top-k normalization layer in our method is replaced by a softmax layer to get q, which means that all of the $M + 1$ context fields are always activated and the output distributions of context fields are merged by attention weights.

Topk-Hard: After top-k normalization layer, K non-zero probabilities in q are set to $\frac{1}{K}$, meaning that the output distributions of K activated context fields are weighted equally.

All-Hard: Regardless of the sequential and speaker dependencies, all the probabilities in q are set as $\frac{1}{M+1}$, which means that all of the $M + 1$ context fields are always activated and the output distributions of context fields are weighted equally.

Topk-Soft: Method in our proposed approach. F1 scores of the test sets are shown in Fig. 5. Compared to All-Hard, AllSoft only has better performance on EmoryNLP. We attribute this to the fact that the attention weights of proper context windows are not significantly larger than those of improper ones. Therefore, directly deactivating improper context fields in our approach is more reasonable than activating them and giving them less attention weights. In response to the above analysis, Topk-Hard outperforms All-Hard nearly across all the datasets, indicating again that we should avoid activating improper context fields. Our top-k normalization layer promotes the attention weights of the K activated context fields, which signified by the superior performance of Topk-Soft over Topk-Hard. According to the above analysis, our generation method of context window distribution not only avoids activating improper context fields but also gives the activated ones more reasonable attention weights. As a result, our method outperforms other generation methods. How to further reduce the attention weights of improper context fields deserves more exploration in future.

| Method | IEMOCAP | DailyDialog | MELD | EmoryNLP |
|---|---|---|---|---|
| *All-Soft* | 65.24 | 60.51 | 63.48 | 37.87 |
| *Topk-Hard* | 65.75 | 60.22 | 63.82 | 38.23 |
| *All-Hard* | 65.42 | 60.56 | 63.79 | 37.11 |
| *Topk-Soft* | **66.35** | **61.22** | **64.42** | **38.93** |

**Fig. 5.** Ablation for the generation method of context window distribution on the test sets of four datasets.

# 6  Conclusion

To alleviate the context scantiness and context redundancy problems in EDC, we present a new EDC approach being capable of recognizing speakers' emotion from variable-length context. In our approach, we first generate a probability distribution over context windows according to sequential and speaker dependencies, where speaker dependencies are explicitly modeled by the newly proposed inner- and inter-speaker units. Then, we introduce a new top-k normalization layer to leverage all top-k best context windows to conduct emotion prediction conditioned on the context window distribution. Elaborate experiments and ablation study demonstrate that our approach can effectively alleviate the context scantiness and context redundancy problems in EDC while achieving competitive performance on three public datasets. In future, we tend to improve the context window distribution by external knowledge or auxiliary tasks. Also, we'll explore more effective mechanisms for the detection of proper context windows.

# References

1. Bosselut, A., Rashkin, H., Sap, M., Malaviya, C., Celikyilmaz, A., Choi, Y.: COMET: commonsense transformers for automatic knowledge graph construction. arXiv preprint: arXiv:1906.05317 (2019)
2. Busso, C., et al.: IEMOCAP: interactive emotional dyadic motion capture database. Lang. Res. Eval. **42**(4), 335–359 (2008). https://doi.org/10.1007/s10579-008-9076-6
3. Devlin, J., Chang, M.W., Lee, K., Toutanova, K.: BERT: pre-training of deep bidirectional transformers for language understanding. arXiv preprint: arXiv:1810.04805 (2018)
4. Ghosal, D., Majumder, N., Gelbukh, A., Mihalcea, R., Poria, S.: COSMIC: commonsense knowledge for emotion identification in conversations. arXiv preprint: arXiv:2010.02795 (2020)
5. Ghosal, D., Majumder, N., Mihalcea, R., Poria, S.: Utterance-level dialogue understanding: An empirical study. arXiv preprint: arXiv:2009.13902 (2020)
6. Ghosal, D., Majumder, N., Poria, S., Chhaya, N., Gelbukh, A.: DialogueGCN: a graph convolutional neural network for emotion recognition in conversation. arXiv preprint: arXiv:1908.11540 (2019)
7. Hazarika, D., Poria, S., Mihalcea, R., Cambria, E., Zimmermann, R.: ICON: interactive conversational memory network for multimodal emotion detection. In: Proceedings of the 2018 Conference on Empirical Methods in Natural Language Processing, pp. 2594–2604 (2018)
8. Hazarika, D., Poria, S., Zadeh, A., Cambria, E., Morency, L.P., Zimmermann, R.: Conversational memory network for emotion recognition in dyadic dialogue videos. In: Proceedings of the Conference Association for Computational Linguistics. North American Chapter. Meeting, vol. 2018, p. 2122. NIH Public Access (2018)
9. Li, J., Ji, D., Li, F., Zhang, M., Liu, Y.: HiTrans: a transformer-based context-and speaker-sensitive model for emotion detection in conversations. In: Proceedings of the 28th International Conference on Computational Linguistics, pp. 4190–4200 (2020)
10. Li, Y., Su, H., Shen, X., Li, W., Cao, Z., Niu, S.: DailyDialog: a manually labelled multi-turn dialogue dataset. arXiv preprint: arXiv:1710.03957 (2017)
11. Liu, B.: Sentiment analysis and opinion mining. In: Synthesis Lectures on Human Language Technologies, vol. 5, no. 1, pp. 1–167. Springer, Cham (2012). https://doi.org/10.1007/978-3-031-02145-9

12. Lu, X., Zhao, Y., Wu, Y., Tian, Y., Chen, H., Qin, B.: An iterative emotion interaction network for emotion recognition in conversations. In: Proceedings of the 28th International Conference on Computational Linguistics, pp. 4078–4088 (2020)

13. Majumder, N., Poria, S., Hazarika, D., Mihalcea, R., Gelbukh, A., Cambria, E.: DialogueRNN: an attentive RNN for emotion detection in conversations. In: Proceedings of the AAAI Conference on Artificial Intelligence, vol. 33, pp. 6818–6825 (2019)

14. Medsker, L.R., Jain, L.: Recurrent neural networks. Des. Appl. **5**, 64–67 (2001)

15. Nasukawa, T., Yi, J.: Sentiment analysis: capturing favorability using natural language processing. In: Proceedings of the 2nd International Conference on Knowledge Capture, pp. 70–77 (2003)

16. Poria, S., Cambria, E., Hazarika, D., Majumder, N., Zadeh, A., Morency, L.P.: Context-dependent sentiment analysis in user-generated videos. In: Proceedings of the 55th Annual Meeting of the Association for Computational Linguistics (volume 1: Long papers), pp. 873–883 (2017)

17. Poria, S., Hazarika, D., Majumder, N., Naik, G., Cambria, E., Mihalcea, R.: MELD: a multimodal multi-party dataset for emotion recognition in conversations. arXiv preprint: arXiv:1810.02508 (2018)

18. Poria, S., Majumder, N., Mihalcea, R., Hovy, E.: Emotion recognition in conversation: research challenges, datasets, and recent advances. IEEE Access **7**, 100943–100953 (2019)

19. Sheng, D., Wang, D., Shen, Y., Zheng, H., Liu, H.: Summarize before aggregate: a global-to-local heterogeneous graph inference network for conversational emotion recognition. In: Proceedings of the 28th International Conference on Computational Linguistics, pp. 4153–4163 (2020)

20. Zahiri, S.M., Choi, J.D.: Emotion detection on tv show transcripts with sequence-based convolutional neural networks. In: Workshops at the Thirty-Second AAAI Conference on Artificial Intelligence (2018)

21. Zhong, P., Wang, D., Miao, C.: Knowledge-enriched transformer for emotion detection in textual conversations. arXiv preprint: arXiv:1909.10681 (2019)

# Multi-relation Word Pair Tag Space for Joint Entity and Relation Extraction

Mingjie Sun, Lisong Wang, Tianye Sheng, Zongfeng He, and Yuhua Huang[✉]

School of Computer Science and Technology, Nanjing University of Aeronautics
and Astronautics, Nanjing, China
{smj_2020,wangls,hezongfeng}@nuaa.edu.cn, hyuhua2k@163.com

**Abstract.** Joint entity and relation extraction from unstructured texts
is a crucial task in natural language processing and knowledge graph
construction. Recent approaches still suffer from error propagation and
exposure bias because most models decompose joint entity and relation
extraction into several separate modules for cooperation. In addition, the
mode of multi-module cooperation to complete the joint extraction task
ignores the information interaction between entities and relations. Most
modeling methods are based on the pattern of token pairs, which leads
to ambiguous information about entities to a certain extent. To address
these issues, in this work, we creatively propose a method to transform
the extraction task of complex triples under multiple relations into a fine-
grained classification problem based on word pairs. Specifically, to fully
utilize entity information and facilitate decoding, the proposed model
uses a tag strategy specific to the feature of the entity itself. Extensive
experiments show that the performance achieved by the proposed model
outperforms public benchmarks and delivers consistent gain on complex
scenarios of overlapping triples.

**Keywords:** Joint entity and relation extraction · Word pairs · Tag
strategy · Overlapping triples

## 1 Introduction

In the form of a triple (subject, relation, object), extracting entity mentions and
judging relation types between them from unstructured texts are basic natural language processing tasks. Some traditional methods named pipelined [2,19]
divide joint extraction into two independent sub-tasks including entity recognition [13] and relation prediction [22]. The pipelined method has the advantage
of low coupling. It is because of low coupling that this method has the disad-
vantages of error propagation, exposure bias, and poor information interaction.
Therefore, building joint extraction frameworks to figure out these problems
becomes increasingly vital.

Recent studies on joint extraction mainly focus on two aspects: multi-task
learning [10,18,21] and single module single step framework [16]. Multi-task

© The Author(s), under exclusive license to Springer Nature Switzerland AG 2023
M. Tanveer et al. (Eds.): ICONIP 2022, LNCS 13624, pp. 207–218, 2023.
https://doi.org/10.1007/978-3-031-30108-7_18

| Yao Ming , chairman of the Chinese Basketball Association , was born in Shanghai , China. The Palace Museum is located in Beijing. | |
| --- | --- |
| *Entity Pair Overlap* *(EPO)* | (Yao Ming, Chairman, Chinese Basketball Association) (Yao Ming, Work, Chinese Basketball Association) |
| *Single Entity Overlap* *(SEO)* | (China, Contains, Shanghai) (Yao Ming, Birthplace, Shanghai) |
| *Subject Object Overlap* *(SOO)* | (Yao Ming, Last name, Yao) |
| *Normal* | (The Palace Museum, Located in, Beijing) |

**Fig. 1.** An example of the Entity Pair Overlap (EPO), Single Entity Overlap (SEO), Subject Object Overlap (SOO), and Normal patterns.

learning is divided into several sub-modules and completed step by step, which is inevitably affected by the propagation of cascading errors between modules and the exposure bias between training and prediction stages. The single module single step framework eliminates these effects. But it falls into the problem that it can not identify overlapping relation triplets. As shown in Fig. 1, it shows the different overlapping patterns of triples under complex relations.

Now more and more work is focused on solving the problem of overlapping triples [10,17,18,21]. Most prevalent architectures adopt the modeling method of token pairs [17,21]. Usually, the entities in a sentence are proper nouns and will be split into several tokens before encoding. For example, "Shijiazhuang" (a place name) contains "Shi", "##ji" , "##az", "##hua", and "##ng" after passing through a tokenizer. However, the context information represented by tokens is very partial. In addition, there are massive redundant calculations existing between token pairs.

To address the aforementioned challenges, in this paper, we propose a simple but effective joint entity and relation extraction model based on multi-relation word pair tag space(MRWPTS). Considering that a token contains less entity information and massive redundant information and calculations exist between token pairs, we employ the max pooling to combine discrete tokens belonging to the same word into word representation by looking up the position of a word in the token sequence. Such as, for the word "Shijiazhuang" and the positions (0, 4) of tokens, the word representation is generated by max pooling of this segment of tokens. In order to solve the problem of overlapping triples, the model assigns meaningful tags to each word pair under all pre-defined relations. The model adopts a tag strategy refined by entity features, which is helpful to improve the training and inference speed. The joint extraction model proposed in this paper achieves outstanding performance with high efficiency in solving the problem of overlapping triples.

## 2   Related Work

Entity and relation extraction have been studied as two separated categories: pipelined extraction and joint extraction. The pipelined extraction strategy [9,20] is to identify named entities first and then predict the possible relations between entities. Because entity recognition and relation prediction are regarded as two isolated tasks in pipelined methods, they still suffer from error propagation and poor information interaction.

The joint extraction model can be subdivided into several different strategies. The joint extraction is regarded as multi-task learning, which is composed of multiple sub-task modules. The first kind of multi-task learning is that all entities in a sentence are recognized first by the entity recognition module, and then the relation classifier module classifies the relation of every entity pair. [1]. This strategy exists a lot of redundancy of entity pairs and relations, resulting in a waste of computing resources. This method also suffers from exposure bias due to different sources of information for entities in the training and prediction stages. The second kind of multi-task learning is that relations in a sentence are predicted first, and then the head entity and tail entity for each corresponding relation are extracted [21]. This strategy can filter out most redundant relations and entity pairs, reducing the burden of computing resources. In the above two multi-task learning strategies, each module is interdependent, so there will be the problem of cascading error propagation.

Another approach is to model joint extraction as a single-stage model [17]. The joint extraction framework proposed by [17] is based on the linking of token pairs of entity boundaries. There is no influence of error propagation and exposure bias in this one-stage model, and it can handle overlapping triples well. However, the token pairs in this method contain little entity boundary information. Second, the decoding process for triples in this method is complicated and inefficient. In contrast, we propose MRWPTS with more comprehensive modeling information and an efficient decoding process.

## 3   Proposed Approach

In this section, we first introduce the definition of the joint entity and relation extraction task, then elaborate on the unique tag strategy and decoding process of MRWPTS, and finally provide the modeling method of the model. An overview illustration of the model is shown in Fig. 2.

### 3.1   Task Definition

Given a pre-defined relation set $R = \{r_1, r_2, ..., r_Q\}$ with $Q$ types and a sentence $W = \{w_1, w_2, ..., w_X\}$ with $X$ words. Note that words and tokens are not equivalent. The joint entity and relation extraction task aims to identify all possible triples $T = \{(h_n, r_n, t_n)\}_{n=1}^{N}$. $h_n$, $t_n$, and $r_n$ represent the head entity, the tail entity, and their relation, respectively. The entity is composed of one or more consecutive words in the sentence.

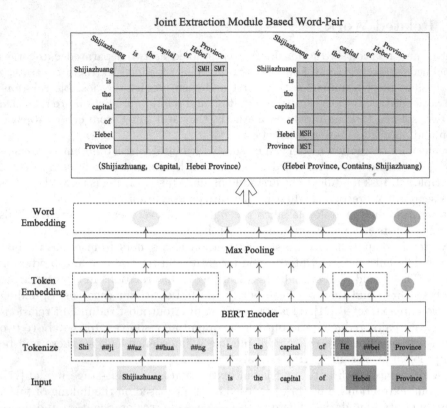

**Fig. 2.** The overall structure of MRWPTS. For a sentence *Input*, perform a tokenization operation on it first and get tokens of the corresponding word. Then, perform a max pooling operation on the token vectors of the corresponding word through the pre-trained BERT model to obtain the word vector. Finally, feed the word sequence to the joint extraction module based on word pairs for word pair tag modeling.

## 3.2    Tag Strategy and Decoding

For the input sentence $W$ and the pre-defined relation set $R$, our classifier model will generate a Q-dimensional tag matrix $TM^{Q \times X \times X}$. Each cell of the matrix $TM$ is assigned a tag with a specific meaning. Each dimension in the matrix $TM$ corresponds to a relation, and the tag in each cell represents the meaning of a word pair in the sentence. The rows and columns in the matrix represent the head entity's words and the tail entity's words, respectively. Decoding refers to extracting triples from the Q-dimensional matrix according to the tags.

We set eight tags according to the length feature of entities and the alignment of entity pairs: **SS, SMH, SMT, MSH, MST, MMH, MMT**, and **A**. In the tags, **S** and **M** indicate that the entity is composed of a single word and multiple words, respectively.(1)**SS.** Both positions of this tag are **S**, meaning that both the head entity and the tail entity consist of a single word. (2)**SMH.** The **S** and **M** in the tag indicate that the head entity consists of a single word, and the tail entity consists of multiple words. **H** means that the first word of the head entity is aligned with the first word of the tail entity. (3)**SMT. T** means that the first word of the head entity is aligned with the end word of the tail entity. (4)**MSH.** The **M** and **S** in the tag indicate that the head entity consists of multiple words, and the tail entity consists of a single word. **H** means that the first word of the head entity is aligned with the first word of the tail entity. (5)**MST. T** means that the end word of the head entity is aligned with the first word of the tail entity. (6)**MMH.** This tag means that the head entity and the tail entity consist of multiple words, and the first word of the head entity is aligned with the first word of the tail entity. (7)**MMT. T** indicates that the end word of the head entity or tail entity appears in the current alignment. (8)**A.** This tag means an empty alignment.

Using this tag strategy, we can make the most of the structural feature of the entity itself and facilitate the decoding. We need to find the correct triples under each relation in the decoding. In other words, we need to find the legal word pair alignment tags in the Q-dimensional matrix. As shown in Fig. 2, the two relations correspond to two sub-matrices. In the *Capital* sub-matrix, we only pay attention to those tags that are not empty. We first get the **SMH** tag, from which we know that the head entity is composed of a single word, the tail entity is composed of multiple words, and the first word of the head entity is aligned with the first word of the tail entity at present. Move to the back of the current line and continue to find the end word of the tail entity. When the tag **SMT** appears, we can extract the triple (Shijiazhuang, Capital, Hebei Province). For the *Contains* sub-matrix, **MSH** tag is obtained first, from which we know that the head entity is composed of multiple words, the tail entity is composed of a single word, and the first word of the head entity is aligned with the first word of the tail entity at present. We need to move down along the current column and find the tag **MST** to extract the triplet (Hebei Province, Contains, Shijiazhuang). For the case where the head entity and the tail entity are composed of multiple words, such as (New York City, City name, New York), the model will assign the **MMH** tag to the (New, New) word pair, the **MMT** tag to the (New, York) word pair, and the (City, York) word pair. The decoding process only needs to find specific tags according to the rules.

## 3.3   Modeling Method

**Encoding Layer.** Given a sentence $W$, we use a pre-trained BERT model [3] to encode the contextualized representation for each token. The output of the encoder is $W_{enc} = \{t_1, t_2, ..., t_n | t_i \in \mathbb{R}^{1 \times d}\}$, where $n$ is the number of the tokens, and $d$ is the embedding hidden dimension.

**Word Embedding Layer.** In the process of tokenization, we maintain a matrix to preserve the token indexes belonging to each word for the subsequent fusion operation. Now we get the token embedding of the sentence. Next, we need to fuse the token embedding belonging to each word to get the word embedding as shown in Fig. 2. We use a max pooling [5] operation to fuse the token embedding. This max pooling method can enable the word embedding representation to contain richer semantic information. The calculation formula of the word embedding is as follows:

$$Index = [(1, n_1)_1, (n_1 + 1, n_2)_2, ..., (n_i, n)_X],$$
$$Emb_i = Maxpool(W_{enc}[n_x : n_y]), \forall (n_x, n_y)_i \in Index, \tag{1}$$

where $Index$ denotes the index set of tokens corresponding to each word in the sentence, $(n_x, n_y)_i$ is the start and end position index of tokens corresponding to the $i$th word in the sentence. : is a slice operation, $Emb_i$ is the word embedding of the $i$th word.

**Joint Extraction Layer.** For an input sentence, we get the word embedding $Emb_i$ for each word in the sentence. We enumerate all possible word pairs ($Emb_i$, $Emb_j$) under all pre-defined relations and design a joint extraction module to assign them high-confidence tags. Inspired by the dependency parsing and the knowledge graph representation, we combine the ideas of Biaffine Attention [4] and HOLE [12] to achieve this goal:

$$h_i = MLP^{head}(Emb_i),$$
$$t_j = MLP^{tail}(Emb_j), \tag{2}$$

where $MLPs \in \mathbb{R}^{d \times d_e}$, $d_e$ denotes the dimension of entity representations. Applying two smaller $MLPs$ maps word embeddings to get entity representations. This approach has two advantages: First, the smaller $MLPs$ can strip away information of high-dimensional word embedding not relevant to the current decision. Second, the low-dimensional entity representations can speed up the subsequent calculation.

$$P(h_i, r_q, t_j)_{q=1}^Q = Softmax(ReLU(drop(h_i + t_j))R^T), \tag{3}$$

where $ReLU$ is the activation function. $drop$ denotes the dropout strategy [15]. $R \in \mathbb{R}^{d_e \times 8Q}$ is a trainable relation projection matrix, where 8 is the number of classification tags. We calculate the tag score for the word pair $P(h_i, r_q, t_j)_{q=1}^Q$ under all relations at once. Here, the entity pair representations obtained by the + operation can also avoid the problem that the head entity and tail entity are not exchangeable under asymmetric relations,i.e., $h_i + t_j \neq h_j + t_i$. The + operation will not cause the expansion of dimension, which will affect the calculation speed.

We optimize the objective function during training time as follows:

$$L_{matrix} = -\frac{1}{Q \times X \times X} \sum_{q=1}^Q \sum_{i=1}^X \sum_{j=1}^X y(h_i, r_q, t_j) log P(h_i, r_q, t_j), \tag{4}$$

where $y(h_i, r_q, t_j)$ is the gold tag.

# 4  Experiments

## 4.1  Experimental Settings

**Datasets.** We evaluate MRWPTS on two benchmark datasets NYT [14] and WebNLG [7]. According to the annotation strategy, both of them have two versions. We use NYT, NYT*, WebNLG, and WebNLG* to distinguish. NYT and WebNLG annotate the whole entity span. NYT* and WebNLG* annotate the last word of entities. In addition, we split the test set into different subsets according to the overlapping patterns. Detailed statistics of datasets are in Table 1.

**Table 1.** Statistics of datasets.

| Dataset | Samples | | | Details of test set | | | | | |
|---|---|---|---|---|---|---|---|---|---|
| | Train | Valid | Test | Normal | SEO | EPO | SOO | #Triples | #Relations |
| NYT* | 56195 | 4999 | 5000 | 3266 | 1297 | 978 | 45 | 8110 | 24 |
| NYT | 56196 | 5000 | 5000 | 3071 | 1273 | 1168 | 117 | 8616 | 24 |
| WebNLG* | 5019 | 500 | 703 | 245 | 457 | 26 | 84 | 1591 | 171 |
| WebNLG | 5019 | 500 | 703 | 239 | 448 | 6 | 85 | 1607 | 216 |

**Evaluation.** For a fair comparison, we follow the prior works. We evaluate the performances with the Precision (Prec.), Recall (Rec.), and F1-score. We use the strict evaluation criteria to measure the quality of extracted triples. An extracted triple (h, r, t) is considered to be correct only if the last word of both the head entity and tail entity (NYT* and WebNLG*) or the whole span of both the head entity and tail entity (NYT and WebNLG) and the relation exactly match with ground truth.

**Implementation Details.** Our model is implemented by PyTorch. The model training is deployed on a server with a Tesla V100-PCIe GPU of 32G memory. We use the BERT base cased English model[1] as the sentence encoder. We use Adam [8] as the optimizer with the initial learning rate of 0.00001. Following the previous work, we limit the max length of sentence tokens to 100. The output dimension $d$ of BERT is 768. We set the dimension $d_e$ of $MLPs$ as 50. To avoid overfitting, the dropout is at a rate of 0.1. We use the batch size of 6/32 for WebNLG(WebNLG*)/NYT(NYT*), respectively. We select six strong models as the baselines for this experiment.

## 4.2  Results and Analysis

We compare MRWPTS with several competing models on two benchmark datasets and the overall results are shown in Table 2. The - in the table indicates that the original paper has no experimental results on this dataset.

---

[1] https://huggingface.co/bert-base-cased.

**Table 2.** Comparison results (%) of our proposed model with other baselines. Note that the experimental results of all baseline models are derived from the original papers.

| Models | NYT | | | WebNLG | | | NYT* | | | WebNLG* | | |
|---|---|---|---|---|---|---|---|---|---|---|---|---|
| | Prec. | Rec. | F1 | Prec. | Rec. | F1 | Prec. | Rec. | F1 | Prec. | Rec. | F1 |
| GraphRel [6] | – | – | – | – | – | – | 63.9 | 60.0 | 61.9 | 44.7 | 41.1 | 42.9 |
| MHSA [10] | – | – | – | – | – | – | 88.1 | 78.5 | 83.0 | 89.5 | 86.0 | 87.7 |
| CasRel [18] | – | – | – | – | – | – | 89.7 | 89.5 | 89.6 | 93.4 | 90.1 | 91.8 |
| TPLinker [17] | 91.4 | 92.6 | 92.0 | 88.9 | 84.5 | 86.7 | 91.3 | 92.5 | 91.9 | 91.8 | 92.0 | 91.9 |
| CasDE [11] | 89.9 | 91.4 | 90.6 | 88.0 | 88.9 | 88.4 | 90.2 | 90.9 | 90.5 | 90.3 | 91.5 | 90.9 |
| PRGC [21] | **93.5** | 91.9 | 92.7 | 89.9 | 87.2 | 88.5 | **93.3** | 91.9 | 92.6 | **94.0** | 92.1 | 93.0 |
| Ours | 93.0 | **92.6** | **92.8** | **92.3** | **89.5** | **90.9** | 92.3 | **93.0** | **92.7** | **94.0** | **94.4** | **94.2** |

Experiments show that MRWPTS achieves the great performance of gaining 92.8%, 90.9%, 92.7%, and 94.2% F1 scores on the benchmark datasets. It can be seen from the table that our model achieves absolute F1 score improvements of 0.8%, 4.2%, 0.8%, and 2.3% compared with the TPLinker model. It achieves absolute F1 score improvements of 0.1%, 2.4%, 0.1%, and 1.2% compared with the PRGC model. In addition, the model significantly gets improvements in both Prec. and Rec. From the dataset type and absolute F1 score improvements, since NYT(NYT*) has much fewer relations and much larger samples than WebNLG(WebNLG*), the boosting performance of the model is relatively small. In other words, the number of samples in the dataset is critical to the performance of models. The above overall results show that our model has outstanding performance.

### 4.3    Detailed Results on Overlapping Triples

To verify the performance of the MRWPTS under different overlapping triple conditions, we split the test set into different types of subsets for detailed experiments. Detailed results are shown in Table 3. From the experimental results, it can be seen that MRWPTS improves the performance in almost all types of overlapping triples. The number of overlapping triples exceeds one-quarter of the total number of triples in both NYT* and WebNLG*. In the case of such a large and complex number of overlapping triples, the performance of our model is outstanding, which is attributed to our modeling method and tag strategy.

First, we employ word pairs to enrich the contextual information of entities, which is crucial for the correctness of model tagging. We conduct a case study shown in Fig. 3, the words "Ampara", "Hospital", "Sri", and "Lanka" in sample1 are tokenized to get a token sequence of 'Am', '##par', '##a', 'Hospital', 'Sri', and 'Lanka'. Next, the demonstration is based on token granularity. Following our proposed model, the model tags the token pairs ('Am', 'Sri'), ('Am', 'Lanka'), and ('Hospital', 'Lanka') with **MMH**, **MMT**, and **MMT**, respectively. According to our decoding, the (Ampara Hospital, country, Sri Lanka) can be easily

**Table 3.** Comparison results (F1 %). We only compare those baselines that set up the detailed experiments.

| Model | NYT* | | | | WebNLG* | | | |
|---|---|---|---|---|---|---|---|---|
| | Normal | SEO | EPO | SOO | Normal | SEO | EPO | SOO |
| CasRel | 87.3 | 91.4 | 92.0 | 77.0 | 89.4 | 92.2 | 94.7 | 90.4 |
| TPLinker | 90.1 | 93.4 | 94.0 | 90.1 | 87.9 | 92.5 | 95.3 | 86.0 |
| PRGC | **91.0** | 94.0 | 94.5 | 81.8 | 90.4 | 93.6 | 95.9 | **94.6** |
| Ours | 90.7 | **94.4** | **94.8** | **85.1** | **91.6** | **94.7** | **96.0** | **94.6** |

---

Sample1:
The 476 bed Ampara Hospital is located in Ampara District , Sri Lanka .
Ground Truth:
(Ampara Hospital, country, Sri Lanka)
Prediction:

(Ampara Hospital, country, Sri Lanka)   √
Sample2:
The 476 bed Am Hospital is located in Ampara District , Sri Lanka .
Ground Truth:
None
Prediction:

(Am Hospital, country, Sri Lanka)   ✗

---

**Fig. 3.** The effect of tokens on the extraction result. The sample is from WebNLG.

extracted. Sample2 is intentionally altered based on sample1. The words "Am", Hospital", "Sri", and "Lanka" in sample2 are tokenized to get a token sequence of 'Am', 'Hospital', 'Sri', and 'Lanka'. However, the model still tags the token pairs ('Am', 'Sri'), ('Am', 'Lanka'), and ('Hospital', 'Lanka') with **MMH**, **MMT**, and **MMT**, respectively. As a result, the wrong triple is extracted. The reason is that tokens contain one-sided contextual information, which brings ambiguity. Our tags incorporate the description of the entity length feature, which will also benefit the model.

## 4.4   The Model Efficiency

Table 4 shows the comparison results of MRWPTS with baselines in the training time and inference time.

It can be seen from the table that the training time of MRWPTS is greatly shortened on both WebNLG* and NYT*. The inference time is also competitive with PRGC. This tag strategy containing entity features makes the difference.

As shown in Fig. 4, we extract the complete entities by finding the start word and the end word of entities in the tag matrix. The prerequisite is to ensure that each entity contains at least two words if we adopt the tag strategy on the right.

**Table 4.** Training Time is the time spent training one epoch. Inference Time is time spent predicting the triples in one sentence.

| Model | WebNLG* | | | NYT* | | |
|---|---|---|---|---|---|---|
| | Training Time | Inference Time | F1 | Training Time | Inference Time | F1 |
| TPLinker | 750 s | 50 ms | 91.9 | 1685 s | 33 ms | 91.9 |
| PRGC | 218 s | **11.6 ms** | 93.0 | 1081 s | **13.5 ms** | 92.6 |
| Ours | **103 s** | 17.8 ms | **94.2** | **576 s** | 17.1 ms | **92.7** |

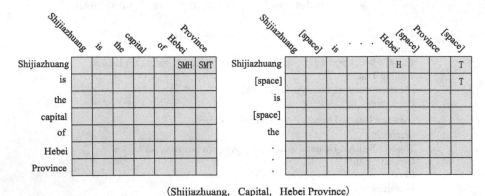

(Shijiazhuang,  Capital,  Hebei Province)

**Fig. 4.** Different tag strategies. The strategy on the left contains entity features, and the strategy on the right does not contain entity features.

Since a large number of single-word entities exist in the dataset, we need to pad an extra space after each word in a sentence. The tag strategy on the right will double the sentence length, which will slow down the training speed and inference speed. However, our tag strategy on the left can distinguish the entity length through **S** and **M**. We do not need to find the end word of a single-word entity during decoding.

## 5    Conclusion

In this paper, we propose a joint entity and relation extraction model based on multi-relation word pair tag space(MRWPTS). MRWPTS uses a tag strategy that integrates the length feature of entities. The model labels all word pairs under different relations at once. We evaluate MRWPTS on two benchmark datasets. The results show that the proposed model outperforms other benchmark models. MRWPTS is competitive in terms of efficiency and has a faster training speed than the state-of-the-art model.

# References

1. Bekoulis, G., Deleu, J., Demeester, T., Develder, C.: Joint entity recognition and relation extraction as a multi-head selection problem. Expert Syst. Appl. **114**, 34–45 (2018)
2. Chan, Y.S., Roth, D.: Exploiting syntactico-semantic structures for relation extraction. In: Proceedings of the 49th Annual Meeting of the Association for Computational Linguistics: Human Language Technologies, pp. 551–560 (2011)
3. Devlin, J., Chang, M.W., Lee, K., Toutanova, K.: Bert: pre-training of deep bidirectional transformers for language understanding. arXiv preprint: arXiv:1810.04805 (2018)
4. Dozat, T., Manning, C.D.: Deep biaffine attention for neural dependency parsing. arXiv preprint: arXiv:1611.01734 (2016)
5. Eberts, M., Ulges, A.: Span-based joint entity and relation extraction with transformer pre-training. arXiv preprint: arXiv:1909.07755 (2019)
6. Fu, T.J., Li, P.H., Ma, W.Y.: GraphRel: modeling text as relational graphs for joint entity and relation extraction. In: Proceedings of the 57th Annual Meeting of the Association for Computational Linguistics, pp. 1409–1418 (2019)
7. Gardent, C., Shimorina, A., Narayan, S., Perez-Beltrachini, L.: Creating training corpora for NLG micro-planners. In: ACL (1) (2017)
8. Kingma, D.P., Ba, J.: Adam: a method for stochastic optimization. arXiv preprint: arXiv:1412.6980 (2014)
9. Lin, Y., Shen, S., Liu, Z., Luan, H., Sun, M.: Neural relation extraction with selective attention over instances. In: Proceedings of the 54th Annual Meeting of the Association for Computational Linguistics (Volume 1: Long Papers), pp. 2124–2133 (2016)
10. Liu, J., Chen, S., Wang, B., Zhang, J., Li, N., Xu, T.: Attention as relation: learning supervised multi-head self-attention for relation extraction. In: Proceedings of the Twenty-Ninth International Conference on International Joint Conferences on Artificial Intelligence, pp. 3787–3793 (2021)
11. Ma, L., Ren, H., Zhang, X.: Effective cascade dual-decoder model for joint entity and relation extraction. arXiv preprint: arXiv:2106.14163 (2021)
12. Nickel, M., Rosasco, L., Poggio, T.: Holographic embeddings of knowledge graphs. In: Proceedings of the AAAI Conference on Artificial Intelligence, vol. 30 (2016)
13. Ratinov, L., Roth, D.: Design challenges and misconceptions in named entity recognition. In: Proceedings of the Thirteenth Conference on Computational Natural Language Learning (CoNLL-2009), pp. 147–155 (2009)
14. Riedel, S., Yao, L., McCallum, A.: Modeling relations and their mentions without labeled text. In: Balcázar, J.L., Bonchi, F., Gionis, A., Sebag, M. (eds.) ECML PKDD 2010. LNCS (LNAI), vol. 6323, pp. 148–163. Springer, Heidelberg (2010). https://doi.org/10.1007/978-3-642-15939-8_10
15. Srivastava, N., Hinton, G., Krizhevsky, A., Sutskever, I., Salakhutdinov, R.: Dropout: a simple way to prevent neural networks from overfitting. J. Mach. Learn. Res. **15**(1), 1929–1958 (2014)
16. Wang, Y., Sun, C., Wu, Y., Zhou, H., Li, L., Yan, J.: UniRE: a unified label space for entity relation extraction. arXiv preprint: arXiv:2107.04292 (2021)
17. Wang, Y., Yu, B., Zhang, Y., Liu, T., Zhu, H., Sun, L.: TPlinker: single-stage joint extraction of entities and relations through token pair linking. arXiv preprint: arXiv:2010.13415 (2020)

18. Wei, Z., Su, J., Wang, Y., Tian, Y., Chang, Y.: A novel cascade binary tagging framework for relational triple extraction. arXiv preprint: arXiv:1909.03227 (2019)
19. Zelenko, D., Aone, C., Richardella, A.: Kernel methods for relation extraction. J. Mach. Learn. Res. **3**(Feb), 1083–1106 (2003)
20. Zeng, D., Liu, K., Chen, Y., Zhao, J.: Distant supervision for relation extraction via piecewise convolutional neural networks. In: Proceedings of the 2015 Conference on Empirical Methods in Natural Language Processing, pp. 1753–1762 (2015)
21. Zheng, H., et al.: PRGC: potential relation and global correspondence based joint relational triple extraction. arXiv preprint: arXiv:2106.09895 (2021)
22. Zhou, P., et al.: Attention-based bidirectional long short-term memory networks for relation classification. In: Proceedings of the 54th Annual Meeting of the Association for Computational Linguistics (volume 2: Short Papers), pp. 207–212 (2016)

# Efficient Policy Generation in Multi-agent Systems via Hypergraph Neural Network

Bin Zhang[1,2], Yunpeng Bai[1,2], Zhiwei Xu[1,2], Dapeng Li[1,2], and Guoliang Fan[1,2(✉)]

[1] Institute of Automation, Chinese Academy of Sciences, Beijing, China
{zhangbin2020,baiyunpeng2020,xuzhiwei2019,lidapeng2020,
guoliang.fan}@ia.ac.cn
[2] School of Artificial Intelligence, University of Chinese Academy of Sciences,
Beijing, China

**Abstract.** The application of deep reinforcement learning in multi-agent systems introduces extra challenges. In a scenario with numerous agents, one of the most important concerns currently being addressed is how to develop sufficient collaboration between diverse agents. To address this problem, we consider the form of agent interaction based on neighborhood and propose a multi-agent reinforcement learning (MARL) algorithm based on the actor-critic method, which can adaptively construct the hypergraph structure representing the agent interaction and further implement effective information extraction and representation learning through hypergraph convolution networks, leading to effective cooperation. Based on different hypergraph generation methods, we present two variants: Actor Hypergraph Convolutional Critic Network (HGAC) and Actor Attention Hypergraph Critic Network (ATT-HGAC). Experiments with different settings demonstrate the advantages of our approach over other existing methods.

**Keywords:** Multi-Agent Reinforcement Learning · Hypergraph Neural Network · Representation Learning

## 1 Introduction

Intelligent decision-making problems has attracted a large number of academics in recent years because of its complexity and extensive application. Deep reinforcement learning (DRL), which combines the function approximation capabilities of deep learning with the trial-and-error learning capabilities of reinforcement learning, is closer to real-world biological learning methods, and it has progressed quickly in many fields, yielding good study outcomes. In the single-agent scenario, DRL's performance in Go [16] and Atari 2600 games [13], for example, has topped that of humans. Simultaneously, academics working on the intelligent decision-making issues in multi-agent systems have produced some impressive

M. Tanveer et al. (Eds.): ICONIP 2022, LNCS 13624, pp. 219–230, 2023.
https://doi.org/10.1007/978-3-031-30108-7_19

outcomes, including intelligent transportation system [14], wireless sensor network management [15], as well as Multiplayer Online Battle Arena (MOBA) and Real-Time Strategy (RTS) games [19]. However, there are still many obstacles in the multi-task and multi-agent setting that significantly restrict the algorithm's deployment and applicability in the real world. When all agents are treated as a single entity, the joint action space grows exponentially with the number of agents [2]. If each agent is individually trained through reinforcement learning, the Markov property of the environment will be invalid. And because the environment is non-stationary, each agent has no way of knowing whether the reward it receives is the result of its own actions or those of others.

Therefore, finding creative training approaches and effectively extracting the attributes of agents is vital to lead the mutual cooperation of agents. MADDPG [11], for example, employs a centralized training with decentralized execution structure to combine the benefits of the two methods. Since the critic is only needed during the training phase, it is convenient to use all agents' input to develop a centralized critic for each independent actor, with each actor relying solely on its own local observations during the execution phase. However, simply concatenate all agents' features may result in information redundancy. Feature extraction and representation from high-dimensional and large-scale data can enhance agents' understanding of complex environments and improve their decision-making level, which is also crucial for MARL. MAAC [7] leverages the attention mechanism [18] to get better results. It allows agents to dynamically and selectively pay attention to the features of other agents. The attention mechanism is also used by ATT-MADDPG [12] to complete the dynamic modeling of teammates. Furthermore, because agents in the system can naturally form graph topological structures depending on their locations, there has been a lot of work combining graph neural network (GNN) [20] with MARL, such as DGN [8] and MGAN [21]. However, the approaches described above need that each agent interact with all other agents in the system. In a complex environment, significant interaction between agents in a neighborhood is usually sufficient, whereas interaction between agents in different neighborhoods can be lessened.

To this end, we discuss the adaptive generation of neighborhoods in the multi-agent system and the cooperation of agents within and between neighborhoods. We explore the application of the hypergraph neural network (HGNN) [3] in multi-agent reinforcement learning and propose Actor Hypergraph Convolutional Critic Network (HGAC) and Actor Hypergraph Attention Critic Network (ATT-HGAC). To achieve efficient state representation learning, the dynamic hypergraph is constructed adaptively and the hypergraph convolution is applied. Despite the complexity of the relationship between agents in the environment, our method is able to extract effective features from large amounts of information to achieve efficient strategy learning. Experiments with different reward settings and different types of collaboration show that our approach outperforms other baselines. And the algorithm's working mechanism is revealed by ablation testing and visualization studies.

# 2 Preliminaries

## 2.1 Markov Game

We employ the framework of Markov Games (also known as Stochastic Games, SG) [10], which is widely used as a standardized game model for sequential decision-making problems in multi-agent systems and can be seen as a multi-agent extension of the single-agent Markov Decision Process (MDP). It is represented by a tuple $\langle \mathcal{S}, \mathcal{A}_1, .., \mathcal{A}_N, r_1, ..., r_N, \mathcal{P}, \gamma \rangle$, where $N$ is the number of agents and $\mathcal{S}$ is the environment state shared by all agents; $\mathcal{A}_i$ is the action set of agent $i$ and the joint action of all agents is described as $\mathcal{A} = \mathcal{A}_1 \times ... \times \mathcal{A}_N$. If agent $i$ performs action $a$ in state $s$ and then transitions to new state $s'$, the environment will reward it with $r_i : \mathcal{S} \times \mathcal{A}_i \times \mathcal{S} \rightarrow \mathbb{R}$; the new state $s'$ is determined by the state transition probability $\mathcal{P} : \mathcal{S} \times \mathcal{A} \times \mathcal{S} \rightarrow [0,1]$. Agent $i$ uses strategy $\pi_i : \mathcal{S} \times \mathcal{A}_i \rightarrow [0,1]$ to take corresponding actions according to its current state, and the joint strategy of all agents is denoted as $\pi = [\pi_1, ..., \pi_N]$. Following the conventional expression of game theory, we use $(\pi_i, \pi_{-i})$ to distinguish the strategy of agent $i$ from all other agents. $\gamma$ represents the discount factor. Under the framework of SG, all agents can move simultaneously in a multi-agent system. If the initial state is $s$, the value function of agent $i$ is expressed as the expectation of discounted return under the joint strategy $\pi$: $v_{\pi_i, \pi_{-i}}(s) = \sum_{t \geq 0} \gamma^t \mathbb{E}_{\pi_i, \pi_{-i}} [r_t^j | s_0 = s, \pi_i, \pi_{-i}]$. According to the Bellman equation, the action-state value function can be written as: $Q_{\pi_i, \pi_{-i}}(s, a) = r_i(s, a) + \gamma \mathbb{E}_{s' \sim p} [v_{\pi_i, \pi_{-i}}(s')]$.

## 2.2 Hypergraph Learning

A hypergraph [23] can be defined as $\mathcal{G} = (\mathcal{V}, \mathcal{E})$, where $\mathcal{V} = \{v_1, ..., v_N\}$ denotes the set of vertices, $\mathcal{E} = \{\epsilon_1, ..., \epsilon_M\}$ denotes the set of hyperedges, $N$ and $M$ are the number of vertices and hyperedges, respectively. Unlike the edges in the graph, the hyperedge can connect any number of vertices in the hypergraph [23]. Hypergraph can be represented by an incidence matrix $\mathcal{H} \in \mathbb{R}^{N \times M}$, with elements specified as:

$$h(v_i, \epsilon_j) = \begin{cases} 1, & \text{if } v_i \in \epsilon_j \\ 0, & \text{if } v_i \notin \epsilon_j \end{cases} \tag{1}$$

where $v_i \in \mathcal{V}, \epsilon_j \in \mathcal{E}$. Each hyperedge is given a weight $w_\epsilon$. All of the weights combine to produce a diagonal hyperedge weight matrix $\mathbf{W} \in \mathbb{R}^{M \times M}$. In addition, the degrees of hyperedges and vertices are defined as $d(\epsilon) = \sum_{v \in V} h(v, \epsilon)$ and $d(v) = \sum_{\epsilon \in E} w_\epsilon h(v, \epsilon)$ respectively, which in turn constitute hyperedge diagonal degree matrix $\mathbf{D_e}$ and vertex diagonal degree matrix $\mathbf{D_v}$ respectively.

In a variety of domains, hypergraph learning is commonly employed. It was first used in semi-supervised learning methods as a propagation process [22]. The learning of distinct modalities is handled in multi-modal learning by building different subhypergraphs and assigning weights [24]. In deep reinforcement learning, it is introduced to model the combined structure of multi-dimensional discrete

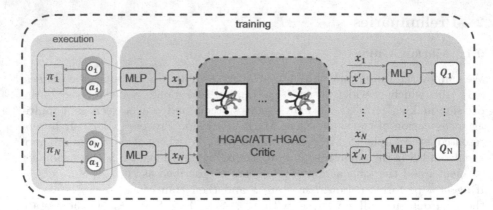

**Fig. 1.** The overall architecture of HGAC/ATT-HGAC.

action space and execute value estimation in a single-agent environment [17]. In the value function decomposition method of multi-agent reinforcement learning, the utility function of each agent is fitted to the global action state value function using a hypergraph neural network [1]. Unfortunately, this method ignores the interaction between agents.

## 3   Method

In this section, we introduce in detail our new methods called Actor Hypergraph Convolutional Critic Network (HGAC) and Actor Hypergraph Attention Critic Network (ATT-HGAC). We begin by discussing adaptive dynamic hypergraph generation. Secondly, we look at how hypergraphs are used in centralized critics to extract and represent information of agents. Finally, we give the overall MARL algorithm.

### 3.1   Hypergraph Generation

Hypergraph, unlike the traditional graph structure, unites vertices with same attributes into a hyperedge. In a multi-agent scenario, if the incidence matrix is filled with scalar 1, as in other works' graph neural network settings, each edge is linked to all agents, then the hypergraph's capability of gathering information from diverse neighborhoods will be lost. Meanwhile, since the states of agents in a multi-agent scenario vary dynamically over time, the incidence matrix should be dynamically adjusted as well.

For the aforementioned reasons, we investigate employing deep learning to dynamically construct hypergraphs. And instead of using a 0–1 incidence matrix, we optimize the elements of the incidence matrix to values in the range of $[0, 1]$, which describe how strong the membership of the vertices in the hyperedge is. In HGAC, we encode each agent's observation and action features and construct

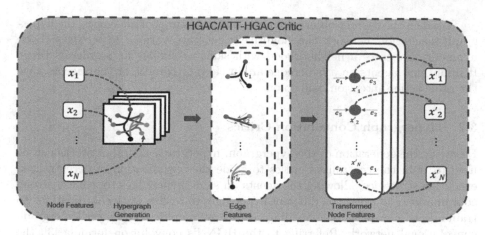

**Fig. 2.** The overall architecture of HGAC/ATT-HGAC. Each agent's feature $h_i$, including observation and action, is used to construct a dynamic hypergraph structure, with each hyperedge $e_i$ representing a neighborhood. In the hypergraph convolution process, the agents in the same neighborhood aggregate information to the hyperedge feature to realize neighborhood cooperation and interaction. Subsequently, the embedding of each agent $h_i'$ is aggregated by the hyperedge information to realize cooperative interaction between different neighbors.

the agent's membership degree to each hyperedge using a Multilayer Perceptron (MLP) model:

$$\mathbf{h}(v_i, E) = Softmax(MLP(concatenate(o_i, a_i))). \tag{2}$$

In addition, rather than utilizing the attention mechanism to aggregate neighbor information in MAAC, we propose using it to construct the hypergraph's incidence matrix. However, calculating the attention weight of a hyperedge to a vertex is unusual since it presupposes that hyperedges are comparable to vertices. To address this issue, we set the number of hyperedges equal to the number of vertices (agents) and assign each hyperedge to a specific agent. On each hyperedge, the membership degree of the specific agent is set to 1. Each hyperedge denotes a neighborhood of high-order attributes centered on the specific agent. Using the attention mechanism to assess the similarity of other agents' attributes to its own, we can create the hypergraph's incidence matrix:

$$h(v_j, \epsilon_i) = \begin{cases} 1, & \text{if } i = j \\ \frac{\exp(f(x_i, x_j))}{\sum_{m=1}^{N} \exp(f(x_i, x_m))}, & \text{if } i \neq j \end{cases} \tag{3}$$

where $x$ represents the feature of vertices, $f(x_i, x_j)$ is the score function used to calculate the correlation coefficient between *query* and *key*. We define $f(x_i, x_j) = x_j^T W_k^T W_q x_i$, where $W_k$ and $W_q$ are learnable parameters as proposed in [18]. Then we normalize the correlation coefficient by *softmax* to obtain the attention coefficient of $i$ to $j$.

Based on the current observation and action features of all agents, the hypergraph generation network can adaptively generate various hyperedges. Each hyperedge indicates a neighborhood with same or similar high-order features. It imply that agents on a hyperedge are in close proximity, or that agents have the same action intention, and so on.

## 3.2   Hypergraph Convolution Critics

Following the generation of the hypergraph, hypergraph neural networks can be used to train the centralized critics to guide the optimization of decentralized execution strategies, allowing the agents on same hyperedges to achieve strong coordination and agents on different hyperedges to realize weak coordination. To train agents' new feature embedding vectors, we employ a two-layer hypergraph convolutional network. Referring to the HGNN's convolution formula [3], the hypergraph convolution operator is defined as:

$$\mathbf{x}^{(l+1)} = \sigma(\mathbf{D}_v^{-1/2}\mathbf{H}\mathbf{W}\mathbf{D}_e^{-1}\mathbf{H}^\top\mathbf{D}_v^{-1/2}\mathbf{x}^{(l)}\mathbf{P}^{(l)}), \tag{4}$$

where $\mathbf{W}$ and $\mathbf{P}$ as learnable parameters represent the hyperedge weight matrix and the linear mapping of the vertices features, respectively. In a convolution process, vertices with the same high-order feature attributes combine their information into the hyperedges to which they belong to generate hyperedges feature vectors. After that, each agent's feature representation will be weighted and aggregated from the hyperedge's feature to which it belongs. Furthermore, as indicated in Fig. 2, we create several hypergraph convolutional neural networks simultaneously to aid the algorithm in gaining a better understanding of crucial information and increasing its robustness. The new characteristics received by each vertex after the convolution operation fuse all of the vertices features required for the agent to collaborate, but naturally, its original attributes are smoothed out. Inspired by DGN [8], we connect the features of original vertices and new features generated by each head of hypergraph convolutional networks and input them into the critic network. The Q-value function of agent $i$ is calculated by:

$$Q_i = ReLU(MLP(concatenate(x_i, x'_{i_1}, ..., x'_{i_K}))), \tag{5}$$

where $x_i$ is the initial feature embedding of agent $i$, and $(x'_{a_1}, ..., x'_{a_K})$ is the new feature embedding generated by hypergraph convolution of $K$ heads. In addition, all parameters of feature embedding and critic networks are shared, considerably reducing training complexity and increasing training efficiency.

## 3.3   Learning with Hypergraph Convolution Critics

To stimulate agent exploration and prevent converging to non-optimal deterministic policies, we advocate employing maximum entropy reinforcement learning [6] for training. In addition, unlike the settings in MADDPG [11], parameter sharing allows us to update all critics together. The loss function of critic networks is defined as:

$$\mathcal{L}(\theta) = \sum_{i=1}^{N} \mathbb{E}_{(o,a,r,o')\sim D}\left[\left(Q_i^\theta(o,a) - target_i\right)^2\right], \tag{6}$$

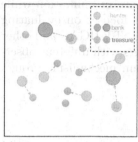 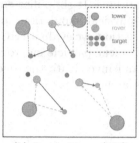 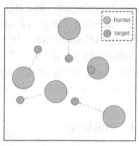

(a) Cooperative Treasure Collection(CTC)   (b) Rover-Tower(RT)   (c) Cooperative Navigation(CN)

**Fig. 3.** multi-agent particle environments used for our evaluating. In CTC and CN, dotted lines point to the target to which the agent needs to go. In RT, dotted lines refers to the rover and target declared by tower, and solid lines indicates the rover's target.

where $o$ represents the observation of the agent, $D$ is the reply buffer used for experience reply, other symbol settings are the same as in Markov Games, and $target_i = r_i(o,a) + \gamma \mathbb{E}_{a'\sim\pi_{\bar{\mu}}(o')}\left[Q_i^{\bar{\theta}}(o',a') - \omega\log\left(\pi_{\bar{\mu}_i}(a_i' \mid o_i')\right)\right]$. $Q_i^\theta(o,a)$ is the Q-value function(parameterized by $\theta$), $target_i$ is the target Q-value function which is calculated by environmental rewards $r_i$, target critics $Q_i^{\bar{\theta}}$ (parameterized by $\bar{\theta}$) and target policies $\pi_{\bar{\mu}_i}$(parameterized by $\bar{\mu}$). $\omega$ is a temperature coefficient to balance the maximization of entropy and rewards. In terms of actor networks, the policy gradient of each agent is expressed as:

$$\nabla_{\mu_i} J(\pi_\mu) = \mathbb{E}_{o\sim D, a\sim\pi}\left[\nabla_{\theta_i}\log\left(\pi_{\mu_i}(a_i \mid o_i)\right) \cdot \right. \\ \left. \left(-\omega\log\left(\pi_{\mu_i}(a_i \mid o_i)\right) + A(o, a_{-i})\right)\right]. \tag{7}$$

Inspired by COMA [5], we use the advantage function $A(o, a_{-i}) = Q_i^\theta(o,a) - \mathbb{E}_{a_i\sim\pi_i(o_i)}\left[Q_i^\theta\left(o, (a_i, a_{-i})\right)\right]$ with a counterfactual baseline, which can achieve the purpose of credit assignment by fixing the actions of other agents and comparing the value function of a specific action with the expected value function so as to determine whether the action lead to an increase or decrease in the expected return.

The whole algorithm adopts the framework of centralized training with decentralized execution (CTDE) [4] and its structure is shown in Fig. 1. It extract the features of agents within and between neighborhoods to guide their value function estimation during training process, so that agents can only follow their own observations during the actual execution process and do not require any other input to complete complex collaboration strategies.

## 4    Experiments

Multi-agent particle environment (MPE) [11] is one of the most commonly used tasks to evaluate MARL algorithms. It simplifies environment animation while still allowing for some basic physical simulation, and it focuses on evaluating strategy effectiveness. In this section, we evaluate HGAC/ATT-HGAC and other baselines in scenarios of multi-agent particle environments with different observation and reward settings, and investigate the algorithm mechanism through ablation and visualization researches.

### 4.1    Settings

We consider environments with continuous observation spaces and discrete action spaces. Specifically, considering different agent types and reward settings, we use three benchmark test environments, including Cooperative Treasure Collection (CTC) and Rover-Tower (RT) proposed in MAAC, and Cooperative Navigation (CN) introduces in MADDPG. They are shown in Fig. 3.

**Cooperative Treasure Collection.** 6 hunters are in charge of gathering treasures, while 2 banks are in charge of keeping treasures, with each bank only storing treasures of a specific hue. Hunters will be rewarded for acquiring treasures on an individual basis. No matter who successfully deposits treasures in the proper bank, all agents will earn global rewards, and if they collide, they will be penalised.

**Rover-Tower.** 4 rovers, 4 towers and 4 landmarks. Rovers and towers are paired randomly in each episode. The tower sends the location signal of the landmark to its paired rover, then the rover receives the signal and heads to the destination. The rewards of each pair are determined by the distance between the rover and the destination.

**Cooperative Navigation.** 5 hunters, 5 landmarks. Hunters need to work together to cover all landmarks and avoid collisions. The environmental rewards are determined by the distance between hunters and landmarks and whether there are collisions.

We choose the well-known MARL methods MADDPG and MAAC as well as completely decentralized independent learning methods DDPG [9] and SAC [6] as baselines to compare with our proposed HGAC/ATT-HGAC approach. Since DDPG and MADDPG are algorithms proposed under continuous control scenarios, we apply the gumbel-softmax reparameterization trick [11] to deal with discrete action scenarios. Furthermore, in order to focus on the enhancement of the experimental effect on the hypergraph convolutional critic network and reduce the impact of the underlying reinforcement learning method, we implement an additional SAC algorithm based on the CTDE framework and named it

MASAC. The hyperparameters common to all algorithms remain the same. We examine the performance of HGAC in the Cooperative Treasure Collection and Cooperative Navigation scenarios, as well as the performance of ATT-HGAC in Rover-Tower, due to characteristics of the experimental environments.

All environments are trained with 60000 episodes, with each episode having 25 time steps and the program having 12 parallel rollouts. During the training process, we keep track of the average return of each episode.

## 4.2   Results and Analysis

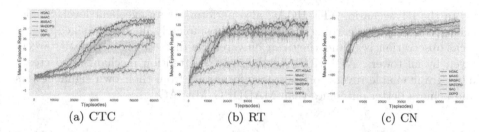

(a) CTC                         (b) RT                         (c) CN

**Fig. 4.** Performance curves with HGAC/ATT-HGAC, MAAC, MADDPG, MASAC, DDPG, SAC for 3 multi-agent particle environments. The solid line represents the median return, and the shadow part denotes the standard deviation.

Figure 4 illustrates the average return of each episode obtained by all algorithms with five random seeds tests in three environments. The results reveal that our methods are quite competitive when compared to other algorithms.

**Experimental Results.** The results show that using SAC as the underlying algorithm has smaller variance and better performance than DDPG. In the CN scenario, since the task is relatively simple, all algorithms achieve good results. But on the whole, MARL methods have better performance than the fully decentralized methods. And it is not hard to see that our HGAC have advantages over other algorithms.

In the CTC scenario, although different types of agents have different reward and observation settings, the convergence result of HGAC is surprising. Single-agent RL algorithms can even yield decent results in this circumstance since all agents can acquire global state information. In contrast, MADDPG and MASAC, which merely concatenating all of the agents' information as the input of the critic networks, perform badly due to the input of excessively redundant data and the lack of feature extraction capability. Correspondingly, HGAC can adaptively split high-order attribute neighborhoods, achieve strong cooperation inside the neighborhood and weak collaboration between neighborhoods to obtain the best performance.

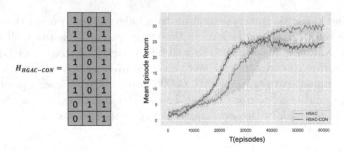

**Fig. 5.** *Left*:Incidence matrix constructed using prior knowledge. *Right*:Performance curve of HGAC and HGAC-CON.

In the RT scenario, ATT-HGAC also performs at an excellent level. Single-agent reinforcement learning algorithms are completely ineffective since rovers' local observation is 0 and they can only receive discrete signals of all targets given by all towers. ATT-HGAC that uses the attention mechanism to create hypergraph enable rovers to focus on information from their own signal towers and achieve better outcomes than MADDPG/MASAC.

**Ablation Studies.** To assess the effectiveness of the hypergraph generating technique, we perform an ablation experiment. We create a static hypergraph based on prior knowledge and utilize it to train the critic network. In the CTC environment, specifically, six hunters are connected using one hyperedge, two banks are connected using another hyperedge, and all agents are connected together using a third hyperedge. We keep other settings the same as HGAC and name it HGAC-CON.

Figure 5 shows the final experimental result. Although HGAC-CON has a faster convergence rate than HGAC, its ultimate performance is inferior to that of the HGAC. This is pretty simple to comprehend. The use of prior knowledge allows the algorithm to skip the stage of hypergraph generation, which speeds up the effect but restricts the hypergraph's expressiveness. As a result, self-adaptive dynamic hypergraph generation has the potential to generate better results.

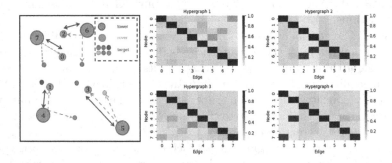

**Fig. 6.** *Left*: Correspondence between towers and rovers. *Right*: Incidence matrixes heat map generated by ATT-HGAC.

**Visualization Research.** We perform a visualization experiment on the hypergraphs generation to investigate the effect of applying the attention mechanism to generate hypergraphs in ATT-HGAC. As we hope, in the absence of clear supervision signals, rovers on different hyperedges successfully find signal towers they need to listen to. As shown in Fig. 6, agents 0–3 indicate rovers and agents 4–7 represent towers. In the *Hypergraph 1*, all of the towers (edge 4–7) successfully notice their rovers (node 0–3). And rovers also successfully notice their corresponding towers in the remaining three hypergraphs. Four hypergraphs can learn the same pairings, they proves and complements each other.

## 5   Conclusion and Future Work

In this paper, we propose HGAC/ATT-HGAC, a novel method for applying hypergraph convolution to the centralized training with decentralized execution paradigm. Our key contribution is to model agents adaptively as hypergraph structures, implement adaptive partition of neighborhoods, as well as efficient information feature extraction and representation to aid actors in forming more effective cooperative policies. We evaluate our algorithms' performance in several multi-agent test scenarios, including various observation and rewards settings. The ablation experiment and visualization verify our method's efficacy and the importance of each component. Facts have proved that HGAC/ATT-HGAC can successfully extract high-order neighborhood information to lead agents to attain efficient collaboration. In the future, we consider making full use of the structural advantages of hypergraphs to carry out related research in the field of multi-agent communication, while improving the efficiency of the algorithm, and increasing its convergence speed and scalability.

## References

1. Bai, Y., Gong, C., Zhang, B., et al.: Value function Factorisation with hypergraph convolution for cooperative multi-agent reinforcement learning (2021)
2. Buşoniu, L., Babuška, R., De Schutter, B.: Multi-agent reinforcement learning: an overview. In: Srinivasan, D., Jain, L.C. (eds.) Innovations in Multi-Agent Systems and Applications - 1. Studies in Computational Intelligence, vol. 310, pp. 183–221. Springer, Heidelberg (2010). https://doi.org/10.1007/978-3-642-14435-6_7
3. Feng, Y., You, H., Zhang, Z., Ji, R., Gao, Y.: Hypergraph neural networks. In: Proceedings of the AAAI Conference on Artificial Intelligence, vol. 33, pp. 3558–3565 (2019)
4. Foerster, J., Assael, I.A., De Freitas, N., Whiteson, S.: Learning to communicate with deep multi-agent reinforcement learning. In: Advances in Neural Information Processing Systems, vol. 29 (2016)
5. Foerster, J., Farquhar, G., Afouras, T., Nardelli, N., Whiteson, S.: Counterfactual multi-agent policy gradients. In: Proceedings of the AAAI Conference on Artificial Intelligence, vol. 32 (2018)
6. Haarnoja, T., Zhou, A., Abbeel, P., et al.: Soft actor-critic: off-policy maximum entropy deep reinforcement learning with a stochastic actor. In: ICML (2018)

7. Iqbal, S., Sha, F.: Actor-attention-critic for multi-agent reinforcement learning. In: International Conference on Machine Learning, pp. 2961–2970. PMLR (2019)
8. Jiang, J., Dun, C., Huang, T., et al.: Graph convolutional reinforcement learning. In: International Conference on Learning Representations (2019)
9. Lillicrap, T.P., Hunt, J.J., Pritzel, A., et al.: Continuous control with deep reinforcement learning. arXiv preprint: arXiv:1509.02971 (2015)
10. Littman, M.L.: Markov games as a framework for multi-agent reinforcement learning. In: Machine Learning Proceedings 1994, pp. 157–163. Elsevier (1994)
11. Lowe, R., WU, Y., Tamar, A., et al.: Multi-agent actor-critic for mixed cooperative-competitive environments. In: Advances in Neural Information Processing Systems, vol. 30, pp. 6379–6390 (2017)
12. Mao, H., Zhang, Z., Xiao, Z., Gong, Z.: Modelling the dynamic joint policy of teammates with attention multi-agent DDPG. arXiv preprint: arXiv:1811.07029 (2018)
13. Mnih, V., Kavukcuoglu, K., Silver, D., et al.: Human-level control through deep reinforcement learning. Nature **518**(7540), 529–533 (2015)
14. Peng, Z., Li, Q., Hui, K.M., Liu, C., Zhou, B.: Learning to simulate self-driven particles system with coordinated policy optimization. In: Advances in Neural Information Processing Systems, vol. 34, pp. 10784–10797 (2021)
15. Sharma, A., Chauhan, S.: A distributed reinforcement learning based sensor node scheduling algorithm for coverage and connectivity maintenance in wireless sensor network. Wireless Netw. **26**(6), 4411–4429 (2020). https://doi.org/10.1007/s11276-020-02350-y
16. Silver, D., Huang, A., Maddison, C.J., et al.: Mastering the game of go with deep neural networks and tree search. Nature **529**(7587), 484–489 (2016)
17. Tavakoli, A., Fatemi, M., Kormushev, P.: Learning to represent action values as a hypergraph on the action vertices. arXiv preprint: arXiv:2010.14680 (2020)
18. Vaswani, A., Shazeer, N., Parmar, N., et al.: Attention is all you need. In: Advances in Neural Information Processing Systems, pp. 5998–6008 (2017)
19. Vinyals, O., Babuschkin, I., Czarnecki, W.M., et al.: Grandmaster level in StarCraft II using multi-agent reinforcement learning. Nature **575**(7782), 350–354 (2019)
20. Wu, Z., Pan, S., Chen, F., Long, G., Zhang, C., Philip, S.Y.: A comprehensive survey on graph neural networks. IEEE Trans. Neural Netw. Learn. Syst. **32**(1), 4–24 (2020)
21. Xu, Z., Zhang, B., Bai, Y., Li, D., Fan, G.: Learning to coordinate via multiple graph neural networks. In: Mantoro, T., Lee, M., Ayu, M.A., Wong, K.W., Hidayanto, A.N. (eds.) ICONIP 2021. LNCS, vol. 13110, pp. 52–63. Springer, Cham (2021). https://doi.org/10.1007/978-3-030-92238-2_5
22. Zhou, D., Huang, J., Schölkopf, B.: Learning with hypergraphs: clustering, classification, and embedding. In: Advances in Neural Information Processing Systems, vol. 19 (2006)
23. Zhou, D., Huang, J., et al.: Learning with hypergraphs: clustering, classification, and embedding. In: Advances in Neural Information Processing Systems, vol. 19, pp. 1601–1608 (2006)
24. Zhu, L., Shen, J., Jin, H., Zheng, R., Xie, L.: Content-based visual landmark search via multimodal hypergraph learning. IEEE Trans. Cybern. **45**, 2756–2769 (2015)

# Bring Ancient Murals Back to Life

Xingeng Zhu$^{(\boxtimes)}$, Ying Yu$^{(\boxtimes)}$, Xiaochao Deng, and Linxia Yang

School of Information Science and Engineering, Yunnan University,
Kunming 650500, China
229933369@qq.com, yuying.mail@163.com

**Abstract.** Digital inpainting of murals has always been a challenging
problem. The damage forms in real murals are complex, such as cracks,
flaking, and fading. There are many difficulties in applying deep learning technology to mural inpainting. First, data sets are often difficult
to obtain. Second, the network based on supervised learning is unfit to
be applied to the real multiple mural damages, which makes the network unpromotable. Third, the output of deep neural network is the
combination of the unmasked area in the label image and the corresponding masked area in the generated image, so there is no change in
the unmasked area. Murals often fade or change color after a hundred
years or more, which leads to the lack of aesthetic feeling in the repaired
images. We propose a mural inpainting model based on the translation
method with three domains, including a SVD block and a dense spatial
attention with mask block. Specifically, the model trains two Variational
Auto-Encoders to respectively map the real mural images and the clean
mural images to two deep spaces, the mapping network learns the transformation between the two deep spaces by paired data. This transformation can well extend to real mural images. Experiments show that the
performance of our model is better than the comparative methods, and
the visual quality is improved.

**Keywords:** Digital inpainting · Mural inpainting · SVD · Dense
spatial attention

## 1 Introduction

As an artistic entity, Chinese ancient mural paintings have rich historical and
scientific values. However, due to natural weathering and destruction by human
factors, ancient murals show considerable signs of deterioration. Digital inpainting of damaged murals can avoid the irreversible defects of manual inpainting
and improve efficiency. Therefore, the digital restoration of ancient murals has
great practical significance for preserving cultural relics.

In digital inpainting solutions for murals, these methods can be divided into
two categories: traditional methods and learnable methods. Traditional methods are always based on diffusion-based methods [2] or patch-based methods [1].

M. Tanveer et al. (Eds.): ICONIP 2022, LNCS 13624, pp. 231–242, 2023.
https://doi.org/10.1007/978-3-031-30108-7_20

Traditional methods are prone to matching errors, blurring, and structural disorder when applied to murals inpainting. Data-driven deep neural network models bring more possibilities. Ren et al. [9] proposed using generalized regression neural networks for the digital inpainting of Dunhuang murals. Cao et al. [3] proposed an enhanced consistency generative adversarial network, which mainly solves the inconsistency between global and local repaired images. Meanwhile, attentional mechanisms are widely used for image inpainting. Yang et al. [10] proposed dilated multi-scale channel attention to perceive image information at different scales. He et al. [5] proposed a residual attention fusion block that enhances the utilization of practical information in the broken image and reduces the interference of redundant information. The inpainting models based on supervised learning only focus on the damaged parts of murals and cannot solve the global color problem. For this reason, we propose a mural inpainting model based on the translation method with three domains [7]. In addition, embedding a SVD block and a dense spatial attention with mask block to the model. The two branches improve the ability of the model to restore murals. Specifically, the main contributions of this paper are as follows:

- For the first time, we apply weak supervised learning to the mural inpainting task, which can restore the missing parts and perfect the overall appearance of murals.
- We design a dense spatial attention with mask block embedded in the mapping net, which further enhances the network's ability to capture the long-distance mapping relationship of deep features; The SVD effectively filters the high-frequency information while retaining most of the structure and detail information, further expanding the overlap between image features.
- Experiments show that our model not only has an excellent ability to repair the cracks, spots, and scratches but also improves the visual effect.

## 2   Proposed Approach

In this section, we first introduce the basic network architecture of the inpainting network. Then we describe the two effective blocks, i.e., the Singular Value Decomposition (SVD) and the Dense Spatial Attention with Mask (DSAWM).

### 2.1   Principle of Inpainting Network

The model formulates mural inpainting as an image transformation process. The training images consist of three parts: the set of real murals $\mathcal{R}$, the synthetic set of $\mathcal{X}$ where images suffer from artificial degradation, and the corresponding set of clean murals $\mathcal{Y}$ that comprises images without degradation. $r \in \mathcal{R}$, $x \in \mathcal{X}$ and $y \in \mathcal{Y}$ represent the image in the three sets. $x$ and $y$ are paired by data synthesizing, i.e., $x$ is degraded from $y$. The artificial degradation forms include holes, blurs, scratches, and low resolution.

First, $r \in \mathcal{R}$, $x \in \mathcal{X}$ and $y \in \mathcal{Y}$ are mapped respectively to the three deep spaces by $E_{\mathcal{R}} : \mathcal{R} \to \mathcal{Z}_{\mathcal{R}}$, $E_{\mathcal{X}} : \mathcal{X} \to \mathcal{Z}_{\mathcal{X}}$ and $E_{\mathcal{Y}} : \mathcal{Y} \to \mathcal{Z}_{\mathcal{Y}}$. Since $r \in \mathcal{R}$ and

$x \in \mathcal{X}$ are both corrupted, we align their spatial features into a shared space by mandatory strategies. As shown in Fig. 1, let the overlap of the two deep spaces (the part between black dotted lines) as large as possible, so there is $\mathcal{Z}_{\mathcal{R}} \approx \mathcal{Z}_{\mathcal{X}}$.

Then we learn the transformation from the spatial features of corrupted murals, $\mathcal{Z}_{\mathcal{X}}$, to the spatial features of clean murals, $\mathcal{Z}_{\mathcal{Y}}$, through the mapping $T_{\mathcal{Z}} : \mathcal{Z}_{\mathcal{X}} \rightarrow \mathcal{Z}_{\mathcal{Y}}$, where $\mathcal{Z}_{\mathcal{Y}}$ can be further reversed to $y$ through generator $G_{\mathcal{Y}} : \mathcal{Z}_{\mathcal{Y}} \rightarrow \mathcal{Y}$. By learning the spatial transformation. real ancient mural $r$ can be restored by sequentially performing the mappings,

$$r_{\mathcal{R} \rightarrow y} = E_{\mathcal{R}}(r) \circ T_{\mathcal{Z}} \circ G_{\mathcal{Y}} \tag{1}$$

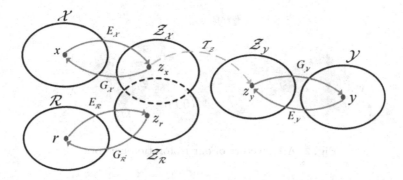

**Fig. 1.** Illustration of the principle of inpainting network.

We use the network shown in Fig. 2 to implement the inpainting process. The model is trained in two stages. In the first stage, two VAEs are trained to recover the input by unsupervised learning, where $\text{VAE}_1$ takes $r$ and $x$ as input, $\text{VAE}_2$ takes $y$ as input. In the second stage, the mapping network is trained by fixing the weight of the VAEs trained in the first stage, the training input is $x$, which first enters the encoder of the $\text{VAE}_1$, then passes through the mapping network, and finally is decoded by $G_{\mathcal{Y}}$ of the $\text{VAE}_2$. The loss function of $\text{VAE}_1$ is defined as

$$\min_{E_{\mathcal{R},\mathcal{X}},G_{\mathcal{R},\mathcal{X}}} \max_{D_{\mathcal{R},\mathcal{X}}} \mathcal{L}_{\text{VAE}_1}(r) + \mathcal{L}_{\text{VAE}_1}(x) + \mathcal{L}_{\text{VAE}_1,\text{GAN}}(r,x) \tag{2}$$

where KL divergence, L1 distance loss and the least-square loss (LSGAN) [6] are included in $\mathcal{L}_{\text{VAE}_1}(r)$. $\mathcal{L}_{\text{VAE}_1}(x)$ and $\mathcal{L}_{\text{VAE}_1}(r)$ are in the same form and will not be repeated. $\mathcal{L}_{\text{VAE}_1,\text{GAN}}(r,x)$ means training another discriminator $D_{\mathcal{R},\mathcal{X}}$ that differentiates $\mathcal{Z}_{\mathcal{R}}$ and $\mathcal{Z}_{\mathcal{X}}$. The loss function of the mapping network can be expressed as

$$\mathcal{L}_T(x,y) = \lambda_1 \mathcal{L}_{T,\text{L}_1} + \mathcal{L}_{T,\text{GAN}} + \lambda_2 \mathcal{L}_{\text{FM}} \tag{3}$$

where $\mathcal{L}_{T,\text{L}_1}$, $\mathcal{L}_{T,\text{GAN}}$ and $\mathcal{L}_{\text{FM}}$ represent L1 distance loss, the least-square loss (LSGAN) [6] and feature matching loss, respectively.

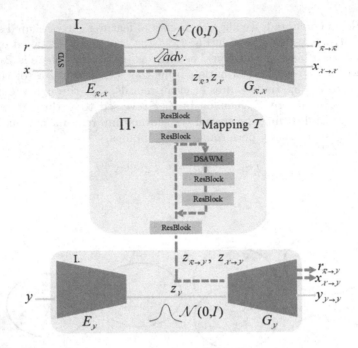

**Fig. 2.** Architecture of our restoration network

## 2.2 Singular Value Decomposition (SVD)

Visually, as shown in Fig. 1, the larger the overlap between spatial features $\mathcal{Z}_{\mathcal{R}}$ and $\mathcal{Z}_{\mathcal{X}}$ (the part between black dotted lines), the better. To achieve the goal, we propose to add the SVD to the encoder of $VAE_1$. SVD is the generalization of eigenvalue decomposition on any matrix. Let a matrix $A \in M_{m \times n}$ with rank $r$, then define the SVD of the matrix as

$$A_{m \times n} = U_{m \times m} \Sigma_{m \times n} V_{n \times n}^{\mathrm{T}} = U_{m \times m} \begin{pmatrix} D_{r \times r} & O \\ O & O \end{pmatrix}_{m \times n} V_{n \times n}^{\mathrm{T}} \tag{4}$$

where, $U_{m \times m} = A \times A^{\mathrm{T}}$, $V_{n \times n} = A^{\mathrm{T}} \times A$, $\Sigma$ is a matrix of $m \times n$, $D_{r \times r} =$

$\begin{pmatrix} \sqrt{\lambda_1} & & & \\ & \sqrt{\lambda_2} & & \\ & & \ddots & \\ & & & \sqrt{\lambda_r} \end{pmatrix}_{r \times r}$, $\lambda_1 \geq \lambda_2 \geq \cdots \lambda_{\min(m,n)} = \lambda_r > 0$ is the non-zero

eigenvalue of $A^{\mathrm{T}} \times A$, so the SVD can be written as follows:

$$A_{m \times n} = U_{m \times m} \Sigma V_{n \times n}^{\mathrm{T}} = (u_1, \ldots, u_m) \begin{pmatrix} \sqrt{\lambda_1} & & \\ & \sqrt{\lambda_2} & \\ & & \ddots \end{pmatrix} \begin{pmatrix} v_1^{\mathrm{T}} \\ \vdots \\ v_n^{\mathrm{T}} \end{pmatrix} = \sqrt{\lambda_1} u_1 \ldots v_1^{\mathrm{T}} + \sqrt{\lambda_2} u_2 v_2^{\mathrm{T}} + \cdots$$

$$(5)$$

at this point, each eigenvector $v_i$ in $V$ is called the right singular vector of $A$, each eigenvector $u_i$ in $U$ is called the left singular vector of $A$. The singular values are ordered from largest to smallest, so the top $N$ larger singular values and its corresponding singular vectors can approximate the matrix.

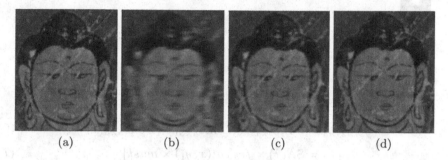

(a)                (b)                (c)                (d)

**Fig. 3.** Effect of SVD: (a) Ground Truth (b) Top 10 singular value composite image, (c) Top 40 singular value composite image, (d) Top 150 singular value composite image.

Using SVD, the gap between the spatial features at high frequencies reduces to a certain extent, and the overlap between the two spatial features further expands. Figure 3 shows the effect of the SVD.

## 2.3  Dense Spatial Attention with Mask

For the image inpainting task, the ordinary spatial attention mechanism is not applicable because the information in the masked region propagates from the adjacent regions, which results in inaccurate attention scores after normalization. Considering this problem, we propose the Spatial Attention with Mask (SAWM), as shown in Fig. 4. The difference between SAWM and ordinary spatial attention is that the mask is added to the feature map before normalization, ensuring that the damaged areas do not affect the attention score, but the output is missing information. To this end, we propose to solve the problem by multi-level fusion of Dense-net. The proposed mechanism is called Dense Spatial Attention with Mask (DSAWM).

Figure 5 shows the DSAWM. Given the input and its mask, three convolution kernels of $1 \times 1$, $3 \times 3$, and $5 \times 5$ are used to extract multi-scale features, the input image and the multi-scale feature maps are densely connected. This process with $x$ as input can be expressed as

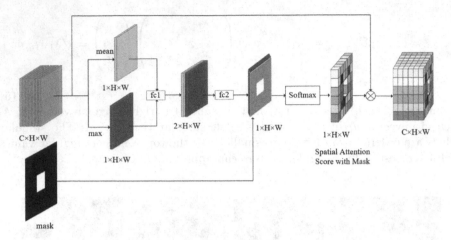

**Fig. 4.** Spatial Attention with Mask (SAWM).

$$y_1, y_2, y_3 = Conv_{1\times1}(x), Conv_{3\times3}(x), Conv_{5\times5}(x) \tag{6}$$

$$h_1 = \text{SA}(x) \times fc_3[cat(x, y_1) \times mask] \tag{7}$$

$$h_2 = \text{SAWM}(h_1) \times fc_4[cat(x, y_1, y_2)] \tag{8}$$

$$h_3 = \text{SA}(h_2) \times fc_5[cat(x, y_1, y_2, y_3)] \tag{9}$$

$$y = \text{SA}(h_3) \times (x) \tag{10}$$

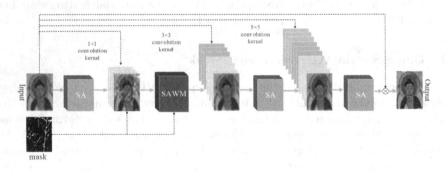

**Fig. 5.** Dense Spatial Attention with Mask (DSAWM).

where, cat is channel concatenating, $fc_3$, $fc_4$, and $fc_5$ indicate that the number of channels will be $\frac{1}{2}$, $\frac{1}{3}$ and $\frac{1}{4}$ of the number of input data channels by the convolution kernel $1\times1$ respectively. The mask is added to SAWM for calculating the attention score; SA calculates the ordinary spatial attention score. The last ordinary spatial attention score times the input image to get the final output.

In this way, the information in the unmasked region of the image is used multiple times, and the advantage of dense connectivity to preserve information is also incorporated.

## 3    Experiments

We select 1535 mural images to form $\mathcal{Y}$, which contains 300 modern murals. $\mathcal{R}$ includes 1632 real damaged murals. The image size is adjusted to 256×256 pixels. The learning rate is set to 0.0002, and batchsize is set to 15. We use the Pytorch framework to train and test the model. The experimental platform equipment configuration: Intel Core i7-6850K 3.60GHz CPU, NVIDIA GeForce GTX 1080Ti GPU.

### 3.1    Artificial Destruction Murals Repair Comparison

**Table 1.** Comparison of PSNR and SSIM of inpainting results.

| | Criminisi | | RN | | DS-net | | Our | |
|---|---|---|---|---|---|---|---|---|
| Image | PSNR/dB | SSIM | PSNR/dB | SSIM | PSNR/dB | SSIM | PSNR/dB | SSIM |
| 1 | 36.43 | 0.9824 | 33.48 | 0.9769 | 38.33 | 0.9848 | 37.21 | 0.9885 |
| 2 | 27.37 | 0.9367 | 21.84 | 0.8941 | 34.16 | 0.9157 | 34.77 | 0.9683 |
| 3 | 25.48 | 0.8661 | 24.72 | 0.8543 | 28.68 | 0.8647 | 31.46 | 0.9364 |
| 4 | 26.35 | 0.9950 | 37.15 | 0.9726 | 34.84 | 0.9792 | 36.41 | 0.9872 |
| 5 | 24.46 | 0.8476 | 19.86 | 0.8375 | 24.02 | 0.8676 | 24.12 | 0.8699 |
| 6 | 23.17 | 0.8285 | 18.39 | 0.8267 | 24.51 | 0.88419 | 24.78 | 0.8493 |
| 7 | 22.02 | 0.8272 | 19.97 | 0.8267 | 23.79 | 0.8363 | 24.10 | 0.8614 |
| 8 | 23.68 | 0.8332 | 19.36 | 0.8394 | 25.03 | 0.8418 | 25.28 | 0.8558 |

Eight murals are selected for inpainting experiments with artificially added masks, random masks are added to the ones numbered 1–4, and the last four murals are masked with center holes. We compare our model to three state-of-the-art models in experiments: Criminisi [4], RN [11] and DS-net [8]. RN and DS-net use the same way of training VAE$_2$ in our model. We used peak signal-to-noise ratio (PSNR) and structural similarity (SSIM) to evaluate the image inpainting quality. Table 1 shows the results of the quantitative analysis.

### 3.2    Experiment in Inpainting Real Damaged Murals

To further verify the effectiveness of our model in restoring the real murals, eight damaged murals are selected for the inpainting experiments. The experimental results are shown in Fig. 6.

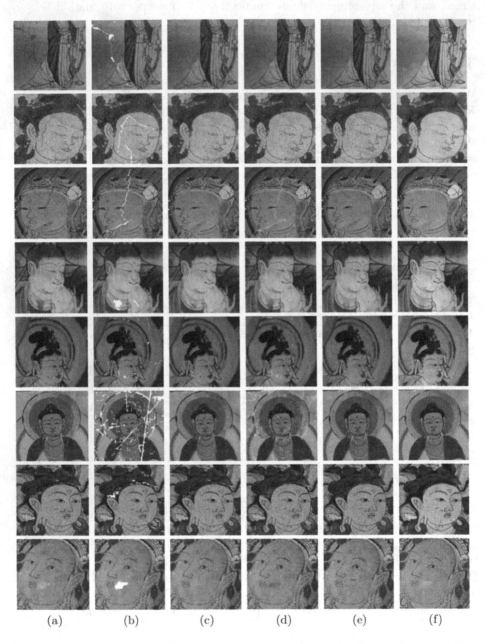

(a)          (b)          (c)          (d)          (e)          (f)

**Fig. 6.** Qualitative comparisons on real damaged murals: (a) Damaged murals, (b) input, (c) Criminisi, (d) RN, (e) DS-net, (f) Ours.

The 1st image has long cracks in the background. It can be observed that the Criminisi algorithm suffers from disruption of the line and fuzzy extension of texture information. RN repair result has blurry area. DS-net has almost no repair ability for the masked area. Note that our repair result is better, which is visually indistinguishable from the inpainting marks. Moreover, the structural consistency is better than the other 3 comparative algorithms. The 2nd and 3rd mural images have some facial cracks. It can be seen that the 3 comparative algorithms have evident artifacts and texture blur in their repair results. Our model achieves better coordination in line fitting, and the contrast of the repaired image is enhanced. For the 4th mural image, some mildew areas can be observed in the lower part. Both the Criminisi and DS-net have varying degrees of structural disorder, and the RN has large fuzzy block effect in its repair result. All these 3 comparative algorithms have obvious repair traces. By comparison, our method can produce better continuity and more reasonable restoration for the deteriorated areas. In the 5th and 6th mural images, there exist lots of scratches. Criminisi, DS-net and our model have visible repair effect, whereas RN performs poorly. Although the Criminisi and DS-net yield noticeable restoration for the scratches, they produced some inpainting errors and residual artifacts. For instance, in the 5th mural image, the restoration result of the Criminisi has a matching error near the corner of the bodhisattva's eye. The 7th mural image has some color falling-off in the hair bun. The Criminisi fails to repair these color falling-offs in this test. RN and DS-net cannot restore color-consistent areas with the surrounding region. By comparison, the repaired areas of our proposed model are visually satisfactory and semantically reasonable. The 8th mural image looks somewhat blurry and has a color inconsistent area in the bodhisattva's face. In this test, The Criminisi can restore this color inconsistency, whereas RN and DS-net cannot produce satisfactory results. Our proposed model can restore the mural image successfully. Moreover, the overall appearance of our result is considerably clearer than the other 3 approaches.

## 3.3   Ablation Study

To verify the utility of the SVD and the DSAWM, the original translation method with three domains [7] is used as the baseline model, "B" denotes it, "S" denotes the model after adding the SVD to the baseline model, "D" denotes the model after adding the DSAWM to the baseline model, "F" denotes our model. Figure 7 shows the variation of the quantitative index with respect to the different number of singularities in "F".

In Fig. 7(a) and Fig. 7(c), the optimal values of PSNR and mean square error (MSE) are received when the number of synthesized singularities is 112; the suboptimal value of SSIM is obtained at the same number 112 in Fig. 7(b). Therefore, the selected number of singularities in all experiments is 112.

The purpose of adding the SVD is to expand the overlap of $\mathcal{Z}_\mathcal{R}$ and $\mathcal{Z}_\mathcal{X}$, so that the network has better generalization performance. The images of building murals not included in the training dataset are selected for visual analysis. Figure 8 and Fig. 9 show qualitative and quantitative evaluation of the ablation

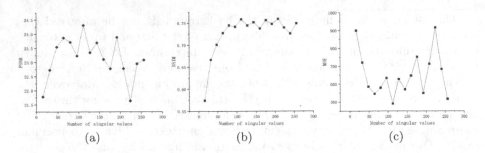

(a)                          (b)                          (c)

**Fig. 7.** (a): PSNR varies with different numbers of singular values. (b): SSIM varies with different numbers of singular values. (c): MSE varies with different numbers of singular values.

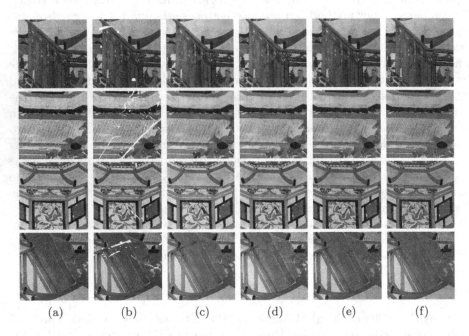

(a)            (b)            (c)            (d)            (e)            (f)

**Fig. 8.** Qualitative comparisons of the inpainting networks. (a) Ground Truth. (b) input. (c) the inpainting results of "B". (d) the inpainting results of "S". (e) the inpainting results of "D". (f) the inpainting results of "F".

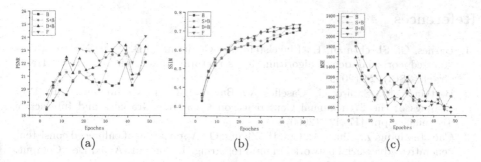

**Fig. 9.** (a): Compare the test PSNR values of the inpainting networks. (b): Compare the test SSIM values of the inpainting networks. (c): Compare the test MSE values of the inpainting networks.

experiment results, respectively. The comparison between Fig. 8(c) and Fig. 8(d) shows that the restoration result is relatively clearer after adding the SVD. From the comparison between Fig. 8(c) and Fig. 8(e), we can see that the output result of adding the DSAWM module has a better ability to capture colour information, structure information and detail information. The output of "F" combines the above two advantages. From Fig. 9(a) and Fig. 9(c), it can be seen that adding the SVD and the DSAWM can improve the PSNR value and reduce the MSE value, which indicates that the above blocks improve the restoration quality at the pixel level and perception level.

## 4    Conclusion

This paper proposed a novel DSAWM block and added the SVD to the inpainting model. The DSAWM enhances the ability of the network to capture the long-distance mapping relationships of deep spatial features, the SVD makes the images decomposed and then reorganized, effectively filtering high-frequency information while retaining most of the structure and detail information, further expanding the overlap of image features. Experiments show that our model based on weak supervised learning not only has good restoration ability for cracks, spots and scratches but also has some improvement in visual effects. However, there are still problems of blurred restoration in large damaged areas, which will be studied from the perspectives of obtaining high-quality data sets, reasonable image enhancement algorithm and optimized network.

**Acknowledgements.** This work was supported by the National Natural Science Foundation of China (Grant No. 62166048, Grant No. 61263048) and by the Applied Basic Research Project of Yunnan Province (Grant No. 2018FB102).

# References

1. Barnes, C., Shechtman, E., Finkelstein, A., Goldman, D.B.: PatchMatch: a randomized correspondence algorithm for structural image editing. ACM Trans. Graph. **28**(3), 24 (2009)
2. Bertalmio, M., Sapiro, G., Caselles, V., Ballester, C.: Image inpainting. In: Proceedings of the 27th annual Conference on Computer Graphics and Interactive Techniques, pp. 417–424 (2000)
3. Cao, J., Zhang, Z., Zhao, A., Cui, H., Zhang, Q.: Application of enhanced consistent generative adversarial network in mural repairing. J. Comput.-Aided Des. Comput. Graph. **32**(8), 1315–1323
4. Criminisi, A., Pérez, P., Toyama, K.: Region filling and object removal by exemplar-based image inpainting. IEEE Trans. Image Process. **13**(9), 1200–1212 (2004)
5. He, P., Yu, Y., Xu, C., Yang, H.: RAIDU-Net: image inpainting via residual attention fusion and gated information distillation. In: Mantoro, T., Lee, M., Ayu, M.A., Wong, K.W., Hidayanto, A.N. (eds.) ICONIP 2021. LNCS, vol. 13108, pp. 141–151. Springer, Cham (2021). https://doi.org/10.1007/978-3-030-92185-9_12
6. Mao, X., Li, Q., Xie, H., Lau, R.Y., Wang, Z., Paul Smolley, S.: Least squares generative adversarial networks. In: Proceedings of the IEEE International Conference on Computer Vision, pp. 2794–2802 (2017)
7. Wan, Z., et al.: Bringing old photos back to life. In: Proceedings of the IEEE/CVF Conference on Computer Vision and Pattern Recognition, pp. 2747–2757 (2020)
8. Wang, N., Zhang, Y., Zhang, L.: Dynamic selection network for image inpainting. IEEE Trans. Image Process. **30**, 1784–1798 (2021)
9. Xiaokang, R., Peilin, C.: Murals inpainting based on generalized regression neural network. Comput. Eng. Sci. **39**(10), 1884–1889 (2017)
10. Yang, H., Yu, Y.: Res2U-Net: image inpainting via multi-scale backbone and channel attention. In: Yang, H., Pasupa, K., Leung, A.C.-S., Kwok, J.T., Chan, J.H., King, I. (eds.) ICONIP 2020. LNCS, vol. 12532, pp. 498–508. Springer, Cham (2020). https://doi.org/10.1007/978-3-030-63830-6_42
11. Yu, T., et al.: Region normalization for image inpainting. In: Proceedings of the AAAI Conference on Artificial Intelligence, vol. 34, pp. 12733–12740 (2020)

# Multi-modal Rumor Detection via Knowledge-Aware Heterogeneous Graph Convolutional Networks

Boqun Li, Zhong Qian, Peifeng Li$^{(\boxtimes)}$, and Qiaoming Zhu

School of Computer Science and Technology, Soochow University, Suzhou, China
20205227046@stu.suda.edu.cn, {qianzhong,pfli,qmzhu}@suda.edu.cn

**Abstract.** With the rapid growth of the number of social media users, a variety of unverified information inevitably spreads on the social platform, which leads to the diffusion of rumors. Although some methods are explored on multi-modal data, they seldom take into account the hidden knowledge behind the text and image, and ignore the widely dispersed structure on multi-modal data in the rumor detection field. To solve the above issues, we propose a novel Multi-Modal Rumor detection model via Knowledge-aware Heterogeneous Graph Convolutional Networks, i.e., M$^3$KHG, which can model a post as a propagation graph, capture the interactive semantic information of image and text at the cross-modal level, and highlight suspicious signals according to the correlation between text-image knowledge in a unified framework. Finally, the "knowledgeable" feature generated by the propagation graph is assigned to debunk rumors. Experimental results on three popular datasets show that our model M$^3$KHG is superior to the state-of-the-art baselines.

**Keywords:** Rumor detection · Text-image knowledge · Heterogeneous graph

## 1 Introduction

Social media has bred the growth of various unverified and misleading information. For example, there was a message that "the cotton swabs used for Covid-19 detection were stained with toxic substances". Such unverified information poses tremendous risks to social stability and security. Rumor detection on social media is of great significance to mitigate the adverse effects of rumors and prevent rumor propagation. Early rumor refuting platforms were mainly reported by users, and invited experts or institutions to confirm. They can achieve the purpose of rumor detection but poor timeliness. Therefore, how to realize automatic and real-time rumor detection has become a key research direction.

With the thriving and rapid rise of multimedia, Twitter has become the focus of rumor detection. To minimize manpower, various neural network methods have come forward. Many earlier studies [2,9,11,18] focused on extracting

© The Author(s), under exclusive license to Springer Nature Switzerland AG 2023
M. Tanveer et al. (Eds.): ICONIP 2022, LNCS 13624, pp. 243–254, 2023.
https://doi.org/10.1007/978-3-031-30108-7_21

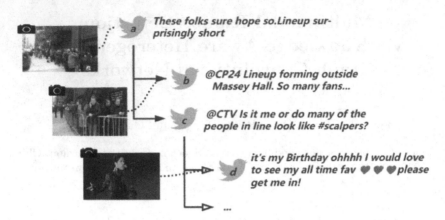

**Fig. 1.** An example of multi-modal tweets.

features from single modality, i.e., text, according to the temporal structure, and a few studies on both texts and images [4,15,19]. With the progressive awareness of the propagation structure on the social platform, researchers began to pay attention to structural information of posts which can construct non-consecutive information for rumor detection [1,12] but mainly focused on the text modality.

Among them, Bian et al. [1] considered the text modality and utilized critical features of rumor propagation and dispersion to acquire the global structure of the rumor tree. Concerning multi-modal modality, Zhang et al. [19] used textual knowledge as semantic supplementary and treated word embedding, visual embedding, and knowledge embedding as multiple stacked channels. Sun et al. [14] focused on source tweets and designed Content-knowledge Inconsistency Subnetwork which spotted inconsistent information among knowledge and texts. However, it is difficult for the above ways to consider background knowledge behind text-image at the same time and introduce the propagation structure on multi-modal data including source tweets and comments.

Take Fig. 1 as an example, the above methods have the following problems: 1) In fact, tweet **d** is not much related to the source tweet **a**, but the temporal structure forces them to the same semantic space, consequently we need to introduce the non-consecutive structural information. 2) We know that the high-dimensional representation learned by images only captured underlying features of the visual layer (such as *"person"*), for better understanding of the image in tweet **a**&**b**, we prefer to acquire semantic knowledge of the concept layer (such as *"crowd"*) which is correlated to the text entity "Lineup" which implies a group of people instead of little people.

To solve the above issues, we propose a Multi-Modal Model Rumor Detection via Knowledge-aware Heterogeneous Graph Convolutional Networks (M³KHG) to detect rumors, which takes multi-modal data into account in the propagation structure and makes use of knowledge in a unified model. Firstly, our Construction of Propagation Graph (CPG) module models a thread as a multi-modal heterogeneous graph structure. Secondly, for each pair of multi-modal posts, we acquire knowledge-level multi-modal representation through Cross-modal Convergence of Knowledge (CCK) module. Then, according to Knowledge-driven Graph Convolutional Network (KGCN), we obtain the feature "debunker" for graph, which selectively obtains the information of each node in the propagation tree. Finally, the "debunker" is assigned to query the event in Knowledgeable Debunker Classification (KDC), and the results are fed into the classifier. To sum up, our contributions are as follows:

(1) We introduce a multi-modal heterogeneous graph structure and creatively evaluate the reliability of each node for the first time.
(2) We exploit both text and image knowledge to supplement semantic information and apply Inter-sequence attention to fuse the knowledge features.

## 2   Related Work

### 2.1   Rumor Detection on Texts

Various methods using text modality have been proposed in rumor detection. Among them, two propagation structures are mainly used.

**Temporal Structure.** Khoo et al. [5] proposed a rumor detection model named PLAN based on attention, in which all tweets were organized into a sequence structure according to the time that tweets were published, and user interaction was achieved by self-attention. Li et al. [7] used user information, attention mechanism, and multi-task learning to detect rumors. They added the user credibility information to the rumor detection layer.

**Propagation Graph Structure.** Ma et al. [12] proposed two recursive neural models based on bottom-up and top-down tree structures for rumor representation learning and classification. EBGCN [16] used graph convolution to acquire node features by aggregating the neighborhood information on the reconstructed edge and paid attention to the uncertainty in the propagation structure. Bi-GCN [1] exploited two GCNs where one GCN used the top-down digraph of rumor propagation to learn the propagation mode of rumor and another GCN used the opposite digraph to capture the spreading pattern of rumors.

### 2.2   Multi-modal Rumor Detection

With the success of deep neural network in multimedia, researchers realized that visual features play an indispensable role in identifying rumors. However, there is little research on introducing propagation structure into multi-modal data.

EANN [15] only straightforwardly concatenated the visual features and text features to obtain multi-modal features. We believe that the concatenation strategy breaks up the connection between textual and visual data. Jin et al. [4] proposed a multi-modal detection model att-RNN and exploited attention mechanism to balance the association between the visual features and the text/social joint features. Zhang et al. [19] proposed a model MKEMN which explored multi-modal information and distilled the background knowledge from the knowledge

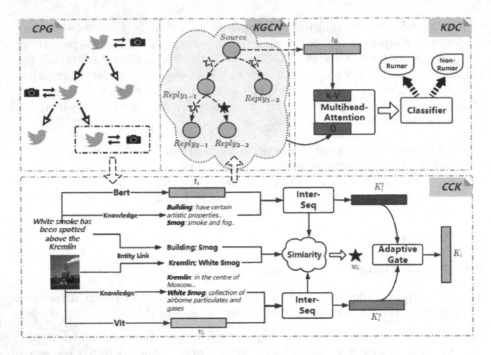

**Fig. 2.** Multi-Modal Rumor detection via Knowledge-aware Heterogeneous Graph Convolutional Networks ($M^3$KHG)

map to supplement the semantic representation of text posts. Different from Zhang et al., Sun et al. [14] took advantage of text knowledge to capture inconsistent features by fusing knowledge with content via attention mechanism.

# 3    Approach

## 3.1    Overview

Rumor detection task can be defined as a binary classification problem, which aims to classify a thread on social media as a rumor or non-rumor. We define a thread, i.e., $thread = \{(T_0, I_0), (T_1, I_1), ..., (T_n, I_n), G\}$, where $T_0$ means source tweet, $T_i(i > 0)$ is the $i$-th relevant responsive post, $I_i$ is the image attached to $T_i$ and it may be null which depends on whether the user published the image, and $G$ refers to the propagation structure. Specifically, $G$ is defined as a graph $<V, E>$ with $T_0$ being the root node.

Figure 2 shows the framework of Multi-Modal Rumor detection via Knowledge-aware Heterogeneous Graph Convolutional Networks ($M^3$KHG), which consists of the following parts: Construction of Propagation Graph (CPG), Cross-modal Convergence of Knowledge (CCK), Knowledge-driven Graph Convolutional Network (KGCN), Knowledgeable Debunker Classification (KDC).

The process is as follows. Given a thread, CPG module firstly constructs a multi-modal heterogeneous graph. For each pair of heterogeneous nodes, CCK module exploits Inter-sequence Attention to fuse textual and visual features with corresponding knowledge, respectively. Then Adaptive Gate is used to balance them as final node representation. KGCN module evaluates the reliability of each node according to entities and knowledge among texts and images, and uses graph convolution to obtain "knowledgeable" features named debunker. In KDC module, the debunker is assigned to debunk the event.

### 3.2   Construction of Propagation Graph(CPG)

For each thread, we build an undirected graph. Specifically, we construct a propagation structure <V, E> for a thread based on the relationship between reply and retweet, where V contains $V_T$ and $V_I$. Given a thread, each text tweet is regarded as a node $n^T$, which is linked with the image node $n^I$, which constitutes a pair of multi-modal heterogeneous node pairs. Edge between $n_j^T$ and $n_k^T$ means adjacent text nodes have a reply or retweet relationship between $T_j$ and $T_k$.

### 3.3   Cross-Modal Convergence of Knowledge (CCK)

**Textual Part.** Text encoder aims to produce the text representation for a given text $T_i$. We feed text into Bert-Base-Uncased [3] to capture the contextual information, "$[CLS]$" is used to represent its features $t_i \in \mathbb{R}^{768}$.

Then, for each tweet, we use the entity linking solution TAGME[1] to link the ambiguous entity mentions to the corresponding entities in Wiki. And we crawl from the Wiki to get the corresponding introduction of the entity as its knowledge. We concatenate each entity with attached knowledge as a sequence and feed it into text encoder to get several sequence features.

**Visual Part.** Our model uses VIT [17] to produce the text representation for a given image $I_i$. We firstly resize the image to 224 × 224 pixels, then feed the image into a vit-base-patch16 model to capture the visual information $v_i \in \mathbb{R}^{768}$.

Also, for each image, using image recognition method to identify image entities is beneficial for model to understand context. The image entities with brief introduction are extracted as image knowledge through object recognition method[2], where only image entities higher than the recognition threshold (i.e., 0.6) will be recognized. Like textual part, we concatenate entity with knowledge to get sequence representation.

**Node Representation.** For text and image knowledge sequences, we use Inter-sequence attention to obtain fusion features. Take the textual part as an example, $M_k$ including $\{t_i, k_1...k_n\}$ represents tweet content and its knowledge. We need to initialize a training parameter $v \in \mathbb{R}^{768}$ to obtain the attention score $\alpha$, the textual knowledge fusion representation $K_i^t \in \mathbb{R}^{768}$ is computed as follows.

$$M_k = [t_i, k_1, ..., k_n] \tag{1}$$

---

[1] https://tagme.d4science.org/tagme/.

[2] The image knowledge consists of the entities with brief introduction and is extracted by an object recognition tool (https://ai.baidu.com/tech/imagerecognition/general).

$$M = tanh(M_k) \tag{2}$$

$$\alpha = softmax(v^T M) \tag{3}$$

$$K_i^t = tanh(M_k \alpha^T) \tag{4}$$

Similarly, we obtain the visual knowledge fusion representation $K_i^v$. To balance the two vectors, we use the adaptive gate to automatically assign weights to them. The final representation $K_i \in \mathbb{R}^{768}$ is computed as follows.

$$e = softmax(W_e(K_i^t \oplus K_i^v) + b_e) \tag{5}$$

$$K_i = (1 - e) \odot K_i^t + e \odot K_i^v \tag{6}$$

where $e$ means weight obtained via training and $W_e, b_e$ are trainable parameters.

## 3.4   Knowledge-Driven Graph Convolutional Network (KGCN)

For each pair of multi-modal heterogeneous nodes, we obtain its final node knowledge representation. In actual conversations on social platforms, some users' tweets and attached images show little obvious correlation as shown in Fig. 1 tweet **d**, for such a propagation node, we don't think it plays an essential role in identifying the authenticity and should have a small proportion in the convergence process. Thus, we calculate the consistency value between text and image knowledge representation and use entity similarity between text and image to revise the value. The consistency value of each node is computed as follows.

$$s_k = \frac{K_i^t \odot K_i^v}{\|K_i^t\| \|K_i^v\|} \tag{7}$$

$$g_e = mean(\sum_{e_t \in E_t} \sum_{e_v \in E_v} tag(e_t, e_v)) \tag{8}$$

$$w_i = s_k * g_e \tag{9}$$

$$W = softmax(w_1, w_2, ..., w_n) \tag{10}$$

where $tag(*, *)$ means obtaining the entity similarity between top-2 confident text entities $(E_t)$ and image entities $(E_v)$ and we calculate the mean value through the method provided by TAGME, $n$ means the count of nodes in propagation tree. Note that for a few posts where the number of entities is less than 2, we make a supplement with pseudo representations with random values.

We introduce feature matrix $X$ associated with consistency value, and adjacency matrix $A$ indicating the edge set. Then we defined $H^{(i)}$ as the input feature matrix of GCN layer ($H^{(0)} = X$). We perform graph convolution as follows.

$$X = [K_1, ..., K_n] * W \tag{11}$$

$$H^{(i+1)} = \sigma(D^{-\frac{1}{2}}(I + A)D^{-\frac{1}{2}}H^{(i)}W) \tag{12}$$

where $\sigma(\cdot)$ refers to a *sigmoid* function, $I$ is the identity matrix, $D$ is diagonal matrix and $W$ is the trainable weight matrix, after two layers of GCN, we choose the root node aggregated features as our final graph representation named "debunker" $d$ which selectively absorbs the information of each node in the propagation tree.

**Table 1.** Distribution of Datasets

| Statistic | Twitter15 | Twitter16 | PHEME |
|---|---|---|---|
| # of tree | 1458 | 818 | 6425 |
| # of tree-depth | 2.8 | 2.7 | 3.2 |
| # of rumor | 1086 | 613 | 2403 |
| # of non-rumor | 372 | 205 | 4022 |
| # of images | 4917 | 1333 | 7239 |
| Average length of tweets | 15.8 | 15.9 | 13.6 |
| Average entities of images | 2.01 | 1.88 | 1.70 |
| Average entities of tweets | 1.82 | 1.72 | 1.77 |

### 3.5 Knowledgeable Debunker Classification (KDC)

The query vector "debunker" $d$ effectively absorbs the information of the whole propagation tree, reducing the influence of the untrusted tweet node. Let this query vector infer the source tweet $t_0$ that we need to judge. The inference process is as follows.

$$R_i = Attention(dW_i^Q, t_0 W_i^K, t_0 W_i^V) \tag{13}$$

$$r = Concat(R_1, ...R_h)W^O \tag{14}$$

$$res = classifier(r) \tag{15}$$

where $h = 8$ is the count of head in attention, $W^O, W_i^Q, W_i^K, W_i^V$ are the trainable parameters, the classifier is a container including (dropout - dense - tanh - dropout - dense).

## 4 Experimentation

### 4.1 Experimental Settings

In this paper, three public datasets are used to evaluate our model, i.e., PHEME [6], Twitter15 [8], and Twitter16 [10]. The data distribution is shown in Table 1. For PHEME, we randomly divided the data and used the same processing method as Sujana et al. [13], 80% of the data was used as the training set, 10% as the validation set, and the remaining 10% as the test set. For Twitter15 and Twitter16, we adopted a 6:2:2 split. Similar to Ma et al. [10], we calculated the accuracy, precision, recall, and F1 score to evaluate the performance.

To better fit the actual situation, the tweets in our datasets contain images randomly, depending on whether the user published the image. For the tweets without images, we uniformly give them a blank image. For the images contained in the tweets, we adjust the size to $224 \times 224$ pixels and normalize them. Adam optimizer is used to update the parameters, some hyperparameters such as tree-depth is 2, entity-num is 2 and the learning rate is $10^{-5}$.

**Table 2.** Results of comparison with different baselines on PHEME

| Dataset | Method | Acc | Pre | Rec | F1 |
|---------|--------|-----|-----|-----|-----|
| PHEME | BiGCN | 0.880 | 0.873 | 0.878 | 0.875 |
| | EANN | 0.824 | 0.813 | 0.833 | 0.818 |
| | Sun2021 | 0.893 | 0.892 | 0.881 | 0.886 |
| | Ours | **0.910** | **0.906** | **0.903** | **0.904** |

**Table 3.** Results of comparison with different baselines on Twitter15 and Twitter16

| Dataset | Method | Acc | Pre | Rec | F1 |
|---------|--------|-----|-----|-----|-----|
| Twitter15 | BiGCN | 0.908 | 0.871 | 0.909 | 0.887 |
| | EANN | 0.866 | 0.824 | 0.886 | 0.843 |
| | Sun2021 | 0.912 | 0.876 | 0.912 | 0.891 |
| | Ours | **0.931** | **0.897** | **0.944** | **0.916** |
| Twitter16 | BiGCN | 0.876 | 0.825 | 0.832 | 0.829 |
| | EANN | 0.839 | 0.827 | 0.689 | 0.722 |
| | Sun2021 | 0.861 | 0.804 | 0.823 | 0.812 |
| | Ours | **0.883** | **0.831** | **0.859** | **0.844** |

## 4.2    Experimental Results

To verify the effectiveness of our proposed $M^3KHG$, the corresponding baselines are conducted for fair comparison as follows.

**BiGCN** [1]. A GCN-based model that uses the two key features of rumor propagation and dispersion to capture the global structure of the rumor tree.
**EANN** [15]. A multi-modal model which uses an event adversarial neural network to extract features.
**Sun2021** [14]. A multi-modal model which explores inconsistency among texts and images and spots information between posts and knowledge.

Considering the different partition methods of datasets and the crawled images may differ, we rerun some experiments according to the source code provided by the authors. Table 2 and Table 3 show the performance of other models, it can be seen that the proposed method achieves the best results both on Acc and F1 and we can draw the following observations:

(1) BiGCN performs well on both datasets, indicating that the propagation structure captures the long-range semantic relations among tweets.
(2) In multi-modal methods, Sun2021 outperforms EANN. Obviously, the simple concatenating strategy implicitly forces textual and visual features into the same dimensional space. Also, external textual knowledge in Sun2021 helps to capture the inconsistent semantics at the cross-modal level.

**Table 4.** Results of comparison among different variants on Datasets

| Text | Image | Graph-based | CCK | KGCN | PHEME | | Twitter15 | | Twitter16 | |
|------|-------|-------------|-----|------|-------|----|-----------|----|-----------|----|
| | | | | | Acc | F1 | Acc | F1 | Acc | F1 |
| ✓ | | | | | 0.866 | 0.856 | 0.904 | 0.884 | 0.854 | 0.800 |
| | ✓ | | | | 0.672 | 0.564 | 0.672 | 0.505 | 0.664 | 0.521 |
| ✓ | ✓ | ✓ | | | 0.824 | 0.817 | 0.851 | 0.757 | 0.752 | 0.719 |
| ✓ | ✓ | ✓ | ✓ | | 0.896 | 0.891 | 0.927 | 0.910 | 0.876 | 0.829 |
| ✓ | ✓ | ✓ | ✓ | ✓ | 0.910 | 0.904 | 0.931 | 0.916 | 0.883 | 0.844 |

(3) Moreover, Sun2021 performs better than BiGCN in PHEME and Twitter15 but worse in Twitter16. It can be attributed to its smaller dataset size seen in Table 1. It also can be found that the removal of visual information leads to the model's declines in PHEME and Twitter15.

(4) Compared with all baselines, our model achieves a promising performance. We attribute the superiority to two properties: a) We make use of propagation structure which can construct non-consecutive information on multi-modal data. b) We used text-image knowledge as complementary information to strengthen the node representation.

### 4.3   Ablation

In this section, we compare the variants of M³KHG with the following two aspects to demonstrate the effectiveness: the usage of multi-modal heterogeneous graph associated with consistency value and the usage of text-image knowledge. We conduct ablation experiments as shown in Table 4 and we can conclude that:

(1) The model based on text modality is superior to the model based on image modality, which is obvious because the textual features contain more abundant semantic information.

(2) The separate graph-network model where each node is represented by concatenating text features and image features performs poorly. Although the propagation structure captures more long-range information among tweets, we need to have a reasonable representation of node and pay attention to the social rule that the reliability of tweets on real platforms is inconsistent.

(3) According to the CCK module, we achieve better results because we obtain the robust representation for each node, and the background knowledge in the image and text is fully utilized which is considered as complementary information for authenticity judgment.

(4) Our KGCN module produces the feature named "debunker" which selectively absorbs the useful information and becomes more "knowledgeable". This "debunker" is more robust than the convergence feature in which each node's reliability is considered the same.

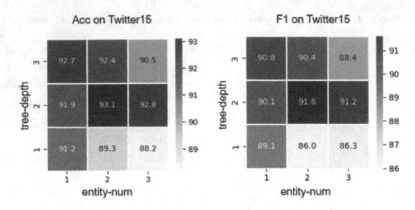

**Fig. 3.** Performance of different parameters on Twitter15

Based on the analysis above, the following conclusions can be drawn, i.e., 1) The multi-modal heterogeneous graph plays an essential role in rumor detection, and the "debunker" which considers the reliability of retweets is more robust; 2) Text-Image knowledge is useful for our model to better understand the context.

Moreover, in our experiments, there are two hyperparameters that need to be manually adjusted. In order to match the "entity-num" and "tree-depth", we took Twitter15 as an example and conducted the following experiments, as shown in the Fig. 3, when we have only a one-layer tree structure, the model performs poorly, because the shallow trees cannot provide enough information. When the number of layers exceeds the average depth shown in Table 1, too many empty nodes also cause interference. Similarly, when we maximize the use of entities, we can achieve the best results.

### 4.4   Case Study

To illustrate the significance of our proposed model. We analyze the rumor case from Fig. 4. In tweet **d**, the extracted image entity *"singer"* (*perform by singing songs...*) shows not much correlation with text entity *"birthday"* (*the anniversary of the birth...*), if we arrange these tweets in chronological order, it is likely that irrelevant posts (image and text are not well-matched) will interfere with the authenticity. In this case, we need to introduce structural information.

The "short lineup" stated by tweet **a** doesn't conform to the attached image. In the image knowledge, from the word *"demonstration"* (*means of modern society public opinion*), we can more intuitively find that this thread belongs to a rumor. Moreover, the tweet and image of tweet **b** like *"fan"* (*sometimes called supporters...*) and *"lineup"* (*a group of people...*) in text knowledge can further prove "short lineup" lacks authenticity.

The case helps to confirm that our model can effectively capture the knowledge information and construct a more practical multi-modal propagation graph.

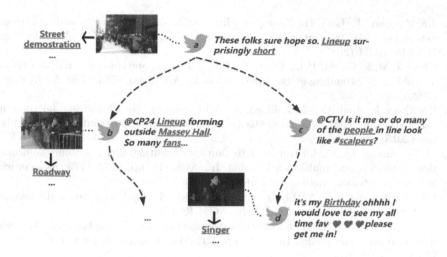

**Fig. 4.** Illustration of case

## 5   Conclusion

In this paper, we propose Multi-Modal Rumor detection via Knowledge-aware Heterogeneous Graph Convolutional Networks (M³KHG), which introduces a multi-modal heterogeneous graph structure and explores the robust feature "debunker" to query the thread, meanwhile semantic knowledge behind images and texts is fully utilized. Specifically, our CPG module constructs the more practical multi-modal propagation graph which is more suitable for social platforms. CCK module offers more intuitive text-image knowledge and utilizes the adaptive gate to balance them which can generalize well. In future work, we will continue to study the fusion method and make full use of knowledge.

**Acknowledgments.** The authors would like to thank the three anonymous reviewers for their comments on this paper. This research was supported by the National Natural Science Foundation of China (Nos. 61836007, 62276177 and 62006167), and Project Funded by the Priority Academic Program Development of Jiangsu Higher Education Institutions (PAPD).

## References

1. Bian, T., et al.: Rumor detection on social media with bi-directional graph convolutional networks. In: AAAI, pp. 549–556. AAAI Press (2020)
2. Chen, W., Zhang, Y., Yeo, C.K., Lau, C.T., Lee, B.: Unsupervised rumor detection based on users' behaviors using neural networks. Pattern Recogn. Lett. **105**, 226–233 (2018)
3. Devlin, J., Chang, M., Lee, K., Toutanova, K.: BERT: pre-training of deep bidirectional transformers for language understanding. In: NAACL-HLT (1), pp. 4171–4186. Association for Computational Linguistics (2019)

4. Jin, Z., Cao, J., Guo, H., Zhang, Y., Luo, J.: Multimodal fusion with recurrent neural networks for rumor detection on microblogs. In: ACM Multimedia, pp. 795–816. ACM (2017)

5. Khoo, L.M.S., Chieu, H.L., Qian, Z., Jiang, J.: Interpretable rumor detection in microblogs by attending to user interactions. In: AAAI, pp. 8783–8790. AAAI Press (2020)

6. Kochkina, E., Liakata, M., Zubiaga, A.: All-in-one: multi-task learning for rumour verification. In: COLING, pp. 3402–3413. Association for Computational Linguistics (2018)

7. Li, Q., Zhang, Q., Si, L.: Rumor detection by exploiting user credibility information, attention and multi-task learning. In: ACL (1), pp. 1173–1179. Association for Computational Linguistics (2019)

8. Liu, X., Nourbakhsh, A., Li, Q., Fang, R., Shah, S.: Real-time rumor debunking on Twitter. In: CIKM, pp. 1867–1870. ACM (2015)

9. Lu, Y., Li, C.: GCAN: graph-aware co-attention networks for explainable fake news detection on social media. In: ACL, pp. 505–514. Association for Computational Linguistics (2020)

10. Ma, J., et al.: Detecting rumors from microblogs with recurrent neural networks. In: IJCAI, pp. 3818–3824. IJCAI/AAAI Press (2016)

11. Ma, J., Gao, W., Wong, K.: Detect rumor and stance jointly by neural multi-task learning. In: WWW (Companion Volume), pp. 585–593. ACM (2018)

12. Ma, J., Gao, W., Wong, K.: Rumor detection on Twitter with tree-structured recursive neural networks. In: ACL (1), pp. 1980–1989. Association for Computational Linguistics (2018)

13. Sujana, Y., Li, J., Kao, H.: Rumor detection on Twitter using multiloss hierarchical BiLSTM with an attenuation factor. In: AACL/IJCNLP, pp. 18–26. Association for Computational Linguistics (2020)

14. Sun, M., Zhang, X., Ma, J., Liu, Y.: Inconsistency matters: a knowledge-guided dual-inconsistency network for multi-modal rumor detection. In: EMNLP (Findings), pp. 1412–1423. Association for Computational Linguistics (2021)

15. Wang, Y., et al.: EANN: event adversarial neural networks for multi-modal fake news detection. In: KDD, pp. 849–857. ACM (2018)

16. Wei, L., Hu, D., Zhou, W., Yue, Z., Hu, S.: Towards propagation uncertainty: edge-enhanced Bayesian graph convolutional networks for rumor detection. In: ACL/IJCNLP (1), pp. 3845–3854. Association for Computational Linguistics (2021)

17. Wu, B., et al.: Visual transformers: token-based image representation and processing for computer vision. CoRR abs/2006.03677 (2020)

18. Wu, L., Rao, Y., Zhao, Y., Liang, H., Nazir, A.: DTCA: decision tree-based co-attention networks for explainable claim verification. In: ACL, pp. 1024–1035. Association for Computational Linguistics (2020)

19. Zhang, H., Fang, Q., Qian, S., Xu, C.: Multi-modal knowledge-aware event memory network for social media rumor detection. In: ACM Multimedia, pp. 1942–1951. ACM (2019)

# DAGKT: Difficulty and Attempts Boosted Graph-Based Knowledge Tracing

Rui Luo[1], Fei Liu[1,2], Wenhao Liang[1], Yuhong Zhang[1], Chenyang Bu[1(✉)], and Xuegang Hu[1(✉)]

[1] Key Laboratory of Knowledge Engineering with Big Data
(The Ministry of Education of China), School of Computer Science
and Information Engineering, Hefei University of Technology, Hefei, China
{chenyangbu,jsjxhuxg}@hfut.edu.cn
[2] Jianzai Tech, Hefei, China

**Abstract.** In the field of intelligent education, knowledge tracing (KT) has attracted increasing attention, which estimates and traces students' mastery of knowledge concepts to provide high-quality education. In KT, there are natural graph structures among questions and knowledge concepts so some studies explored the application of graph neural networks (GNNs) to improve the performance of the KT models which have not used graph structure. However, most of them ignored both the questions' difficulties and students' attempts at questions. Actually, questions with the same knowledge concepts have different difficulties, and students' different attempts also represent different knowledge mastery. In this paper, we propose a difficulty and attempts boosted graph-based KT (DAGKT) (https://github.com/DMiC-Lab-HFUT/DAGKT), using rich information from students' records. Moreover, a novel method is designed to establish the question similarity relationship inspired by the F1 score. Extensive experiments on three real-world datasets demonstrate the effectiveness of the proposed DAGKT.

**Keywords:** Educational Data Mining · Knowledge Tracing · Graph Neural Network

## 1 Introduction

In recent years, with the development of intelligent tutoring systems, more users choose online education because it is more convenient to provide personalized and high-quality education than traditional classrooms [2]. Knowledge tracing (KT), which evaluates students' knowledge mastery based on their performance on coursework, has attracted great attention and in-depth research.

This work is supported by the National Natural Science Foundation of China (under grants 61806065, 62120106008, 62076085, and 61976077), and the Fundamental Research Funds for the Central Universities (under grants JZ2022HGTB0239).

M. Tanveer et al. (Eds.): ICONIP 2022, LNCS 13624, pp. 255–266, 2023.
https://doi.org/10.1007/978-3-031-30108-7_22

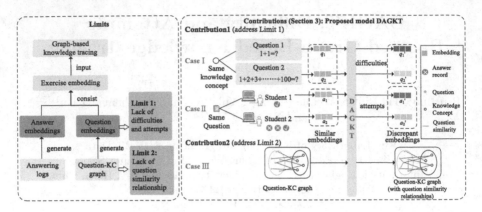

**Fig. 1.** The limits and contributions. The limit of lack of difficulties and attempts is addressed shown in Case I and II, and the limit of lack of question similarity relationship is addressed shown in Case III.

Nowadays, KT models based on graph neural networks (GNNs) present satisfied performance, because there are natural graph structures among knowledge concepts (KCs) and questions in KT [9]. Nakagawa et al. [13] proposed the graph-based KT (GKT) to learn the graph relations among KCs using the GNN. Graph-based interaction model for KT (GIKT) [18] focuses on the relationships between questions and KCs, obtaining higher-order embeddings of questions and KCs by the graph convolutional network (GCN) [7]. Question embeddings and answer embeddings in KT task [15] are integral parts of exercise embeddings. Among them, question and answer embeddings represent the information of questions and students' performance on questions, respectively. These GNN-based KT models obtain satisfied performance because better exercise embeddings are achieved through question embeddings with graph relationships using GNNs.

Exercise embedding plays an important role in KT task, because cognition evaluation in KT relies on students' performance on exercises. There is rich information involved in exercises such as stem texts [17] and student behaviors features [12]. There is still room for improvement for both embeddings, analyzed from the following aspects, as shown in Case I–III of Fig. 1.

First, most existing GNN-based KT models ignore the question difficulties in question embeddings as well as attempts in answer embeddings. Difficulties and the number of attempts are critical as question embeddings and answer embeddings which reasons are as follows. When two questions $q_1$ and $q_2$ examine the same KC, student $s_1$ may give different answers because $q_1$ and $q_2$ have different difficulties (shown in Case I of Fig. 1). And if the number of attempts is not considered, the model will think that student $s_2$ who has tried 10 times to get it right, and student $s_3$ who got it right after only one attempt have the same experience (shown in Case II of Fig. 1). So if difficulties and attempts are not considered models can't discriminate between questions with the same KCs, or between answers with different attempts.

Furthermore, the question embedding is achieved by GNN aggregating the information of the surrounding nodes in the question-KC graph, so the question-KC graph is very important. Most existing graph-based KT models perform convolution on bipartite graphs and there is no question-question relationship in the graph (shown in III of Fig. 1). Gao et al. [5] hold the view that there are two kinds of relationships between questions: prerequisite relationships and similarity relationships. In the field of GNN-based KT, few studies put the relationships between questions into the convolution process (most of them only use the question similarities in the prediction process, such as [18]). Tong et al. [16] designed a method of constructing prior support relationships between questions from students' answer results illustrating the effectiveness of constructing relations from students' answer results. However, most existing studies construct the question similarity relationship through question text information or problem embedding distances, without using the students' answer results. There is still a need for a method that can use students' answer results to build similarity relationships.

To address these two problems, we propose the DAGKT model. Specifically, to solve the first problem, we design a fusion module to fuse two types of information: difficulty and attempts. We get the difficulties of the questions and the students' number of attempts from the datasets and encode them into embeddings through the encoder. After that, we put them with question embeddings and answer embeddings to the fusion module to obtain exercise embeddings that contain enormous information. Secondly, to address the second question and obtain a good question embedding, we design a relationship-building module that enriches the question-KC graph so that GCN can generate question embeddings that combine the information of the question relationships. We use statistical information combined with the calculation method of the F1 score to calculate similarity relationships between questions. It is assumed that the two questions may have a close relationship when students always obtain similar answering results (correct/incorrect) on the two questions. The F1 score is an indicator used in statistics to measure the accuracy of binary models. Another way to say, the F1 score infers to the degree of similarity between predicted and target values [10]. Therefore, the similarity of questions in this study is calculated according to the F1 score.

Finally, extensive experiments on real world datasets demonstrate the effectiveness of DAGKT and each module. In summary, our main contributions are as follows:

- To address the problem that most graph-based KT models cannot clearly discriminate between questions with same KCs, or between answers with different attempts, DAGKT is proposed with a fusion module. In this module, the question and answer embeddings are fused with difficulty and attempts.
- Furthermore, the relationship-building module is designed to construct the similarity relationship between questions, inspired by the F1 score. The con-

structed relationship enhances the representation of questions and improves the performance of KT.

– Several experiments are conducted on three public datasets. The results show that our model outperforms the baseline models and the effectiveness of the above two contributions is demonstrated.

## 2    Related Work

First, the KT models are reviewed. Then, the GNN-based KT model, i.e., GIKT, is introduced.

### 2.1    Knowledge Tracing

KT is the task of estimating the dynamic changes in students' knowledge state based on their exercise records. Existing KT models can be categorized into two main types: Bayesian-based KT and deep learning KT models [6]. BKT is based on the hidden Markov model which is the first model proposed to solve the KT task. Several studies have integrated some other information into BKT, such as student's prior knowledge [14], slip and guess probabilities [1], and student individualization [19]. Due to the powerful ability to achieve non-linearity and feature extraction making it well suited to modeling the complex learning process, deep neural networks have been leveraged in many KT models. DKT [15] uses a recurrent neural network to trace the knowledge state of the students which is the first deep KT model. DKVMN [20] introduces an external memory module to store the KCs and accurately points out students' specific knowledge state on KCs. Based on these two models, several studies consider adding more information to the models to improve their performance [8], such as the forgetting behavior of students [12] and student individualization [11].

With the development of GNN, it has been found that it can work well when dealing with graph-structured data. Nakagawa et al. [13] presented the GKT, which uses GNNs to handle the complex graph-related data, such as knowledge concepts. GIKT [18] using GCN [7] to obtain higher-order embeddings of questions through relations between questions and KCs.

### 2.2    GIKT

In this subsection, we introduce the student state evolution module and prediction part in GIKT [18]. Our work is inspired by GIKT, an effective graph-based KT model, and we refer readers to the reference [18] for more details about GIKT.

**Student State Evolution Module:** For each time step $t$, GIKT concatenates the question and answer embeddings and transforms them into the representation of exercises through nonlinear layers:

$$\mathbf{e}_t = \text{ReLU}\left(\mathbf{W}_1\left([\widetilde{\mathbf{q}}_t, \mathbf{a}_t]\right) + \mathbf{b}_1\right), \tag{1}$$

where [,] denotes embedding concatenation. GIKT models the whole exercise process to capture the students' state changes and to learn the potential relationships between exercises. To model behaviors of students doing exercises, GIKT uses LSTM which can capture coarse-grained dependency like potential relationships between KCs to learn students' states from input exercise embeddings. And GIKT learns the hidden state as the current student state, which contains the coarse-grained mastery state of KCs.

**Prediction:** To improve the performance of model, GIKT designed a history recap module that can select relevant history exercises (question-answer pair) to better represent a student's ability on a specific question. GIKT chooses history questions sharing the same skills with the new question for prediction. After that, GIKT uses the interaction of cognitive state and questions, the interaction of cognitive states with related skills and interaction of the cognitive state at the time step of the relevant history exercise with the new question and its skills to predict, GIKT calculates the attention weights of all relevant interaction terms and computes the weighted sum as the prediction.

## 3   The Proposed Model DAGKT

In this section, our model is introduced in detail. We use the GIKT model as our base model because it is one of the state-of-the-art models in the graph-based KT field and can make good use of exercise embeddings to predict.

### 3.1   Framework

The framework of DAGKT is shown in Fig. 2. First, we establish the similarity relationships between questions from the records to enrich the question-KC graph and generate the embeddings of questions through GCNs. Then, we extract the number of attempts students made on each question and the difficulty of each question from the records and encoder them into embeddings. After that, we put the question embedding, difficulty embedding, answer embedding and attempts embedding into the fusion module to obtain exercise embedding which denotes information about this exercise. Finally we put exercise embeddings into LSTM to obtain the knowledge mastery of students at each time step. And we make predictions through the prediction module.

### 3.2   Embedding Module

In this subsection, we will introduce how the model produces three important parts of the four components of the exercise: question embedding, difficulty embedding and attempts embedding. The relationship-building module is used to generate similarity relationships between questions, and then the question embedding propagation is used to generate question embedding. Finally, attempts embeddings and difficulty embeddings are generated through the difficulty and attempts encoder module.

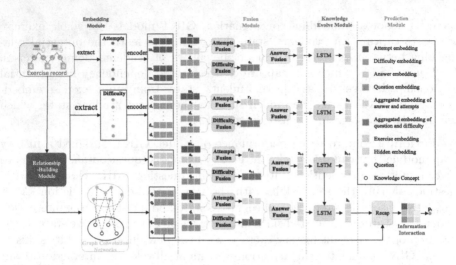

**Fig. 2.** The framework of DAGKT, including four modules: embedding module (detailed in Sect. 3.2), fusion module (detailed in Sect. 3.3), knowledge evolution module (detailed in Sect. 3.4) and prediction module (detailed in Sect. 3.4).

**Relationship-Building Module:** In this module, we introduce how to establish similarity relationships between questions. The same person's answers will be similar when the questions have similarity relationships. And the F1 score is a good indicator used in statistics to measure the performance of binary models which is used to measure how well the predicted results match the real results. Because the F1 score is a good indicator to measure the similarity of two sets of binary data, we use the true response result of the previous question as the prediction of the response result of the latter question, and calculate the F1 score between them as the similarity between them:

$$\text{Sim}\,(q_1, q_2) = \frac{(F_1\,(q_1, q_2) + F_1\,(q_2, q_1))}{2}, \tag{2}$$

where $F_1\,(q_1, q_2)$ denotes when $q_1$ is answered before $q_2$, the similarity of the two answers. And $F_1\,(q_1, q_2)$ is calculated by:

$$F_1\,(q_1, q_2) = 2 \cdot \frac{P\,(q_1, q_2) \cdot R\,(q_1, q_2)}{P\,(q_1, q_2) + R\,(q_1, q_2)}, \tag{3}$$

where $P\,(q_1, q_2)$ and $R\,(q_1, q_2)$ are the parts used to calculate $F_1\,(q_1, q_2)$. $P\,(q_1, q_2)$ and $R\,(q_1, q_2)$ are calculated by Eqs. (4) and (5).

$$P\,(q_1, q_2) = \frac{\text{Count}\,((q_1, q_2) = (1, 1)) + \lambda}{\sum_{a_1=0,1} \text{Count}\,((q_1, q_2) = (a_1, 1)) + \lambda}, \tag{4}$$

$$R\,(q_1, q_2) = \frac{\text{Count}\,((q_1, q_2) = (1, 1)) + \lambda}{\sum_{a_2=0,1} \text{Count}\,((q_1, q_2) = (1, a_2)) + \lambda}, \tag{5}$$

where Count $((q_i, q_j) = (a_i, a_j))$ denotes the number of question sequences that reply $q_i$ with answer $a_i$ before $e_j$ with an answer $a_j$. Besides, to prevent the denominator from becoming too small, we introduced the laplacian smoothing parameter $\lambda = 0.01$ in Eqs. (4) and (5). After generating similarity between the two questions, we add the edges between the questions $q_1, q_2$ to the question-knowledge concept graph when Sim $(q_1, q_2)$ is larger than hyperparameter $\omega$. So far, we have completed the construction of the question-KC graph.

**Question Embedding Propagation:** We put the initialized question embeddings and question-KC graph into GCNs to obtain better question embeddings that have higher-order information between questions and KCs.

**Difficulty and Attempts Encoder Module:** Since the difficulty and attempts play an important role in KT. In this module, we incorporate difficulties and attempts into exercise embeddings. Firstly, we obtain the number of attempts from the real dataset $M = \{m_{1,1}, m_{1,2}, \ldots, m_{i,j}\}$ where $i$ denotes student ID and $j$ denotes question ID. Then we count the accuracy of all students on each question and calculate the questions' difficulties as

$$d_i = \text{function}(\frac{\text{n}(correct(q_i))}{\text{n}(correct(q_i)) + \text{n}(false(q_i))}), \tag{6}$$

where $\text{n}(\cdot)$ denotes the number of $[\cdot]$. After obtaining the attempts and difficulties, we use encoders such as three nonlinear fully connected layers to encode them into 100-dimensional embeddings $\mathbf{m}$ and $\mathbf{d}$:

$$\mathbf{m_{i,j}} = \sigma(\sigma(\tanh(m_{i,j}))), \mathbf{d_i} = \sigma(\sigma(\tanh(d_i)))). \tag{7}$$

And three more nonlinear layers are used to transform the 100-dimensional embeddings into numerical values:

$$\tilde{m}_{i,j} = \sigma(\sigma(\tanh(\mathbf{m_{i,j}}))), \tilde{d}_i = \sigma(\sigma(\tanh(\mathbf{d_i}))). \tag{8}$$

As the optimizer optimizing parameters, we obtain embeddings of attempts $\mathbf{m_{i,j}}$ and difficulties $\mathbf{d_i}$ when the $\tilde{m}_{i,j}$ and $\tilde{d}_i$ come in close to the original inputs $m_{i,j}$ and $d_i$.

### 3.3   Fusion Module

After embedding module, we can obtain question embedding $\mathbf{q}$, difficulty embedding $\mathbf{d}$, attempts embedding $\mathbf{m}$ and answer embedding $\mathbf{a}$. In this module, we fuse the four embeddings. First, we fuse the difficulties and questions through difficulty fusion module to obtain the aggregated embeddings of question and difficulty. Then we fuse the attempts and answers through attempts fusion module to obtain the aggregated embeddings of answers and attempts. Finally, the two aggregated embeddings are passed through one more layer of the nonlinear neural network to obtain the final exercise embeddings we need. The formula for the whole process can be expressed as follows:

$$\mathbf{x}_t = \text{Relu}(\mathbf{W}_3([\mathbf{W}_1([\mathbf{q}_t, \mathbf{d}_t]) + \mathbf{b}_1, \mathbf{W}_2([\mathbf{a}_t, \mathbf{m}_{i,j}]) + \mathbf{b}_2]) + \mathbf{b}_3), \tag{9}$$

where $\mathbf{W}_1, \mathbf{W}_2, \mathbf{W}_3$ and $\mathbf{b}_1, \mathbf{b}_2, \mathbf{b}_3$ are trainable matrices and parameters.

### 3.4 Knowledge Evolution Module and Prediction Module

In this subsection, we describe how the model utilizes exercise embeddings to generate the cognitive state of students, how to use the cognitive state to generate predictions, and how to optimize the model.

**Knowledge Evolution Module:** From the previous step, exercise embeddings absorb the relationships between questions and the relationships between questions and KCs, representing the behavior of students doing exercises. In each history step, to model the sequential behavior of students doing exercises, we put exercise embeddings into LSTM to learn the knowledge mastery changes of students where the hidden state denotes the current student state.

**Prediction Module:** When students do exercises, it is easy for them to associate the experience of doing similar questions to help them to solve the current questions. Like GIKT [18], we select questions with similar KCs from historical questions to help the model make predictions. And we use the interaction of cognitive state and question, the interaction of cognitive state and related skills, and the interaction of the cognitive state at the time step of the relevant historical question with the current question to make predictions.

**Optimization:** To optimize our model, we choose the method of Adam optimization, the parameters in the model can be updated by minimizing the loss function which contains three parts: (1) cross-entropy between the probabilities that the students will answer the question correctly $p_t$ and the true labels of the students' answer $a_t$, (2) mean square error between difficulties before encoder $d_i$ and difficulties after decoder $\tilde{d}_i$ and (3) mean square error between attempts before encoder $m_{i,j}$ and attempts after decoder $\tilde{m}_{i,j}$:

$$
\mathcal{L} = -\sum_t \left( a_t \log p_t + (1 - a_t) \log (1 - p_t) \right)
$$
$$
+ \sum_n \sum_t ((m_{i,j} - \tilde{m}_{i,j})^2) + \sum_t ((d_i - \tilde{d}_i)^2). \tag{10}
$$

## 4 Experiments

In this section, we conduct several experiments to investigate the performance of our model. We evaluate the prediction by comparing our model with other baselines on three public datasets. Then we make ablation studies show our modules' effectiveness in Sect. 4.3.

### 4.1 Setup

Datasets, baselines and implementation details are introduced in this subsection.

**Datasets:** To evaluate our model DAGKT, We have conducted extensive experiments on three public available datasets, i.e., ASSIST09[1], JUNYI [3][2] and

---

[1] https://sites.google.com/site/assistmentsdata/home/assistment-2009-2010-data/skill-builder-data-2009-2010.

[2] https://pslcdatashop.web.cmu.edu/DatasetInfo?datasetId=1198.

**Table 1.** Dataset statistics

| Numbers→ | Students | Questions | Skills | Logs | Questions/skills | Skills/questions |
|---|---|---|---|---|---|---|
| ASSIST09 | 3,852 | 17,737 | 167 | 282,619 | 173 | 1.12 |
| JUNYI | 4,872 | 835 | 41 | 569,111 | 1 | 20.36 |
| CSEDM | 343 | 50 | 18 | 32,082 | 14.5 | 5.22 |

CSEDM[3]. In the ASSIST09 dataset, we remove the duplicated records and scaffolding problems. We randomly select 5000 students in JUNYI dataset as the whole dataset is too big [18]. Table 1 illustrates the statistics of the datasets. We conduct a comprehensive examination of the models' performance in different fields and under different questions on these three datasets: ASSIST09, which is a record of mathematical questions with both single and multiple knowledge concepts; JUNYI, which is a collection of mathematical questions with a single KC, and CSEDM, which is a collection of programming questions with multiple KCs. For each dataset, we take at least sequences with lengths greater than 3, as it is meaningless to be too short.

**Baselines:** We select the following models as the baselines: DKT [15][4], DKVMN [20][5], GKT [13][6] and GIKT [18][7] (detailed in Sect. 2.1).

**Implementation Details:** We initialize the 100-dimensional embeddings of KCs, questions. In the LSTM part, we use an LSTM with two hidden layers where the sizes of the memory cells are set to 200 and 100. For GCNs, we set the maximal aggregate layer number $L = 3$. To avoid overfitting, we use a dropout with a keep probability of 0.8 for GCNs. We use the Adam optimizer to optimize the parameters of the model with the learning rate at 0.001 and the batch size of 32. We use Bayesian optimization to choose appropriate values for the other hyperparameters including the number of related exercises to the new question, skills related to the new question, skill neighbors in GCNs and question neighbors in GCNs.

Five-fold cross-validation is used to obtain a stable experimental results, setting 80% of the sequences as the training set and 20% as the test set. We take the average of the best results for each fold as the final result. The comparison models use their own parameters, and each model is trained for 50 epochs. To evaluate the performance of each model on each dataset, we use the AUC as the evaluation metric, and the larger AUC the better performance of model.

## 4.2   Overall Performance

Table 2 shows the AUC results of all the compared methods. Figure 3(a–c) shows the boxplots on three datasets. From these results, it is found that our DAGKT

---

[3] https://pslcdatashop.web.cmu.edu/Files?datasetId=3458.
[4] https://github.com/chrispiech/DeepKnowledgeTracing.
[5] https://github.com/lucky7-code/DKVMN.
[6] https://github.com/jhljx/GKT.
[7] https://github.com/ApexEDM/GIKT.

**Table 2.** Performance comparison on the datasets.

| Model→ | DKT [15] | DKVMN [20] | GKT [13] | GIKT [18] | DAGKT |
|---|---|---|---|---|---|
| ASSIST09 | 0.6838 | 0.7253 | 0.7214 | 0.7641 | **0.7759** |
| JUNYI | 0.8142 | 0.8398 | 0.8663 | 0.9066 | **0.9168** |
| CSEDM | 0.7321 | 0.7138 | 0.7328 | 0.7378 | **0.7719** |

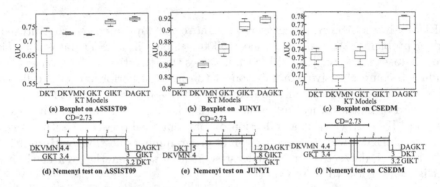

**Fig. 3.** The boxplots and Nemenyi tests on three datasets. Our model performs best and the distribution is more concentrated.

model performs best on three datasets, which proves the effectiveness of our model. To be specific: (1) GKT, GIKT and DAGKT achieve better results than DKT and DKVMN, which shows the effectiveness of GNN in handling the graph structure between question and KC. (2) The results of GIKT and DAGKT are better than those of GKT, which indicates that relationships of questions and KCs should be considered when constructing graph neural networks, and it is better to use relevant historical questions' information when predicting new questions. (3) DAGKT performs better than GIKT, which demonstrates the effectiveness of considering questions' difficulties, attempts and establishing similarity relationships between question. Moreover, the performance of our model is relatively concentrated, indicating that our model can obtain more stable results.

To better illustrate the superiority of our model, we performed Nemenyi test [4] on three datasets, results as Fig. 3(d–f), the smaller the value, the better the performance of the model.

### 4.3    Ablation Studies

To demonstrate the effectiveness of the modules in DAGKT, we design the following models:

- DAGKT-R removes relationship-building module and not fuse difficulties and attempts into exercise embeddings.
- DAGKT-D only adds difficulty information to help the model predict.
- DAGKT-A only adds attempts information to help the model predict.
- DAGKT-DA adds information on both difficulty and number of attempts to help model predict.

**Table 3.** Ablation studies of DAGKT. The relationship-building module, difficulty and attempts fusion module all improve model performance.

| Model→ | DAGKT-R | DAGKT-D | DAGKT-A | DAGKT-DA | DAGKT-G | DAGKT |
|---|---|---|---|---|---|---|
| ASSIST09 | 0.76414 | 0.7690 | 0.77146 | 0.7729 | 0.7647 | **0.7759** |
| JUNYI | 0.9066 | 0.9099 | 0.9079 | 0.9135 | 0.9106 | **0.9168** |
| CSEDM | 0.7378 | 0.7570 | 0.7560 | 0.7688 | 0.7577 | **0.7719** |

- DAGKT-G uses the relationship-building module to construct similarity relationships between questions so that it can produce better exercise embeddings to help model predict.

Table 3 shows that DAGKT achieve best performance considering similarity relationships between questions, difficulty of the question and the number of attempts. The results of DAGKT-D and DAGKT-A are better than DAGKT-R, which shows the effectiveness of fusing difficulties and attempts. The results of DAGKT-D and DAGKT-A are better than DAGKT-R, which shows that the attempts and difficulty have different improvement effects and the improvement of the effects are more stable when used at the same time. The results of DAGKT-G are better than DAGKT-R, which shows the effectiveness of relationship-building module which constructs similarity relationships between questions.

## 5 Conclusion

In this paper, to solve the problem that most existing graph-based KT models do not consider difficulty and attempts and do not establish similarity relationships between questions, we propose the DAGKT model which digs into questions' difficulties and number of attempts. Moreover, we design the relationship-building module to calculate the similarity between questions through the F1 score calculation method to establish similarity relationships between questions. Several experiments on three datasets demonstrate the effectiveness of the proposed model and the modules in the model.

## References

1. Baker, R.S.J., Corbett, A.T., Aleven, V.: More accurate student modeling through contextual estimation of slip and guess probabilities in Bayesian knowledge tracing. In: Woolf, B.P., Aïmeur, E., Nkambou, R., Lajoie, S. (eds.) ITS 2008. LNCS, vol. 5091, pp. 406–415. Springer, Heidelberg (2008). https://doi.org/10.1007/978-3-540-69132-7_44
2. Bu, C., et al.: Cognitive diagnostic model made more practical by genetic algorithm. IEEE Trans. Emerg. Top. Comput. Intell. **7**(2), 447–461 (2023). https://ieeexplore.ieee.org/document/9812476
3. Chang, H.S., Hsu, H.J., Chen, K.T.: Modeling exercise relationships in e-learning: a unified approach. In: Proceedings of International Conference on Educational Data Mining, pp. 532–535 (2015)

4. Demšar, J.: Statistical comparisons of classifiers over multiple data sets. J. Mach. Learn. Res. **7**, 1–30 (2006)
5. Gao, W., et al.: RCD: relation map driven cognitive diagnosis for intelligent education systems. In: Proceedings of the International ACM SIGIR Conference on Research and Development in Information Retrieval, pp. 501–510 (2021)
6. Hu, X., Liu, F., Bu, C.: Research advances on knowledge tracing models in educational big data. J. Comput. Res. Dev. **57**(12), 2523 (2020)
7. Kipf, T.N., Welling, M.: Semi-supervised classification with graph convolutional networks. In: Proceedings of International Conference on Learning Representations (2017)
8. Liu, F., Hu, X., Bu, C., Yu, K.: Fuzzy Bayesian knowledge tracing. IEEE Trans. Fuzzy Syst. **30**(7), 2412–2425 (2021)
9. Liu, Q., Shen, S., Huang, Z., Chen, E., Zheng, Y.: A survey of knowledge tracing. arXiv preprint (2021)
10. Ma, W., Cai, L., He, T., Chen, L., Cao, Z., Li, R.: Local expansion and optimization for higher-order graph clustering. IEEE Internet Things J. **6**(5), 8702–8713 (2019)
11. Minn, S., Yu, Y., Desmarais, M.C., Zhu, F., Vie, J.J.: Deep knowledge tracing and dynamic student classification for knowledge tracing. In: Proceedings of IEEE International Conference on Data Mining, pp. 1182–1187. IEEE (2018)
12. Nagatani, K., Zhang, Q., Sato, M., Chen, Y.Y., Chen, F., Ohkuma, T.: Augmenting knowledge tracing by considering forgetting behavior. In: Proceedings of the World Wide Web Conference, pp. 3101–3107 (2019)
13. Nakagawa, H., Iwasawa, Y., Matsuo, Y.: Graph-based knowledge tracing: modeling student proficiency using graph neural network. In: Proceedings of IEEE/WIC/ACM International Conference on Web Intelligence (2019)
14. Pardos, Z.A., Heffernan, N.T.: Modeling individualization in a Bayesian networks implementation of knowledge tracing. In: De Bra, P., Kobsa, A., Chin, D. (eds.) UMAP 2010. LNCS, vol. 6075, pp. 255–266. Springer, Heidelberg (2010). https://doi.org/10.1007/978-3-642-13470-8_24
15. Piech, C., et al.: Deep knowledge tracing. In: Advances in Neural Information Processing Systems, vol. 28 (2015)
16. Tong, H., Wang, Z., Liu, Q., Zhou, Y., Han, W.: HGKT: introducing hierarchical exercise graph for knowledge tracing. arXiv preprint arXiv:2006.16915 (2020)
17. Xiao, R., Zheng, R., Xiao, Y., Zhang, Y., Sun, B., He, J.: Deep knowledge tracking based on exercise semantic information. In: Jia, W., et al. (eds.) SETE 2021. LNCS, vol. 13089, pp. 278–289. Springer, Cham (2021). https://doi.org/10.1007/978-3-030-92836-0_24
18. Yang, Y., et al.: GIKT: a graph-based interaction model for knowledge tracing. In: Hutter, F., Kersting, K., Lijffijt, J., Valera, I. (eds.) ECML PKDD 2020. LNCS (LNAI), vol. 12457, pp. 299–315. Springer, Cham (2021). https://doi.org/10.1007/978-3-030-67658-2_18
19. Yudelson, M.V., Koedinger, K.R., Gordon, G.J.: Individualized Bayesian knowledge tracing models. In: Lane, H.C., Yacef, K., Mostow, J., Pavlik, P. (eds.) AIED 2013. LNCS (LNAI), vol. 7926, pp. 171–180. Springer, Heidelberg (2013). https://doi.org/10.1007/978-3-642-39112-5_18
20. Zhang, J., Shi, X., King, I., Yeung, D.Y.: Dynamic key-value memory networks for knowledge tracing. In: Proceedings of the International Conference on World Wide Web, pp. 765–774 (2017)

# AMRE: An Attention-Based CRNN for Manchu Word Recognition on a Woodblock-Printed Dataset

Zhiwei Wang[1], Siyang Lu[2(✉)], Mingquan Wang[1], Xiang Wei[1], and Yingjun Qi[3]

[1] School of Software Engineering, Beijing Jiaotong University, Beijing, China
[2] School of Computer and Information Technology, Beijing Jiaotong University, Beijing, China
sylu@bjtu.edu.cn
[3] School of Japanese studies, Dalian University of Foreign Languages, Dalian, China

**Abstract.** Ancient minority language character recognition could be challenging due to limited documentation, but it is always critical to better understand history and conduct social science researches. As an important minority language, Manchu language is confronted with the similar challenges due to the lack of systematic document studies. Recently, more researches focus on solving this problem through different approaches, such as document digitalization or character image segmentation. However, there are still some limitations. On one hand, existing digitalized Manchu documents are carried out based upon machine-printed style, which is not common in real historical documents and can cause severe recognition bias. On the other hand, most of Manchu identification methods are based on coarse image segmentation and may result in recognition error since it is difficult to consistently cut the words accurately. To tackle these two challenges, we propose a segmentation-free method for Manchu recognition with a medium scale dataset of Woodblock-printed Manchu Words (WMW). We first develop WMW based-upon woodblock-printed Manchu words, which are more common in ancient documents. With the developed dataset, we conduct document mining and carry out a framework, namely AMRE, with Attention-based Convolutional Recurrent Neural Network. AMRE leverages attention mechanism by weighted aggregation of the convolution results from differently sized kernels and more effectively mine the valid information of morphed words in recognition process. By implementing our proposed AMRE, the digitalized characters can be more accurately recognized. The experiment results show that the word recognition accuracy of AMRE exceeds the baseline by more than 5%.

**Keywords:** Manchu word recognition · Woodblock-printed dataset · Optical character recognition · Deep learning

## 1 Introduction

Manchu language is the lingua of Manchu people and belongs to the Manchu-Tungus language branch. During the Qing Dynasty of China, the Manchu language is leveraged as one of the official languages. However, Manchu becomes a

© The Author(s), under exclusive license to Springer Nature Switzerland AG 2023
M. Tanveer et al. (Eds.): ICONIP 2022, LNCS 13624, pp. 267–278, 2023.
https://doi.org/10.1007/978-3-031-30108-7_23

rarely used language cased by the culture assimilation, thus few people can recognize Manchu words in China now. Generally, Manchu is a phonetic language consisting of multiple letters. Moreover, according to different positions, Manchu letters can be divided into four forms: beginning, middle, end and alone, as illustrated in Fig. 1. In China, a part of ancient Manchu documents is preserved in historical archives and cultural institutions. These documents contain substantial values from both historical and cultural aspects, and are of great importance to Manchu studies. However, during the long storage time, these materials have undergone serious degradation, such as surface damages, ink bleedings, ink fadings, and paper stains. In order to preserve the contents of these deteriorated documents, digitizations are made to transform them into image format. Figure 1 shows a fragment scanned image of ancient Manchu documents. To effectively recognize woodblock-printed Manchu words in these digitized images still face three main obstacles: (1) Woodblock-printed Manchu words have special morphological variations because of various woodcut techniques and complex inking relationships. Therefore, the same letter in different positions have different shapes. At the same time, there are differences in the shape of the same word. As shown in Fig. 1 and Fig. 2. (2) Due to the long storage time, the ancient documents in Manchu have problems such as paper deformation and blurred fonts. (3) There are additional intervening characters in the ancient documents because of the manual woodcut technique.

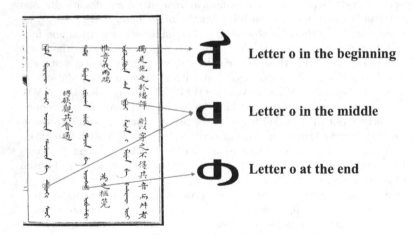

**Fig. 1.** Illustration of a Manchu document and primary characteristics of a Manchu word.

In this paper, we introduce a new dataset of woodblock-printed Manchu words. The new dataset consists of 21,245 woodblock-printed Manchu words exacted from a set of books called the Imperially-Published Revised and Enlarged mirror of Qing (han i araha nonggime toktobuha manju gisun buleku bithe). Moreover, to address the above challenges, we propose AMRE, a

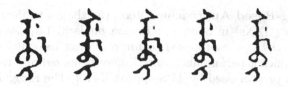

Fig. 2. An example of different glyphs of the same Manchu word "sembi".

segmentation-free method based on attention mechanism, combining Convolutional Neural Networks (CNN), Bi-directional Long short-term memory (Bi-LSTM), and Connectionist Temporal Classification (CTC) to achieve high-precision recognition of woodblock-printed Manchu words. Firstly, we leverage an Attention-based CNN for feature extraction on Manchu images. The Attention-based CNN could calculate the weights between different convolutional layers efficiently and focus more on the valid feature information in the images. Next, a Bi-LSTM is used to make the features extracted from CNN focus on the contextual semantic relationships. Finally, CTC is used to decode the feature vectors and output each recognized Manchu word. We validate our proposed framework on the new dataset, and the experimental results show the effectiveness of our method in Manchu word recognition.

Our work has the following two contributions:

- A new woodblock-printed Manchu words dataset called WMW (**W**oodblock-printed **M**anchu **W**ords) is introduced, which provides a criteria for comparing different woodblock-printing Manchu recognition algorithms.
- A novel method, namely AMRE (**A**ttention-based CRNN for **M**anchu Word **RE**cognition), is proposed to tackle Manchu word recognition task. What's more, AMRE achieves significant performances on WMW dataset, where word recognition accuracy exceeds the baseline CRNN by more than 5%.

## 2   Related Work

In this section, traditional text recognition approaches, deep learning-based approaches and Manchu recognition approaches are introduced respectively.

**Traditional Text Recognition Approaches.** In the traditional works of ancient document recognition, features in the images are extracted manually. Gradient analysis [1], stroke analysis [2], and connected component analysis [3] are the commonly used approaches for extracting features. Next, the extracted features are fed into the classifier to output the results. Previous classifiers including Support Vector Machine (SVM) [4] and Hidden Markov Model (HMM) [5] obtain better performance in small sample dateset, but have difficulty identifying multiple classes of similar features.

**Deep Learning-Based Approaches.** Recently, thanks to the success of deep neural networks (DNNs) in many application fields [6–14], several approaches integrated with DNNs are proposed to improve text recognition performance and achieve significant performance gain. Before being fed into the network, the distorted text is preprocessed by TPS approach [15]. The recognition of text in images is divided into Segmentation-based schemes [16] and Non-segmentation-based schemes [17]. In Segmentation-based schemes, the accuracy of text recognition depends heavily on the effectiveness of the specific segmentation. Due to the problem of data imbalance, Wei et al. [18] expanded the dataset with SMOTE method. Kass et al. [19] introduce the concept of Transfer Learning, which uses optimized pre-trained models for feature migration.

Among the frameworks for deep learning, Recurrent Neural Network (RNN) is a generally adopted network structure that is good at handling temporal correlation tasks. Kang et al. [17] leverage LSTM to recognize text in ancient documents. Mei et al. [20] improve this approach by jointly training CNN and RNN, and analyzing the effect of different networks on text recognition. Later, Shi et al. [21] incorporate CTC [22] into this model. It outputs predictions by efficiently counting the highest conditional probabilities of input sequences. Liao et al. [23,24] use a deep neural network to localize text in images before recognition.

In recent years, attention mechanism becomes a current research hotspot. It ignores unimportant information and highlights some distinct features in the data. Inspired by this, Bahdanau et al. [25] first apply attention to text recognition. Soon after, Vaswani et al. [26] introduce a new attention concept. It discards the previous ways of using CNN or RNN to obtain character location information, making the model better able to capture feature information in the context. In addition, Jiang et al. [27] fuse attention with CTC to recognize the input sequence with joint decoding. Cui et al. [28] propose a triple attention network for Mongolian recognition.

**Manchu Recognition Approaches.** In the previous studies of Manchu recognition, Zhang et al. [29] locate the baseline in Manchu words and use a wavelet neural network to recognize Manchu characters on both sides of the baseline. Depending on the baseline selected, this way will result in different recognition results. Subsequently, Xu et al. [30] adapt the method by splitting the Manchu words into separate components. Each separate component is represented by a letter, and SVM is used for identification. This approach is limited in the number of components and does not classify them well. To overcome these problems, Li et al. [31] use a Segmentation-free scheme. Manchu words are overall recognized by CNN and it has 666 classes in total. This approach avoids the problem of inaccurate component segmentation but does not prevail in parameter tuning.

## 3   WMW Dataset Construction

In this section, a novel dataset called WMW is proposed. WMW contains a total of 21,245 images of woodblock-printed Manchu words, taking 4,506 vocabularies into consideration. The construction of WMW dataset is comprised of three parts: Manchu word separation, Manchu word annotation, and dataset statistic analysis.

## 3.1    Manchu Word Separation

Manchu letters can form various kinds of Manchu words. However, collecting all Manchu words from existing ancient Manchu materials is a challenging task. Therefore, we choose a set of books which named Imperially-Published Revised and Enlarged mirror of Qing as word vocabulary. This set of books compose a Manchu-Manchu dictionary with great value in Manchu studies.

To establish the dataset, firstly, we implement a splitting algorithm to segment single Manchu words from each page of the books. The word locating process is shown in Fig. 3. Each Manchu word is marked with red boxes. The words which have additional parts need to be cut again. In this way, a total of 21,245 images of woodblock-printed Manchu words are obtained.

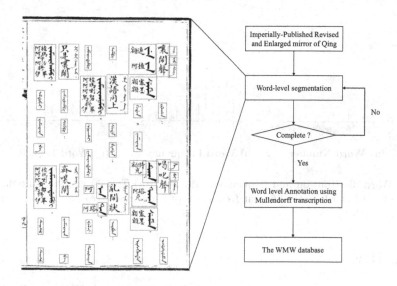

**Fig. 3.** The overview of dataset preparation pipeline. (Color figure online)

## 3.2    Manchu Word Annotation

In the next step, each woodblock-printed Manchu word is annotated. In the existing Manchu datasets, the entire word is usually annotated with a unique identifier. It causes not only the problem of over-classification of the model, but also the loss of the ability of generalization. Therefore, we annotate each letter in the word. The Möllendorff transliteration scheme proposed by German linguist Paul Georg von Möllendorff is leveraged to annotate each different character. The scheme is a widely adopted international transliteration scheme for Manchu words. It contains 36 different glyph codes (a, e, i, o, u, ū, n, ng, k, g, h, b, p, s, š, t, d, l, m, c, j, y, r, f, w, k', g', h', ts', ts, dz, dzi, ž, sy, c'y, jy). The overall creation process is given in Fig. 3.

### 3.3   Dataset Statistic Analysis

Finally, we analyze some features of WMW dataset. Figure 4 shows the distribution of dataset. The top 20 words are shown in Fig. 4a. Specifically, in our dataset, the word "sembi" has the highest count of 1,262. Figure 4b shows the distribution of the word frequency, where the number of words with 1 occurrence is 2503. As the frequency of occurrence increases, the number of words decreases sharply. Figure 4c shows the distribution of word lengths. The largest number of words with length 5 is 5713. The number of letters in the longest word is 15, and 6 words stands in that length.

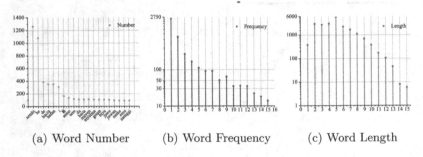

| (a) Word Number | (b) Word Frequency | (c) Word Length |

**Fig. 4.** Word distribution and frequency of dataset. (a) Top 20 words in the dataset. (b) Number of words with different frequency of occurrence. (c) Number of words of different lengths.

## 4   AMRE

In this section, our proposed AMRE is described in detail. The overall process of the model is shown in Fig. 5. The architecture of the model consists of three stages: feature extraction, sequence modeling, and prediction.

### 4.1   Feature Extraction Stage

Before being input into the network, all the images should be preprocessed. Each image is normalized to $60 \times 120$ while maintaining the aspect ratio. Since the Manchu is written from top to bottom, all the images need to be rotated 270° clockwise.

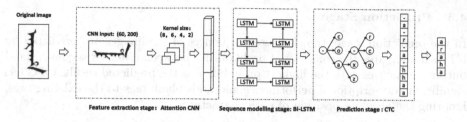

**Fig. 5.** The network architecture of our AMRE model.

In the feature extraction stage, the input images are encoded into two-dimensional feature vectors by using CNN. The convolution layer mainly contains four independent convolution operations. The sizes of the convolutional kernels are 8, 6, 4, and 2, respectively. To increase the computational nonlinearity in the convolution process, we set up three convolution layers.

In order to accurately extract features through convolution operation, we propose an attention model on CNN, named CNN-layer Attention. It focuses on the relationship among the outputs of each convolutional layer. Suppose the output of CNN is $D$, and it is divided into $n$ independent vectors. These vectors are denoted by $d_1, d_2 \ldots d_n$, where $d_i$ represents the $i_{th}$ row of $D$. For each different layer of convolution outputs, the relationship between the input vector $d_i$ and the output vector $c_i$ can be expressed as follows:

$$c_i = \lambda_i d_i \tag{1}$$

where $\lambda_i$ is the coefficient of CNN-layer Attention. It is calculated as follows:

$$e_i = \xi^T conv\left(W d_i + b\right) \tag{2}$$

$$\lambda_i = \frac{exp\left(e_i\right)}{\sum_{j=1}^{n} exp\left(e_j\right)} \tag{3}$$

where $\xi^T$, $W$ and $b$ are trainable parameters. $conv$ is the convolution function. $e_i$ is the intermediate value.

## 4.2 Sequence Modeling Stage

In the sequence modeling stage, Bi-LSTM is used to capture contextual information in the feature vector. Recognizing letters in word is more stable and accurate than recognizing each letter individually. Bi-LSTM is good at handling time-sensitive problems. It contains a special gating unit structure to select effective memory information and utilize it. Compared to a single LSTM, Bi-LSTM can obtain information containing both directions. At the same time, the deep network structure allows for a higher level of abstraction.

## 4.3   Prediction Stage

In the prediction stage, the character sequences of the images are predicted by CTC. It compares the conditional probabilities between different sequences and outputs sequences with the highest probability as the predicted result. CTC has significant transcriptional performance and adds blank tags to the existing tags, ignoring the absolute position of each character.

## 5   Experiments

This section describes the evaluation protocols, implementation details and the experimental results.

### 5.1   Evaluation Protocols

Character recognition accuracy (CRA) and word recognition accuracy (WRA) are used as the evaluation protocols for recognizing Manchu words. CRA is defined by:

$$CRA = \frac{\sum_{z=1}^{M} \{x_z = \hat{x}_z\}}{M} \tag{4}$$

where $M$ is the total number of recognized characters. $x_Z$ represents the ground truth characters and $\hat{x}_Z$ denotes the predicted characters. WRA is given as:

$$WRA = \frac{\sum_{z=1}^{N} Dis\,(y_z, \hat{y}_z)}{N} \tag{5}$$

where $N$ is the total number of words. $y_z$ represents the ground truth labels and $\hat{y}_z$ represents the predicted words. $Dis(\cdot)$ is the identification function and returns a value in $\{0, 1\}$. 1 is returned only when $y_z$ has the same value as $\hat{y}_z$.

### 5.2   Results

We conduct experiments on our proposed WMW dataset. The training data includes a total of 11,252 words, and a total of 9,993 words are selected as test data. The edit distance is adopted in the model to determine whether the predicted words and ground truth words are identical or not. The edit distance refers to the number of different letters between two Manchu words. Table 1 shows the proportion of correctly recognized words at different edit distances. If the edit distance is 0, the predicted word and ground truth word is the same. A smaller edit distance means that the predicted word is more similar to the ground truth word.

Figure 6 shows the experiment results. Our proposed AMRE model reaches 97.44% and the baseline method reaches 95.74% in CRA. In Fig. 6b, AMRE reaches 90.92% and the baseline method reaches 85.52% in WRA. Compared with the baseline method, AMRE is 1.70% higher in CRA and 5.40% higher in

**Table 1.** The comparison of proportion of correctly identified words at different edit distances.

| Edit Distance | 0 | $\leq 1$ | $\leq 2$ | $\leq 3$ |
|---|---|---|---|---|
| Proportion | 90.92% | 97.53% | 99.22% | 99.58% |

WRA. The accuracy of both methods stabilized after 60 rounds. The baseline way reached the convergence condition at 60 rounds, and the accuracy fluctuated more during the convergence process. Our proposed AMRE reaches the convergence condition at 30 rounds, and the convergence process is more stable. It shows that AMRE is better than the baseline method in terms of performance and stability. In Fig. 6c, the loss value of AMRE is always lower than that of the baseline method. As the training epochs increase, the loss values of the two methods converge to the same. It indicates that our proposed AMRE has better robustness.

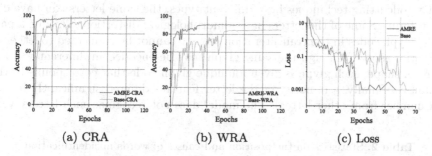

| (a) CRA | (b) WRA | (c) Loss |

**Fig. 6.** The comparison results between AMRE and baseline method. (a) CRA performance comparison of AMRE and baseline. (b) WRA performance comparison of AMRE and baseline. (c) Loss performance comparison of AMRE and baseline.

## 5.3   Discussion

To further explain the possible causes of the model's errors, 100 Manchu words are further analyzed. Figure 7 illustrates the recognition results of a part of Manchu words. The left part of the figure shows some examples of the correctly recognized words. The model can recognize complex words such as 'becunure' because our proposed framework could focus on the valid information in words to perform word identification. The right part of the figure shows examples of the incorrectly recognized words. For example, 'ne' is recognized as 'na' since the distorted letter 'e' is visually similar to letter 'a'.

Table 2 lists the misidentification causes and positions of these 100 Manchu words. The main causes are incomplete letters, distorted letters, additional part, semantic collocation, etc. The specific analysis of the word misidentification is

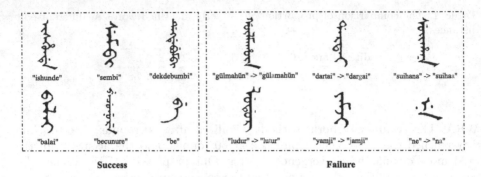

**Fig. 7.** The recognition results of some woodblock-printed Manchu words.

discussed as follows. Firstly, some letters are incomplete as a result of degradation, stains, and artificial factors in the documents. At the same time, the process of recognition is affected by additional part. Secondly, due to the fact that woodcutting techniques have different types, the same letters can have different forms. Some of the letters have a morphological distortion. Thirdly, a part of the glyph codes in the transliteration scheme cannot be connected together in semantics aspect. Though the same glyph may be encoded in different ways in terms of shape, each glyph is given a unique glyph code that corresponds to the semantics. Finally, a minority of the letter features differed significantly in size and number from other letters. The model does not recognize such letters well.

**Table 2.** Statistics on the position and causes of words misidentification

|            | Beginning | Middle | End | Alone | Total |
|------------|-----------|--------|-----|-------|-------|
| Incomplete | 5         | 10     | 1   | 0     | 16    |
| Distorted  | 9         | 29     | 9   | 0     | 47    |
| Additional | 2         | 4      | 0   | 0     | 6     |
| Semantic   | 3         | 4      | 3   | 0     | 10    |
| Other      | 5         | 10     | 6   | 0     | 21    |
| **Total**  | 24        | 57     | 19  | 0     | 100   |

# 6   Conclusion

In this paper, an Attention-based CRNN for Manchu Word Recognition method, namely AMRE, is proposed for woodblock-printed Manchu words. At the same time, an innovative dataset is introduced and named as WMW. It contains a total of 21,245 images of woodblock-printed Manchu words. The Manchu words are annotated by using the widely adopted Möllendorff transliteration scheme. The proposed method is validated on WMW dataset with character recognition accuracy and word recognition accuracy, reaching 97.44% and 90.92% respectively. The experimental results show that our proposed AMRE is superior to the baseline method.

**Acknowledgement.** This research was supported by NSF of China (62006016).

# References

1. Shivakumara, P., Phan, T.Q., Tan, C.L.: A gradient difference based technique for video text detection. In: 2009 10th International Conference on Document Analysis and Recognition, pp. 156–160. IEEE (2009)
2. Mosleh, A., Bouguila, N., Hamza, A.B.: Image text detection using a bandlet-based edge detector and stroke width transform. In: BMVC, pp. 1–12 (2012)
3. Pan, Y.F., Hou, X., Liu, C.L.: A hybrid approach to detect and localize texts in natural scene images. IEEE Trans. Image Process. **20**(3), 800–813 (2010)
4. Ye, Q., Huang, Q., Gao, W., et al.: Fast and robust text detection in images and video frames. Image Vis. Comput. **23**(6), 565–576 (2005)
5. Rodríguez-Serrano, J.A., Perronnin, F.: A model-based sequence similarity with application to handwritten word spotting. IEEE Trans. Pattern Anal. Mach. Intell. **34**(11), 2108–2120 (2012)
6. Wang, D., Li, Y., Wang, L., Gong, B.: Neural networks are more productive teachers than human raters: active mixup for data-efficient knowledge distillation from a blackbox model. In: Proceedings of the IEEE/CVF Conference on Computer Vision and Pattern Recognition, pp. 1498–1507 (2020)
7. Wang, D., Zhang, S., Wang, L.: Deep epidemiological modeling by black-box knowledge distillation: an accurate deep learning model for COVID-19. In: Proceedings of the AAAI Conference on Artificial Intelligence, vol. 35, no. 17, pp. 15424–15430 (2021)
8. Wang, D., Liu, Q., Wu, D., et al.: Meta domain generalization for smart manufacturing: tool wear prediction with small data. J. Manuf. Syst. **62**, 441–449 (2022)
9. Yang, Y., Xing, W., Wang, D., et al.: AEVRNet: adaptive exploration network with variance reduced optimization for visual tracking. Neurocomputing **449**, 48–60 (2021)
10. Wei, X., Wei, X., Kong, X., et al.: FMixCutMatch for semi-supervised deep learning. Neural Netw. **133**, 166–176 (2021)
11. Lu, S., Rao, B.B., Wei, X., et al.: Log-based abnormal task detection and root cause analysis for spark. In: 2017 IEEE International Conference on Web Services (ICWS), pp. 389–396. IEEE (2017)
12. Lu, S., Wei, X., Rao, B., et al.: LADRA: log-based abnormal task detection and root-cause analysis in big data processing with Spark. Future Gener. Comput. Syst. **95**, 392–403 (2019)
13. Yao, J., Wang, D., Hu, H., Xing, W., Wang, L.: ADCNN: towards learning adaptive dilation for convolutional neural networks. Pattern Recogn. **123**, 108369 (2022)
14. Yao, J., Xing, W., Wang, D., Xing, J., Wang, L.: Active dropblock: method to enhance deep model accuracy and robustness. Neurocomputing **454**, 189–200 (2021)
15. Shi, B., Yang, M., Wang, X., et al.: ASTER: an attentional scene text recognizer with flexible rectification. IEEE Trans. Pattern Anal. Mach. Intell. **41**(9), 2035–2048 (2018)

16. Xu, Y., Wang, Y., Zhou, W., et al.: TextField: learning a deep direction field for irregular scene text detection. IEEE Trans. Image Process. **28**(11), 5566–5579 (2019)
17. Kang, Y., Wei, H., Zhang, H., et al.: Woodblock-printing Mongolian words recognition by bi-LSTM with attention mechanism. In: 2019 International Conference on Document Analysis and Recognition (ICDAR), pp. 910–915. IEEE (2019)
18. Wei, H., Gao, G.: A holistic recognition approach for woodblock-print Mongolian words based on convolutional neural network. In: 2019 IEEE International Conference on Image Processing (ICIP), pp. 2726–2730. IEEE (2019)
19. Kass, D., Vats, E.: AttentionHTR: handwritten text recognition based on attention encoder-decoder networks. arXiv preprint arXiv:2201.09390 (2022)
20. Mei, J., Dai, L., Shi, B., et al.: Scene text script identification with convolutional recurrent neural networks. In: 2016 23rd International Conference on Pattern Recognition (ICPR), pp. 4053–4058. IEEE (2016)
21. Shi, B., Bai, X., Yao, C.: An end-to-end trainable neural network for image-based sequence recognition and its application to scene text recognition. IEEE Trans. Pattern Anal. Mach. Intell. **39**(11), 2298–2304 (2016)
22. Graves, A., Fernández, S., Gomez, F., et al.: Connectionist temporal classification: labelling unsegmented sequence data with recurrent neural networks. In: Proceedings of the 23rd International Conference on Machine Learning, pp. 369–376 (2006)
23. Liao, M., Shi, B., Bai, X., et al.: TextBoxes: a fast text detector with a single deep neural network. In: Thirty-First AAAI Conference on Artificial Intelligence (2017)
24. Liao, M., Shi, B., Bai, X.: TextBoxes++: a single-shot oriented scene text detector. IEEE Trans. Image Process. **27**(8), 3676–3690 (2018)
25. Bahdanau, D., Cho, K., Bengio, Y.: Neural machine translation by jointly learning to align and translate. arXiv preprint arXiv:1409.0473 (2014)
26. Vaswani, A., Shazeer, N., Parmar, N., et al.: Attention is all you need. In: Advances in Neural Information Processing Systems, vol. 30 (2017)
27. Jiang, Y., Jiang, Z., He, L., et al.: Text recognition in natural scenes based on deep learning. Multimedia Tools Appl. **81**(8), 10545–10559 (2022). https://doi.org/10.1007/s11042-022-12024-w
28. Cui, S.D., Su, Y.L., Ji, Y.T.: An end-to-end network for irregular printed Mongolian recognition. Int. J. Doc. Anal. Recogn. (IJDAR) **25**, 41–50 (2022). https://doi.org/10.1007/s10032-021-00388-y
29. Zhang, G., Li, J., Wang, A.: A new recognition method for the handwritten Manchu character unit. In: 2006 International Conference on Machine Learning and Cybernetics, pp. 3339–3344. IEEE (2006)
30. Xu, S., Qi, G.Q., Li, M., et al.: An improved Manchu character recognition method (2016)
31. Li, M., Zheng, R., Xu, S., et al.: Manchu word recognition based on convolutional neural network with spatial pyramid pooling. In: 2018 11th International Congress on Image and Signal Processing, BioMedical Engineering and Informatics (CISP-BMEI), pp. 1–6. IEEE (2018)

# High-Accuracy and Energy-Efficient Action Recognition with Deep Spiking Neural Network

Jingren Zhang[✉], Jingjing Wang, Xie Di, and Shiliang Pu

Hikvision Research Institute, Hangzhou Hikvision Digital Technology Co., Ltd., Hangzhou, China
{zhangjingren,wangjingjing9,xiedi,pushiliang}@hikvision.com

**Abstract.** In recent years, spiking neural networks (SNNs) received significant attention as the third generation of networks and have successfully been employed in energy-efficient image classification tasks. However, typical SNN construction methods still suffer from problems such as high inference latency or incompatibility with complicated models. Thus, applications of SNNs are limited to relatively simple tasks. In this paper, we establish an SNN-based action recognition model which aims at a more challenging video classification task. Specifically, the action recognition SNN model with a deep two-stream architecture is constructed with a hybrid conversion method combining channel-wise normalization and tandem learning. A skipping-step rate decoder is applied to decrease the conversion errors and improve the transmission accuracy. A new conversion and inference method for recurrent spiking neural network (RSNN) is introduced into the framework. The tandem learning method with bounded ReLU (bReLU) function is employed to fine-tune the normalized SNN parameters, decreasing the inference latency while still preserving high accuracy. Experiments on UCF-101 show that our proposed model obtains an accuracy of 88.46% with only 200 time steps, which achieves a high-accuracy and energy-efficient performance in SNN.

**Keywords:** spiking neural network · action recognition · ANN-to-SNN conversion · tandem learning

## 1 Introduction

In recent years, many researchers focus attention on spiking neural networks (SNNs) which are considered as the third-generation of networks with biological plausibility [1]. With supports of neuromorphic platforms and dynamic vision sensors, SNNs are allowed to work in asynchronous mode with lower power and computational consumption, while still providing acceptable accuracy [2,3].

Supported by the National Key Research and Development Program of China under Grant 2020AAA0109004.

Despite the promising potential, it remains a challenge to bridge the gap between the continuous-valued ANNs and event-based SNNs. Training methods with supervised or unsupervised paradigms are effective approaches to yield SNNs. Despite the comparable accuracy and inference latency, training methods suffer from unconvergence or constraints of specific architectures [7,8,12]. ANN-to-SNN conversion is a more straightforward approach that maps pre-trained ANNs into equivalent SNNs [4–6]. Layer-wise and Channel-wise normalization are typical data-based conversion methods. While implementations of normalization processes are compatible with complicated deep architectures [6], these methods suffer from approximation or quantization errors and high inference latency [10]. To minimize the conversion errors, tandem learning as a specific conversion framework composed of an SNN and a coupled ANN is proposed [9,10]. The coupled ANN shares parameters with the SNN and replaces the output activations of each layer with the corresponding spike counts of the SNN. The tandem learning achieves better performance in inference latency compared with the normalization methods. However, the tandem learning method has lower compatibility with deep architectures [10], which limits the application.

The network architectures are also important research issues for SNN constructions. Multi-layer feedforward SNNs are mostly applied to existing conversion and training processes [13]. In order to leverage the temporal characteristics of SNNs further, many researchers focus on reservoir networks constructed by STDP methods [7]. However, the constraint of architecture is still an obstacle to their applications. Hybrid feedforward/recurrent SNNs have also shown promising performances in temporal signal processing tasks such as speech recognition [13]. However, the training process relies on specific spike-train-level learning techniques and suffers from gradient explosion or vanishment.

Because of the problems mentioned above, SNNs are still limited to relatively simple models or tasks. In this paper, we establish an SNN-based action recognition model which aims at a more challenging video classification task. The total framework consists of a deep two-stream SNN and a recurrent spiking neural network (RSNN) fusion module. In order to achieve a high-accuracy and low-latency performance for the relatively complicated architecture, a hybrid conversion method combining channel-wise normalization and tandem learning is applied to the model. To decrease the conversion errors and improve the transmission efficiency in the deep architecture, we introduce a new kind of skipping-step rate decoder and its corresponding clipped proportional gradient into the framework. Because the RSNN module after conversion preserves independent recurrence and time step loops, the inference process is different from previous RSNNs and remains a problem. We propose an RSNN inference process with the skipping-step rate decoder and optimize the required memory of hidden and potential states. Despite the acceptable accuracy after channel-wise normalization, there is still scope for the inference latency to improve further. Thus the SNN model is also fine-tuned by the tandem learning method. The tandem learning method utilizes SNN firing rates as the hidden-layer outputs of the coupled ANN model, and the normalized model can be directly loaded as a pre-trained model. Bounded ReLU functions are introduced in the coupled ANN model as activations. The gradient modification for saturated neurons helps improve the performance of tandem training. Our proposed model is evaluated on the popular action recognition dataset UCF-101, and obtains an accuracy of 88.46% with only 200 time steps. High-accuracy and energy-efficient performance in video action recognition has been achieved.

Our main contributions can be summarized as follows: 1) An SNN-based action recognition model is constructed with a hybrid conversion method combining channel-wise normalization and tandem learning. A new type of skipping-step rate decoder is applied to decrease the conversion error. And a state-of-the-art performance in SNN is achieved on the UCF-101 dataset. 2) A new conversion and inference method for RSNN is introduced into the framework. And the required memory during the inference process is analyzed and optimized. 3) Tandem learning with a bounded ReLU module is employed to fine-tune the normalized deep SNN model. The inference latency is reduced by 75% after tandem learning.

## 2   Related Work

### 2.1   ANN-to-SNN Conversion

ANN-to-SNN conversion is a straightforward method to construct SNNs with pre-trained ANN parameters. To enable fast and efficient information transmission in deep SNNs, channel-wise normalization (channel-norm) is applied and converges several times faster than the layer-norm method [5]. Parameters are normalized corresponding to the maximum activations of each channel. Layer-norm and channel-norm can both be applied to deep network architectures. However, these conversion processes still suffer from quantization or approximation errors that lead to high inference latency. To minimize the conversion errors, Wu et al. proposed a tandem learning framework which is comprised of an SNN coupled layerwise with an ANN through shared weights [9]. During the forward pass, the output activations of the ANN are replaced with the spike counts of the SNN. During the backward pass, the gradients in SNN are back-propagated through the coupled ANN and the shared weights are updated. Competitive performances on a number of frame- and event-based benchmarks have been demonstrated. However, the direct applicability of tandem learning to deep SNNs is limited because of the difference of characteristics between spiking neurons and analog activation functions. While an adaptive training scheduler was proposed to overcome this obstacle [10], the layer-wise fine-tuning process complicates the learning process and increases the computational resource.

In this study, our hybrid conversion method combines the advantage of channel-wise normalization in deep architectures and that of tandem learning in fine-tuning together. A bounded ReLU (bReLU) module is introduced into the tandem learning framework to correct the back-propagated gradients of over-activated neurons [11], which improves the learning performance.

## 2.2  Recurrent Spiking Neural Network

Most existing conversion and training methods are aimed at constructions of feedforward SNNs. Different from feedforward SNNs, recurrent spiking neural networks with additional recurrent connections are more capable of extracting temporal features of time series data such as video or speech signals [13]. Reservoir networks with input and readout layers are special implementations of RSNNs, which can be trained by unsupervised STDP methods [7,8]. However, the constraint of the specific network architecture and training process limits the performance of reservoir networks. There are also deep RSNNs such as the hybrid feedforward/recurrent networks. Training processes of these RSNNs are challenged by the problems of back-propagation through recurrent connections and vanishing/exploding gradients. These RSNNs can be trained by the spike-train level back-propagation, which relies on specific aggregated effects calculation techniques for recurrent neurons [13].

Different from previous RSNNs, our proposed RSNN in this study is converted from an pretrained RNN and includes an independent recurrence loop. The tandem learning method can be applied to fine-tune the parameters, and the learning techniques for traditional RNNs can be easily leveraged.

## 2.3  Action Recognition

Action recognition tasks aim to recognize human actions in videos. In recent studies, deep learning methods have achieved great success and outperformed traditional methods. Existing deep learning methods mainly include 3D CNNs and two-stream networks [15,16]. 3D CNN-based methods extract spatial and temporal features simultaneously with the extended temporal dimension in convolutions. However, the high training complexity requires massive data sets or 3D convolution kernel factorization [15]. Two-stream networks exploit spatiotemporal information with RGB and stacked optical flow images as inputs. They are able to achieve competitive performance in spite of limited training data [14].

There are also researches of SNN-based video action recognition models constructed with unsupervised or supervised learning methods. Constrained by the network architecture and learning performance, the gap of accuracies between these methods and ANN-based methods still remains [17,18].

In this study, we focus on the SNN-based video action recognition task. A deep two-stream architecture constructed with the hybrid conversion method is applied to the framework, which achieves comparable performance to the original ANN model.

## 3  Methods

### 3.1  SNN Framework

Our SNN framework mainly includes two-stream SNNs with ResNet50 architecture and an RSNN fusion module. The total SNN framework is shown in Fig. 1.

ResNet backbones and an ordinary RNN in a pretrained ANN are both converted with the channel-wise normalization method. Then, the tandem learning method fine-tunes the parameters, which only requires few learning epoches. The total framework combines the compatibility with deep architectures in the normalization process and the advantage of fine-tuning in tandem learning. A new type skipping-step decoder is introduced to decrease the inference latency.

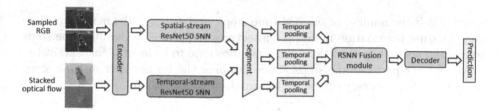

**Fig. 1.** The total SNN framework for action recognition.

**SNN Structure.** In our SNN framework, the inputs consist of sampled RGB and stacked optical flow images. We sample 12 frames evenly for each video regardless of its length. The frames are then converted into spike signals by a rate-based encoder. Spatial and temporal features are extracted by the two-stream ResNet50 SNN. These features in frame sequence are divided into $N_s$ temporal segments to learn their distinct feature representations [14]. Each temporal segment contains features corresponding to $12/N_s$ input frames, and the segment number $N_s$ is set to 3 in this study. The features in each temporal segment are then down-sampled via temporal max-pooling operation. The outputs of $N_s$ temporal segments are fed into the RSNN sequentially. Finally, the outputs of the two-layer RSNN fusion module are decoded into analog types and utilized for prediction.

Different from previous SNNs, ResNet50 SNN in our model includes no batch normalization (BN) or batch normalization through time (BNTT) module during the following SNN conversion process [6], which simplifies the construction. The spiking neurons are chosen as integrate-and-fire (IF) neurons reset by substraction [4,5]. Except the modules mentioned above, the entire architecture and total layer number are the same as ResNet50 ANN.

Different from previous RSNNs constructed by training methods, the proposed RSNN in this study is converted from an ordinary RNN which recurrence loop is preserved independently after the normalization process. So the RSNN consists of the independent recurrence and SNN time step loop during inference.

**Skipping-Step Decoder.** To increase the accuracy of information transferred in the deep architecture, a new kind of skipping-step decoder is proposed and applied to the decoding operations in the total framework. The input spike train of the skipping-step decoder is divided into the former and latter part along the temporal axis. And the firing rate of the spike train is evaluated according to the latter part of time steps. A certain number of time steps are skipped and the output of the decoder is described as:

$$r_o = \sum_{t=T_{sk}}^{T_{s\max}} o(t) \tag{1}$$

where $T_{sk}$ is the number of skipped time steps, and $T_{s\max}$ is the maximum time step. Feature information in deep architectures requires considerable time steps to transfer accurately. And the spiking neurons tend to be insufficiently activated at the former part of time steps. Thus skipping the former part of time steps leads to higher accuracy of feature information transmission.

### 3.2   RSNN Inference

**RSNN with Independent Recurrence Loops.** Different from previous RSNNs, the RSNN in our model includes the independent recurrence loop and time step loop. The inference process remains a problem. In this part, we propose two different inference settings and discuss the transfer modes of hidden states.

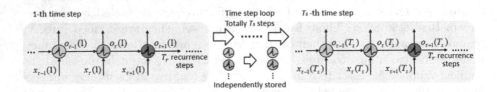

**Fig. 2.** RSNN inference process with the recurrence loop as inner loop

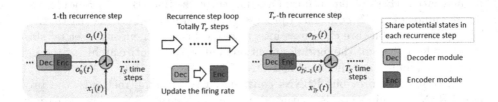

**Fig. 3.** RSNN inference process with the recurrence loop as outer loop

During RSNN inference, input spikes of a spiking neuron are generated from the preceding layer, while the hidden state spikes are generated from the previous recurrence step. According to the transmission modes of hidden state spikes, two different RSNN inference settings can be built up. When we choose the time step loop as the outer loop, the recurrence loop is set as the inner loop (shown in Fig. 2). In the $t$-th time step, the RSNN can be expanded along the recurrence loop and the hidden state spike $o_\tau(t)$ is fed into the spiking neuron at the $(\tau + 1)$-th recurrence step. For the inference alone the time step loop, the membrane states at different recurrence steps should be stored independently. Thus the memory of the membrane potential and hidden states (for one neuron) during inference can be evaluated as:

$$M_1 = T_r m_f + m_i \tag{2}$$

where $T_r$ is the total step number in a recurrence loop, $m_f$ is the memory of an analog value for the membrane potential, and $m_i$ is the memory of a spike signal for the hidden state.

We can also choose the time step loop as the inner loop, and the recurrence loop as the outer loop (shown in Fig. 3). In the $\tau$-th recurrence step, the inference of the spiking neuron lasts for $T_s$ time steps. And the hidden state $o_\tau(.)$ should be restored at the $(\tau + 1)$-th recurrence step. When the whole spike train is stored in each recurrence step, the memory of the potential and hidden states during inference can be evaluated as:

$$M_2 = T_s m_i + m_f \tag{3}$$

where $T_s$ is the total step number in a time step loop.

**Improvement of RSNN Inference.** It can be seen that when the total step number $T_s$ and $T_r$ are high, both of the inference settings require large memory. To overcome this problem, we optimize the restoration of hidden states based on the second inference setting (shown in Fig. 3). In each recurrence step, the spike train of the hidden state is decoded according to the firing rate. The decoding output is received by an encoder at the next RNN recurrence step, and then the spike train of the hidden state is restored via rate-based encoding. To improve the restoration accuracy, the skipping-step decoder is also applied to the decoding operation. The memory of the potential and hidden states can be written to:

$$M_2' = m_i + 2m_f \tag{4}$$

Compared with Eq. 5 and 6, the required memory turns into a fixed value that reduces the risk of insufficient memory during inference.

### 3.3   Tandem Learning

After channel-wise normalization, the normalized SNN model is fine-tuned with tandem learning, which is shown in Fig. 4. Different from previous tandem learning frameworks [9], the outputs of the coupled ANN layers are replaced with the firing rates of corresponding SNN layers. And the normalized SNN model is directly employed as the pre-trained model in tandem learning.

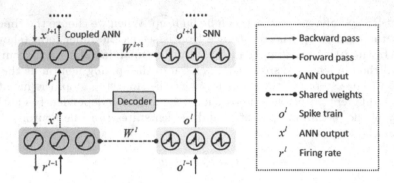

**Fig. 4.** Tandem learning framework with bReLU in coupled ANN

**Tandem Learning for RSNN.** In this study, the tandem learning method is not only applied to the feedforward SNN backbone but also to the RSNN fusion module. The RSNN is coupled with an ordinary RNN with shared parameters. Because of the specific inference process of RSNN with the independent recurrence loop, the firing rates of each layer in the RSNN are allowed to replace the corresponding outputs of RNN at each recurrence step. Then the problem of RSNN learning can be transformed into the training of RNN. The BPTT technique for traditional RNN can be leveraged by expanding alone the recurrence loop instead of the time step loop. In this study, the total number of recurrence steps is far lower than that of time steps, and the problem of gradient explosion during RSNN training can be avoided. On the other hand, different modifications to RNN architectures can also be introduced.

**Bounded ReLU in Tandem Learning.** Previous tandem learning methods employ ordinary ReLU functions in the coupled ANN. However, the difference of upper bound between the ReLU function and IF neuron degrades the accuracy of back-propagated gradients that may lead to failure of training.

Considering the difference between the spiking neuron and the coupled activation function, we introduce a bounded ReLU function (bReLU) in the coupled ANN [11]. The bReLU keeps the maximum output consistent with the maximum firing rates of the IF neuron. Because the maximum firing rate is always 1, the bReLU function is described as:

$$f_{sat}(x) = \begin{cases} 0, x < 0 \\ x, 0 \leq x < 1 \\ 1, x \geq 1 \end{cases} \tag{5}$$

When the firing rate of the IF neuron is saturated, the gradient of corresponding bReLU becomes zero. Then the updates of synaptic weights and input spike rates corresponding to the saturated neuron can be inhibited, which is especially essential for RSNN training. When an IF neuron appears saturated in RSNN, the other neurons are affected through connections of hidden-hidden weights during training. The zero gradient of bReLU clears up the gradient for other neurons that back-propagated from the saturated one. This improves the accuracy of back-propagated gradients and prevents failure of training caused by increasing occurrence of saturated IF neurons.

**Proposed Gradient of the Skipping-Step Decoder.** On the other hand, the skipping-step decoder for hidden states in RSNN is also an obstacle for training. The gradient of the corresponding module in the coupled ANN is related to the input and output firing rate. We can simply define the gradient as 1, which means that the output of the decoder varies synchronously with the input. The gradient with this incremental mode can be described as:

$$\frac{\partial r_{post}}{\partial r_{pre}} \approx 1 \tag{6}$$

where $r_{post}$ are the firing rate of the latter part of the spike train, $r_{pre}$ is the firing rate of the whole spike train. However, the incremental gradient can't reflect the difference between input and output firing rate caused by skipping steps.

Thus, we proposed a clipped proportional gradient for the skipping-step decoder, which is described as:

$$\frac{\partial r_{post}}{\partial r_{pre}} \approx clip\left(\frac{r_{post}}{r_{pre}}, 1 - \varepsilon, 1 + \varepsilon\right) \tag{7}$$

where $\varepsilon$ is a hyper-parameter, which is chosen as 0.5. If the former part of the spike train achieves the same firing rate as the latter part, the proportional gradient equals to 1. If the IF neuron is insufficiently activated during the former part of time steps, the proportional gradient is more than 1. The proportional gradient reflects the transmission efficiency of SNN.

## 4    Experiments

### 4.1    Experimental Setup

Action recognition aims to detect human actions in videos. In experiments, we report the Top-1 accuracy for each video to evaluate the performance of our method. The experiments are based on PyTorch and conducted on NVIDIA Tesla-P40 GPUs. Our action recognition SNN model is evaluated on the popular benchmark dataset UCF-101, which consists of 13,320 videos with 101 action classes.

### 4.2    Implementation Details

Firstly, the pretrained ANN model is converted to the SNN model with the channel- norm method. The SNN time step is set as 800 and the recurrence step is 3. Then, the normalized SNN model is fine-tuned with the tandem learning method. The SNN time step is fixed as 200. During tandem learning, the learning rates of the spatial stream, temporal stream and RNN module are all set to 1e−5, and divided by 10 when the accuracies are saturated. The total number of learning epoch for the coupled ANN is set to 15.

## 4.3   Ablation Studies

We first present ablation studies to explore the effectiveness of our proposed methods. Through rigorous experiments, we conclude that: (i) SNN with the skipping-step decoder performs better than that with the ordinary rate-based decoder in accuracy. (ii) bReLU helps improve the performance of tandem learning for both feedforward and recurrent SNN compared with the ordinary ReLU module. (iii) The gradient of the skipping-step decoder in clipped proportional mode outperforms that in incremental mode.

**Table 1.** Conversion performance with skipping-step rate decoders

| Model | Fusion | Skipping step | Total step | Accuracy |
|-------|--------|---------------|------------|----------|
| ANN | RNN | – | – | 88.42% |
| SNN (channel norm) | RSNN | 0 | 800 | 84.05% |
|  |  | 400 | 800 | 87.90% |
| SNN (channel norm + tandem learning) | RSNN | 0 | 200 | 86.35% |
|  |  | 100 | 200 | 88.46% |

**Table 2.** Training performance of bReLU for feedforward SNN

| Model | Fusion | Activation | Total step | Accuracy |
|-------|--------|-----------|------------|----------|
| SNN (channel norm + tandem learning) | Avg. | ReLU | 200 | 87.32% |
|  |  | bReLU | 200 | 87.63% |
|  |  | ReLU | 50 | 81.61% |
|  |  | bReLU | 50 | 83.67% |

**Table 3.** Training performance of bReLU and the proportional gradient for RSNN (with 200 time steps in total and 100 skipping steps)

| Mmodel | Fusion | Activation | Skipping-step decoder | Accuracy |
|--------|--------|-----------|----------------------|----------|
| SNN (channel norm + tandem learning) | RSNN | ReLU | Incremental gradient | 79.30% |
|  |  | bReLU | Incremental gradient | 88.14% |
|  |  | bReLU | Proportional gradient | 88.46% |

**Impact of the Hybrid Conversion Method.** In Table 1, our original ANN model achieves an accuracy of 88.42%. The converted SNN with channel-wise normalization achieves 4.37% accuracy loss at 800 time steps with an ordinary rate decoder (0 skipping step). And the SNN model with channel-wise normalization and tandem learning achieves 2.07% accuracy loss with only 200 time steps. Thus our hybrid conversion method improves both the accuracy and inference latency compared with the normalization method.

**Impact of Skipping-Steps Rate Decoder.** The skipping-step decoder is employed at the output layers of the SNN backbones and RSNN. We simply set the skipping step as half of the total time step. Performance of the skipping-step decoder is compared with the ordinary rate decoder. In Table 1, compared with the ordinary rate decoder, the skipping-step decoder obtains 3.85% and 2.11% accuracy boost for the SNN with the channel-wise normalization and the hybrid conversion method respectively. It indicates that the spikes at the former part of time steps with insufficient activation degrade the transmission accuracy in deep SNNs. And the skipping-step decoder can avoid the accuracy degradation by skipping the former part of spike trains.

**Impact of bReLU.** We then evaluate the impact of the bReLU module in tandem learning. The SNN model is evaluated with average or RSNN fusion module. When the average fusion module is employed, the SNN becomes a feed-forward network. In Table 2, it can be seen that the bReLU in coupled ANN outperforms the ordinary ReLU for both 200 and 50 time steps. The accuracy improvement is 0.31% when the total time step is 200. When the total time step decreases, the transmission accuracy of rate coding decreases and the gap between SNN and the coupled ANN grows larger. Then the inaccuracy gradients corresponding to the saturated neurons with ordinary ReLU increase the performance degradation. While bReLU provides necessary gradient modification for saturated neurons. Thus the accuracy improvement reaches 2.06% when the total time step reduces to 50. When the RSNN fusion module is employed, it can be seen that the accuracy of the coupled ANN model with ordinary ReLU functions is only 79.30%, as shown in Table 3. It is much lower compared with the employment of average fusion. The training process is affected by the problem of increasing saturated neurons. While the coupled ANN model with bReLU achieves an accuracy improvement of 8.84% compared with the ordinary ReLU. bReLU clears up the gradients back-propagated from the saturated neurons and avoids the problem of increasing saturated neurons. Thus bReLU performs better in RSNN training.

**Impact of the Clipped Proportional Gradient for the Skipping-Step Decoder.** Our proposed clipped proportional gradient for the skipping-step decoder is compared with the incremental gradient. In Table 3, it can be seen that the proportional gradient for skipping-step decoders in tandem learning achieves an accuracy of 88.46%. Though there are only 4 skipping-step decoders employed in the whole SNN model, the clipped proportional gradient still achieves an accuracy improvement of 0.32%.

### 4.4   Comparison with State-of-the-Art Methods

Our proposed model is compared with previous state-of-the-art SNN models for video action recognition, as shown in Table 4. Existing SNN models mainly include STS-ResNet and D/A model [17,18]. Our implementation combines channel-wise normalization and tandem learning. The normalization process enables the construction of relatively complicated and deep SNN architecture, and then the tandem learning process fine-tunes the normalized SNN model. Our SNN model finally reaches an accuracy of 88.46% with only 200 time steps and achieves SOTA performance in SNN. The high-accuracy and energy-efficient action recognition task is implemented.

**Table 4.** Comparison with state-of-the-art methods on UCF-101

| Methods | Type | Accuracy |
|---------|------|----------|
| Our original ANN | ANN | 88.42% |
| STS-ResNet | SNN | 42.10% |
| D/A model | | 81.30% |
| Our proposed SNN | | 88.46% |

**Table 5.** Energy consumption of ANN (GPU) and proposed SNN (neuromorphic chip)

| Model | Device | FLOPs | Time steps | GFLOPS/W | Energy (J) |
|-------|--------|-------|------------|----------|------------|
| ANN | Titan V100 | 2.13E+11 | – | 56 | 3.803 |
| SNN (channel norm) | Tianjic | 1.64E+09 | 800 | 650 | 2.014 |
| SNN (proposed) | | 1.52E+09 | 200 | 650 | 0.468 |

### 4.5   Energy Consumption

According to [19] and [3], the computing performance (GFLOPS/W) of a GPU (Titan V100) and a neuromorphic chip (Tianjic) can be obtained respectively. Based on these measurements, we evaluated the energy consumptions of the ANN and SNN, shown in Table 5. Compared with the ANN model, the total energy consumption of our proposed SNN model is approximately 8 times lower. On the other hand, the tandem learning process reduces the total time step from 800 to 200 compared with the channel-norm SNN, that reduces the energy consumption by 76.7%.

# 5   Conclusion

In this study, an action recognition model composed of a two-stream deep SNN and an RSNN fusion module is constructed with channel-wise normalization and tandem learning successively. A new kind of skipping-step decoder and a tandem learning method with bReLU function are proposed to improve the inference accuracy and latency. Experiments on UCF-101 show that our proposed model obtains an accuracy of 88.46% with only 200 time steps, and achieves state-of-the-art performance in SNN.

# References

1. Maass, W.: Networks of spiking neurons: the third generation of neural network models. Neural Netw. **10**(9), 1659–1671 (1997)
2. Merolla, P.A., Arthur, J.V., Icaza, R.A., et al.: A million spiking-neuron integrated circuit with a scalable communication network and interface. Science **345**(6197), 668–673 (2014)
3. Pei, J., Deng, L., Song, S., et al.: Towards artificial general intelligence with hybrid Tianjic chip architecture. Nature **572**, 106–124 (2019)
4. Rueckauer, B., Lungu, I.A., Hu, Y., Pfeiffer, M., Liu, S.: Conversion of continuous-valued deep networks to efficient event-driven networks for image classification. Front. Neurosci. **11**, 682–693 (2017)
5. Kim, S., Park, S., Na, B., Yoon, S.: Spiking-YOLO: spiking neural network for energy-efficient object detection. In: 2020 AAAI Conference on Artificial Intelligence, New York, pp. 11270–11277, February 2020
6. Zhang, J., Wang, J., Yan, J., Wang, C., Pu, S.: Deep spiking neural network for high-accuracy and energy-efficient face action unit recognition. In: 2021 International Joint Conference on Neural Networks, Virtual Event, pp. 1–7, July 2021
7. George, A.M., Banerjee, D., Dey, S., Mukherjee, A., Balamurali, P.: A reservoir-based convolutional spiking neural network for gesture recognition from DVS input. In: 2020 International Joint Conference on Neural Networks, pp. 1–9, July 2020
8. Priyadarshini, P., Kaushik, R.: Learning to generate sequences with combination of hebbian and non-hebbian plasticity in recurrent spiking neural networks. Front. Neurosci. **11**, 69 (2017)
9. Wu, J., Chua, Y., Zhang, M., et al.: A tandem learning rule for effective training and rapid inference of deep spiking neural networks. IEEE Trans. Neural Netw. Learn. Syst. **34**(1), 446–460 (2023)
10. Wu, J., Xu, C., Zhou, D., et al.: Progressive tandem learning for pattern recognition with deep spiking neural networks. IEEE Trans. Pattern Anal. Mach. Intell. **44**(11), 7824–7840 (2022)
11. Liew, S.S., Khalil-Hani, M., Bakhteri, R.: Bounded activation functions for enhanced training stability of deep neural networks on visual pattern recognition problems. Neurocomputing **216**, 718–734 (2016)
12. Lee, J.H., Delbruck, T., Pfeiffer, M.: Training deep spiking neural networks using backpropagation. Front. Neurosci. **10**, 508 (2016)
13. Zhang, W., Li, P.: Spike-train level backpropagation for training deep recurrent spiking neural networks. In: 2019 Conference on Neural Information Processing Systems, Montreal, Vancouver, pp. 1–12, December 2019

14. Ma, C., Chen, M., Kira, Z., AlRegib, G., et al.: TS-LSTM and temporal-inception: exploiting spatiotemporal dynamics for activity recognition. Sig. Process. Image Commun. **71**, 76–87 (2018)
15. Tran, D., Bourdev, L., Fergus, R., Torresani, L., Paluri, M.: Learning spatiotemporal features with 3D convolutional networks. In: 2015 Proceedings of the IEEE International Conference on Computer Vision, Santiago, pp. 4489–4497, December 2015
16. Simonyan, K., Zisserman, A.: Two-stream convolutional networks for action recognition in videos. In: 2014 International Conference on Neural Information Processing Systems, Montreal, pp. 1–11, December 2014
17. Samadzadeh, A., Far, F., Javadi, A., et al.: Convolutional spiking neural networks for spatio-temporal feature extraction, pp. 1–10, January 2021
18. Priyadarshini, P., Narayan, S.: Learning to recognize actions from limited training examples using a recurrent spiking neural model. Front. Neurosci. **12**, 1–15 (2018)
19. NVIDIA Tesla V100 GPU architecture (2017)

# Sequence Recommendation Based on Interactive Graph Attention Network

Qi Liu(✉) [ID], Jianxia Chen [ID], Shuxi Zhang [ID], Chang Liu [ID], and Xinyun Wu [ID]

School of Computer Science, Hubei University of Technology, Wuhan, China
260129443@qq.com

**Abstract.** Sequence recommendation aims to model the dynamic preferences of users from their historical interactions and accurately predict the next item that the users may be interested. Sequence recommendation models based on graph neural networks (GNNs) have become popular in academic research recently with remarkable results. However, it is difficult for existing GNNs-based models to learn the rapidly changing patterns of the user interests. Therefore, this paper proposes a novel GNNs-based model with a graph attention network (GAT) for the sequence recommendation, named Interactive Graph Attention Network Sequence Recommendation, IGANSR in short. In particular, the proposed IGANSR model constructs the user attributes graph and item attributes graph respectively to acquire the dynamic characteristics of both users and items. In addition, the IGANSR model utilizes a multi-layer graph attention network to dynamically learn the higher-order features and the representations of new nodes. Afterward, the IGANSR model can aggregate various information of each user's neighbors' graph and capture the embedding of similar users. Lastly, the proposed IGANSR model combines the dynamic item representations with the user representations together and projected onto multiple scales for the augmented learning. Experimental results carried out on three public datasets demonstrate that the IGANSR model outperforms other existing recommendation models.

**Keywords:** Recommendation Systems · Graph Neural Networks · Graph Interaction · Information Transfer

## 1  Introduction

Sequence recommendation (SR) aim to model the dynamic preferences of users from their historical interactions and accurately predict the next item that the users may be interested. Compared with the traditional recommender systems, the purpose of SR is to obtain the intention and preference information of different users from different behavior types and hidden behavior objects. Consequently, it is necessary for the sequential recommend to extract as much effective information as possible from the sequence to learn the user's interest in the

This work is supported by National Natural Science Foundation of China (Grant No. 61902116).

sequence, including short-term interest, long-term interest, dynamic interest, etc. In recent years, there are some models that have been utilized for sequence modeling, such as Markov chain-based and factorization-based, or deep neural networks (DNNs). However, it is difficult for these advanced techniques to be well leveraged for short-term, dynamic interest modeling to satisfy the requirements of SR.

Since GNNs [17] are so good at learning graph representations that they have been widely used recently in the area of SR. Li et al. [8] and Su et al. [12] utilize GNNs-based methods to convert sequence structures to graph structures. However, existing GNNs-based models still might not provide a satisfactory joint decision. One reason is that these models make final prediction judgments lacking enough information such as the internal information, cross-interaction information. For example, the graph convolutional network (GCN) [7] considers only the user-item interactions; KGAT [15] cannot distinguish the internal interaction and cross-interaction; A-PGNN [11] cannot capture useful structural information about attribute interactions. Another reason is that these models often rarely consider the feature interaction information. For example, GMCF [13] didn't consider the various effects resulted from different node attributes.

To address the above issues, this paper proposes a novel GNNs-based model with a GAT [14] for the sequence recommendation, named Interactive Graph Attention Network Sequence Recommendation, IGANSR in short. The main contributions of this paper are introduced as follows:

(1) Construct the user and item attributes graph respectively to acquire the dynamic characteristics of both them.
(2) Utilize a multi-layer graph attention network to dynamically learn the higher-order features and the representations of new nodes.
(3) Aggregate various information of each user's neighbors' graph and capture the embedding of similar users.
(4) Combine the dynamic item representations with the user representations together and projected onto multiple scales for the augmented learning.

## 2 Methods

As shown in Fig. 1, the model architecture includes three important modules, namely: the graph constructing module of users and items; an interaction module; and an aggregation and matching module. We will describe the main idea of each module in detail as follows.

### 2.1 Problem Definition

For each input data sample $X_n$, node $i$ of the user or item attribute graph represents the attribute features of that user or item. Thus, the model can capture the potential associations between users-users, items-items and users-items, as well as complex interactions between different feature pairs. The most favorable

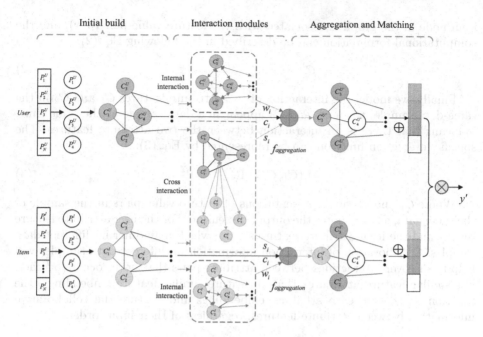

**Fig. 1.** IGANSR framework

pairwise feature interactions are detected by the prediction model $F(x_n)$, and the final predicted value $y_n$ is output by graph matching.

For each training data $(X_n, y_n)$, it is contained in a dataset D with input-output pairs: $D = \{(X_n, y_n)\}_{1 \leq n \leq N}$, $y_n$ is the predicted value calculated by the model and represents the probability of a user clicking, the data sample $X_n$ is denoted in the following Eq. (1):

$$X_n = \left\{ P_n^U = (p, x) \right\}_{p \in J_n^U}, \left\{ P_n^I = (p, x) \right\}_{p \in J_n^I} \tag{1}$$

The value of $X_n$ is 1 when the user has an interaction record with the item i.e., it is recorded in this data sample, otherwise it is 0. Where $p$ denotes the attribute feature and $x$ is its corresponding value, $P_n^U, P_n^I$ represent the user and item attribute feature value pairs respectively, $J_n^U \subseteq J^U$, $J_n^I \subseteq J^I$ represent the set of user attributes and the set of item attributes, respectively, and J is the index set of all features in datasets $D$.

For each attribute feature $p \in J^U \cup J^I$, it is firstly represented as an initial embedding vector $v_p^e$ in the d-dimensional space $R^d$ for interaction modelling. A parameter matrix is constructed as an embedding query table and the embedding vector $v_p^e$ for that attribute is shared for data samples with the same attribute $p$; Secondly, for each feature value pair $(p, x)$, the user, item attribute graphs are constructed separately. The feature vector corresponding to each user or item node in the graph is associated with the feature value pair $(p, x)$, and the computational formulation can be described in the following Eq. (2):

$$C^U = x \cdot v_p^e, C^I = x \cdot v_p^e \tag{2}$$

Finally, we model the interaction of two attribute features $C_1$ and $C_2$ in the same data sample, when $C_1$ and $C_2$ appear in a data sample at the same time, it means that there is an interaction between the two attribute features. The specific interaction function can be represented by Eq. (3):

$$f(C_1, C_2) : R^{2 \times d} \to R^\ell \tag{3}$$

Where $C_1$ and $C_2$ are representations of feature-value pairs in the sample of the same data, and $l$ denotes the output dimension. For the user attribute feature set $J_n^U$ and the item attribute feature set $J_n^I$, which may differ in different data samples. Collaboration of interaction information in different data samples will help to discover interactions between attribute pairs that never occur together. For similar feature attributes that constitute similar feature embeddings, the function $f(\cdot, \cdot)$ can learn attribute embeddings and capture the collaborative information between attribute features, regardless of their input order.

## 2.2    User and Item Diagram Constructing Blocks

This paper constructs the graph structure on the sequence of inputs, representing each user and each item as an attribute graph, and explicitly incorporating the user and item information into the GAT. For a data sample N and a graph $G_N(V_N, E_N)$ can be constructed:

Vertices: $V_N = \{v_i\}_{1 \le i \le N}$ is a set of N nodes, each node is initialized to the corresponding sequence-encoded feature vector $v_i$. The first-level state vector of nodes can be represented as $V = (v_1, \ldots, v_N)$, each node $i$ representing an attribute in the user or item sequence is expressed as a node feature representation $C_i^U$ or $C_i^I$ in the user or item attribute graph. Thus, the set of nodes of the user, item attribute graph can be denoted in the following Eq. (4):

$$V^U = \left\{C_i^U\right\}_{i \in J^U}, V^I = \left\{C_i^I\right\}_{i \in J^I} \tag{4}$$

Edges: The weight of each edge reflects the importance of different feature interactions. $E = (e_1, \ldots, e_N)$, $E \in R^{N \times N}$ denotes the set of edges containing adjacency information between nodes. $E_{ij} = 1$ when an edge connects nodes $i$ and $j$ (there are beneficial feature interaction), otherwise $E_{ij} = 0$. $E_{ij}$ denotes a set of edge interaction values, which is the set of edges with neighborhood information between nodes. Nodes $i$, $j$ are also called neighborhood nodes.

For the user attribute graph, which is denoted as $G^U = \langle V^U, E^U \rangle$, where $E^U$ contains the set of edges of all edges in the user attributes graph. By performing the same transformation, the item attribute graph can be constructed as $G^I = \langle V^I, E^I \rangle$, where $E^I$ contains the set of edges of all edges in the user attribute graph.

## 2.3   Information Interaction Module

**Internal Information Transfer Between Nodes.** Inspired by papers [1,12], the Multilayer Perceptron (MLP) are utilized to model each internal interaction and aggregates the interaction modeling results into the internal messaging. This paper uses an MLP function $f(\cdot,\cdot) : R^{2\times d} \to R^{\ell}$, which takes as input the dimension d of a pair of node embeddings utilized for edge prediction and outputs a binary value indicating whether the two nodes are connected by an edge. The formulation can be described in the following Eq. (5):

$$w_{ij} = f_{\mathrm{M}}(C_i, C_j) \tag{5}$$

Where $C_i = x_i v_i$, $w_{ij}$ is the result of modeling the interaction of node pair $(i,j)$, and $W_i$ is the internal transfer information. The formulation can be expressed in the following Eq. (6):

$$W_i = \sum_{j \in N_i} w_{ij} \tag{6}$$

Where $W_i \in R^d$, $N_i$ denotes the set of neighboring nodes within the graph including node $i$. The method makes use of the characteristics of graph structure to enable the iterative transfer, updating and enrichment of information between nodes.

**Interaction Graph Attention Network.** The core idea of the IGANSR model is to use GAT to integrate nodes, edges, and their interaction information in a unified graph interaction architecture. For a graph with N nodes, the initial node characteristics of a single-layer GAT can be described as $H = [h_1, h_2, \ldots, h_N]$, with $h_i \in R^d$ as input, to obtain a more abstract feature representation $H^t = [h_1^t, h_2^t, \ldots, h_N^t]$, with $h_i^t \in R^{d'}$ as the model output. Where t denotes the number of messages passed between nodes. Since the interactions on each edge has different effect, we aim to achieve interactions along the edges, which requires each edge to have a unique weight and transformation function. The graph attention operation implemented on the node representations can be written in the following Eq. (7):

$$h_i^t = \sigma \left( \sum_{j \in N_i} \alpha_{ij} W_h h_j \right) \tag{7}$$

Where $N_i$ is a set containing node $i$ and its neighborhood nodes in the graph, d and $d'$ represent the input and output dimensions, respectively, $W_h \in R^{d' \times d}$ represents the trainable weight matrix and $\sigma$ is the nonlinear activation function utilized to make the weights easily comparable between different nodes. Since we need to infer the importance of interactions between different nodes, the weights $\alpha_{ij}$ in the Eq. (7) are used to learn the weights of the edges via an attention mechanism, which calculates the weight value of each neighboring node $h_j$ with respect to node $h_i$, as expressed in the following Eq. (8):

$$\alpha_{ij} = \frac{\exp\left(\mathcal{F}\left(h_i, h_j\right)\right)}{\sum\limits_{j' \in N_i} \exp\left(\mathcal{F}\left(h_i, h_{j'}\right)\right)} \tag{8}$$

Where $\mathcal{F}$ is an attention function, which can be expressed as: $\mathcal{F}\left(h_i, h_j\right) = LeakyReLU\left(a^\top\left[W_h h_i \parallel W_h h_j\right]\right)$, $a \in R^{2d'}$ is a trainable weight matrix. Furthermore, to stabilize the process of self-attentive learning, GAT extends the above mechanism to employ a multi-headed attentional implementation, denoted in the following Eq. (9):

$$h_i^t = \Vert_{k=1}^K \sigma\left(\sum_{j \in N_i} \alpha_{ij}^k W_h^k h_j\right) \tag{9}$$

Where K is the number of multi-headed attention heads, $\alpha_{ij}^k$ is the k-headed normalized attention weight, and $\parallel$ is the serial operation. Finally, the state vector of the node is updated by the initial node feature $h_i$ and its hidden state $h_i^{t-1}$ at the last step to implement the gated recurrent neural network GRU operation. This is expressed as $h_i^{t'} = GRU\left(h_i^{t-1}, h_i\right)$. The detailed formulation is as follows (10):

$$\begin{aligned} z_i^t &= \sigma\left(W_z h_i + U_z h_i^{t-1} + b_z\right) \\ r_i^t &= \sigma\left(W_r h_i + U_r h_i^{t-1} + b_r\right) \\ \widetilde{h}_i^t &= \tanh\left(W_h h_i + U_h\left(r_i^t \odot h_i^{t-1}\right) + b_h\right) \\ h_i^{t'} &= \widetilde{h}_i^t \odot z_i^t + h_i^{t-1} \odot\left(1 - z_i^t\right) \end{aligned} \tag{10}$$

Where $W_z, W_r, W_h, b_z, b_r, b_h$ are the weights and deviations of the update function gating recursive unit GRU [2], $z_i^t$ and $r_i^t$ are the update and the reset gate vector respectively. Based on this, node matching between two graphs can be expressed in the following Eq. (11):

$$s_{ij} = h_i^{t'} \odot h_j^{t'} \tag{11}$$

Where $h_i^{t'}$ represents the embedding feature of node $i$ in the user graph and $h_i^{t'} \in J_n^U$, $\widetilde{h}_j^{t'}$ represents the embedding feature of node $j$ in the item graph and $\widetilde{h}_j^{t'} \in J_n^V$, $s_{ij}$ represents the node matching result of two nodes from different graphs. Indicates the corresponding multiplication operation of each element. Similar to intra-node information transfer, the node matching results from different graphs are summed to obtain the final information transfer result $S_i$ for the interaction of the two graphs: $S_i = \sum_{j \in J_n^V} s_{ij}$.

**Information Aggregation and Graph Matching.** To fuse the own information $C_i$ of each node in the graph, the internal association information $W_i$, and the node interaction information $S_i$ between the two graphs, this paper uses the fusion function $f_{\text{aggregation}} \in R^{3 \times d} \rightarrow R^d$ to integrate the parts of the information. The input information is passed through GRU to update its node state based on the aggregated information and historical information to obtain $c'_i = f_{\text{aggregation}} (C_i, W_i, S_i)$ and use the element summation method to aggregate the nodes represented as $f_G$. Aggregating nodes via element-wise addition can be described in the following Eq. (12):

$$f_G\left(G, J_n^V\right) = \sum_{i \in I_n^U} c'_i \tag{12}$$

For graph matching, this paper obtains vector representations of both user and item attribute graphs by using the $f_G(\cdot, \cdot)$ function: $v_G^U = f_G\left(G^U, V^I\right), v_G^I = f_G\left(G^I, V^U\right)$. Due to the graph matching with the dot product function $f(\cdot, \cdot)$ on the two graphs, we can get the predicted output $y' = v_G^{U^\top} v_G^I$.

**Training Process.** We normalize all parameters of the model using $L_2$ regularization in the training process. Thus, the loss function is expressed in the following Eq. (13):

$$\mathcal{R}(\theta) = \tfrac{1}{N} \sum_{n=1}^{N} \mathcal{L}\left(F_{DIGAT}\left(X_n; \theta\right), y_n\right) + \lambda\left(\|\theta\|_2\right)$$
$$\theta^* = \operatorname{argmin}_\theta \mathcal{R}(\theta) \tag{13}$$

Where $F_{DIGAT}$ is the prediction function of this model, $y'$ is the output, $\lambda$ is the regularized weight coefficient, $L(\cdot)$ is the binary cross-entropy loss function, $\theta$ contains all the parameters of the model, and $\theta^*$ is the final parameter representative.

## 3   Experiments

### 3.1   The Dataset and Its Sources

The proposed IGANSR model is evaluated on the following three benchmark datasets, the statistical details of which are summarized in Table 1.

**MovieLens 1M [4]:** is a set of movies rating data, containing movie data, user data, etc. We convert the display feedback from this dataset into implicit data by marking the user's rating of the item as 0 or 1. Each data sample is a graph containing a user and a movie with the corresponding attributes.

**Bookcrossing [19]:** contains ratings of books by users implicitly and explicitly. Each data sample includes the user, the book and its attributes, such as title, book name, author, etc.

**Taobao [18]:** contains data about clicks on advertisements displayed on the page by Taobao. Each data sample contains the information of a specific user and their corresponding attributes.

**Table 1.** Dataset statistics.

| Datasets | Data | User | Item | User features | Item features |
|---|---|---|---|---|---|
| MovieLens 1M | 1 144 739 | 6060 | 3952 | 30 | 6049 |
| Bookcrossing | 1 050 834 | 4873 | 53 168 | 87 | 43 157 |
| Taobao | 2 599 463 | 4532 | 371 760 | 36 | 4 344 254 |

## 3.2  Parameter Setting

In this paper, binary cross-entropy is used as the loss function, and each dataset is randomly divided into a training set, a validation set and a test set in a ratio of 6:2:2. Adam [6] is utilized to combine the first-order momentum and the second-order momentum to automatically adjust the learning rate of the parameters. AUC and NDCG@k are the evaluation metrics. In detail, the specific hyper-parameter settings are shown in Table 2.

**Table 2.** Hyperparameters of model.

| Hyperparameters | Description | Value |
|---|---|---|
| $d$ | Nodal dimension | 128 |
| $\lambda$ | Regularization factor | $1 \times 10^{-5}$ |
| $lr$ | Learning rate | $1 \times 10^{-3}$ |
| $n\_epoch$ | Number of training iterations | 50 |
| $Batch\ size$ | Number of batches | 512 |
| $n\_hidden\_layer$ | Number of hidden layers | 1 |
| $hidden\_layer$ | Hidden layer dimension | 256 |

## 3.3  Correlation Comparison Model

In this paper, the following benchmark model is compared with the IGANSR model proposed in this paper.

- FM [9]: uses point multiplication to calculate and model the interaction of each feature and has a good learning ability for sparse data.
- NFM [5]: combines FM with neural networks to improve FM's ability to capture multi-order interaction information between features.
- AutoInt [10]: map raw sparse high-dimensional feature vectors to a low-dimensional space for explicit modeling of feature interactions.
- Fi-GNN [8]: each feature map consists of data samples, and each node in the graph represents a feature field, which is modeled using a multi-head self-attention method.
- NGCF [16]: is a graph-based CF approach that broadly follows the standard GCN [3], aiming to model neighborhood information by up to three orders.

- GMCF [13]: is a graph-based CF approach that uses the MLP structure of interaction modeling for model internal interaction modeling.
- KGAT [15]: modeling based on collaborative knowledge graphs with a form of graph convolution.

## 3.4  Experimental Results and Analysis

From the Table 3, we find that IGANSR improves the performance of books and Taobao better than the Movie dataset, suggesting that IGANSR can handle sparse data well.

The FM model has the worst effect because only the use of dot product calculation is not enough to effectively capture the information features; NFM is inspired by FM and uses MLP instead of point multiplication, and the experimental effect is significantly improved compared with the FM results, which demonstrate the effectiveness of the MLP calculation method; AutoInt and Fi-GNN models combine multi-head self-attention neural network to model attribute features, which effectively improves the ability of the model to capture multi-order interactive information between features; Both NGCF and GMCF implement the graph modeling process.

For the above models, although high-level feature extraction is achieved to a certain extent, the weights are not calculated for the relevant influencing factors of information between nodes and their adjacent nodes. Therefore, the IGANSR model utilizes the GAT to analyze the relationship between each node and its neighbors in the graph. Different weights are allocated between domain nodes, and the weights of different neighbors are adaptively allocated based on extracting high-order features of information. The experimental results show that the proposed model can effectively improve the accuracy and expressive ability of recommend results.

**Table 3.** Comparison of Models.

|        | MovieLens 1M | | | Book-Crossing | | | Taobao | | |
|--------|--------|---------|----------|--------|---------|----------|--------|---------|----------|
|        | AUC    | NDCG @5 | NDCG @10 | AUC    | NDCG @5 | NDCG @10 | AUC    | NDCG @5 | NDCG @10 |
| FM     | 0.8758 | 0.8278  | 0.8457   | 0.7486 | 0.7694  | 0.8059   | 0.6115 | 0.8824  | 0.1119   |
| NFM    | 0.8893 | 0.8537  | 0.8843   | 0.7934 | 0.7733  | 0.8315   | 0.6514 | 0.0922  | 0.1249   |
| AutoInt| 0.8922 | 0.8721  | 0.8947   | 0.8125 | 0.8079  | 0.8455   | 0.6376 | 0.0898  | 0.1207   |
| Fi-GNN | 0.8946 | 0.8827  | 0.9077   | 0.8131 | 0.8124  | 0.8593   | 0.6425 | 0.0942  | 0.1254   |
| NGCF   | 0.8958 | 0.9145  | 0.9258   | 0.8197 | 0.8319  | 0.8612   | 0.6457 | 0.0958  | 0.1286   |
| GMCF   | 0.8991 | 0.9373  | 0.9416   | 0.8231 | 0.8689  | 0.8957   | 0.6567 | 0.1018  | 0.1371   |
| KGAT   | 0.9008 | 0.9364  | 0.9429   | 0.8292 | 0.8725  | 0.8981   | 0.6544 | 0.1027  | 0.1393   |
| IGANSR | 0.9035 | 0.9437  | 0.9438   | 0.8443 | 0.9030  | 0.9149   | 0.6604 | 0.1121  | 0.1485   |

## 3.5  Research on Node Dimension and Network Layers

In this section, we attempt to analyze the performance of the proposed IGANSR with different network layers and different dimensions node representations.

As shown in the Fig. 2, a line graph is utilized to feed back the recommendation accuracy of the IGANSR model on the node representation dimensions of 64, 128, and 256. Each subgraph is distinguished by three colors: red, green, and blue. Different numbers of network layers (GATlayer_1, GATlayer_2, GATlayer_3) recommended results.

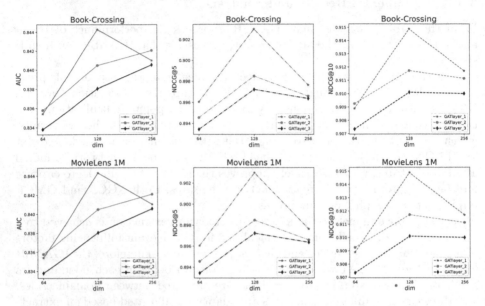

**Fig. 2.** Comparison of Recommendation Accuracy of Different Datasets in Different Number of Network Layers and Different Node Attribute Embedding Dimensions. (Color figure online)

In the Fig. 2, that when node dimension is set to 128, the model achieves the best performance. Low-dimensional may have problems such as insufficient parameters, while high-dimensional needs to fit more parameters that may lead to other problems such as overfitting. This indicates that the model needs to match the appropriate node dimensions for optimal performance. In addition, the model does not achieve better performance with the deepening of the GAT network layers, and there may also be an overfitting problem. It also shows that a single-layer GAT is sufficient to meet the information transfer requirements of cross-interaction between nodes.

## 4    Conclusion

In this work, we propose a novel IGANSR model incorporating GAT that facilitates model generalization and interpretability, exploring higher-order connectivity and internal association of two-attribute graphs in a graph neural network structure. In particular, the IGANSR models dual-attribute interactions between the users and items in an end-to-end manner according to different ways of computing interactions for different interaction patterns. In addition, the IGANSR model uses GAT to per-form preference matching in the cross-interactions between dual-attribute graphs with adaptively propagating embedded representations. The experimental results demon-strate the robustness and effectiveness of the IGANSR model.

In the future, we will explore the higher-order interactions in the model architecture and reveal the user's decision-making process of selecting items. In addition, we will study the prevalence of sparsity problems in recommender systems through the self-supervised learning methods.

# References

1. Battaglia, P., Pascanu, R., Lai, M., Jimenez Rezende, D., et al.: Interaction networks for learning about objects, relations and physics. In: Advances in Neural Information Processing Systems, vol. 29 (2016)
2. Cho, K., et al.: Learning phrase representations using RNN encoder-decoder for statistical machine translation. arXiv preprint arXiv:1406.1078 (2014)
3. Defferrard, M., Bresson, X., Vandergheynst, P.: Convolutional neural networks on graphs with fast localized spectral filtering. In: Advances in Neural Information Processing Systems, vol. 29 (2016)
4. Harper, F.M., Konstan, J.A.: The movielens datasets: history and context. ACM Trans. Interact. Intell. Syst. (TIIS) 5(4), 1–19 (2015)
5. He, X., Chua, T.S.: Neural factorization machines for sparse predictive analytics. In: Proceedings of the 40th International ACM SIGIR Conference on Research and Development in Information Retrieval, pp. 355–364 (2017)
6. Kingma, D.P., Ba, J.: Adam: a method for stochastic optimization. arXiv preprint arXiv:1412.6980 (2014)
7. Kipf, T.N., Welling, M.: Semi-supervised classification with graph convolutional networks. arXiv preprint arXiv:1609.02907 (2016)
8. Li, Z., Cui, Z., Wu, S., Zhang, X., Wang, L.: Fi-GNN: modeling feature interactions via graph neural networks for CTR prediction. In: Proceedings of the 28th ACM International Conference on Information and Knowledge Management, pp. 539–548 (2019)
9. Rendle, S.: Factorization machines. In: 2010 IEEE International Conference on Data Mining, pp. 995–1000. IEEE (2010)
10. Song, W., et al.: AutoInt: automatic feature interaction learning via self-attentive neural networks. In: Proceedings of the 28th ACM International Conference on Information and Knowledge Management, pp. 1161–1170 (2019)
11. Spinelli, I., Scardapane, S., Uncini, A.: Adaptive propagation graph convolutional network. IEEE Trans. Neural Netw. Learn. Syst. 32(10), 4755–4760 (2020)
12. Su, Y., Zhang, R., Erfani, S., Xu, Z.: Detecting beneficial feature interactions for recommender systems. In: Proceedings of the AAAI Conference on Artificial Intelligence, vol. 35, pp. 4357–4365 (2021)
13. Su, Y., Zhang, R., Erfani, S.M., Gan, J.: Neural graph matching based collaborative filtering. In: Proceedings of the 44th International ACM SIGIR Conference on Research and Development in Information Retrieval, pp. 849–858 (2021)
14. Veličković, P., Cucurull, G., Casanova, A., Romero, A., Lio, P., Bengio, Y.: Graph attention networks. arXiv preprint arXiv:1710.10903 (2017)

15. Wang, X., He, X., Cao, Y., Liu, M., Chua, T.S.: KGAT: knowledge graph attention network for recommendation. In: Proceedings of the 25th ACM SIGKDD International Conference on Knowledge Discovery & Data Mining, pp. 950–958 (2019)
16. Wang, X., He, X., Wang, M., Feng, F., Chua, T.S.: Neural graph collaborative filtering. In: Proceedings of the 42nd International ACM SIGIR Conference on Research and Development in Information Retrieval, pp. 165–174 (2019)
17. Wu, S., Sun, F., Zhang, W., Xie, X., Cui, B.: Graph neural networks in recommender systems: a survey. ACM Comput. Surv. (CSUR) **55**(5), 1–37 (2022)
18. Zhou, G., et al.: Deep interest network for click-through rate prediction. In: Proceedings of the 24th ACM SIGKDD International Conference on Knowledge Discovery & Data Mining, pp. 1059–1068 (2018)
19. Ziegler, C.N., McNee, S.M., Konstan, J.A., Lausen, G.: Improving recommendation lists through topic diversification. In: Proceedings of the 14th International Conference on World Wide Web, pp. 22–32 (2005)

# Imbalanced Equilibrium: Emergence of Social Asymmetric Coordinated Behavior in Multi-agent Games

Yidong Bai[(✉)] and Toshiharu Sugawara

Computer Science and Engineering, Waseda University, Tokyo 169-8555, Japan
ydbai@moegi.waseda.jp, sugawara@waseda.jp

**Abstract.** Multi-agent deep reinforcement learning (MADRL) has made remarkable progress but usually requires delicate and fragile reward engineering. Modeling other agents (MOA) is an effective method for compensating for the absence of efficient reward signals. However, existing MOA methods often assume that only one agent can model other non-learning agents. In this study, we propose *continuous mutual modeling* (CMM), which constantly models other agents that also learn appropriate behaviors from their viewpoints to facilitate the coordination among agents in complex MADRL environments. We then propose a CMM framework referred to as *predictor-actor-critic* (PAC) in which every agent determines its actions by estimating those of other agents through mutual modeling. We experimentally show that the proposed method enables agents to realize other agents' activities and promotes the emergence of better-coordinated behaviors in agent society.

**Keywords:** Control and decision theory · Modeling other agents · Multi-agent deep reinforcement learning · Coordination

## 1 Introduction

Multi-agent deep reinforcement learning (MADRL) has gained much attention recently and has made remarkable progress in many challenging areas, such as video games [13] and robot control [2]. To effectively use multiple intelligent agents, they must learn sophisticated social behaviors through cooperation, coordination, or competition. A large part of the successful learning of social behaviors has been accomplished by reward engineering. Carefully hand-designed rewards are used to encourage or punish specific behaviors of agents so that they can establish a desired social interaction pattern.

Unfortunately, these achievements are often delicate and fragile for several reasons. First, the rewards are usually parameter-sensitive and may lead to unexpected results, even minor modifications. Second, appropriate rewards are scenario-specific; thus, when the scenario changes, rewards should be re-tuned or they do not work. Finally, such a reward scheme is often integrated with

M. Tanveer et al. (Eds.): ICONIP 2022, LNCS 13624, pp. 305–316, 2023.
https://doi.org/10.1007/978-3-031-30108-7_26

the intrinsic worth of individuals and non-local information, such as allowing agents to see the rewards obtained by other agents [9]. However, these intertwined rewards make it difficult to achieve decentralized training of MARL and partially push aside the primary challenge of MARL, i.e., how agents can learn to coordinate using limited observations with only available reward signals [6].

A unique and effective method for compensating for the absence of comprehensive observations and efficient reward signals is *modeling other agents* (MOA) [1]. An agent is trained to interact with other agents effectively by constructing their models to predict their actions and the reason behind their intentions. MOA is a well-known and remarkable study in adversarial games like Chess and Go [10,18], but there are also prior studies on the issue committed to promoting coordination [3,15,19]. For example, Barrett and Stone [3] proposed *planning and learning to adapt swiftly to teammates to improve cooperation – policy* (PLASTIC–Policy) to reuse policies that cooperate with past teammates to adapt to new teammates quickly. Bowling and McCracken [4] proposed two techniques for coordination between agents that join an impromptu soccer team with other unfamiliar agents. However, they rely on the assumption that only *one* agent can model other *non-learning* agents [1,3–5,8], but this is impractical in actual situations. We believe that agents must mutually model other agents continuously only from locally available information and rewards to adapt their policy to others' behaviors. However, it remains largely unexplored whether and how such *continuous mutual modeling* (CMM) facilitates social behaviors in complex MARL environments.

Therefore, we propose a CMM framework, referred to as *predictor-actor-critic* (PAC), in which each agent determines its actions by estimating those of other agents through mutual modeling. To evaluate the effectiveness and performance of PAC, we use the *cleanup game* [9] generating a situation of *public goods dilemma* [12] in which an agent must pay a personal cost to provide resources that will be shared with all agents. It is usually difficult for agents to learn cooperative policies independently. They are likely to be trapped at a socially deficient equilibrium where no agents are willing to pay personal costs based on individual rationality, converging in situations where no one can obtain reasonable rewards.

Unlike the naive methods, we experimentally found that our approach avoided such a socially deficient equilibrium, and established another type of equilibrium, we named it *imbalanced equilibrium* due to its imbalanced reward distribution among agents. We further analyzed the behaviors of individual agents after learning and found that (1) one agent played as a *farmer*, which continuously paid the personal cost to create a large number of public goods, and (2) other agents played as *robbers*, which paid a few personal costs but consumed most of the public goods, and (3) each *robber* acted with its range of activities to reduce unnecessary competition among robbers. Although such a coordination structure seems imbalanced, agents with PAC successfully obtained higher rewards from a social viewpoint than those obtained by balanced and fair behaviors. Notably, the emergence of social structure is not caused by a reward engineering such as external rewards, nor did by manually dividing agents into groups. Experimental results indicate that the proposed framework enabled agents to infer other agents' activities and promoted the emergence of better-coordinated behaviors in the agent society.

## 2    Background

### 2.1    Multi-agent Partially Observable Markov Games

Let $A = \{1, \ldots, N\}$ be the set of $N$ agents. Our framework for MADRL is based on a *decentralized partially observable Markov decision process* (Dec-POMDP), $\langle A, S, O, A, P, R \rangle$, tuple of state space $S$, the set of observations $O = O_1 \times \cdots \times O_N$, the joint action space $A = A_1 \times \cdots \times A_N$, the transition probability $P$ and the joint reward $R = R_1 \times \cdots \times R_N$. At time step $t$, agent $i \in A$ observes the state and get the local observation $o_t^i = o_t^i(s_t) \in O_i$. It then chooses an action $a_t^i \in A_i$ according to its local stochastic policy $\pi^i(a_t^i | o_t^i)$. As a result of the joint action $a_t = (a_t^1, \ldots, a_t^N)$, the state $s_t$ transits to another state $s_{t+1}$ according to the state transition probability $P(s_{t+1} | s_t, a_t)$, while $i$ may receive its individual reward $r_t^i$. At the end of episodes, $i$ updates its behavior policy $\pi^i(a_t^i | o_t^i)$ though its own experience to attempt to maximize its total expected discounted future reward $R_i = \sum_{t=0}^{T} \gamma^t r_t^i$, where $\gamma$ is the discount factor, and $T > 0$ is the horizon parameter.

### 2.2    Policy Gradient, Actor-Critic and Proximal Policy Optimization

*Policy gradient* [14] aims to maximize agent $i$'s expected rewards $J(\theta) = \mathbb{E}(R_i)$ by directly updating its policy with respect to parameters $\theta$, using the policy gradient:

$$\nabla_\theta J(\theta) = \nabla_\theta \log(\pi_\theta^i(a_t^i | o_t^i)) \sum_{t'=t}^{T} \gamma^{t'-t} r_{t'}^i. \tag{1}$$

As rewards may vary dramatically over episodes, the term $\sum_{t'=t}^{T} \gamma^{t'-t} r_{t'}^i$ leads to high variance, so actor-critic algorithms [16,17] replace it with a function approximation and rewrite the objective as

$$J(\theta) = \mathbb{E}_t[\log \pi_\theta^i(a_t^i | o_t^i) A_t], \tag{2}$$

where $A_t$ is the advantage function [16,17].

*Proximal policy optimization* (PPO) [17] uses constraints to guarantee monotonic improvement in a simple form so that the actor network is trained to maximize:

$$J(\theta) = \mathbb{E}_t[\min(r_t(\theta)A_t, clip(r_t(\theta), 1 - \epsilon, 1 + \epsilon)A_t) + \sigma S[\pi_\theta^i(o_t^i)]], \tag{3}$$

where $r_t(\theta)$ is probability ratio $\frac{\pi_\theta^i(a_t^i | o_t^i)}{\pi_{\theta_{old}}^i(a_t^i | o_t^i)}$, $\theta_{old}$ is the collection of policy parameters before the update, $S$ is the policy entropy, $\sigma$ is the entropy coefficient parameter, and $A_t$ is computed via *Generalized Advantage Estimation* (GAE) [16].

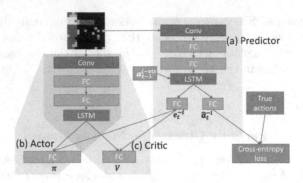

**Fig. 1.** Architecture of predictor-actor-critic.

# 3  Proposed Method

## 3.1  Policy with Actions of Other Agents

A general form of Dec-POMDP described in the previous section has uncertainty owing to the partial observability, i.e., agent $i$ updates its policy $\pi^i(a_t^i|o_t^i)$ to maximize the cumulative reward $R_i$, but reward is a function of state and joint action $r_t^i(s_t, \boldsymbol{a}_t, i)$. Although many existing studies showed that their algorithms in this setting could converge and achieve somewhat acceptable performance, we believe that if agents can predict the actions of other agents, they might be able to derive better results. Thus, ideally, we would like to redefine the policy as

$$a_t^i \sim \pi_\theta(a_t^i|o_t^i, \boldsymbol{a}_t^{-i}), \tag{4}$$

where $\boldsymbol{a}_t^{-i}$ is the joint action of all agents except $i$, and $\sim$ means that $i$'s action $a_t^i$ is sampled according to the conditional probability distribution $\pi_\theta(a_t^i|o_t^i, \boldsymbol{a}_t^{-i})$. This is an anti-causal definition, because agent $i$ needs the actions of other agents as a precondition, but currently, other agents are also waiting for $i$'s action. In centralized algorithms this problem can be solved by parallel processing, while in our fully decentralized environment, a straightforward solution is to rollout $\boldsymbol{a}_t^{-i}$ using a neural network and rewrite Eq. 4 as

$$a_t^i \sim \pi_\theta(a_t^i|o_t^i, P_\omega(o_t^i)), \tag{5}$$

where $P_\omega$ is the *predictor network*, which is fed the local observation $o_t^i$ as input and output prediction $\hat{\boldsymbol{a}}_t^{-i}$ of joint action except $i$.

## 3.2  Predictor Network

Our network structure with the predictor network is shown in Fig. 1. To reduce the noise caused by partial observability [8], we use an LSTM [7] for the predictor network (Fig. 1(a)). As $i$ can observe other agents only in the local observable area, all visible agents' joint action (include the action of itself) at the last time step, $a_{t-1}^{vis-i}$, is also input to the LSTM to predict other agents' action $\hat{a}_t^{-i}$,

$$(e_t^{-i}, \hat{a}_t^{-i}) = P_\omega(o_t^i, a_{t-1}^{vis-i}), \tag{6}$$

where $e_t^{-i}$ is an action feature representation adjusted by a fully-connected (FC) layer for actor and critic. Notably, each visible agent's action in $a_{t-1}^{vis-i}$ is represented in a one-hot encoding, while the actions of agents out of $i$'s field-of-view are masked as zeros.

### 3.3  Predictor-Actor-Critic (PAC)

At each time step, agent $i$'s predictor (Fig. 1(a)) outputs one one-step simulated joint action, $\hat{a}_t^{-i}$, and its representation $e_t^{-i}$ according to Eq. 6. Then, $e_t^{-i}$ is connected with the features in the actor network (Fig. 1(b)) and critic network (Fig. 1(c)). Agent $i$'s action is finally sampled according to

$$a_t^i \sim \pi_\theta(a_t^i | o_t^i, e_t^{-i}). \tag{7}$$

The pseudocode of the training procedure is provided in Algorithm 1. The predictor is trained together with the actor network using a weighted sum loss function:

$$L^{PAC} = L^{PPO} + \lambda L^{pred}, \tag{8}$$

where $L^{PPO}$ is the loss function of PPO [17], $\lambda$ is a weight parameter and $L^{pred}$ is the cross-entropy loss for agent $i$'s predictor network,

$$L^{pred} = \frac{1}{M} \sum_{j=1}^{M} H(a_t^j) + D_{KL}(a_t^j || \hat{a}_t^j), \tag{9}$$

where $M = N - 1$, is the number of all agents except $i$, $H(a_t^j)$ is entropy of $a_t^j$, and $D_{KL}(a_t^j || \hat{a}_t^j)$ is the Kullback-Leibler divergence of $a_t^j$ from $\hat{a}_t^j$.

## 4    Experimental Evaluation

### 4.1  Experimental Setting

To evaluate our method, we used a cleanup game [9] generating a situation of *public goods dilemma* [12], wherein an agent must pay a personal cost to provide resources that are shared with all agents. The cleanup game is performed in a 2D partially observable grid environment. An example snapshot is shown in Fig. 2, where five agents, expressed by red, brown, red-violet, dark-blue, and sky-blue nodes, move around the environment to collect apples (green nodes) to earn rewards; one reward for one apple. The left grayish-blue area is the river; the right black area is the land. At each time step, a piece of waste (khaki nodes) may spawn randomly in the river area, and apples may spawn randomly on the land. The spawn probability of a piece of waste, $P_{wst}$, is constant $C_{wst}$ if the

---

**Algorithm 1.** Training Procedure of PAC

---

1: Initialize $\theta$, parameters for actor $\pi$ for each agent,
2: Initialize $\phi$, parameters for critic $V$ for each agent,
3: Initialize $\omega$, parameters for predictor $P$ for each agent.
4: **while** $step \leq step_{max}$ **do**
5:　　reset data buffer for each agent $D^i = \{\}$
6:　　**for** $t = 1$ **to** $T$ **do**
7:　　　　**for** agent $i = 1$ **to** $N$ **do**
8:　　　　　　Get a local observation: $o_t^i$
9:　　　　　　Observe previous actions $a_{t-1}^{vis-i}$
10:　　　　　　Predict others' actions $(e_t^{-i}, \hat{a}_t^{-i}) = P(o_t^i, a_{t-1}^{vis-i})$
11:　　　　　　Choose an action $a_t^i \sim \pi(o_t^i, e_t^{-i})$
12:　　　　**end for**
13:　　　　Execute the joint action $a_t = [a_t^1, a_t^2, \cdots, a_t^N]$
14:　　　　Update buffer for each agent $D^i += [o_t^i, a_t^i, r_t^i, o_{t+1}^i, a_{t-1}^{vis-i}, \hat{a}_t^{-i}]$
15:　　**end for**
16:　　**for** agent $i = 1$ **to** $N$ **do**
17:　　　　**for** mini-batch $k = 1$ **to** **K** **do**
18:　　　　　　Sample random mini-batch date $b_k^i \leftarrow D^i$
19:　　　　　　Compute $L^{PAC} = L^{PPO} + \lambda L^{pred}$ with $b_k^i$
20:　　　　　　Compute gradient based on $L^{PAC}$
21:　　　　　　Update $\theta^i$ and $\omega^i$
22:　　　　　　Update $\phi^i$
23:　　　　**end for**
24:　　**end for**
25: **end while**

---

ratio of the number of waste to the number of nodes in the river area, $R_{wst}$, is lower than the threshold $T_{wst}$ $(0 < T_{wst} < 1)$; otherwise $P_{wst} = 0$. Meanwhile, the spawn probability of apples, $P_{apl}$, is lowered by $R_{wst}$ as

$$P_{apl} = \left(1 - \frac{R_{wst}}{T_{wst}}\right) \cdot C_{apl}, \qquad (10)$$

if $R_{wst} < T_{wst}$ and otherwise $P_{apl} = 0$. $C_{apl}$ is a constant.

At the start of each episode, there are no apples on the land by setting $R_{wst}$ is slightly larger than $T_{wst}$. To spawn apples, at least one agent must clean the waste using the *cleaning beam* whose width is three nodes and reaches five nodes; this is shown as a yellow beam in Fig. 2. However, cleaning does not provide any direct reward; thus, this situation results in a social dilemma, i.e., which agent should clean them. Furthermore, all agents are equipped with a fining beam (cyan nodes), which costs $-1$ as a negative reward when used but incurs a $-50$ reward as damage to the agents hit. Agents are trained to maximize their rewards by collecting as many apples as possible, but, of course, the collective rewards of all agents also indicate the quality of their cooperation and coordination from the social viewpoint.

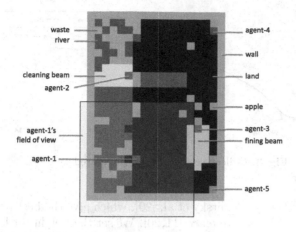

**Fig. 2.** Cleanup game environment. (Color figure online)

## 4.2   Comparative Methods

In all experiments, each agent has its neural network for learning and thus is trained in a decentralized way to maximize its rewards without access to non-local information, including other agents' policies, rewards, and observations. The only assumption is that agents can observe the actions of other agents at the last time step. We compare our method PAC with the following methods:

**PPO-baseline.** Vinitsky et al. [20] proposed this by implementing a CNN-LSTM network (only the (b) and (c) part of Fig. 1) and performing fully decentralized training using *proximal policy optimization* (PPO).

**DQN-baseline.** It uses the same structure of convolution layers, fully connected layers, and LSTM as that in the PPO-baseline. The difference from PPO-baseline is that its last layer outputs Q-values rather than logits and values and is trained as *deep Q-network* (DQN) instead of PPO.

**DRPIQN.** *Deep recurrent policy inference Q-network* (DRPIQN) [8] used *adaptive loss*, and policies are trained as DQN. Its loss function is defined as $L = \lambda L^Q + L^{pred}$, where $\lambda = 1/\sqrt{L^{pred}}$, and $L^Q$ is the loss function for deep Q-network. In the earlier training phases for DRPIQN, $L^{pred}$ is large and $\lambda L^Q$ is small, which encourages the network to focus on other agents' behaviors [8]. As the training progresses, the $L^{pred}$ becomes smaller, and $\lambda L^Q$ becomes dominant.

**DRPI-PPO.** *Deep recurrent policy inference-PPO* (DRPI-PPO) is a variant of DRPIQN [8]. It uses the network structure and adaptive loss of DRPIQN but is trained as PPO rather than DQN. The loss function in Eq. 8 is modified as $L = \lambda L^{PPO} + L^{pred}$, with $\lambda = 1/\sqrt{L^{pred}}$ for DRPI-PPO.

We did not fine-tune specific parameters for PAC. The learning rate is set at 0.00126 initially and annealed linearly to 0.000012 until $2 \times 10^7$ time steps. The *entropy coefficient* ($\sigma$ in Eq. 3) is set to 0.00176 for PPO. These parameter values

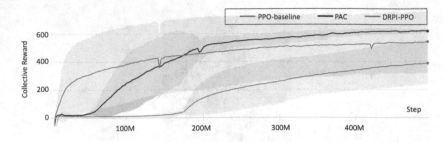

**Fig. 3.** Collective rewards for compared methods.

are identical to those by Vinitsky et al. [20], which give the best performance for PPO-baseline and *social influence* [11,20]. We set the weight for loss function of PAC in Eq. 8, $\lambda = 0.06663557$. DQN-baseline and DRPIQN failed to converge; therefore, we omit their results.

### 4.3   Performance Comparison and Analysis

To compare the performance of three methods, PPO-baseline, DRPI-PPO, and PAC, we plotted the mean collective rewards of five agents every 1000 time steps until 500M time steps in five runs with different random seeds in Fig. 3. The light-colored areas above and below the curves show their standard deviation. At the beginning of training, the collective rewards were negative (minimally $-4000$) because agents started shooting the fining beam recklessly. However, they could soon learn to stop shooting during 3M to 9M time steps. Thus, we only show the part of curves of collective rewards over 0 for readability.

Figure 3 indicates that first, PAC achieved the highest collective rewards finally, and its variance was much smaller than those of other methods. This suggested that the predictor network in PAC provided helpful information to the owner agents to achieve better-coordinated activities. Second, the PPO-baseline quickly increased the collective rewards in the earlier stage (until 30M steps). In contrast, PAC made little learning progress during this stage, gradually increasing the collective rewards and achieving the highest around 200M steps. We think that the rapid convergence of the PPO-baseline was owing to its simple network structure, which only requires a small amount of data to converge. In contrast, other methods with more complicated structures needed more data to learn.

Finally, DRPI-PPO failed to achieve good performance and learned very slowly. This is probably because the adaptive loss function does not work correctly. In the original environment [8], the opponent agent uses a manual non-learning policy. Therefore, the prediction loss could be reduced gradually as the training progressed, and the adaptive weights allowed the algorithm to focus on the opponent's behavior in the earlier stage and then on getting higher scores later.

We can see that all agents' policies continued to change during the whole training process from Fig. 4, which plotted the prediction loss of PAC and DRPI-PPO. The loss would first decrease, then increase, and then gradually decrease

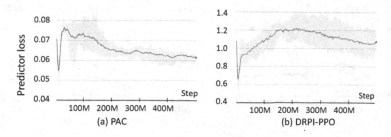

**Fig. 4.** Prediction loss for PAC and DRPI-PPO (Note the difference in vertical axes).

to a relatively stable range. In this case, the adaptive weights of the DRPI-PPO may not have appropriately adjusted the algorithm's attention and may instead have become a burden on training.

### 4.4 Analyzing Learned Coordinated Behavior

We believe it is crucial to understand the learned coordination by analyzing the agent's behaviors in these methods. Figure 5 shows the rewards of individual agents during the training process in five experimental runs. We can find that among the five agents trained by PAC (Fig. 5(b-1) to (b-5)), one agent earned low rewards, while the other four agents gained almost identical high rewards in all runs. This phenomenon always appeared in PAC but was often not apparent in the PPO-baseline and DRPI-PPO.

We further analyzed the movements based on the learned policies by testing its last checkpoint (saved at step 499.2M time step) in a new cleanup game and recording the agents' actions and locations in the first 1000 steps. Their locations are shown in Fig. 6 as heat maps in which darkness indicates the frequencies of staying locations on the map. Figure 6(a) was generated using the PPO-baseline policies of the first run (i.e., Fig. 5(a-1)) and Fig. 6(b) was generated using the PAC policies (i.e., Fig. 5(b-1)).

We found clear patterns in the PAC agents' behaviors in Fig. 6(b). First, *agent-2* (agents whose ID is 2) continued to move clockwise. Furthermore, when moving in the river area (Fig. 6(f)), it sometimes fired the cleaning beam into the river. At the same time, many apples were generated on the land, and the other four agents moved around to collect apples greedily. Second, after *agent-2* left the river area to search for the remaining apples (Fig. 6(g)), the other four agents were still looking for apples greedily and would not intentionally leave any apples for *agent-2*. When the number of apples in the environment became small, *agent-2* moved to the river area again. Here, the agent that cleaned the river area, like *agent-2*, is called a *farmer* and other agents that collected apples are called *robbers*. Finally, each robber built their territory to collect apples (Fig. 6(b-1)(b-3)(b-4) and (b-5)); this may reduce unnecessary competition among robbers. In this agent society, no one can gain more rewards by changing its own strategy, that is, they reach an equilibrium. However, there are obvious differences between

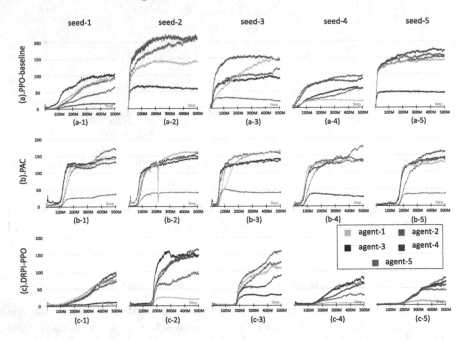

**Fig. 5.** Earned rewards of agents in individual runs.

agents in their strategies and rewards. Therefore, we call this social coordinated structure imbalanced equilibrium.

Table 1 lists the numbers of apples collected and cleaned nodes with wastes in the same episode in Fig. 6. *Agent-2* in PAC cleaned up almost all the waste in the episode but collected the smallest number of apples. The other four agents rarely cleaned up the waste but divided the land into four subregions (Fig. 6), collecting a roughly equal number of apples. Note that in Fig. 6, *agent-2* was a farmer but another agent was a farmer in a different episode (see Fig. 5).

In contrast, for agents trained by PPO-baseline, *agent-4* cleaned up in the river but acted as a part-time farmer because it also collected some apples by occupying a lower-right region as the territory (see (h) in Fig. 6(a-4)). However, it could collect only a few apples there, so resulting in the lower rewards rather than that of *agent-2* in PAC (Table 1). Furthermore, the fact that *agent-4* has such a region to stay there and collect apples reduced (1) the sizes of occupied regions of the other four agents and (2) the efficiency of cleaning, which in turn lowered the number of apples spawned and so overall rewards to earn.

We believe that the ability of the agents with PAC to infer the activities of other agents is critical to establishing a sophisticated and effective social structure and promoting better coordination by establishing an imbalanced but socially efficient equilibrium. Robbers will not clean the river area because they recognize that the farmer will clean the river so that the apples will respawn, and thus they can easily collect apples to receive many rewards. In contrast, the farmer will give up fighting the robbers for occupying a part of the land

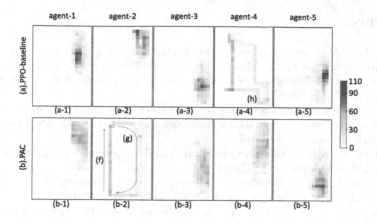

**Fig. 6.** Locations of each agent.

**Table 1.** Number of consumed apples and cleaned wastes in an episode.

| Agent ID | | 1 | 2 | 3 | 4 | 5 | Sum |
|---|---|---|---|---|---|---|---|
| Apple | PPO-baseline | 82 | 63 | 101 | **22** | 91 | 359 |
| | PAC | 138 | **38** | 163 | 141 | 168 | 648 |
| Waste | PPO-baseline | 0 | 0 | 0 | 574 | 0 | 574 |
| | PAC | 1 | 629 | 7 | 0 | 0 | 637 |

area because it realizes that the other four robbers are greedy and will not clean up the river, and thus the farmer attempts to focus on actions for cleaning that suit other agents' activities; otherwise, its rewards will also be decreased. Agents with the PPO-baseline also built a similar equilibrium but were incomplete and earned lower rewards from a social perspective.

## 5 Conclusion

This study proposed a CMM framework called PAC. Each agent constructed a model to estimate the actions of all other agents and then decided their actions considering these actions. We examined the framework in a public goods dilemma called the cleanup game. Our experiments indicate that the trained agents spontaneously establish an effective social structure for coordination and can avoid socially deficient equilibrium.

We experimentally found the emergence of imbalanced equilibrium in this study, and we plan to conduct more mathematical and theoretical research on it in the future. We also plan to extend our method to mixed cooperative-competitive environments.

**Acknowledgements.** This work was partly supported by JST KAKENHI and SPRING, Grant Numbers 20H04245 and JPMJSP2128.

# References

1. Albrecht, S.V., Stone, P.: Autonomous agents modelling other agents: a comprehensive survey and open problems. Artif. Intell. **258**, 66–95 (2018)
2. Baker, B., et al.: Emergent tool use from multi-agent autocurricula. arXiv preprint arXiv:1909.07528 (2019)
3. Barrett, S., Stone, P.: Cooperating with unknown teammates in complex domains: a robot soccer case study of ad hoc teamwork. In: Twenty-Ninth AAAI Conference on Artificial Intelligence (2015)
4. Bowling, M., McCracken, P.: Coordination and adaptation in impromptu teams. In: AAAI, vol. 5, pp. 53–58 (2005)
5. He, H., et al.: Opponent modeling in deep reinforcement learning. In: International Conference on Machine Learning, pp. 1804–1813. PMLR (2016)
6. Heess, N., et al.: Emergence of locomotion behaviours in rich environments. arXiv preprint arXiv:1707.02286 (2017)
7. Hochreiter, S., Schmidhuber, J.: Long short-term memory. Neural Comput. **9**(8), 1735–1780 (1997)
8. Hong, Z.W., et al.: A deep policy inference q-network for multi-agent systems. In: Proceedings of the 17th International Conference on Autonomous Agents and MultiAgent Systems, pp. 1388–1396 (2018)
9. Hughes, E., et al.: Inequity aversion improves cooperation in intertemporal social dilemmas. In: Advances in Neural Information Processing Systems, vol. 31 (2018)
10. Iida, H., Handa, K.I., Uiterwijk, J.: Tutoring strategies in game-tree search. ICGA J. **18**(4), 191–204 (1995)
11. Jaques, N., et al.: Social influence as intrinsic motivation for multi-agent deep reinforcement learning. In: International Conference on Machine Learning, pp. 3040–3049. PMLR (2019)
12. Kollock, P.: Social dilemmas: the anatomy of cooperation. Ann. Rev. Sociol. **24**(1), 183–214 (1998)
13. OpenAI: OpenAI five (2018). https://blog.openai.com/openai-five/
14. Peters, J., Schaal, S.: Reinforcement learning of motor skills with policy gradients. Neural Netw. **21**(4), 682–697 (2008)
15. Rovatsos, M., Weiß, G., Wolf, M.: Multiagent learning for open systems: a study in opponent classification. In: Alonso, E., Kudenko, D., Kazakov, D. (eds.) AAMAS 2001-2002. LNCS (LNAI), vol. 2636, pp. 66–87. Springer, Heidelberg (2003). https://doi.org/10.1007/3-540-44826-8_5
16. Schulman, J., Moritz, P., Levine, S., Jordan, M., Abbeel, P.: High-dimensional continuous control using generalized advantage estimation. arXiv preprint arXiv:1506.02438 (2015)
17. Schulman, J., Wolski, F., Dhariwal, P., Radford, A., Klimov, O.: Proximal policy optimization algorithms. arXiv preprint arXiv:1707.06347 (2017)
18. Silver, D., et al.: Mastering the game of Go with deep neural networks and tree search. Nature **529**(7587), 484–489 (2016)
19. Stone, P., Kaminka, G.A., Kraus, S., Rosenschein, J.S.: Ad hoc autonomous agent teams: collaboration without pre-coordination. In: Twenty-Fourth AAAI Conference on Artificial Intelligence (2010)
20. Vinitsky, E., et al.: An open source implementation of sequential social dilemma games (2019). GitHub repository. https://github.com/eugenevinitsky/sequential_social_dilemma_games/issues/182

# Model-Based Reinforcement Learning with Self-attention Mechanism for Autonomous Driving in Dense Traffic

Junjie Wen[1,2], Zuoquan Zhao[1], Jinqiang Cui[2(✉)], and Ben M. Chen[1]

[1] Department of Mechanical and Automation Engineering,
The Chinese Uinversity of Hong Kong, Hong Kong, People's Republic of China
[2] Department of Mathematics and Theories, Peng Cheng Laboratory, Shenzhen 518055, China
cuijq@pcl.ac.cn

**Abstract.** Reinforcement learning (RL) has recently been applied in autonomous driving for planning and decision-making. However, most of them use model-free RL techniques, requiring a large volume of interactions with the environment to achieve satisfactory performance. In this work, we for the first time introduce a solely self-attention environment model to model-based RL for autonomous driving in a dense traffic environment where intensive interactions among traffic participants may occur. Firstly, an environment model based solely on the self-attention mechanism is proposed to simulate the dynamic transition of the dense traffic. Two attention modules are introduced: the Horizon Attention (HA) module and the Frame Attention (FA) module. The proposed environment model shows superior predicting performance compared to other state-of-the-art (SOTA) prediction methods in dense traffic environment. Then the environment model is employed for developing various model-based RL algorithms. Experiments show the higher sample efficiency of our proposed model-based RL algorithms in contrast with model-free methods. Moreover, the intelligent agents trained with our algorithms all outperform their corresponding model-free methods in metrics of success rate and passing time.

**Keywords:** Reinforcement Learning · Attention Mechanism · Autonomous Driving

## 1 Introduction

Decision-making in a dense environment with dynamically moving obstacles has been a hot topic in the mobile robotics community. One typical problem is passing through dense traffic like unsignalized urban intersections for autonomous vehicles.

Although there are several algorithms that have already been successfully applied to autonomous vehicles [19,25], most of them still rely on hand-crafted rule-based methods which are only applicable in simple driving circumstances. Situations as complex as passing through intersections would dramatically increase the complexity of rule-based algorithm. In contrast, learning-based algorithms can better deal with complex driving situations without complexity increase. Some researchers applied imitation learning (IL) to autonomous driving by training the intelligent agent in a supervised manner [7,30]. However, IL suffers from tedious data labeling and poor generalizability.

M. Tanveer et al. (Eds.): ICONIP 2022, LNCS 13624, pp. 317–330, 2023.
https://doi.org/10.1007/978-3-031-30108-7_27

Unlike IL, RL-based methods can better generalize to unseen scenarios without data labeling by interacting with the environment. Decision-making in dense traffic environment with RL-based methods have already been studied [11,16]. However, most of them are based on model-free RL algorithms, which only train the intelligent agent by exhaustively interacting with the environment. Model-based RL algorithms, on the other hand, explicitly learn the environment transitions as well as interact with the environment, leading to an improvement in sample efficiency.

Researchers [9,31] have already applied model-based RL to autonomous driving. However, the environments they are dealing with are simple driving scenarios. In a complex environment where interactions across traffic participants are unavoidable, the environment model should be carefully designed. In such an environment, one participant's action not only determines its own future state but also others' behaviors. Compared to other model architectures, the self-attention mechanism is more capable of extracting the interactive dependencies across all inputs.

In this work, we for the first time apply a solely self-attention environment model to model-based RL for autonomous driving in a dense traffic environment. Our contributions are listed as follows:

- We propose an environment model solely on the self-attention mechanism to simulate the dynamic transitions in a dense traffic environment and it shows better performance in predicting the future states of traffic participants than other SOTA methods.
- We introduce two self-attention modules: the Horizon Attention module to encode the spatially interactive information and the Frame Attention module to encode sequential state features.
- The developed environment model has been introduced to model-based RL for autonomous driving in dense traffic. Testing results show the superiority of our proposed model-based RL algorithms over their model-free RL counterparts.

## 2    Related Work

### 2.1    Attention Mechanism

Attention Mechanism is first introduced in natural language processing (NLP) by Bahdanau et al. [3] as an improvement over the encoder-decoder neural machine translation system based on recurrent neural network (RNN) or long short-term memory (LSTM) [10]. However, due to the vanishing/exploding gradient problem, RNN suffers from long-range dependency. In addition, the sequential nature of RNN/LSTMs makes the parallelization of training examples extremely hard. By calculating the embeddings of all input words and the weighted sum of hidden states, the attention mechanism not only allows for the modeling of dependencies without regard to the distances but also accelerates the training process by data parallelization.

Based on the attention-mechanism, many breakthroughs have been made in NLP and computer vision tasks. The Transformer [26] differs from previous neural translation models in that it is based on self-attention mechanisms to draw global dependencies between input and output. It gets SOTA performance in translation with less training time and significantly more parallelization. BERT [5] also achieves SOTA performances in a wide range of NLP tasks by pretraining deep bidirectional representations from unlabeled text based on Transformers and finetuning one additional task-specific output layer. ViT [6] directly applied a pure Transformer architecture [26] to sequences of image patches and showed excellent results on image classification tasks. Swin-Transformer [17] surpassed previous SOTA networks on a broad range of vision tasks by a large margin with a hierarchical Transformer whose representation is computed with shifted windows.

In this work, we equip the environment model of dense traffic with self-attention mechanisms to extract the dependencies across both different traffic participants and different state histories.

## 2.2 Traffic Trajectory Prediction Methods

Predicting the future trajectories of other traffic participants in dense traffic is a complex problem for the autonomous driving community. Early researchers [21,25] mostly focused on rule-based Finite/Hybrid State Machines to encode the desired behavior of the vehicle in the encountered urban scenarios. However, the complexity of these methods could dramatically increase as the driving scenarios become more general. Besides, the constant velocity assumption that is usually employed in those systems ignores the surrounding vehicles' reactions, leading to potential risks in dense urban traffic scenario [8].

The learning-based prediction method, on the other hand, does not show a complexity increase even in complicated situations and is more capable of modeling the participants' intentions. Altche et al. [2] introduced an LSTM-based network that predicted the future longitudinal and lateral trajectories for vehicles on the highway. Ma et al. [18] also proposed an LSTM-based network that contains instance and category layers to predict the trajectories of heterogeneous traffic participants. Tang et al. [24] introduced a probabilistic framework to efficiently model the future motions of agents with a dynamic attention-based state encoder.

However, most previous learning-based prediction methods are based upon RNN or combine RNN with the attention mechanism, which could be time-consuming during inference due to RNN's poor parallelization capability. In this work, we propose a prediction network that is solely based on the self-attention mechanism for forecasting the future motions of all the participants involved in dense traffic.

## 2.3 Reinforcement Learning

RL involves interactions between the intelligent agent and the environment. It is usually formulated as a Markov decision process $< S, A, P, R, \gamma >$, with the measurable space $S$, action space $A$, state transition dynamics $P$, reward function $R$ and discount factor $\gamma$. The training purpose for solving RL task is just to find a policy $\pi$ that maximizes the expected $\gamma$-discounted cumulative reward.

In the RL paradigm, based on whether the intelligent agents predict the environment responses, there are two kinds of RL methods: model-free and model-based RL. Model-free RL indicates that the agent learns policies by directly interacting with the

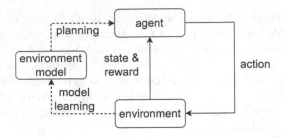

**Fig. 1.** Illustration of model-free and model-based RL. The loop formed by the solid line denotes the model-free RL while the loop of dotted line represents the model-based RL.

environment, while model-based RL constructs a predictive model of the environment dynamics and integrates both learning and planning, as illustrated in Fig. 1.

Among various model-free algorithms, Q-learning [28] is a typical value optimization method by learning a near optimal state-action value function with a sufficient number of learning samples. Deep Q-Networks (DQN) [20] achieved SOTA performances in many Atari games by introducing deep neural network (DNN) as a non-linear value approximator and experience replay technique. The Q value updating rule is:

$$Q(S_t, A_t) \leftarrow Q(S_t, A_t) + \alpha(R_{t+1} + \gamma \max_a Q(S_{t+1}, a) - Q(S_t, A_t)), \qquad (1)$$

where $Q(S_t, A_t)$ is the estimated action value of state $S$ when choosing action $A$ at time $t$, $\alpha \in [0, 1]$ is the learning rate.

Instead of evaluating a value function before getting the optimal policy, policy optimization methods aim to obtain an optimal policy directly. In policy optimizations, the policy function is usually parameterized as a learnable neural network $\pi_\theta$. The REINFORCE [29] algorithm is a simple straightforward policy optimization algorithm, whose policy parameters are updated by:

$$\theta \leftarrow \theta + \alpha \gamma^t G \nabla_\theta \ln \pi(A_t | S_t, \theta), \qquad (2)$$

where $\theta$ denotes the parameter of policy network, $\alpha$ is the learning rate and $\gamma^t G$ is the $\gamma$-discounted cumulative reward. Actor-Critic [14] is a policy optimization method which integrates the benefits of value optimization. Deep Deterministic Policy Gradient (DDPG) [22] is an actor-critic algorithm that can learn policies for continuous action space with DNN approximation function.

However, model-free methods usually suffer from low sample efficiency. For intelligent agents which need to be applied to practical situations, interacting with real-world environments would be limited due to cost and safety concerns. Model-based RL [12,23] is advantageous in reducing the required number of interactions by simultaneously learning the environment transitions. In this work, we develop various model-based RL algorithms with our proposed environment model for autonomous driving in dense traffic.

## 3   Methodology

### 3.1   Problem Definition

The problem of developing model-based RL algorithms for passing through dense traffic could be decomposed into two tasks: one is constructing an environment model for dense traffic, and the other is solving the passing problem with RL algorithm with the environment model.

We first assume the states of agents involved in the traffic are preprocessed, including their spatial coordinates, velocities, and headings. At time $t$, the feature vector of any agent $A_i^t$ is denoted as $f_i^t = (p_x, p_y, v_x, v_y, \cos\phi, \sin\phi)_i^t$, where $p_x/p_y$ is the coordinate in $x$-/$y$- axis, $v_x/v_y$ is the velocity in $x$-/$y$- axis, and $\phi$ is the agent heading. In the following paragraphs, we denote 'ego-agent' as the traffic participant controlled by RL algorithms and 'other-agent' as the participants in the environment except 'ego-agent'. The task of formulating an environment model is just to observe the states of all the agents during the time interval $[1, T_{obs}]$ and predict their states and rewards from the environment at time step $t = T_{obs} + 1$.

With the environment model, the problem of autonomous driving in dense traffic could be viewed as the ego-agent navigating to a target position without collision with other-agents. It is an optimization problem whose objective is to maximize the expected cumulative reward. For an environment with one ego-agent and $N$ other-agents, the optimization objective should be:

$$\underset{\pi_\theta}{\mathrm{argmax}} \quad \mathbb{E}[R|\pi_\theta, S_e, S_{o,1:N}],$$
$$s.t. \quad T_e \in [0, T_{\max}], P_e \in P_f, V_e \in V_f \tag{3}$$

where $\pi_\theta$ is the policy taken by the intelligent ego agent, $R$ is the cumulative reward, $S_e$ is the ego agent state, $s_{o,1:N}$ is the states for other agents. $T_e$ is the passing time of ego agent, with the maximum allowable time as $T_{\max}$. $P_e$ and $V_e$ are the ego agent position and velocity, which should be limited in the feasible position region $P_f$ and velocity region $V_f$, respectively. The collision restriction is not shown in Eq.(3) because it is integrated in the cumulative reward $R$.

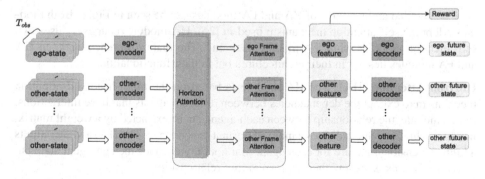

**Fig. 2.** The overall network structure for traffic environment modeling with HA precedes FA (HA-FA model).

## 3.2   Environment Model

The transition dynamics of a dense traffic environment is complex due to the interactions between different agents. The action taken by one agent results in both its own
state transition and others' decision makings. Hence, the environment model for a dense
traffic should consider not only the state histories of each agent, but also the interaction dependencies on others. We propose two self-attention modules to deal with both
the state histories and the interactions: the Frame Attention (FA) module to encode
the sequential state features for each agent and the Horizon Attention (HA) module to
encode the interactive information with others.

The overall structure of our environment model can be seen in Fig. 2. The sequential
states of each agent within the time interval $T_{obs}$ are first embedded by encoders. The
parameters of the ego-encoder differ from the other-encoders because it is used for
embedding ego-agent's state. The embedded features are then fed into the HA module
to extract the interactive information across agents at each time step $t \in [1, T_{obs}]$. After
that, the FA module draws the time dependencies of sequential features with length $T_{obs}$
for each agent, with ego-FA for ego-agent and other-FA for other-agents. Decoders are
finally used to get the state of each agent in the next time step. The reward from the
environment is regressed by a feed-forward layer after concatenating all the FA features.

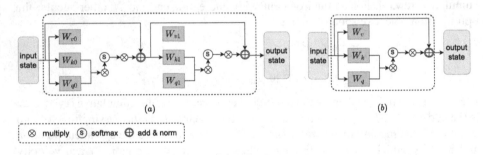

(a)

(b)

⊗ multiply   ⓢ softmax   ⊕ add & norm

**Fig. 3.** (a) The detailed architecture of single-head HA-module. (b) The detailed architecture of
single-head FA-module.

The detailed architectures of HA and FA modules can be seen in Fig. 3. Both modules adopt the self-attention mechanism used in [26]. HA-module recursively uses two
self-attention modules while FA-module uses one. The main difference between HA
and FA modules lies not in their architectures but in their functionalities.

**Horizon Attention (HA):** The HA module is based on the multi-head self-attention
mechanism to extract the dependencies between different agents that have interactions.
In this module, the relationship between each agent can be extracted by a weight matrix,
in which a higher value indicates a more important dependency. The HA module outputs
a high-dimensional feature for each agent that not only encodes the ego-agent state but
also other-agents' state and their corresponding relationship.

**Frame Attention (FA):** The FA module is proposed to encode the sequential features of
each agent in the scene. Compared to RNN/LSTM architecture, the FA module is advan-

tageous in data parallelization and dependency extraction. Besides, taking an agent's full history of states gives it a better understanding of the dynamic environment.

---

**Algorithm 1.** Procedure for model-based value optimization in dense traffic

---
1: Initialize value function $Q(s, a)$, environment model $Model(s, a)$ and an empty experience buffer $B$
2: While not done:
3:     Observe $S$ from real environment
4:     Get action $A \leftarrow \varepsilon$-greedy$(S, Q)$
5:     Execute action $A$, observe reward $R$ and state $S'$
6:     $B \leftarrow B \cup (S, A, R, S')$
7:     Update $Q(S, A)$ with Eq.(1) and $Model(s, a)$ with $(S, A, R, S')$
8:     Repeat $n$ times:
9:         $S \leftarrow$ random previously observed state
10:        $A \leftarrow$ random previous action taken in $S$
11:        $R, S' \leftarrow Model(S, A)$
12:        Update $Q(S, A)$ with Eq.(1)

---

### 3.3  Model-Based RL for Dense Traffic

Model-based RL is more advantageous than model-free RL in its higher sample efficiency. With an environment model, the intelligent agent could learn how the environment transfers and then predict the action that would lead to desirable outcomes. In this work, we use our proposed environment model to generate imagined future rollouts and add them to the experience buffer from which the intelligent agent could learn a passing policy.

The procedure for model-based value optimization RL that combines direct RL, model learning, and planning is presented in Algorithm 1. The environment model is trained online with the real experience from the environment, and the value function is updated by both the real environment experience and the imagined rollouts from the environment model.

---

**Algorithm 2.** Procedure for model-based policy optimization in dense traffic

---
1: Initialize parameters $\theta$ of policy $\pi$, parameters $\alpha$ of environment model $M$ and an empty experience buffer $B$
2: While not done:
3:     Observe $S$ from real environment
4:     Get action $A \leftarrow \varepsilon$-greedy$(\pi_\theta)$
5:     Execute action $A$, observe reward $R$ and state $S'$
6:     $B \leftarrow B \cup (S, A, R, S')$
7:     Update environment model $M_\alpha$ with $(S, A, R, S')$
8:     Update policy $\pi_\theta$ with $M_\alpha$ and $B$

---

The general algorithm for model-based policy optimization is shown in Algorithm 2. The policy is also updated with both real and imagined experiences. The policy updating method is general such that it could be implemented by any policy optimization method such as TD3 [4], DDPG [22], etc.

## 4   Experiments

### 4.1   Environment Setup

The simulated environment is modified upon the intersection environment presented in [15]. The environment range is set to $[-40\,\text{m}, 40\,\text{m}]$. The ego-agent is rewarded by 5 if it reaches its destination, $-5$ if it collides with other-agents or violates traffic rules, and a positive reward of 0.5 is used to encourage driving in a desired speed range. There are 3 levels of difficulty for the environment based on the average number of other-agents, i.e. easy, medium, and hard.

The agents in the simulated environment are controlled in a hierarchical manner, including a low-level lateral controller and a high-level longitudinal controller. The lateral action is implemented by a low-level lateral tracker for both ego- and other- agent. The longitudinal acceleration is realized by our learned policies for the ego-agent while implemented by a rule-based intelligent driver model (IDM) [13] for other-agents.

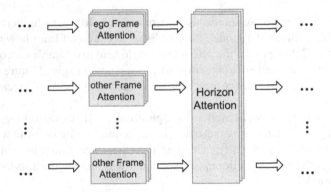

**Fig. 4.** FA-HA environment model. (Dots indicate the same with HA-FA model.)

### 4.2   Environment Model Evaluation

In order to evaluate the effectiveness of our environment model, we first collect experience data from the simulated environment through random policy and then train environment models in a supervised manner. The training set contains 100k data while the testing set contains 10 k. The models are trained on a computer with the configurations: NVIDIA GTX1080Ti graphics card, Intel i7 6800k, 48 GB RAM. During training, the batch size is 32, the learning rate is set to 0.0005, and maximum training epoch is set to 50.

Besides the proposed HA-FA model, we also implement an FA-HA model as shown in Fig. 4, where the only difference is the order of HA and FA modules. The HA/FA module is also replaced with multi-layer perceptron (MLP) to get MLP-FA/HA-MLP models. Other implemented environment models include: Social-LSTM (SL) [1], Social-Attention (SA) [27], GMM [31], WM [9] and a naive MLP model.

**Table 1.** Testing Displacement Error of Different Models

| Models | MLP | SA | SL | GMM | WM | MLP-FA | HA-MLP | FA-HA | HA-FA |
|---|---|---|---|---|---|---|---|---|---|
| Average Error (m) | 0.650 | 0.325 | 0.329 | 0.401 | 0.457 | 0.359 | 0.190 | 0.436 | **0.167** |
| Straight Road Error (m) | 0.276 | 0.151 | **0.141** | 0.191 | 0.188 | 0.183 | 0.192 | 0.204 | 0.143 |
| Intersection Error (m) | 0.782 | 0.388 | 0.397 | 0.477 | 0.555 | 0.425 | 0.187 | 0.520 | **0.172** |

The positional displacement error of these methods with respect to the testing data is shown in Table 1. It is noted that our proposed HA-FA model achieves the best average prediction performance than other methods. All these methods achieve good results in straight road prediction, indicating the effectiveness of common prediction methods in simple driving scenarios. However, in complex situations like intersections, the performances of these methods differ a lot. Our proposed HA-FA model performs best in such situation, mainly contributed to the adoption of HA module as compared with MLP-FA model. It should be noted that FA-HA model shows a dramatic performance drop even though it contains both HA and FA modules. The reasons are discussed in Sect. 5.

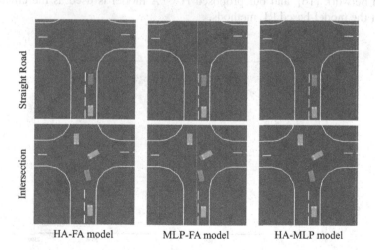

**Fig. 5.** Visualization of environment transition predictions. Green rectangle: ego-agent, blue rectangle: other-agents, transparent yellow rectangle: predicted states. (Color figure online)

In order to better understand the effectiveness of HA/FA modules, the visualization results of the predicted future positions for different agents of HA-FA/MLP-FA/HA-MLP models are shown in Fig. 5. It is seen that our HA-FA model has superior prediction

**Table 2.** Average Inference Time of Different Models

| Models | MLP | SA | SL | GMM | WM | MLP-FA | HA-MLP | FA-HA | HA-FA |
|---|---|---|---|---|---|---|---|---|---|
| Average Inference Time (s) | **0.042** | 0.102 | 0.097 | 0.059 | 0.115 | 0.047 | 0.051 | 0.052 | 0.055 |

performance for agents in both straight roads and intersections. The MLP-FA model, however, has similar performance in straight road scenario while shows bad predictions when agents are in the intersection region. In contrast, the HA-MLP model shows better results than MLP-FA in intersections but slightly worse when agents are at straight road.

The average inference time of each model is computed in Table 2. Although the inference time of our proposed HA-FA model is not the shortest, it's still much less than the time used by other RNN/LSTM- based models like SA/SL/WM, validating the effectiveness of parallelization for the self-attention mechanism.

### 4.3 Model-Based and Model-Free RL

We implement the training of different RL algorithms in both model-based and model-free manners. Model-free RL methods have been tried with both value optimization like DQN and policy optimization methods like DDPG and TD3. For DQN, the ego-agent output three discrete actions "slow-down","idle", and "speed-up". For policy optimization algorithms, the ego-agent produces continuous acceleration ranging in $[-5, 5]$. The ego-agents for all the implemented algorithms have the same structure as the Ego-Attention network [16], and our proposed HA-FA model is used as the environment model for the model-based RL methods.

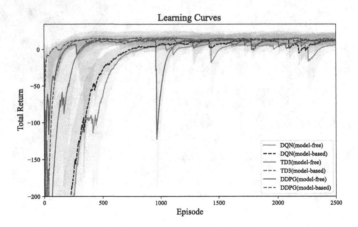

**Fig. 6.** Learning curves of model-based and model-free methods

The learning curves are shown in Fig. 6 and the average cumulative reward for each methods is shown in Table 3. If we define "Rising Time" as the used episodes when

**Table 3.** Cumulative Reward and Rising Time of Different Methods (MF: model-free, MB: model-based)

| Models | DQN (MF) | DQN (MB) | TD3 (MF) | TD3 (MB) | DDPG (MF) | DDPG(MB) |
|---|---|---|---|---|---|---|
| Cumulative Reward | 15.3 | 20.9 | 20.8 | **22.3** | 21.1 | 21.3 |
| Rising Time | 1215 | 1038 | 381 | 219 | 238 | **137** |

**Table 4.** Evaluated result of success rate and passing time (MF: model-free, MB: model-based)

| Success Rate | Easy | Medium | Hard | Passing Time | Easy | Medium | Hard |
|---|---|---|---|---|---|---|---|
| IDM (baseline) | 97.02% | 95.05% | 88.12% | IDM (baseline) | 132.46 | 147.19 | 173.84 |
| DQN (MF) | 95.34% | 92.23% | 86.78% | DQN (MF) | 128.36 | 140.34 | 175.53 |
| DQN (MB) | 96.81% | 92.13% | 87.09% | DQN (MB) | 127.53 | 139.28 | 175.42 |
| TD3 (MF) | 96.32% | 95.31% | 88.19% | TD3 (MF) | 125.55 | 138.54 | 172.33 |
| TD3 (MB) | **97.57%** | **96.87%** | **89.77%** | TD3 (MB) | **120.21** | **136.91** | **170.85** |
| DDPG (MF) | 95.44% | 94.98% | 88.09% | DDPG (MF) | 125.38 | 139.33 | 173.95 |
| DDPG (MB) | 96.87% | 95.87% | 88.59% | DDPG (MB) | 123.29 | 137.65 | 171.32 |

the cumulative reward reaches 90% of the final average reward, then the method with less rising time has better sample efficiency, as shown in Table 3. It can be seen that all the model-based methods equipped with our proposed environment model show higher cumulative reward and better sample efficiency than their corresponding model-free methods. The model-based TD3 method achieves the highest cumulative reward and the model-based DDPG method has the best sample efficiency.

The testing result of each method with respect to the success rate and passing time is shown in Table 4. The baseline metrics are implemented upon the IDM method. It is seen that the model-based methods with our proposed environment model achieve a relatively higher success rate and less passing time than their corresponding model-free methods. Model-based TD3 and DDPG algorithms could even outperform the rule-based IDM method in our evaluated metrics.

## 5  Discussion

In this section, we qualitatively discuss the performance drop by the order exchange of HA and FA modules. As shown in Fig. 7(a), the HA-FA model first extracts the interactive dependencies across agents in each frame by HA, and then the decoded features of all frames are fed into FA to get the time dependencies. In contrast, as shown in Fig. 7(b), the FA-HA model first gets the time dependencies of each agent and then extracts the interactive dependencies with the encoded features, which has fewer connections than the HA-FA model.

Hence, the performance drop of the FA-HA model might be caused by two factors: (1) The environmental information is not fully utilized by the FA-HA model. As shown in Fig. 7(a)(b), the HA-FA model fully utilizes the interactive information of all frames,

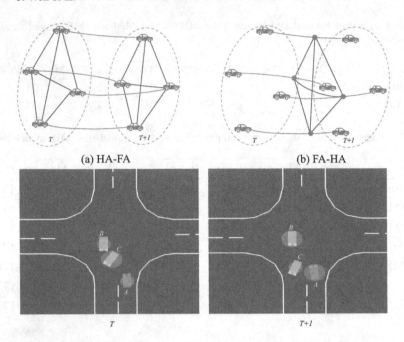

**Fig. 7.** (a) Network connections for HA-FA model, (b) Network connections for FA-HA model, Blue Lines: Horizon Attention, Green Lines: Frame Attention. (c) Illustration about the different attention importance for different agents at different time steps. (Color figure online)

while the FA-HA model only uses the encoded features from the FA module with time dependency while ignoring the interactive information at most time frames; (2) The time dependencies extracted by the FA module for each agent could vary for different agents such that the HA module could not get useful interactive information. As shown in Fig. 7(c), the encoded feature for agent $C$ might have higher importance at frame $T$ while agents $A$ and $B$ higher at frame $T + 1$.

## 6    Conclusion

In this work, an environment model for dense traffic based solely on the self-attention mechanism has been proposed and model-based RL algorithms upon the proposed environment model have been developed. The FA and HA modules are presented: the FA module is used to encode temporal features of one agent while the HA module is to understand the interactive dependencies among all agents spatially at each time frame. The HA-FA model shows superior performance over other SOTA methods in predicting future states of traffic participants. The performance drop of the FA-HA model could be caused by less information utilization and a mismatch of temporal dependencies for different agents. Model-based RL algorithms for both value and policy optimizations have been developed for dense traffic environments. With our proposed HA-FA model, we successfully trained intelligent agents in a model-based manner for various RL algorithms like DQN, DDPG, and TD3 with better sample efficiency and higher cumulative reward. The agents trained with the proposed model-based algorithms also surpass the corresponding agents trained in a model-free manner with a higher success rate and less passing time.

# References

1. Alahi, A., Goel, K., Ramanathan, V., et al.: Social LSTM: human trajectory prediction in crowded spaces. In: Proceedings of the IEEE Conference on Computer Vision and Pattern Recognition, pp. 961–971 (2016)
2. Altché, F., de La Fortelle, A.: An LSTM network for highway trajectory prediction. In: 2017 IEEE 20th International Conference on Intelligent Transportation Systems (ITSC), pp. 353–359. IEEE (2017)
3. Bahdanau, D., Cho, K., Bengio, Y.: Neural machine translation by jointly learning to align and translate. arXiv preprint arXiv:1409.0473 (2014)
4. Dankwa, S., Zheng, W.: Twin-delayed ddpg: a deep reinforcement learning technique to model a continuous movement of an intelligent robot agent. In: Proceedings of the 3rd International Conference on Vision, Image and Signal Processing, pp. 1–5 (2019)
5. Devlin, J., Chang, M.W., Lee, K., Toutanova, K.: Bert: pre-training of deep bidirectional transformers for language understanding. arXiv preprint arXiv:1810.04805 (2018)
6. Dosovitskiy, A., Beyer, L., Kolesnikov, S., et al.: An image is worth 16x16 words: transformers for image recognition at scale. arXiv preprint arXiv:2010.11929 (2020)
7. Eraqi, H.M., Moustafa, M.N., Honer, J.: End-to-end deep learning for steering autonomous vehicles considering temporal dependencies. arXiv preprint arXiv:1710.03804 (2017)
8. Fletcher, L., Teller, S., Olson, E., et al.: The MIT-cornell collision and why it happened. J. Field Robot. 25(10), 775–807 (2008)
9. Ha, D., Schmidhuber, J.: World models. arXiv preprint arXiv:1803.10122 (2018)
10. Hochreiter, S., Schmidhuber, J.: Long short-term memory. Neural Comput. 9(8), 1735–1780 (1997)
11. Isele, D., Nakhaei, A., Fujimura, K.: Safe reinforcement learning on autonomous vehicles. In: 2018 IEEE/RSJ International Conference on Intelligent Robots and Systems (IROS). pp. 1–6. IEEE (2018)
12. Kaiser, L., Babaeizadeh, M., Milos, P., et al.: Model-based reinforcement learning for atari. arXiv preprint arXiv:1903.00374 (2019)
13. Kesting, A., Treiber, M., Helbing, D.: Enhanced intelligent driver model to access the impact of driving strategies on traffic capacity. Philos. Trans. R. Soc. A: Math. Phys. Eng. Sci. 368(1928), 4585–4605 (2010)
14. Konda, V., Tsitsiklis, J.: Actor-critic algorithms. Adv. Neural Inf. Process. Syst. 12, 1008–1014 (1999)
15. Leurent, E.: An environment for autonomous driving decision-making. GitHub (2018)
16. Leurent, E., Mercat, J.: Social attention for autonomous decision-making in dense traffic. arXiv preprint arXiv:1911.12250 (2019)
17. Liu, Z., Lin, Y., Cao, Y., et al.: Swin transformer: hierarchical vision transformer using shifted windows. arXiv preprint arXiv:2103.14030 (2021)
18. Ma, Y., Zhu, X., Zhang, S., et al.: Trafficpredict: trajectory prediction for heterogeneous traffic-agents. In: Proceedings of the AAAI Conference on Artificial Intelligence, vol. 33, pp. 6120–6127 (2019)
19. Minderhoud, M.M., Bovy, P.H.: Extended time-to-collision measures for road traffic safety assessment. Accid. Anal. Prev. 33(1), 89–97 (2001)

20. Mnih, V., Kavukcuoglu, K., Silver, D., et al.: Human-level control through deep reinforcement learning. Nature **518**(7540), 529–533 (2015)
21. Montemerlo, M., Becker, J., Bhat, S., et al.: Junior: the stanford entry in the urban challenge. J. Field Robot. **25**(9), 569–597 (2008)
22. Silver, D., Lever, G., Heess, N., et al.: Deterministic policy gradient algorithms. In: International Conference on Machine Learning, pp. 387–395. PMLR (2014)
23. Sutton, R.S.: Dyna, an integrated architecture for learning, planning, and reacting. ACM Sigart Bull. **2**(4), 160–163 (1991)
24. Tang, C., Salakhutdinov, R.R.: Multiple futures prediction. Adv. Neural Inf. Process. Syst. **32**, 15424–15434 (2019)
25. Urmson, C., Anhalt, J., Bagnell, D., et al.: Autonomous driving in urban environments: boss and the urban challenge. J. Field Robot. **25**(8), 425–466 (2008)
26. Vaswani, A., Shazeer, N., Parmar, N., et al.: Attention is all you need. In: Advances in Neural Information Processing Systems, pp. 5998–6008 (2017)
27. Vemula, A., Muelling, K., Oh, J.: Social attention: modeling attention in human crowds. In: 2018 IEEE international Conference on Robotics and Automation (ICRA), pp. 4601–4607. IEEE (2018)
28. Watkins, C.J., Dayan, P.: Q-learning. Mach. Learn. **8**(3–4), 279–292 (1992)
29. Williams, R.J.: Simple statistical gradient-following algorithms for connectionist reinforcement learning. Mach. Learn. **8**(3), 229–256 (1992)
30. Xu, H., Gao, Y., Yu, F., Darrell, T.: End-to-end learning of driving models from large-scale video datasets. In: Proceedings of the IEEE Conference on Computer Vision and Pattern Recognition, pp. 2174–2182 (2017)
31. Zhuo, X., Jianyu, C., Masayoshi, T.: Guided policy search model-based reinforcement learning for urban autonomous driving. arXiv preprint arXiv:2005.03076 (2020)

# CLCDR: Contrastive Learning for Cross-Domain Recommendation to Cold-Start Users

Yanyu Chen⑩, Yao Yao⑩, and Wai Kin Victor Chan[(✉)]⑩

Tsinghua-Berkeley Shenzhen Institute, Tsinghua Shenzhen International Graduate
School, Tsinghua University, Shenzhen 518055, China
{cyy20,y-yao19}@mails.tsinghua.edu.cn, chanw@sz.tsinghua.edu.cn

**Abstract.** Recent advances in cross-domain recommendation have shown great potential in improving sample efficiency and coping with the challenge of data sparsity via transferring the knowledge from the source domain to the target domain. Previous cross-domain recommendation methods are generally based on extracting information from overlapping users, which limits their performance when there are not sufficient overlapping users. In this paper, we propose a contrastive-based cross-domain recommendation framework for cold-start users that simultaneously transfers knowledge about overlapping users and user-item interactions to optimize the user/item representations. To this end, two contrastive loss functions and two specific learning tasks are proposed. The proposed framework can make fuller use of the information on the source domain and reduce the demand for overlapping users, while maintaining or even enhancing recommendation performance. Experimental results on a real-world dataset demonstrate the efficacy and effectiveness of our framework on the top-N cross-domain recommendation task.

**Keywords:** Cross-Domain Recommendation · Contrastive Learning · Cold-Start · Neighborhood Inference

## 1 Introduction

Recommendation systems, a tool to assist in retrieving helpful information, have applications covering all aspects of daily life. Limited by the data-driven property, there are two challenges in recommendation systems: data sparsity and the cold-start problem. Data sparsity refers to the few-shot case [1], while the cold-start problem corresponds to the zero-shot case [2,3]. To address the challenges, researchers are devoted to introducing auxiliary knowledge from other domains,

This research was funded by the Shenzhen Science and Technology Innovation Commission (JCYJ20210324135011030, WDZC20200818121348001), Guangdong Pearl River Plan (2019QN01X890), and National Natural Science Foundation of China (Grant No. 71971127).

M. Tanveer et al. (Eds.): ICONIP 2022, LNCS 13624, pp. 331–342, 2023.
https://doi.org/10.1007/978-3-031-30108-7_28

which has given rise to a new research hotspot, namely cross-domain recommendation (CDR) [4–7]. CDR aims to leverage data from the source domain to improve recommendation performance in the target domain.

Most existing CDR approaches transfer knowledge from the source domain to alleviate the data sparsity issue, but fail to enhance the cold-start user experience. With specific consideration for cold-start users, [8] proposed the Embedding and Mapping approach (EMCDR) framework, which consists of three steps: (1) Encode users and items to obtain the corresponding representations (a.k.a. embedding in some literature) in each domain. (2) Learn a mapping function to transfer the overlapping users' information from the source domain to the target domain. (3) Based on the output representation of the mapping function, items on the target domain are filtered to be recommended. Along the lines of EMCDR, there are many EMCDR-based approaches [9–12]. However, these methods focus on optimizing the second or third step of EMCDR yet neglect the optimization of the first step. It is well known that user and item representations serve as inputs to downstream recommendation models, and their quality plays a nontrivial role in the ultimate recommendation performance. Therefore, it is highly desired to design a mechanism for learning better representations.

The effectiveness of the EMCDR framework relies on the ability to efficiently transfer information about overlapping users, but with this comes the sensitivity to the number of overlapping users, i.e., the performance of CDR deteriorates significantly if there are not sufficient overlapping users. According to the analysis of the Amazon dataset[1] [13] in [10], the largest public real-world dataset for CDR research, overlapping users (those with records on multiple domains) account for less than 20% of all users. In order to reduce the dependency on overlapping users, recent studies additionally consider user-item interactions when transferring information. For instance, SSCDR [10] proposed a semi-supervised learning framework to transfer the knowledge of user-item interactions. TMCDR [14] designed a transfer stage and a meta stage to prevent the learned mapping function from becoming biased to work well only for overlapping users. Following these studies, we aim to leverage user-item interactions to enhance CDR performance while relaxing the assumption of sufficient overlapping users. Different from existing approaches, our work achieves the objective from the perspective of learning better representations. Inspired by the extraordinary success of contrastive learning [15–17], a paradigm for learning representations of unlabeled samples on the latent space by teaching the model which data points are similar (positive samples) or different (negative samples), we propose two contrastive loss functions that are applied to the first and second steps of the EMCDR framework to optimize the input representation for the third step.

In this paper, we propose a **C**ontrastive **L**earning for **C**ross-**D**omain **R**ecommendation (CLCDR) framework that is geared towards cold-start users. Akin to EMCDR [8], the CLCDR framework is divided into three steps: (1) Learn user and item representations with a proposed contrastive loss function to distill domain-specific information on each domain. (2) Calibrate the representation with another proposed contrastive loss function to transfer information from the

---

[1] https://nijianmo.github.io/amazon/index.html.

source domain to target domain. Two tasks are involved in this step, where the main task is in charge of extracting information from overlapping users while the auxiliary task corresponds to transferring user-item interaction knowledge. (3) Infer the representation of cold-start users from the neighborhood of the users. The contributions of this paper are three-fold and are summarized below.

- We propose the CLCDR framework whereby two complementary tasks account for both overlapping user information and user-item interactions to optimize representations and improve the experience of cold-start users.
- We propose two kinds of contrastive learning-based loss functions to extract domain-specific and inter-domain knowledge effectively. These loss functions are plug-and-play and compatible with most EMCDR-based frameworks.
- Extensive experiments show that CLCDR outperforms several baselines by large margins and demonstrate the effectiveness of CLCDR.

## 2    The Proposed CLCDR Framework

### 2.1    Notations

Without loss of generality, we consider the most basic case, i.e., cross-domain recommendations from a source domain $X$ to a target domain $Y$. In this paper, the "source domain" refers to the domain where we transfer the knowledge of user preferences from, while the "target domain" refers to the domain in which we make recommendations. Each domain has its corresponding item set $\mathcal{V}$, a user set $\mathcal{U}$, and a binary matrix $R \in \{0,1\}^{|\mathcal{U}| \times |\mathcal{V}|}$ whose elements indicate whether there is an interaction between a user and an item. Let $\mathcal{U}^X, \mathcal{V}^X, R^X$ and $\mathcal{U}^Y, \mathcal{V}^Y, R^Y$ denote the user sets, item sets, and user-item interactions of the source and target domain, respectively. Specifically, the overlapping users are denoted by $\mathcal{U}^O = \mathcal{U}^X \cap \mathcal{U}^Y$. Note that it is commonly assumed that the set of items in the source and target domains do not overlap. In the source domain $X$, let $M(u_i^X)$ denote the subset of items that user $u_i^X$ has ever interacted with, $M(v_j^X)$ denote the subset of users who have interaction with item $v_j^X$, and $\neg M(u_i^X) = \mathcal{V}^X - M(u_i^X)$ denote the subset of items for which user $u_i^X$ has no interaction record. In the following, users and items are represented by vectors if not otherwise stated.

### 2.2    Overview of CLCDR

As presented in the Fig. 1, this CLCDR framework is composed of three steps. (1) In the encoding step, CLCDR aims to model the user and item representations of the source and target domains respectively with a newly proposed contrastive loss. In this way, the interactions between users and items can be represented by the distances in the latent space. (2) In the transferring step, CLCDR sets two specific tasks to learn a cross-domain mapping function that can extract useful information from both domains and get better representations by utilizing both overlapping users and user-item interactions. (3) In the recommendation step, given a cross-domain mapping function, for a cold-start user in the target domain, CLCDR can obtain an approximation of its expected representation in the latent space of the target domain based on its neighbors in the source domain, then return a recommendation item list via preference prediction.

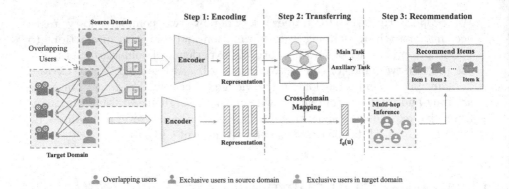

**Fig. 1.** The architecture of CLCDR. The first step is to encode users and items by contrastive loss. The second step is to learn a mapping function through combining main task with auxiliary task to obtain better representations. The third step is to recommend items via multi-hop neighborhood inference.

**User and Item Representation Encoding.** To encode users and items in each domain, deep neural networks (DNNs) are utilized as encoders to extract information. Note that the DNN model can be replaced by any other kinds of models which can learn the representation well. Different from many EMCDR-based methods, we aim to optimize representation learning by proposing a novel contrastive loss function that minimizes the distance between positive samples while maximizing the distance between negative samples. We regard the items that a user has interacted with as positive samples for that user, and otherwise as negative samples. Compared with SSCDR [10], there is no need in CLCDR to add the unit sphere constraint to prevent user and item vectors from spreading too widely. The proposed contrastive loss function is formulated as follows:

$$\mathcal{L} = \sum_{\substack{u_i \in \mathcal{U}^O \\ v_j \in M(u_i) \\ v_k \in \neg M(u_i)}} -\frac{1}{n} \log \frac{\exp\left[s\left(u_i, v_j\right)/\tau\right]}{\exp\left[s\left(u_i, v_j\right)/\tau\right] + \exp\left[s\left(u_i, v_k\right)/\tau\right]}, \quad (1)$$

where $n$ is the number of training examples, $\tau > 0$ is a tunable temperature hyperparamete, and $s(a, b)$ is the cosine similarity between vector $a$ and $b$, e.g., $s(u_i, v_j)$ can be calculated as follows:

$$s\left(u_i, v_j\right) = \frac{\langle u_i, v_j \rangle}{\|u_i\| \cdot \|v_j\|}. \quad (2)$$

**Learning the Cross-Domain Mapping Function for Knowledge Transfer.** The second step of CLCDR is to transfer the knowledge from the source domain to the target domain by learning a cross-domain mapping function. Unlike EMCDR [8] only exploits the information of overlapping users by formulating a supervised problem, CLCDR accounts for both overlapping users and user-item interactions. To this end, we design two specific tasks, namely the main task and the auxiliary task.

The main task is designed to minimize the distance of overlapping users between two domains. Specifically, the objective of the main task is to learn a cross-domain mapping function such that the representations of overlapping users in the source domain can, after mapping, be close to their corresponding representations in the target domain.

The loss function is formulated as follows.

$$\mathcal{L}_{Main} = -\sum_{u_i \in \mathcal{U}^O} s\left(f_\theta\left(u_i\right), u_i\right). \tag{3}$$

where $f_\theta\left(\cdot\right)$ is the cross-domain mapping function parameterized by $\theta$.

According to collaborative filtering [18], users who have interacted with the same item (i.e., users with similar preferences) in the source domain tend to share the preferences in the target domain to some extent. Therefore, we set up an auxiliary task to extend our model to scenarios with fewer overlapping users by additionally transferring user-item interaction information. To this end, we propose a contrastive loss to extract user-item interaction knowledge from the source domain to the target domain, in which the set of items $(M(u_i^X))$ that one user $u_i^X$ (simplified as $u_i$ if it is an overlapping user) has interacted with in the source domain is considered as the positive sample set for this user, while the others $(\neg M(u_i^X))$ are considered as the negative sample set. Specifically, the proposed contrast loss is given the following definition.

$$\mathcal{L}_{CL} = \sum_{\substack{u_i \in \mathcal{U}^O \\ v_j^X \in M(u_i) \\ v_k^X \in \neg M(u_i)}} -\frac{1}{n} \log \frac{\exp\left[s\left(f_\theta\left(v_j^X\right), u_i\right)/\tau\right]}{\exp\left[s\left(f_\theta\left(v_j^X\right), u_i\right)/\tau\right] + \exp\left[s\left(f_\theta\left(v_k^X\right), u_i\right)/\tau\right]}. \tag{4}$$

Combining the two losses together and adjusting the relative importance of the two tasks by hyperparameter $\lambda$, the complete loss function used in the second step is defined as follows.

$$\mathcal{L} = \mathcal{L}_{Main} + \lambda \cdot \mathcal{L}_{CL}. \tag{5}$$

**Top-$N$ Items Recommendation.** Following the Multi-hop Neighborhood Inference method proposed in SSCDR [10], which can fully utilize the information of users' interacted items and items' interacted users, we first get the $H$-th aggregated representation of cold-start users through aggregating their neighbors' representation in the source domain. Given the initial representation

**Algorithm 1.** The *Contrastive Learning for Cross-Domain Recommendation (CLCDR)* Framework

---

**Require:** Given the user and item sets of each domain, $\mathcal{U}^X, \mathcal{U}^Y, \mathcal{V}^X, \mathcal{V}^Y$.
Given the binary matrix of user-item interaction, $R^X, R^Y$.
Given the overlapping user set, $U^O$.
**Ensure:** Recommend Top-$N$ items to the cold-start users.
**Step 1:** User and Item Representation Encoding
1: Train a source-domain model to get representation of $u^X$ and $v^X$ with Eq.(1).
2: Train a target-domain model to get representation of $u^Y$ and $v^Y$ with Eq.(1).
**Step 2:** Learning the Cross-domain Mapping function for Knowledge Transfer.
3: Learn a cross-domain mapping function $f_\theta(\cdot)$ parameterized by $\theta$ with Eq.(4).
**Step 3:** Top-$N$ Items Recommendation
4: Calculate the $H$-th aggregated representation of $u_i^{X,H}/v_j^{X,H}$ with Eq.(6)/(7).
5: Map the aggregated representation of $u_i^{X,H}$ through $f_\theta(\cdot)$ with Eq.(8).
6: Get Top-$N$ Recommendation items.

---

as $u_i^{X,0} = u_i^X, v_j^{X,0} = v_j^X$ ($\forall u_i^X \in \mathcal{U}^X$ and $v_j^X \in \mathcal{V}^X$), the $h$-hop ($h \in \{1, \cdots, H\}$) aggregated representation of a user ($u_i^{X,h}$) and an item ($v_i^{X,h}$) can be calculated by iteration as follows.

$$u_i^{X,h} = \frac{1}{|M(u_i^X)| + 1}\left(u_i^{X,h-1} + \sum_{v_j^X \in M(u_i^X)} v_j^{X,h-1}\right), \tag{6}$$

$$v_j^{X,h} = \frac{1}{|M(v_j^X)| + 1}\left(v_j^{X,h-1} + \sum_{u_i^X \in M(v_j^X)} u_i^{X,h-1}\right), \tag{7}$$

where $|\cdot|$ is the number of elements in a given set.

Then, the practical representation of cold-start users in the target domain is obtained with the cross-domain mapping function as follows.

$$\hat{u}_i^Y = f_\theta\left(u_i^{X,H}\right) \tag{8}$$

Finally, we obtain the top-$N$ nearest items for a cold-user $u_i^Y$, i.e., the top-$N$ recommendation results, by calculating the cosine similarity between $\hat{u}_i^Y$ and each $v_j \in V^Y$ in the target domain.

## 3   Experiments

Extensive experiments are conducted to answer the following questions:
**Q1:** How does our CLCDR perform compared with the competitive baselines?
**Q2:** How does the number of overlapping users impact the model performance?
**Q3:** How does the contrastive loss affect the performance?

## 3.1  Experimental Settings

**Datasets.** To evaluate our approach, we conduct experiments on Amazon datasets, which is the largest public CDR dataset. And we define three CDR scenarios as *scenario 1*: Book → Movie, *scenario 2*: Book → Music, *scenario 3*: Movie → Music. The detailed statistics of each domain are listed in Table 1.

**Table 1.** Statistics of cross-domain scenarios (Overlap. denotes overlapping users).

| Datasets | #Users | #Items | #Interactions | Density | #Overlap. |
|---|---|---|---|---|---|
| Book Movie | 12,0918,231 | 40,12219,890 | 583,801357,656 | 0.12%0.22% | 5,644 |
| BookMusic | 11,4206,031 | 38,34112,501 | 566,309107,376 | 0.13%0.14% | 3,921 |
| MovieMusic | 8,6805,425 | 20,45811,539 | 360,231102,213 | 0.20%0.16% | 2,374 |

**Baselines.** In our experiments, we compare with the following baselines.

– **CMF** [4]: CMF generalizes Matrix Factorization to cross-domain problems by sharing the user latent factors and factorizing the joint rating matrix.
– **BPR** [19]: BPR model the latent vector by pairwise ranking loss, which optimizes the order of the inner product of user and item latent vectors.
– **EMCDR** [8]: EMCDR is a widely used CDR framework. It first learns user and item representations, and then uses a network to bridge the representations from the source domain to the target domain.
– **SSCDR** [10]: SSCDR is a self-supervised bridge-based method that gets the final item list by multi-hop neighborhood inference. This method combines the EMCDR method with CML [20]. And CML is the state-of-the-art collaborative filtering method.
– **DAN** [11]: DAN captures high-order relationships to learn user preferences by utilizing the user-item interaction graph end-to-end.
– **TMCDR** [14]: TMCDR is the state-of-the-art cross-domain recommendation method for cold-start users.

Note that, several approaches are proposed to recommend items to users who have interactions in the target domain [9,21]. They are not included in baselines because they can not recommend items to cold-start users. In addition, CDR methods utilizing side information such as review texts and item details [23] are also not included in our baselines as they require more types of data to enhance the performance. And graph-based methods are also not included.

**Evaluation Metrics.** To evaluate the performance of our CLCDR, we use three general metrics: AUC, HR and NDCG, which are widely used in the Top-$N$ recommendation task [6–8,22].

**Implementation Details.** We convert explicit data to implicit data (i.e., convert rating data to binary data). In each scenario, 50% of the overlapping users are randomly selected, and their interaction information is removed in the target domain to regard these users as cold-start users for evaluating the model's performance. To study the performance of CLCDR with respect to the number of overlapping users, we extract four training sets with a certain proportion $\eta \in \{5\%, 20\%, 50\%, 100\%\}$ of overlapping users that are not the test users.

## 3.2 Experimental Results and Analysis

Table 2 shows the model performance with four different proportions $\eta$ of overlapping users on three scenarios. We evaluate the performance of recommendation models by HR@10, HR@20, NDCG@10, NDCG@20. In summary, our approach CLCDR obtains the best performance compared with the baselines for all scenarios and all $\eta$. We analyze the results from several perspectives.

**Cold-Start Experiments (Q1).** We evaluate the effectiveness of CLCDR on three scenarios, comparing with the existing cross-domain recommendation methods. The experimental results are shown in the Table 2, and the best performance is shown in boldface. From the experiments, our approach CLCDR performs best. We have several findings: (1) The CMF performs worst in scenarios 1 and 3 because it combines data from different domains by regarding the data from the different domains as the same, which ignores the potential domain knowledge transferring. Our method transfers the representation of user and item from the source domain to the target domain, which can fully use the source domain knowledge and alleviate the influence of domain shift. (2) We find that CLCDR could outperform the EMCDR. The EMCDR only transfers the overlapping users' knowledge and ignores the user-item interaction. While our approach CLCDR considers user-user and user-item interaction, which can transfer more knowledge from the source domain to get a better recommendation. (3) We can also find CLCDR could outperform the SSCDR, which uses Euclidean distance to reconstruct latent space. While in our method, we design a novel contrastive loss with excellent properties of uniformity and alignment that can effectively calibrate the representation of the target domain. (4) Our approach CLCDR could also outperform the state-of-the-art method TMCDR, because we design an effective contrastive loss and auxiliary task to better transfer the domain knowledge. In summary, CLCDR is effective and performs very well for cold-start user recommendations.

**The Impact of Overlapping Users (Q2).** In the Table 2, we find that the performance of baselines is affected by the number of overlapping users. When there are few overlapping users, that is $\eta < 20\%$, the performance of baselines deteriorates as the number of overlapping users decreases. Especially the models which only transfer the information of overlapping users from the source domain to the target domain. According to the statistic of [10], the average number

**Table 2.** Performance comparison on three scenarios. We report the mean result over ten runs with outliers removed. CMF, BPR and EMCDR do not utilize the user-item interaction knowledge while others do. Best results are in boldface. Best baseline is underlined. *Improve.* denotes relative improvement over the best baseline.

| | $\eta$ | Metrics | CMF | BPR | EMCDR | SSCDR | DAN | TMCDR | CLCDR | Improve. |
|---|---|---|---|---|---|---|---|---|---|---|
| Scenario 1: | 5% | HR@10 | 0.1421 | 0.1480 | 0.1563 | 0.1849 | 0.1924 | <u>0.2010</u> | **0.2135** | 6.22% |
| Book → Movie | | HR@20 | 0.1862 | 0.1902 | 0.2037 | 0.2269 | 0.2344 | <u>0.2373</u> | **0.2441** | 2.87% |
| | | NDCG@10 | 0.1274 | 0.1295 | 0.1346 | 0.1577 | 0.1620 | <u>0.1689</u> | **0.1840** | 8.94% |
| | | NDCG@20 | 0.1546 | 0.1597 | 0.1683 | 0.1870 | 0.1921 | <u>0.1962</u> | **0.2107** | 7.39% |
| | 20% | HR@10 | 0.1447 | 0.1509 | 0.1585 | 0.1873 | 0.1989 | <u>0.2062</u> | **0.2201** | 6.74% |
| | | HR@20 | 0.1900 | 0.1941 | 0.2082 | 0.2304 | 0.2373 | <u>0.2398</u> | **0.2474** | 3.17% |
| | | NDCG@10 | 0.1316 | 0.1332 | 0.1407 | 0.1624 | 0.1665 | <u>0.1733</u> | **0.1871** | 7.96% |
| | | NDCG@20 | 0.1587 | 0.1648 | 0.1717 | 0.1900 | 0.1978 | <u>0.2015</u> | **0.2182** | 8.29% |
| | 50% | HR@10 | 0.1621 | 0.1628 | 0.1699 | 0.1918 | 0.2022 | <u>0.2111</u> | **0.2286** | 8.29% |
| | | HR@20 | 0.2011 | 0.2131 | 0.2289 | 0.2467 | 0.2491 | <u>0.2503</u> | **0.2647** | 5.75% |
| | | NDCG@10 | 0.1445 | 0.1481 | 0.1535 | 0.1782 | 0.1810 | <u>0.1821</u> | **0.1903** | 4.50% |
| | | NDCG@20 | 0.1647 | 0.1692 | 0.1796 | 0.1931 | 0.2027 | <u>0.2044</u> | **0.2258** | 10.47% |
| | 100% | HR@10 | 0.1737 | 0.1748 | 0.1821 | 0.2035 | 0.2100 | <u>0.2238</u> | **0.2402** | 7.33% |
| | | HR@20 | 0.2108 | 0.2188 | 0.2354 | 0.2503 | 0.2532 | <u>0.2564</u> | **0.2733** | 6.59% |
| | | NDCG@10 | 0.1501 | 0.1537 | 0.1623 | 0.1838 | <u>0.1862</u> | 0.1857 | **0.1939** | 4.14% |
| | | NDCG@20 | 0.1695 | 0.1716 | 0.1864 | 0.2062 | <u>0.2168</u> | 0.2147 | **0.2325** | 7.24% |
| Scenario 2: | 5% | HR@10 | 0.1282 | 0.1275 | 0.1310 | 0.1332 | 0.1338 | <u>0.1352</u> | **0.1535** | 13.54% |
| Book → Music | | HR@20 | 0.1721 | 0.1680 | 0.1803 | 0.1891 | 0.1879 | <u>0.1892</u> | **0.2004** | 8.29% |
| | | NDCG@10 | 0.0910 | 0.0913 | 0.1045 | 0.1137 | <u>0.1146</u> | 0.1145 | **0.1226** | 6.98% |
| | | NDCG@20 | 0.1131 | 0.1168 | 0.1238 | 0.1368 | 0.1373 | <u>0.1385</u> | **0.1503** | 8.52% |
| | 20% | HR@10 | 0.1301 | 0.1286 | 0.1344 | 0.1368 | 0.1374 | <u>0.1401</u> | **0.1594** | 13.78% |
| | | HR@20 | 0.1797 | 0.1724 | 0.1848 | 0.1967 | <u>0.1979</u> | 0.1977 | **0.2131** | 7.68% |
| | | NDCG@10 | 0.0995 | 0.0982 | 0.1138 | 0.1158 | 0.1155 | <u>0.1187</u> | **0.1269** | 6.91% |
| | | NDCG@20 | 0.1171 | 0.1195 | 0.1341 | 0.1452 | 0.1471 | <u>0.1484</u> | **0.1551** | 4.51% |
| | 50% | HR@10 | 0.1344 | 0.1301 | 0.1487 | 0.1501 | 0.1491 | <u>0.1510</u> | **0.1620** | 7.28% |
| | | HR@20 | 0.1805 | 0.1762 | 0.1883 | 0.2050 | 0.2072 | <u>0.2083</u> | **0.2187** | 4.99% |
| | | NDCG@10 | 0.1001 | 0.1003 | 0.1174 | 0.1224 | 0.1230 | <u>0.1254</u> | **0.1361** | 8.53% |
| | | NDCG@20 | 0.1258 | 0.1261 | 0.1502 | 0.1512 | 0.1539 | <u>0.1547</u> | **0.1631** | 5.43% |
| | 100% | HR@10 | 0.1387 | 0.1323 | 0.1525 | 0.1611 | 0.1603 | <u>0.1623</u> | **0.1689** | 4.07% |
| | | HR@20 | 0.1854 | 0.1793 | 0.1902 | 0.2109 | 0.2123 | <u>0.2141</u> | **0.2234** | 4.34% |
| | | NDCG@10 | 0.1099 | 0.1108 | 0.1406 | 0.1428 | 0.1433 | <u>0.1457</u> | **0.1476** | 1.30% |
| | | NDCG@20 | 0.1291 | 0.1303 | 0.1582 | 0.1633 | 0.1668 | <u>0.1679</u> | **0.1720** | 2.44% |
| Scenario 3: | 5% | HR@10 | 0.0921 | 0.1010 | 0.1075 | 0.1132 | 0.1101 | <u>0.1172</u> | **0.1219** | 4.01% |
| Movie → Music | | HR@20 | 0.1438 | 0.1492 | 0.1580 | 0.1645 | 0.1611 | <u>0.1675</u> | **0.1734** | 3.52% |
| | | NDCG@10 | 0.0770 | 0.0621 | 0.0839 | 0.0911 | 0.0910 | <u>0.0924</u> | **0.1001** | 8.33% |
| | | NDCG@20 | 0.0811 | 0.0663 | 0.0909 | 0.1007 | 0.1081 | <u>0.1100</u> | **0.1197** | 8.82% |
| | 20% | HR@10 | 0.0946 | 0.1021 | 0.1121 | 0.1201 | 0.1166 | <u>0.1199</u> | **0.1270** | 5.92% |
| | | HR@20 | 0.1471 | 0.1517 | 0.1601 | 0.1682 | 0.1673 | <u>0.1691</u> | **0.1785** | 5.56% |
| | | NDCG@10 | 0.0781 | 0.0628 | 0.0866 | 0.1024 | 0.1031 | <u>0.1044</u> | **0.1091** | 4.50% |
| | | NDCG@20 | 0.0839 | 0.0670 | 0.0921 | 0.1139 | 0.1152 | <u>0.1168</u> | **0.1231** | 5.39% |
| | 50% | HR@10 | 0.0967 | 0.1038 | 0.1219 | 0.1298 | 0.1317 | <u>0.1321</u> | **0.1475** | 11.66% |
| | | HR@20 | 0.1481 | 0.1547 | 0.1681 | 0.1772 | 0.1799 | <u>0.1840</u> | **0.1996** | 8.48% |
| | | NDCG@10 | 0.0794 | 0.0635 | 0.0929 | 0.1064 | 0.1085 | <u>0.1088</u> | **0.1128** | 3.68% |
| | | NDCG@20 | 0.0845 | 0.0681 | 0.0976 | 0.1202 | 0.1210 | <u>0.1247</u> | **0.1284** | 2.97% |
| | 100% | HR@10 | 0.0983 | 0.1065 | 0.1375 | 0.1411 | 0.1458 | <u>0.1462</u> | **0.1583** | 8.28% |
| | | HR@20 | 0.1502 | 0.1565 | 0.1796 | 0.1888 | 0.1902 | <u>0.1910</u> | **0.2067** | 4.50% |
| | | NDCG@10 | 0.0810 | 0.0711 | 0.1138 | 0.1205 | 0.1214 | <u>0.1249</u> | **0.1288** | 3.12% |
| | | NDCG@20 | 0.0866 | 0.0825 | 0.1222 | 0.1365 | 0.1350 | <u>0.1377</u> | **0.1425** | 3.48% |

of overlapping users is less than 20%, so it is necessary to try to transfer other knowledge. In our approach, we not only transfer the overlapping user knowledge but also transfer the user-item interaction by designing two kinds of more effective loss inspired by contrastive learning. One is for inner-domain representation. Another one is for cross-domain calibration.

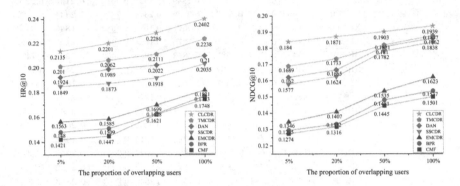

**Fig. 2.** The impact of the proportion $\eta$ of the overlapping users on scenario 1. (left: HR@10, right: NDCG@10)

The Fig. 2 shows the impact of the proportion $\eta$ of overlapping users on the model. Here we only compare with CMF, BPR and EMCDR, which only model overlapping user information, to verify that it is practical to model the user-item interaction. We can see that our approach CLCDR and other approach (SSCDR, DAN, TMCDR), which also transfer the knowledge of user-item interaction, work much better than other models, and the performance is smoother. Even with very few overlapping users, our model still works very well.

**The Ablation Study of Contrastive Learning (Q3).** There are many EMDCR based methods that focus on the mapping function itself. While the performance to some extent may rely on the encoding step. Many researchers apply their methods upon MF to conduct experiments. While in the real-world recommendations, the MF is hard to achieve excellent performance by just modeling the user-item interaction as the inner product. Thus, we use a neural-based model to replace the MF and design the contrastive loss to better model the representations. Contrastive loss is to minimize the distance between positive pairs, and maximize the distance between negative pairs. And this loss function can be applied in other networks, which means our approach can be applied upon various models, e.g., YouTube DNN, DSSM.

Figure 3 shows the ablation study of the contrastive learning. In our representation and calibration step, we use MF to replace the contrastive learning, and the performance of "without CL" is shown as a blue one. The purple one is the AUC of our approach CLCDR which is "with CL". The result verifies the effectiveness of the contrastive learning.

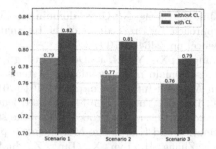

**Fig. 3.** The ablation study (AUC) of w/o contrastive learning on three scenarios.

## 4 Conclusion

In this paper, we studied the cross-domain recommendation (CDR) for cold-start users. Many existing EMCDR-based methods focus on learning a better mapping function. However, the representation quality plays a nontrivial role in the ultimate recommendation performance. Thus, we propose a novel approach called CLCDR, to effectively retrieve the top-$N$ items to cold-start users. The main contribution comes from enhancing the qualities of user and item representations by domain-specific contrastive loss and inter-domain calibration loss. We empirically demonstrated that our CLCDR learns the cross-domain knowledge more accurately and is practical even with few overlapping users according to the comprehensive experiments on the Amazon dataset.

## References

1. Yue, X., et al.: Prototypical cross-domain self-supervised learning for few-shot unsupervised domain adaptation. In: Proceedings of the IEEE/CVF Conference on Computer Vision and Pattern Recognition, pp. 13834–13844 (2021)
2. Li, J., Jing, M., Lu, K., Zhu, L., Yang, Y., Huang, Z.: From zero-shot learning to cold-start recommendation. In: Proceedings of the AAAI Conference on Artificial Intelligence, pp. 4189–4196 (2019)
3. Wu, T., Chio, E.K.I., Cheng, H.T., Du, Y.: Zero-shot heterogeneous transfer learning from recommender systems to cold-start search retrieval. In: Proceedings of the Conference on Information & Knowledge Management, pp. 2821–2828 (2010)
4. Singh, A.P., Gordon, G.J.: Relational learning via collective matrix factorization. In: Proceedings of the 14th ACM SIGKDD International Conference on Knowledge Discovery and Data Mining, pp. 650–658 (2008)
5. Mirbakhsh, N., Ling, C.X.: Improving top-n recommendation for cold-start users via cross-domain information. In: ACM Transactions on Knowledge Discovery from Data (TKDD), pp. 1–19 (2015)
6. Gao, C., et al.: Cross-domain recommendation without sharing user-relevant data. In: The World Wide Web Conference, pp. 491–502 (2019)
7. Hu, G., Zhang, Y., Yang, Q.: Conet: collaborative cross networks for cross-domain recommendation. In: Proceedings of the 27th ACM International Conference on Information and Knowledge Management, pp. 667–676 (2018)

8. Man, T., Shen, H., Jin, X., Cheng, X.: Cross-domain recommendation: an embedding and mapping approach. In: Proceedings of the International Joint Conference on Artificial Intelligence, pp. 2464–2470 (2017)

9. Fu, W., Peng, Z., Wang, S., Xu, Y., Li, J.: Deeply fusing reviews and contents for cold start users in cross-domain recommendation systems. In: Proceedings of the AAAI Conference on Artificial Intelligence, pp. 94–101 (2019)

10. Kang, S., Hwang, J., Lee, D., Yu, H.: Semi-supervised learning for cross-domain recommendation to cold-start users. In: Proceedings of the 28th ACM International Conference on Information and Knowledge Management, pp. 1563–1572 (2019)

11. Wang, B., Zhang, C., Zhang, H., Lyu, X., Tang, Z.: dual autoencoder network with swap reconstruction for cold-start recommendation. In: Proceedings of the Conference on Information and Knowledge Management, pp. 2249–2252 (2020)

12. Zhao, C., Li, C., Xiao, R., Deng, H., Sun, A.: CATN: cross-domain recommendation for cold-start users via aspect transfer network. In: Proceedings of the 43rd International ACM SIGIR Conference on Research and Development in Information Retrieval, pp. 229–238 (2020)

13. Ni, J., Li, J., McAuley, J.: justifying recommendations using distantly-labeled reviews and fine-grained aspects. In: Proceedings of the Conference on Empirical Methods in Natural Language Processing (EMNLP-IJCNLP), pp. 188–197 (2019)

14. Zhu, Y., et al.: Transfer-meta framework for cross-domain recommendation to cold-start users. In: Proceedings of the 44th International ACM SIGIR Conference on Research and Development in Information Retrieval, pp. 1813–1817 (2021)

15. Chen, T., Kornblith, S., Norouzi, M., Hinton, G.: A simple framework for contrastive learning of visual representations. In: International Conference on Machine Learning, pp. 1597–1607 (2020)

16. Chen, X., He, K.: Exploring simple siamese representation learning. In: Proceedings of the IEEE/CVF Conference on Computer Vision and Pattern Recognition, pp. 15750–15758 (2021)

17. He, K., Fan, H., Wu, Y., Xie, S., Girshick, R.: Momentum contrast for unsupervised visual representation learning. In: Proceedings of the IEEE/CVF Conference on Computer Vision and Pattern Recognition, pp. 9729–9738 (2020)

18. Schafer, J.B., Frankowski, D., Herlocker, J., Sen, S.: collaborative filtering recommender systems. In: The Adaptive Web, pp. 291–324 (2007)

19. Rendle, S., Freudenthaler, C., Gantner, Z., Schmidt-Thieme, L.: BPR: bayesian personalized ranking from implicit feedback. In: Proceedings of the Twenty-Fifth Conference on Uncertainty in Artificial Intelligence, pp. 452–461 (2009)

20. Hsieh, C.K., Yang, L., Cui, Y., Lin, T.Y., Belongie, S., Estrin, D.: Collaborative Metric Learning. In: The World Wide Web Conference, pp. 193–201 (2017)

21. He, J., Liu, R., Fuzhen, Z., Fen, L., Qing, H.: A general cross-domain recommendation framework via bayesian neural network. In: IEEE International Conference on Data Mining (ICDM) (2018)

22. He, X., Liao, L., Zhang, H., Nie, L., Hu, X., Chua, T.S.: Neural collaborative filtering. In: The World Wide Web Conference, pp. 173–182 (2017)

23. Zhu, F., Wang, Y., Chen, C., Liu, G., Zheng, X.: A graphical and attentional framework for dual-target cross-domain recommendation. In: Proceedings of the International Joint Conference on Artificial Intelligence, pp. 3001–3008 (2020)

# Medical Visual Question Answering via Targeted Choice Contrast and Multimodal Entity Matching

Hui Guo, Lei Liu, Xiangdong Su[✉], and Haoran Zhang

College of Computer Science, Inner Mongolia University, Hohhot, China
cssxd@imu.edu.cn

**Abstract.** Although current methods have advanced the development of medical visual question answering (Med-VQA) task, two aspects remain to be improved, namely extracting high-level medical visual features from small-scale data and exploiting external knowledge. To strengt-hen the performance of Med-VQA, we propose a **pre-training** model called **T**argeted **C**hoice **C**ontrast (TCC) and a **M**ultimodal **E**ntity **M**atc-hing (MEM) module, and integrate them into an end-to-end framework. Specifically, the TCC model extracts deep visual features on the small-scale medical dataset by contrastive learning. It improves model robustness by a targeted selection of negative samples. The MEM module is dedicated to embedding knowledge representation into the framework more accurately. Besides, we apply a mixup strategy for data augmentation during the framework training process to make full use of the small-scale images. Experimental results demonstrate our framework outperforms state-of-the-art methods.

**Keywords:** Medical Visual Question Answering · Contrastive Learning · Multimodal Entity Matching · Knowledge Graph

## 1 Introduction

Medical visual question answering (Med-VQA) aims to answer the clinical questions based on the visual information of medical images. Currently, most Med-VQA methods [4,7,10] leverage transfer learning to obtain better performance, where the initial weights of the visual feature extractor are derived from the pre-trained model with large-scale unannotated radiology images. Further, concerning the fact that understanding complex clinical questions depends on not only the Image-Question (I-Q) pairs but also prior knowledge, the recent approach [8] introduces the knowledge graph (KG) to improve the Med-VQA performance. These methods advance Med-VQA from the embryonic stage to the development stage.

However, there are two problems with the methods described above. The first problem is that these methods excessively rely on external large-scale medical images for pre-training. The pre-trained models MEVF [10], MTPT [4] and

---

L. Liu—Equal technical contribution.

M. Tanveer et al. (Eds.): ICONIP 2022, LNCS 13624, pp. 343–354, 2023.
https://doi.org/10.1007/978-3-031-30108-7_29

CPRD [7] capture visual features from 11,779, 11,014 and 22,995 unlabeled images, respectively. In fact, it is very costly and time-consuming to obtain large-scale medical image data. Besides, these approaches require that the data for the downstream tasks are similar in statistical distribution to most of the data used in the pre-training phase [20]. Otherwise, their performance will be degraded. Therefore, it is necessary to develop a pre-training model to mine deeper visual information using existing small-scale data as much as possible.

The second problem is that how to accurately embed the knowledge from the KG into the Med-VQA framework is under-explored. To our best knowledge, only Liu et al. [8] propose a pipeline method to embed the external knowledge, which introduces irrelevant information in some cases. This irrelevant knowledge misleads the generation of correct answer, thereby degrading the performance of model. We take the question **"which organs/organ in the picture belong to the respiratory system"** in Fig. 1 as an example to describe their approach. First, they use two LSTMs to predict the "tail" entity, "relation" of the question, and obtain **"respiratory system"** and **"belong to"**. Second, they combine the "tail" entity, "relation" through TransE [1] method to get the question-related "head" entity, which is the **"larynx"** entity. Finally, the **"larynx"** embedding and the I-Q pair fusion information are combined to get the answer. However, there are multiple "head" entities obtained through the "tail" entity "respiratory system" and the relation "belong to", including **"lung"**, **"larynx"**, **"pharynx"** et. Therefore, the way that Liu et al. [8] only utilize the question information to obtain the "head" entity is inaccurate. A better way to get the "head" entity is to combine multimodal information in a matching manner.

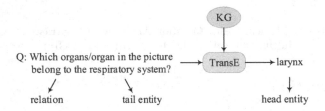

**Fig. 1.** An example of embedding knowledge from the knowledge graph. The embedding of each entity in the KG is reinforced by the TransE [1] method.

In this paper, we propose the following measures to address the two problems above. For the **first** problem, inspired by the superior ability of contrastive learning in visual extraction, we propose a **pre-training** model called Targeted Choice Contrast (TCC). It captures high-level visual features on small-scale data through contrastive learning. Specifically, the main difference between TCC and existing contrastive learning methods [2,3,5] is the selection of negative samples. When images most similar to the positive sample are used for the negative samples, it is harder for the model to distinguish between positive and negative samples since the same domains should be more similar, resulting in a more robust model [13]. Considering the existence of various domains in medical images, such as different body parts and modalities (CT, MRI, X-rays), in order to identify the more difficult negative samples, we choose samples that are identical to the positive samples in modalities and body parts as the negative samples. Notably, we only use the **450** images in SLAKE [8] to train the TCC model. For the **second** problem, we design a Multimodal Entity Matching (MEM) module which combines semantic and visual information to match the most relevant "head" entity, helping to predict answers more accurately. We integrate TCC and MEM into our Med-VQA framework to enhance the robustness and accuracy of the framework. In addition, we apply the mixup [19] technology in our framework training to exploit small-scale data where possible. Our contributions in this work are as follows:

- A effective pre-training model TCC dedicates to mining high-level visual features from small-scale data by contrastive learning. It strengthens the robustness of the model with targeted selection of negative samples.
- A novel MEM module combines multimodal information to obtain the "head" entity from the KG that is most relevant to the I-Q pair. This representation of knowledge grants positive guidance to the framework.
- Our framework outperforms the state-of-the-art baselines on the SLAKE [8] benchmark dataset. Figure 5 provides visualization examples showing our framework can predict the answer more accurately by locating and identifying the special region associated with the question.

## 2   Method

### 2.1   The Proposed Med-VQA Framework

In the proposed Med-VQA framework shown in Fig. 2. First, to extract medical visual features $F_v$, every input image is passed through the ResNet-50 (initialization weights from TCC), forming the 256-D enhanced image feature $F_v$. Further, each input question is trimmed into a 20-word sentence, and each word is represented a 300-D vector by a pre-trained GloVe [12]. And the question is zero-padded in case its length is less than 20. Then the word embeddings are fed into a 1024-D LSTM to generate the question embedding $F_q$.

Second, the MEM module is introduced to obtain the "head" entity embedding $\hat{e}$ from KG that is most relevant to the multimodal representation $F_a$. Within the MEM module, the image features $F_v$ and question embedding $F_q$ are fed to two fully connected layers, forming two different length vectors $F_v'$ and $F_q'$, respectively. Then, $F_v'$ and $F_q'$ are passed into BAN [6] to generate the multimodal representation $F_a$. In the end, we compute the relevance scores between the $F_a$ and the embedding $R_e$ of each "head" entity found from KG, and select the most relevant "head" entity embedding $\hat{e}$.

Finally, to predict the answer with the image feature, question embedding and the most relevant "head" entity embedding, the image features $F_v$ and question embedding $F_q$ are fed into another BAN [6] to obtain the joint multimodal

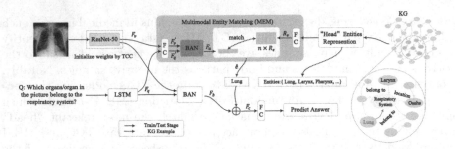

**Fig. 2.** Overview of our proposed framework

representation $F_b$. It is worth noting that both BANs in our proposed framework have the same structure and different parameters. The embedding of the most relevant "head" entity $\hat{e}$ and $F_b$ are summed to obtain the feature vector $F_c$, which is used as the input of the answer predictor. We employ the cross-entropy loss as the classification loss of Med-VQA.

## 2.2 Targeted Choice Contrast (TCC)

In order to learn high-level visual features extraction from small-scale images, we design the pre-training TCC model using contrastive learning. Considering the images most similar to the positive samples for the negative samples will enhance the robustness of the model and improve its ability to mine visual information [13], in order to identify the more difficult negative samples, we choose samples that are identical to the positive samples in modalities and body parts as the negative samples. Notably, we only use the 450 images in SLAKE [8] to train the TCC. Its structure is shown in Fig. 3a.

First, we randomly sample an image $v_{bm}$ labeled with a given body part $b$ and modality $m$, and $N$ images with the same body part $b$ and modality $m$. These $N$ images form the list $l = \left\{ v_j^- \right\}_{j=1}^N$. $N$ equals to the batch size subtracting 1. Second, we employ the data augmentation (denoted as $Aug$) on $v_{bm}$ twice and $l$ once, respectively. The augmented views from $v_{bm}$ and $l$ are as

$$\hat{v_{bm}} = Aug(v_{bm}), \; \hat{v_{bm}}^+ = Aug(v_{bm}), \; \hat{l} = \left\{ \hat{v}_j^- = Aug\left(v_j^-\right) \right\}_{j=1}^N \quad (1)$$

where, $\hat{v_{bm}}^+$ considered positive sample for $\hat{v_{bm}}$, $\hat{l}$ considered negative samples for $\hat{v_{bm}}$. Next, the first view $\hat{v_{bm}}$ is passed through the encoder $E_q$ to extract visual features $f_{bm} = E_q(\hat{v_{bm}})$. The other view $\hat{v_{bm}}^+$ and the $\hat{l}$ are fed to the encoder $E_k$ to produce representations $f_{bm}^+$ and $f^- = \left\{ f_1^-, f_2^-, \ldots, f_N^- \right\}$. Finally, since $f_{bm}$ and $f_{bm}^+$ come from different view features for $v_{bm}$, $f_{bm}$ should be similar to $f_{bm}^+$, but dissimilar to the other $N$ representations in $f^-$.

This learning process can be guided with the contrastive loss InfoNCE [11]:

$$\mathcal{L}_{f_{bm},f_{bm}^+,\{f_j^-\}} = -\log \frac{\exp\left(f_{bm} \cdot f_{bm}^+ / \tau\right)}{\exp\left(f_{bm} \cdot f_{bm}^+ / \tau\right) + \sum_{j=1}^{N} \exp\left(f_{bm} \cdot f_j^- / \tau\right)} \tag{2}$$

where $\tau$ is a temperature parameter and $\cdot$ is dot product.

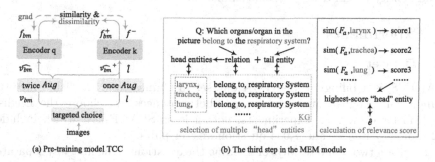

(a) Pre-training model TCC

(b) The third step in the MEM module

**Fig. 3.** (a) Our proposed TCC. (b) The third step in the MEM module.

## 2.3  Multimodal Entity Matching (MEM)

To extract more accurate information from the KG, we propose the MEM module which exploits multimodal information to obtain the "head" entity from the KG that is most relevant to the I-Q pair. This module takes the following three steps to find the most relevant "head" entity, unlike the method [8] in Fig 1.

In the first step, in order to learn the implicit semantic representation of images and questions, we use fully connected layers to encode the image features $F_v$ and question embedding $F_q$, respectively. We denote the two resulting vectors of different lengths as $F_v'$ and $F_q'$. In the second step, to utilize given vision-language information seamlessly, we employ BAN [6] to fuse $F_v'$ and $F_q'$ to generate the joint representation $F_a$. In the third step, we calculate the relevance scores between $F_a$ and multiple "head" entities embeddings found by KG, and choose the highest-score "head" entity embeddings $\hat{e}$.

The third step is extremely important as it determines the accuracy of the entity matching. The third step of the MEM module is shown in Fig. 3b. It contains two parts: the selection of multiple "head" entities and the calculation of relevance score. For the first part, the extraction of the "tail" entity and "relation" in the question, we follow the method in QA-GNN [16]. After that, we combine the "tail" entity with the "relation" to find the multiple "head" entities from the KG. For the second part, to obtain the most relevant "head" entity to $F_a$ from multiple "head" entities, we use cosine similarity as a method of calculating the relevance score after the ablation experiment in Sect 3.4. The embedding of each "head" entity and $F_a$ calculate the cosine similarity to obtain the relevance score. We select the "head" entity embedding with the highest-score, and denote it as $\hat{e}$. The calculation is as follows:

$$\hat{e} = \begin{cases} \underset{e \in \mathbb{G}}{\arg\max} \left( \text{sim} \left( F_a, R_e \right) \right) & \text{if } \mathbb{G} \neq \varnothing \\ 0 & \text{else} \end{cases} \tag{3}$$

where $\mathbb{G}$ denotes the multiple "head" entities set, $R_e$ denotes the embedding of "head" entity $e$, sim deontes cosine similarity.

## 3   Experiments

### 3.1   Dataset

SLAKE [8] is a bilingual Med-VQA dataset in Chinese and English. Since the Chinese and English questions are parallel sentences, we choose the English question for evaluation experiments, noting it as SLAKE-EN, which includes 642 radiology images and 7,033 question-answer pairs.

There are two different ways of dividing the questions. The first way separates the questions into "closed-ended" questions where the answers are restricted choices such as "yes", "no" and "lung", and "open-ended" questions where the answers are free-form text. The second way splits the questions into vision-only questions and knowledge-based questions based on semantic labels. We evaluate our proposed Med-VQA framework based on the first dividing way on SLAKE-EN. Besides, to validate the superiority of the MEM module, we experiment on the knowledge-based questions.

### 3.2   Experimental Setup

The experiments of the pre-training TCC are performed on a single NVIDIA Tesla P100 with 16 GB Graphic Memory. The experiments with the Med-VQA framework are conducted on two NVIDIA Tesla P100 with 16 GB Graphic Memory. We implement all models with PyTorch.

**Targeted Choice Contrast (TCC).** We use ResNet-50 followed by two fully connected layers to instantiate $E_q$ and $E_k$ (Sect 2.2) and train 200 epochs. We set the batch size as 16 in each epoch. The temperature parameter $\tau$ in Eq. 2 is set to 0.4. We optimize the model using the AdamW [9] method, with the initial value of the learning rate set to 0.01. The learning rate is dynamically adjusted by the cosine schedule. Furthermore, we use weight decay to prevent over-fitting due to small-scale data.

**The Proposed Med-VQA Framework.** The initialization weights of the visual extractor in Med-VQA are derived from the pre-trained TCC. Then, we train the whole Med-VQA framework in an end-to-end way on SLAKE [8] for 600 epochs. The visual extractor RseNet-50 is fine-tuned in this process. We employ the mixup [19] technology in the framework training stage, with the hyperparameters $\alpha$ set to 5 and $\beta$ set to 1 for the Beta$(\alpha, \beta)$ distribution. We use AdamW [9] optimizer with an initial learning rate of $5 * 10^{-4}$ for model optimization. Like the normal Med-VQA, we use accuracy to evaluate our model.

## 3.3   Comparison with the State-of-the-arts

To demonstrate the effectiveness of the proposed Med-VQA framework, we compare it with the general Med-VQA model MFB [17], SAN [15] and BAN [6]. Besides, we compare it with the latest models involving MEVF [10,18] and CPRD [7] , all of which are trained with extra radiology images. Table 1 shows the results on SLAKE-EN [8]. The initialization weights of the visual feature extractor in MFB [17], SAN [15] and BAN [6] are derived from the pre-trained model ResNet-50 in ImageNet.

**Table 1.** Test accuracy of our method and baselines.

| # | Methods | SLAKE-EN [8] | | |
|---|---------|-----------|---------|----------|
| | | Overall(%) | Open(%) | Close(%) |
| 1 | MFB fw [17] | 73.3 | 72.2 | 75.0 |
| 2 | SAN fw [15] | 76.0 | 74.0 | 79.1 |
| 3 | BAN fw [6] | 76.3 | 74.6 | 79.1 |
| 4 | **BAN fw** [6][†] | 79.9 | 76.9 | 84.6 |
| 5 | MEVF + SAN [10]* | 76.5 | 75.3 | 78.4 |
| 6 | MEVF + BAN [10]* | 78.6 | 77.8 | 79.8 |
| 7 | CPRD + BAN [7]* | 81.1 | 79.5 | 83.4 |
| 8 | **TCC + BAN (ours)** | 81.9 | 79.1 | **86.3** |
| 9 | MEVF + BAN + CR [18]* | 80.0 | 78.8 | 82.0 |
| 10 | CPRD + BAN + CR [7]* | 82.1 | **81.2** | 83.4 |
| 11 | **TCC + BAN + MEM (ours)** | **82.2** | 79.2 | **86.8** |

[1] * indicates that the method of extra medical images is used.
[2] † indicates the result of our re-implementation.

The following observations can be made from Table 1. (1) Our TCC+BAN+ MEM outperforms the state-of-the-art baselines. (2) The TCC+BAN+ MEM achieves state-of-the-art results on the "closed-ended" question with a 3.4% improvement over the latest model CPRD+BAN+CR [7]. This demonstrates that our proposed pre-training TCC model can deliver high-level visual features for the downstream task Med-VQA, and the MEM module can provide positive guidance for the Med-VQA framework.

## 3.4   Ablation Studies

In this section, we first study the contribution of the pre-training TCC model, the MEM module and the mixup technique to our framework. Next, we discuss the contribution of negative example selection in TCC and $\tau$ in Eq. 2 to the framework. Finally, we explore the impact of different relevant score calculation methods in MEM on our framework.

**Med-VQA Framework.** To demonstrate the effectiveness of our proposed framework, we conduct ablation study with the BAN (row #1–4), which is shown in Table 2.

**Table 2.** Ablation studies of mixup, TCC, MEM in Med-VQA framework.

| # | mixup | TCC | MEM | SLAKE-EN [8] Overall(%) | Open(%) | Close(%) |
|---|-------|-----|-----|------------|---------|----------|
| 1 |       |     |     | 79.9 | 76.9 | 84.6 |
| 2 | ✓ |     |     | 80.3 | 76.6 | 86.1 |
| 3 | ✓ | ✓ |     | 81.9 | 79.1 | 86.3 |
| 4 | ✓ |     | ✓ | 81.8 | 78.8 | 86.5 |
| 5 | ✓ | ✓ | ✓ | **82.2** | **79.2** | **86.8** |

In Table 2, the BAN network is set as the baseline and its accuracy is 79.9%. Comparing to the baseline, we utilize pre-trained TCC model to provide initialization weights for the visual extractor, apply MEM module to learn more accurate knowledge representation and introduce the mixup strategy into framework (row #2–5), which brings an accuracy gain of 2.3%. Bedides, **mixup** brings a 0.4% boost, which is used in **all subsequent ablation experiments**.

**Selection of Negative Samples and $\tau$ in TCC.** To obtain a better negative samples selection method and $\tau$ in Eq. 2, we perform experiments in the pre-training model TCC. Specifically, we use different negative sample selection methods and $\tau$ in TCC to obtain the pre-trained models. Then, these pre-trained models are fine-tuned and evaluated by our Med-VQA framework (MEM module is missing). The results are shown in Table 3.

**Table 3.** Selection of negative samples and $\tau$ in TCC

(a) Experiments for negative samples choice on TCC

| # | modality | body part | SLAKE-EN [8] Overall(%) | Open(%) | Close(%) |
|---|----------|-----------|------------|---------|----------|
| 1 |   |   | 80.9 | 78.1 | 85.3 |
| 2 | ✓ |   | 81.4 | 78.1 | 86.5 |
| 3 |   | ✓ | 81.1 | 78.0 | 85.8 |
| 4 | ✓ | ✓ | **81.8** | **79.1** | 85.8 |
| 5 | × |   | 81.4 | 78.0 | **86.8** |
| 6 |   | × | 80.9 | 77.7 | 85.9 |
| 7 | × | × | 81.2 | 78.0 | 86.1 |

(b) The choice of parameter $\tau$ in Eq.2

| # | t | SLAKE-EN [8] Overall(%) | Open(%) | Close(%) |
|---|---|------------|---------|----------|
| 1 | 0.07 | 81.8 | 79.1 | 85.8 |
| 2 | 0.1 | 81.4 | 78.6 | 85.6 |
| 3 | 0.2 | 81.5 | 78.9 | 85.6 |
| 4 | 0.3 | 81.2 | 78.0 | 86.3 |
| 5 | 0.4 | **81.9** | **79.1** | 86.3 |
| 6 | 0.5 | 81.9 | 78.6 | **87.0** |
| 8 | 0.6 | 81.5 | 79.1 | 85.3 |

In Table 3a, the $\tau$ is set to 0.07, the row #3–1 indicates that the selection of negative samples is random. Row #3–2 show that negative samples have the same modalities as positive samples. Row #3–5 indicates that the negative samples are different in modalities from the positive samples. In the TCC model, we observe that the framework performs best when the negative and positive samples have the same modalities and body parts (row #3–4). For small-scale image data, we argue that this selection method enhances the robustness of the model and improves its ability to capture high-level visual features.

Based on the selection in the fourth row of Table 3a, we experiment on $\tau$ in Eq. 2, and the results are shown in Table 3b. We can find the temperature $\tau$ is a key parameter to control the strength of penalties on difficult negative samples. When $\tau$ is set to 0.4 (row #3–5), it improves the feature quality.

**Relevance Score Calculation in MEM Module.** In the MEM module, in order to choose a good method for calculating the relevance score, we experiment with our proposed framework, which does not use TCC to provide initialization weights for the visual extractor. We compare Euclidean Distance, dot product and cosine similarity. The result is shown in Table 4.

**Table 4.** Relevance Score Calculation in MEM Module

| Methods | SLAKE-EN [8] | | |
| --- | --- | --- | --- |
| | Overall(%) | Open(%) | Close(%) |
| Euclidean Distance | 80.7 | 77.4 | 86.1 |
| dot | 81.3 | 77.4 | 86.0 |
| cos | **81.8** | **78.8** | **86.5** |

As shown in Table 4, in the MEM module, the improvement in using cosine similarity is 1.1% compared to the Euclidean distance calculation method. This is because the cosine similarity method is a better measure of individual differences in scientific research. Therefore, we choose the cosine similarity method in the MEM module.

### 3.5 Comparison of Contrastive Learning Methods

In this part, we compare the structure and accuracy of TCC with existing contrastive learning methods [2,3,5]. Figure 4 provides an abstraction of these method structures.

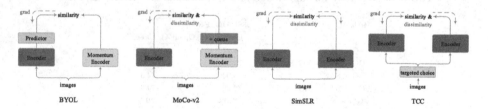

**Fig. 4.** Comparison of Contrastive Learning Methods. Predictor denotes the MLP. In MoCo-v2 [3], the queue is used to hold lots of negative examples. In TCC, we select negative samples in a targeted manner.

BYOL [5] use positive samples only to learn model representations. MoCo-v2 [3] and SimCLR [2] utilizes negative samples to avoid model collapse. To extract deeper visual features, we select negative samples with the same modalities and body parts as the positive samples in a targeted manner. To compare the performance of these contrastive learning methods on small-scale data, these methods are first pre-trained on 450 images and then fine-tuned in our Med-VQA framework (MEM module is missing, mixup [19] is used). The results are shown in Table 5.

**Table 5.** Comparison of different contrastive learning methods on Med-VQA fine-tuned on SLAKE-EN [8].

| Methods | SLAKE-EN [8] | | |
|---|---|---|---|
| | Overall(%) | Open(%) | Close(%) |
| BYOL [5] | 80.4 | 76.6 | 86.3 |
| MoCo-v2 [3] | 80.7 | 77.2 | 86.1 |
| SimCLR [2] | 80.9 | 78.1 | 85.4 |
| **TCC** | **81.9** | **79.1** | **86.3** |

In Table 5, we can find the following: (1). In small-scale datasets, our proposed TCC can provide better visual features for the downstream task Med-VQA. (2). For a small-scale data set, methods using negative samples perform better than approaches using positive samples only. (3). Compared with SimCLR [2], our TCC achieves better performance on the downstream task Med-VQA, which benefits from the targeted choice of negative samples.

### 3.6 Superiority of MEM Module

In order to assess the superiority of the MEM module in terms of knowledge embedding, we compare it with the latest KG embedding approach [8]. Based on the second way of dividing the dataset, the performance of the methods is evaluated in knowledge-based questions. For the fairness of comparison, following Liu et al. [8], we use VGG as visual extractor, LSTM as text extractor and SAN [15] as fusion module. The results of the experiments are shown in Table 6. On the Table 6, the method of Liu et al. [8] to embed knowledge representation is denoted by KG.

**Table 6.** Accuracy for knowledge-based questions on SLAKE-EN [8].

| Methods | SLAKE-EN [8] | | |
|---|---|---|---|
| | Overall(%) | Open(%) | Close(%) |
| VGG + SAN [8] | 70.27 | – | – |
| VGG + SAN + KG [8] | 72.30 | – | - |
| VGG + SAN$^\dagger$ | 70.94 | 69.72 | 74.35 |
| VGG + SAN + KG$^\dagger$ | 72.29 | 71.56 | 74.36 |
| **VGG + SAN + MEM (ours)** | **77.03** | **76.15** | **79.5** |

$^{1\ \dagger}$ indicates the result of our re-implementation.

In Table 6, our proposed MEM module is superior to the knowledge embedding method used in [8], bringing a 4.74% improvement. This suggests that the most relevant "head" entity obtained by combining semantic and visual information has a more positive guiding effect on Med-VQA.

### 3.7 Visualization

In this subsection, we compare our proposed framework with the baseline model MEVF+BAN+CR [18]. Figure 5 visualizes the test results of some questions in SLAKE-EN [8]. We leverage Grad-CAM [14] to highlight the crucial regions for images. Our proposed framework can predict the answer more accurately by locating and identifying the special region associated with the question.

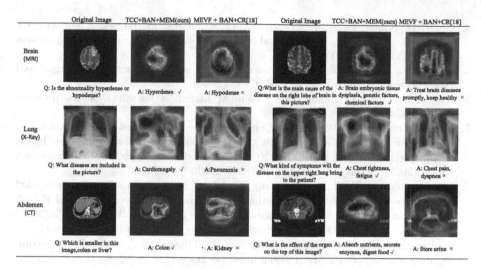

**Fig. 5.** Visualization of our proposed framework and baseline MEVF+BAN+CR [18] results. ✓ and × denote the correctness of what each model gives for the answer.

## 4   Conclusion

In this paper, we propose the pre-training TCC model to tackle the poor visual feature extraction ability on small-scale dataset. Further, we design the MEM module to improve the accuracy of embedding knowledge graph in the Med-VQA task. Experimental results demonstrate that our framework outperforms the state-of-the-art baselines on the benchmark dataset SLAKE, especially in "closed-ended" and knowledge-based questions.

**Acknowledgements.** This work was funded by National Natural Science Foundation of China (Grant No. 61762069), Key Technology Research Program of Inner Mongolia Autonomous Region (Grant No. 2021GG0165), Key R&D and Achievement Transformation Program of Inner Mongolia Autonomous Region (Grant No. 2022YFHH0077), Big Data Lab of Inner Mongolia Discipline Inspection and Supervision Committee (Grant No. 21500–5206043).

## References

1. Bordes, A., Usunier, N., Garcia-Duran, A., Weston, J., Yakhnenko, O.: Translating embeddings for modeling multi-relational data. In: Advances in Neural Information Processing Systems, vol. 26 (2013)
2. Chen, T., Kornblith, S., Norouzi, M., Hinton, G.: A simple framework for contrastive learning of visual representations. In: International Conference on Machine Learning, pp. 1597–1607. PMLR (2020)
3. Chen, X., Fan, H., Girshick, R., He, K.: Improved baselines with momentum contrastive learning. arXiv preprint arXiv:2003.04297 (2020)

4. Gong, H., Chen, G., Liu, S., Yu, Y., Li, G.: Cross-modal self-attention with multi-task pre-training for medical visual question answering. arXiv preprint arXiv:2105.00136 (2021)
5. Grill, J.B., et al.: Bootstrap your own latent: a new approach to self-supervised learning. arXiv preprint arXiv:2006.07733 (2020)
6. Kim, J.H., Jun, J., Zhang, B.T.: Bilinear attention networks. arXiv preprint arXiv:1805.07932 (2018)
7. Liu, B., Zhan, L.-M., Wu, X.-M.: Contrastive pre-training and representation distillation for medical visual question answering based on radiology images. In: de Bruijne, M., et al. (eds.) MICCAI 2021. LNCS, vol. 12902, pp. 210–220. Springer, Cham (2021). https://doi.org/10.1007/978-3-030-87196-3_20
8. Liu, B., Zhan, L., Xu, L., Ma, L., Yang, Y., Wu, X.M.: Slake: a semantically-labeled knowledge-enhanced dataset for medical visual question answering. In: IEEE 18th International Symposium on Biomedical Imaging (ISBI), pp. 1650–1654 (2021)
9. Loshchilov, I., Hutter, F.: Fixing weight decay regularization in adam (2018)
10. Nguyen, B.D., Do, T.-T., Nguyen, B.X., Do, T., Tjiputra, E., Tran, Q.D.: Overcoming data limitation in medical visual question answering. In: Shen, D., et al. (eds.) MICCAI 2019. LNCS, vol. 11767, pp. 522–530. Springer, Cham (2019). https://doi.org/10.1007/978-3-030-32251-9_57
11. Oord, A.v.d., Li, Y., Vinyals, O.: Representation learning with contrastive predictive coding. arXiv preprint arXiv:1807.03748 (2018)
12. Pennington, J., Socher, R., Manning, C.D.: Glove: global vectors for word representation. In: Proceedings of the 2014 Conference on Empirical Methods in Natural Language Processing (EMNLP), pp. 1532–1543 (2014)
13. Robinson, J., Chuang, C.Y., Sra, S., Jegelka, S.: Contrastive learning with hard negative samples. arXiv preprint arXiv:2010.04592 (2020)
14. Selvaraju, R.R., Cogswell, M., Das, A., Vedantam, R., Parikh, D., Batra, D.: Gradcam: visual explanations from deep networks via gradient-based localization. In: Proceedings of the IEEE International Conference on Computer Vision, pp. 618–626 (2017)
15. Yang, Z., He, X., Gao, J., Deng, L., Smola, A.: Stacked attention networks for image question answering. In: Proceedings of the IEEE Conference on Computer Vision and Pattern Recognition (CVPR), June 2016
16. Yasunaga, M., Ren, H., Bosselut, A., Liang, P., Leskovec, J.: QA-GNN: reasoning with language models and knowledge graphs for question answering. arXiv preprint arXiv:2104.06378 (2021)
17. Yu, Z., Yu, J., Fan, J., Tao, D.: Multi-modal factorized bilinear pooling with co-attention learning for visual question answering. In: Proceedings of the IEEE International Conference on Computer Vision (ICCV), October 2017
18. Zhan, L., Liu, B., Fan, L., Chen, J., Wu, X.M.: Medical visual question answering via conditional reasoning. In: Proceedings of the 28th ACM International Conference on Multimedia, pp. 2345–2354 (2020)
19. Zhang, H., Cisse, M., Dauphin, Y.N., Lopez-Paz, D.: mixup: beyond empirical risk minimization. arXiv preprint arXiv:1710.09412 (2017)
20. Zhang, W., Wang, H., Lai, Z., Hou, C.: Constrained contrastive representation: classification on chest x-rays with limited data. In: 2021 IEEE International Conference on Multimedia and Expo (ICME), pp. 1–6 (2021). https://doi.org/10.1109/ICME51207.2021.9428273

# Boosting StarGANs for Voice Conversion with Contrastive Discriminator

Shijing Si[1,2], Jianzong Wang[1](✉), Xulong Zhang[1], Xiaoyang Qu[1],
Ning Cheng[1], and Jing Xiao[1]

[1] Ping An Technology (Shenzhen) Co., Ltd., Shenzhen, China
jzwang@188.com
[2] School of Economics and Finance, Shanghai International Studies University,
Shanghai, China

**Abstract.** Nonparallel multi-domain voice conversion methods such as the StarGAN-VCs have been widely applied in many scenarios. However, the training of these models usually poses a challenge due to their complicated adversarial network architectures. To address this, in this work we leverage the state-of-the-art contrastive learning techniques and incorporate an efficient Siamese network structure into the StarGAN discriminator. Our method is called SimSiam-StarGAN-VC and it boosts the training stability and effectively prevents the discriminator overfitting issue in the training process. We conduct experiments on the Voice Conversion Challenge (VCC 2018) dataset, plus a user study to validate the performance of our framework. Our experimental results show that SimSiam-StarGAN-VC significantly outperforms existing StarGAN-VC methods in terms of both the objective and subjective metrics.

**Keywords:** Contrastive Learning · Nonparallel Voice Conversion · StarGAN · Siamese Networks · Data Augmentation · Training Stability

## 1 Introduction

Voice conversion (VC) is a speech processing task that converts an utterance from one speaker to that of another [19,25,32,33]. VC can be useful to various scenarios and tasks such as speaker-identity modification for text-to-speech (TTS) systems [16], speaking assistance [30], and speech enhancement [1].

Voice contains significant information of the speaker [23], so increasingly complicated models are employed to capture the feature of voice. Statistical methods based on Gaussian mixture models (GMMs) [7,29] have been quite successful in VC task. Recently, deep neural networks (DNNs), including feed-forward deep NNs [21], recurrent NNs [26], and generative adversarial nets (GANs) [11], have also achieved promising results on VC task. Most of these conventional VC methods require accurately aligned parallel source and target speech data. However, in many scenarios, it may be impossible to access parallel utterances. Even if we could collect such data, we typically need to utilize time alignment procedures,

© The Author(s), under exclusive license to Springer Nature Switzerland AG 2023
M. Tanveer et al. (Eds.): ICONIP 2022, LNCS 13624, pp. 355–366, 2023.
https://doi.org/10.1007/978-3-031-30108-7_30

which becomes relatively difficult when there is a large acoustic gap between the source and target speech [8,11]. These challenges motivate how to train high-quality VC models with non-parallel data.

Many research study non-parallel VC methods, because they require no parallel utterances, transcriptions, or time alignment procedures. Currently, two representative methods of this type are CycleGAN-VCs [12,14] and StarGAN-VCs [10]. These methods are first developed by the computer vision (CV) community for style transfer of figures [4,34]. The main difference between CycleGAN-VCs and StarGAN-VCs lies in the multi-domain cases. CycleGAN-VCs are specialized to two domain cases, while StarGAN-VCs can handle multi-domains by taking account of the latent code for each domain [10]. Other researchers also investigate how to perform voice coversion in few-shot cases, such as, [27,28]. However, the training of GAN-like models is a challenge due to their non-convex nature. Therefore, the training stability of StarGAN-VCs is poor and can consume a significantly large amount of time.

In this paper, we focus on how to boost the training stability of StarGAN-VCs that utilize a StarGAN architecture to perform VC tasks. Due to the non-convex/stationary nature of the mini-max game, however, training StarGANs in practice is often very unstable and extremely sensitive to many hyperparameters [5,22]. Data augmentation techniques have recently proven beneficial to stabilizing GAN-like adversarial models [31]. Researchers also have applied contrastive learning methods to the basic GAN as an auxiliary task upon the GAN loss [9,17]. From the literature, contrastive methods can strengthen the discriminator of GAN models, thereby improving the capability of the entire GAN model. However, little attempt has been done to the complicated StarGAN models, let alone for the VC tasks. In this paper, we leverage the efficient simple Siamese (SimSiam) representation learning [3], one kind of contrastive learning, to train the discriminator of the StarGAN-VC model, and our method is called SimSiam-StarGAN-VC. We evaluated the performance of the proposed SimSiam-StarGAN-VC on the commonly used multi-speaker VC dataset Conversion Challenge 2018 (VCC 2018) [18]. We observe that SimSiam-StarGAN-VC presents better stability of the training process and better naturalness of converted voices, compared with the original StarGAN-VC2.

Our contributions are summarized as follows:

- We propose a SimSiam-StarGAN-VC method, which incorporates a Siamese network into StarGAN-VCs and stabilizes the training of StarGAN-VCs.
- We empirically investigate the performance of SimSiam-StarGAN-VC and show its superiority over StarGAN-VCs in terms of both subjective and objective metrics.

## 2    Background

Prior to the introduction of our SimSiam-StarGAN-VC, we elaborate the StarGAN-VC2 and SimSiam methods in this section.

## 2.1  StarGAN-VC2 Method

Inspired by the success of StarGAN in the computer vision community, [10] proposed to leverage its power to train a single generator $G$ that converts voices among multiple speakers or domains. For each speaker, StarGAN-VC posits a domain code (e.g., a speaker identifier or embedding). The generator $G$ of StarGAN-VC takes a real acoustic feature map $x$ and the target domain code $c'$ as input and produces a feature map $x'$ of the target speaker domain $c'$. The mathematical notations are presented in Table 1. Specifically, We denote $x$ as a 2-dimensional acoustic feature map (like MFCC). We use $c \in \{1, \ldots, N\}$ to denote the domain code of a speaker, where the number of domains or speakers is $N$.

To further enhance the conversion performance of StarGAN-VC, StarGAN-VC2 [13] introduces the source-and-target conditional adversarial loss to replace the classification loss and target conditional adversarial loss in StarGAN-VC. Both the generator and discriminator in StarGAN-VC2 take the source $(c)$ and target $(c')$ codes as input, i.e., $G(x, c, c') \rightarrow x'$. The training objectives of StarGAN-VC2 is the source-and-target adversarial loss which is shown as follows.

**Source-and-Target Adversarial Loss:** the most significant contribution of StarGAN-VC2

$$\mathcal{L}_{st-adv} = \mathbb{E}_{(x,c)\sim P(x,c),c'\sim P(c')} \left[\log D\left(x, c', c\right)\right] + \\ \mathbb{E}_{(x,c)\sim P(x,c),c'\sim P(c')} \left[\log(1 - D\left(G\left(x, c, c'\right), c, c'\right))\right], \tag{1}$$

lies in that both the generator $G$ and discriminator $D$ takes the acoustic feature map $(x)$, source domain $(c)$ and target domain codes $(c')$ as input. The striking difference between StarGAN-VC2 and StarGAN-VC is that StarGAN-VC ignores the source code $c$. $D(x, c', c)$ outputs the probability that an acoustic feature $x$ is real from the target domain $c$, and its range is from 0 to 1. Similar to other GAN models, it is a min-max game: maximizing the loss in Eq. (1) with respect to $D$ leads to a powerful fake voice detector, but minimizing the loss with regards to $G$ will train a generator to mimic the true acoustic features.

[13] also explored to deploying a conditional instance normalization (CIN) module [6] inside the network architecture, which proceeds as in Eq. 2.

$$\text{CIN}\left(f; c'\right) = \gamma_{c'} \left(\frac{f - \mu(f)}{\sigma(f)}\right) + \beta_{c'}, \tag{2}$$

In Eq. 2, $f$ represents a feature map of input audio, $\mu(f)$ and $\sigma(f)$ are the average and standard deviation of $f$ that are computed for each training sample. $\gamma_{c'}$ and $\beta_{c'}$ are speaker(domain)-specific scale and bias parameters for the speaker (i.e., domain) $c$. In the training process, we train and learn these speaker-specific parameters with other network parameters/weights. For the source and target generator loss in Eq. (1), these domain specific parameters $\gamma$ and $\beta$ are dependent on both the source $(c)$ and target speakers $(c')$, i.e., $\gamma_{c'}$ and $\beta_{c'}$ are replaced by $\gamma_{c,c'}$ and $\beta_{c,c'}$, respectively.

## 2.2  Simple Siamese Representation Learning

SimSiam is one kind of contrastive learning [3], which requires none of the following: 1.) negative sample pairs, 2.) large batches, and 3.) momentum encoders. It utilizes two random data augmentations of each audio as input, and extracts features via the same encoder network $f$ and a multi-layer perceptron (MLP) projection header $h$. More specifically, augmented speeches $x_1$ and $x_2$ come from $x$, with their high-level features $z_i = f(x_i)$ and $p_i = h(f(x_i))$. The SimSiam loss for each real speech is

$$\mathcal{L}_{SimSiam}(x_1, x_2, f) = \frac{1}{2}\mathcal{D}(p_1, \qquad (z_2))$$
$$+ \frac{1}{2}\mathcal{D}(p_2, \qquad (z_1)), \qquad (3)$$

where

$$\mathcal{D}(p_1, z_2) = -\frac{p_1}{||p_1||_2} \cdot \frac{z_2}{||z_2||_2}$$

with $|| \cdot ||_2$ the $\ell_2$-norm and          represents the stop-gradient operation.

# 3  Methodology

In this section, we illustrate how our SimSiam-StarGAN-VC works. We utilize the same network architecture as StarGAN-VC2 in [13], but our framework is compatible to many existing GAN based VC architectures. The overall architecture is shown in Fig. 1, where the $G$ and $D$ are the generator and discriminator, respectively. The source and target domain codes are $c$ and $c'$, respectively, and they are embedded to latent vectors before being fed into the generator and discriminator. For clarity, we list the mathematical notations in Table 1.

## 3.1  Contrastive Learning for Real Samples

In this part, we describe how to train the discriminator $D$ with contrastive learning. We denote the encoder part of the discriminator $D$ as $D_e$, and it can extract high-level features (a real vector) from an input speech, i.e., $D_e : (x) \rightarrow \mathbb{R}^{d_e}$.

Overall, the encoder network $D_e$ of SimSiam-StarGAN-VC is trained by minimizing two different contrastive losses: (a) the SimSiam loss in Eq. 3 on the real speech samples, and (b) the supervised contrastive loss [15] on fake speech samples. Figure 1 displays the loss functions used in our SimSiam-StarGAN-VC. We elaborate these two contrastive losses in detail.

**Contrastive Learning with Real Speech Samples.** Here, we attempt to simply follow the SimSiam training scheme for each real sample $x$, the loss function is

$$L_{Sim}(x, D_e) = L_{SimSiam}(t_1(x), t_2(x), D_e), \qquad (4)$$

where $t_1$ and $t_2$ are augmentation methods for audio data [20].

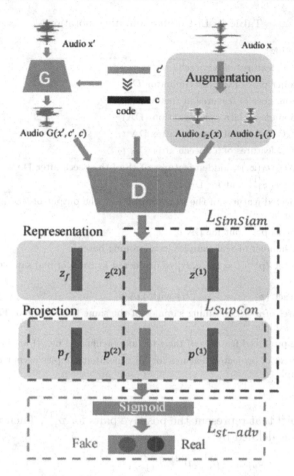

**Fig. 1.** The overall architecture of SimSiam-StarGAN-VC. The source and target speaker (domain) codes are $c$ and $c'$, respectively.

## 3.2   Supervised Contrastive Learning for Fake Speech Samples

In order for the encoder of $D_e$ to keep necessary information to discriminate real and fake speech samples, we consider an auxiliary loss $L_{con}$. Specifically, we employ the supervised contrastive loss [15] over fake (generated) speech samples. This loss is an extended version of contrastive loss to support supervised learning by allowing more than one sample to be positive, so that samples of the same label can be attracted to each other in the embedding space. On a mini-batch, we treat the real samples and their augmented versions as positive, and the generated fake speech samples as negative. For a mini-batch of real samples, we denote $p^{(1)}$ and $p^{(2)}$ as the projected representations (after a MLP) of two kinds of data augmentation. $p_f$ is the set of projected representations for a batch of converted (generated by $G$) fake audio samples. Formally, for each $p_i^{(1)}$, let $P_{i+}^{(2)}$

**Table 1.** List of mathematical notations

| Notation | Meaning |
|---|---|
| $x$ | MFCC features of a speech |
| $D_e$ | Encoder part of the discriminator $D$ |
| $d_e$ | The number of features in the output of $D_e$ |
| $t_1, t_2$ | Two kinds of data augmentation (DA) |
| $t_1(x)$ | MFCC features of a speech after DA $t_1$ |
| $t_2(x)$ | MFCC features of a speech after DA $t_2$ |
| $z_i^{(1)}$ | i.e., $D_e(t_1(x_i))$, hidden features of the $i$-th speech after DA $t_1$ |
| $z_i^{(2)}$ | Same as $z_i^{(1)}$, but for DA $t_2$ |
| $p_i^{(1)}$ | Projected features of the i-th sample, e.g., the output of feeding $z_i^{(1)}$ to a linear layer |
| $p_i^{(2)}$ | Same as $p_i^{(1)}$, but for DA $t_2$ |
| $B$ | The number of real samples in a training batch |
| $p^{(1)}$ | The set $\{p_i^{(1)}, i = 1, \ldots, B\}$ of projected features of real samples after DA $t_1$ |
| $p^{(2)}$ | The set $\{p_i^{(2)}, i = 1, \ldots, B\}$ with DA $t2$ |
| $P_{i+}^{(2)}$ | Subset of $p^{(2)}$ containing samples of the same label (True or Fake) as the $i$-th sample |
| $p_{f,i}$ | The projected features of the $i$-th fake (generated by $G$) audio sample |
| $p_{f,-i}$ | The set of projected features for all generated samples except the $i$-th sample |

be a subset of $p^{(2)}$ that represent the positive pairs for $p_i^{(1)}$. Then the supervised contrastive loss is defined by:

$$L_{SupCon}(p_i^{(1)}, p^{(2)}, P_{i+}^{(2)}) =$$
$$-\frac{1}{|P_{i+}^{(2)}|} \sum_{p_{i+}^{(2)} \in P_{i+}^{(2)}} \log \frac{\exp(s(p_i^{(1)}, p_{i+}^{(2)}))}{\sum_j \exp(s(p_i^{(1)}, p_j^{(2)}))}, \tag{5}$$

where $s(\cdot, \cdot)$ is the inner product used in SimCLR [2].

Using this notation, we define the loss for fake samples as follows:

$$L_{Con} = \frac{1}{B} \sum_{i=1}^{B} L_{SupCon}(p_{f,i}, [p_{f,-i}; p^{(1)}; p^{(2)}], [p_{f,-i}]), \tag{6}$$

where $B$ is the batch size and $[p_{f,-i}; p^{(1)}; p^{(2)}]$ is the union of three sets of projected features.

The loss function of SimSiam-StarGAN-VC for the generator is $L_{st-adv}$ in Eq. (1), and for discriminator the loss is defined as follows:

$$L_D = -L_{st-adv} + \lambda_1 \cdot L_{Sim} + \lambda_2 \cdot L_{Con}, \tag{7}$$

where $\lambda_1$ and $\lambda_2$ are the strength parameters for the SimSiam and supervised contrastive loss.

# 4  Experiments

## 4.1  Experimental Setup

**Dataset:** We utilized data from the most popular VCC 2018 dataset [18] in a similar manner to the experiments in StarGAN-VC2 [13]. We describe how we conduct the experiments briefly. To perform both inter-gender and intra-gender VC, we randomly selected two male and two female speakers from VCC 2018, denoted as SF1, SF2, SM1, and SM2, short for "Speaker of Female/Male 1 or 2". Therefore, we have $N = 4$ as the number of domains (speakers). To ensure the non-parallel setting, there is no overlapping content between the training and evaluation datasets. For a thorough comparison, we conduct all $4 \times 3 = 12$ combinations intra-gender and inter-gender conversions. Each speaker has approximately 80 utterances for model training and 30 for model evaluation.

**Implementation Details:** For StarGAN-VC and StarGAN-VC2, we employ the same network architecture as shown in the Fig. 3 of [13]. For the data augmentation methods in the SimSiam-StarGAN-VC, we utilized time masking and frequency masking as $t_1$ and $t_2$, respectively. The upper limit of training epochs is set to be $1 \times 10^5$, early stopping is deployed, and the learning rate parameters for $G$ and $D$ is tuned by carefully by closely monitoring the loss of discriminators and generators.

## 4.2  Objective Evaluation

As common in the literature, an objective evaluation is done to verify the benefits of our SimSiam-StarGAN-VC over other existing StarGAN-VCs. Similar to [13], we also utilized the Mel-cepstral distortion (MCD) and the modulation spectra distance (MSD). Essentially, these two metrics measure the overall and local structural differences between the target and converted Mel-cepstral coefficients (MCEPs). For both MCD and MSD metrics, smaller values indicate better voice conversion performance.

**Table 2.** Comparison of MCD and MSD among three different models.

| Method | MCD [dB] | MSD [dB] |
|---|---|---|
| StarGAN-VC | $7.11 \pm .10$ | $2.41 \pm .13$ |
| StarGAN-VC2 | $6.90 \pm .07$ | $1.89 \pm .03$ |
| SimSiam-StarGAN-VC | $\mathbf{6.35 \pm .12}$ | $\mathbf{1.48 \pm .10}$ |

Table 2 displays the performance of 3 different VC approaches in terms of two objective metrics (MCD and MSD). To show the statistical significance, we have computed the mean scores by taking average over models trained with five different initializations and reported the standard deviations in this table. From Table 2, our SimSiam-StarGAN-VC (MCD: 6.35, MSD: 1.48) significantly outperforms both StarGAN-VC (MCD: 7.11, MSD: 2.41) and StarGAN-VC2 (MCD: 6.90, MSD: 1.89) in terms of both metrics. This indicates that contrastive losses ($L_{Sim}$ and $L_{Con}$) are useful for improving the feature extraction capability of the discriminator, which further boosts the quality of converted speeches.

### 4.3    Subjective Evaluation

To analyze the effectiveness of SimSiam-StarGAN-VC, we conducted listening tests to cpmpare it with StarGAN-VC2. We collected 36 generated (converted) sentences (12 source-target combinations ×3 sentences, where the first one is the real target utterance and the other two are generated by SimSiam-StarGAN and StarGAN-VC2). Eight well-educated Chinese native speakers participated in the tests as audiences. We conducted a mean opinion score (MOS) test to evaluate the naturalness of generated speeches, from 5 (for excellent quality) to 1 (for poor quality). In these tests we presented the target speech as a reference for audiences (the average MOS for target speeches is about 4.5), so that the audiences can evaluate the generated speeches properly.

We also implemented an XAB test to evaluate speaker similarity by randomly selecting 30 sentences from the evaluation set. Here we denote "X" the target speech, and "A" and "B" were converted utterances from StarGAN-VC2 and SimSiam-StarGAN-VC, respectively. When presenting each set of speeches, we display "X" first, then "A" and "B" randomly. After the audiences heard one set of speeches, we asked them to choose which speech ("A" or "B") is closer to the target ("X"), or to be "Fair".

Figure 2 and Fig. 3 display the main findings of naturalness and the preference scores of StarGAN-VC2 and SimSiam-StarGAN-VC, respectively. In Fig. 2, the pink and orange bars represent the MOS of SimSiam-StarGAN-VC and StarGAN-VC2, respectively. These results empirically demonstrate that SimSiam-StarGAN-VC (overall MOS: 3.7) outperforms StarGAN-VC2 (overall MOS: 3.1) on naturalness for every category. In Fig. 3, the pink, light blue and orange colors the preference scores for SimSiam-StarGAN-VC, fair, and StarGAN-VC2, respectively. The SimSiam-StarGAN-VC (overall preference: 75.0%) outperforms StarGAN-VC2 (overall: 5.6%) significantly on speaker similarity. We also highlight that our SimSiam-StarGAN-VC takes only 100 training epochs to converge, which is shown in Fig. 4. However, StarGAN-VC2 still oscillates significantly after 400 epochs. This demonstrates the effectiveness of contrastive training of the discriminator.

### 4.4    Training Stability

To show the training stability of SimSiam-StarGAN-VC, we have produced a figure of discriminator loss and mean opinion score (MOS) for speech naturalness along training epochs, which is presented in Fig. 4.

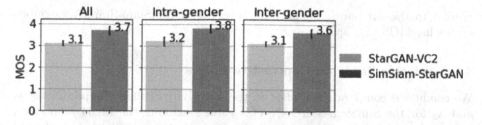

**Fig. 2.** The average MOS values of all, intra-gender and cross-gender conversion of StarGAN-VC2 (orange bars) and SimSiam-StarGAN-VC (pink bars) (Color figure online)

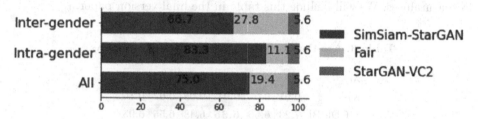

**Fig. 3.** The preference of all, intra-gender and cross-gender conversion of StarGAN-VC2 (orange) and SimSiam-StarGAN-VC (pink) (Color figure online)

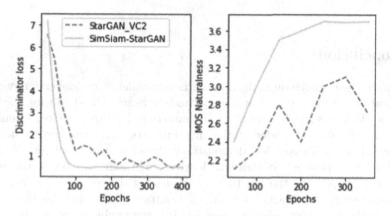

**Fig. 4.** The discriminator loss and MOS values along training epochs. Left panel shows the discriminator loss of StarGAN-VC2 (orange dashed line) and SimSiam-StarGAN (pink solid line) versus the training epochs; and the right panel displays the MOS for naturalness of the two approaches. (Color figure online)

Figure 4 displays the discriminator loss traces and MOS for naturalness of StarGAN-VC2 (orange dashed lines) and SimSiam-StarGAN (pink solid lines). The discriminator loss of StarGAN-VC2 oscillates over training epochs, while the discriminator loss of SimSiam-StarGAN converges steadily. The right panel illustrates the MOS for naturalness of generated speech by these two methods.

Similar to the left panel, the speech converted by SimSiam-StarGAN exhibits increasing MOS over epochs.

### 4.5 Ablation Study on Contrastive Losses

We conducted comparative studies on the sensitivity of the hyperparametes $\lambda_1$ and $\lambda_2$ for the SimSiam-StarGAN-VC. Table 3 exhibits the MCD scores over different combinations of $\lambda_1$ and $\lambda_2$, and $\lambda_1 = \lambda_2 = 0.01$ is the best choice for the VCC 2018 dataset. We have recorded the MOS values for the ablation study in the experiments. The extended table in shown in Table 3. When $\lambda_1 = \lambda_2 = 0.01$, the SimSiam-StarGAN-VC performs best in terms of both the MCD and MOS for naturalness. We will include this table in the final version paper.

**Table 3.** Ablation study of hyper-parameters $\lambda_1$ and $\lambda_2$.

| $\lambda_1$ | 0.0 | 0.01 | 0.01 | 0.02 | 0.05 | 0.1 |
|---|---|---|---|---|---|---|
| $\lambda_2$ | 0.01 | 0.0 | 0.01 | 0.05 | 0.02 | 0.1 |
| MCD[dB] | 7.23 | 6.56 | **6.35** | 6.48 | 6.55 | 6.95 |
| MOS | 3.05 | 3.56 | **3.70** | 3.68 | 3.65 | 3.45 |

## 5   Conclusion

To advance the research on multi-domain non-parallel voice conversion, we have incorporated the contrastive learning methods in StarGAN-VC during the training stage. We leveraged the SimSiam and supervised contrastive loss to enhance the capability of the encoder of the discriminator. The empirical studies on non-parallel multi-speaker VC demonstrate the effectiveness of our SimSiam-StarGAN-VC. Therefore, contrastive learning methods can boost the performance of StarGANs on the VC task by improving the convergence and stability of the complicated StarGAN training. Contrastive learning has shown good promise in the computer vision community. It is reasonable to believe that it will advance the speech processing area in many aspects. In the next step, we may attempt to employ the variational information bottleneck [24] with contrastive learning to disentangle the speaker identity information from the input speech, which may improve the controllability of VC models.

**Acknowledgment.** This paper is supported by the Key Research and Development Program of Guangdong Province under grant No.2021B0101400003.

# References

1. Chen, C.Y., Zheng, W.Z., Wang, S.S., Tsao, Y., Li, P.C., Li, Y.: Enhancing intelligibility of dysarthric speech using gated convolutional-based voice conversion system. In: IEEE Interspeech (2020)
2. Chen, T., Kornblith, S., Norouzi, M., Hinton, G.: A simple framework for contrastive learning of visual representations. In: ICML, pp. 1597–1607. PMLR (2020)
3. Chen, X., He, K.: Exploring simple siamese representation learning. In: Proceedings of CVPR, pp. 15750–15758 (2021)
4. Choi, Y., Choi, M., Kim, M., Ha, J.W., Kim, S., Choo, J.: Stargan: unified generative adversarial networks for multi-domain image-to-image translation. In: Proceedings of the IEEE Conference on Computer Vision and Pattern Recognition, pp. 8789–8797 (2018)
5. Choi, Y., Uh, Y., Yoo, J., Ha, J.W.: Stargan v2: diverse image synthesis for multiple domains. In: Proceedings of CVPR, pp. 8188–8197 (2020)
6. Dumoulin, V., Shlens, J., Kudlur, M.: A learned representation for artistic style. In: ICLR (2017). https://openreview.net/forum?id=BJO-BuT1g
7. Helander, E., Virtanen, T., Nurminen, J., Gabbouj, M.: Voice conversion using partial least squares regression. IEEE Trans. Audio Speech Lang. Process. 18(5), 912–921 (2010)
8. Hsu, C.C., Hwang, H.T., Wu, Y.C., Tsao, Y., Wang, H.M.: Voice conversion from non-parallel corpora using variational auto-encoder. In: Proceedings of APSIPA, pp. 1–6. IEEE (2016)
9. Jeong, J., Shin, J.: Training gans with stronger augmentations via contrastive discriminator. In: ICLR (2021)
10. Kameoka, H., Kaneko, T., Tanaka, K., Hojo, N.: Stargan-vc: non-parallel many-to-many voice conversion using star generative adversarial networks. In: SLT, pp. 266–273. IEEE (2018)
11. Kaneko, T., Kameoka, H.: Cyclegan-vc: non-parallel voice conversion using cycle-consistent adversarial networks. In: Proceedings of EUSIPCO, pp. 2100–2104. IEEE (2018)
12. Kaneko, T., Kameoka, H., Tanaka, K., Hojo, N.: Cyclegan-vc2: improved cyclegan-based non-parallel voice conversion. In: Proceedings of ICASSP, pp. 6820–6824. IEEE (2019)
13. Kaneko, T., Kameoka, H., Tanaka, K., Hojo, N.: Stargan-vc2: rethinking conditional methods for stargan-based voice conversion. In: Proceedings of INTERSPEECH (2019)
14. Kaneko, T., Kameoka, H., Tanaka, K., Hojo, N.: Cyclegan-vc3: examining and improving cyclegan-vcs for mel-spectrogram conversion. In: Proceedings of Interspeech, pp. 2017–2021 (2020)
15. Khosla, P., et al.: Supervised contrastive learning. In: NeurIPS, vol. 33 (2020)
16. Kim, T.H., Cho, S., Choi, S., Park, S., Lee, S.Y.: Emotional voice conversion using multitask learning with text-to-speech. In: Proceedings of ICASSP, pp. 7774–7778. IEEE (2020)
17. Lee, K.S., Tran, N.T., Cheung, N.M.: InfoMax-GAN: improved adversarial image generation via information maximization and contrastive learning. In: Proceedings of WACV, pp. 3942–3952 (2021)
18. Lorenzo-Trueba, J., et al.: The voice conversion challenge 2018: promoting development of parallel and nonparallel methods. In: Proceedings of Speaker Odyssey (2018)

19. Nercessian, S.: Zero-shot singing voice conversion. In: Proceedings of ISMIR (2020)
20. Park, D.S., et al.: Specaugment: a simple data augmentation method for automatic speech recognition. In: Proceedings of Interspeech, pp. 2613–2617 (2019)
21. Saito, Y., Takamichi, S., Saruwatari, H.: Voice conversion using input-to-output highway networks. IEICE Trans. Inf. Syst. **100**(8), 1925–1928 (2017)
22. Salimans, T., Goodfellow, I., Zaremba, W., Cheung, V., Radford, A., Chen, X.: Improved techniques for training GANs. NeurIPS **29**, 2234–2242 (2016)
23. Si, S., et al.: Speech2video: cross-modal distillation for speech to video generation. In: Proceedings of the Annual Conference of the International Speech Communication Association, INTERSPEECH (2021)
24. Si, S., et al.: Variational information bottleneck for effective low-resource audio classification. In: Proceedings of the Annual Conference of the International Speech Communication Association, INTERSPEECH, p. 31 (2021)
25. Stylianou, Y., Cappé, O., Moulines, E.: Continuous probabilistic transform for voice conversion. IEEE Trans. Speech Audio Process. **6**(2), 131–142 (1998)
26. Sun, L., Kang, S., Li, K., Meng, H.: Voice conversion using deep bidirectional long short-term memory based recurrent neural networks. In: Proceedings of ICASSP, pp. 4869–4873. IEEE (2015)
27. Tang, H., Zhang, X., Wang, J., Cheng, N., Xiao, J.: Avqvc: one-shot voice conversion by vector quantization with applying contrastive learning. In: ICASSP 2022–2022 IEEE International Conference on Acoustics, Speech and Signal Processing (ICASSP), pp. 4613–4617. IEEE (2022)
28. Tang, H., et al.: Tgavc: improving autoencoder voice conversion with text-guided and adversarial training. In: 2021 IEEE Automatic Speech Recognition and Understanding Workshop (ASRU), pp. 938–945. IEEE (2021)
29. Toda, T., Black, A.W., Tokuda, K.: Voice conversion based on maximum-likelihood estimation of spectral parameter trajectory. IEEE Trans. Audio Speech Lang. Process. **15**(8), 2222–2235 (2007)
30. Urabe, E., Hirakawa, R., Kawano, H., Nakashi, K., Nakatoh, Y.: Electrolarynx system using voice conversion based on wavernn. In: Proceedings of ICCE, pp. 1–2. IEEE (2020)
31. Zhang, H., Zhang, Z., Odena, A., Lee, H.: Consistency regularization for generative adversarial networks. In: ICLR (2020). https://openreview.net/forum?id=S1lxKlSKPH
32. Zhang, M., Zhou, Y., Zhao, L., Li, H.: Transfer learning from speech synthesis to voice conversion with non-parallel training data. IEEE/ACM Trans. Audio Speech Lang. Process. **29**, 1290–1302 (2021)
33. Zhou, K., Sisman, B., Liu, R., Li, H.: Seen and unseen emotional style transfer for voice conversion with a new emotional speech dataset. In: Proceedings of ICASSP, pp. 920–924. IEEE (2021)
34. Zhu, J.Y., Park, T., Isola, P., Efros, A.A.: Unpaired image-to-image translation using cycle-consistent adversarial networks. In: Proceedings of the IEEE International Conference on Computer Vision, pp. 2223–2232 (2017)

# Next POI Recommendation with Neighbor and Location Popularity

Xianxian Li[1,2], Tianran Liu[2], Li-e Wang[1,2]($\boxtimes$), Zhigang Sun[1,2], and Huachang Zeng[2]

[1] Guangxi Key Lab of Multi-source Information Mining and Security,
Guangxi Normal University, Guilin 541004, China
`lixx@gxnu.edu.cn`
[2] School of Computer Science and Engineering, Guangxi Normal University,
Guilin 541004, China
`wanglie@gxnu.edu.cn`

**Abstract.** Next point-of-interest (POI) recommendation aims to predict the next destination for users. In the past, most POI recommendation models were based on the user's historical check-in trajectory to achieve recommendations. However, when these models are trained with sparse historical trajectory data, the learned user's sequence patterns are unstable, which is difficult to obtain good recommendations. In view of the above problem, we propose the next POI recommendation approach that combines neighbor information with location popularity to alleviate the sparsity of data. Specifically, we construct User-POI graph and POI-POI graph, and use graph neural networks (GNN) to capture neighbor information of effective users on these two graphs. In addition, considering that location popularity is influenced by different times and distances, we design a dynamic method to measure the impact of location popularity on the user's check-in preferences. In evaluating the experimental performance of two real-world datasets, our approach outperforms several classical next POI recommendation approaches.

**Keywords:** POI Recommendation · Graph Neural Network · Data Fusion · Location Popularity

## 1 Introduction

With the rapid development of Internet technology, location-based social networks (LBSN) are overwhelmingly popular in our society. Users share the experience of POI by check-in records. More and more location information is being collected and used to improve user experience on LBSN, which provides a valuable opportunity to explore the user's POI. Machine learning and deep learning have recently become very common in the recommendation field. Markov chains [1] captured the sequence correlation of the user's check-in records, but it didn't obtain the effective impact of different check-in records on the user. Recurrent Neural Networks (RNN) [2] have emerged to model the correlation between

© The Author(s), under exclusive license to Springer Nature Switzerland AG 2023
M. Tanveer et al. (Eds.): ICONIP 2022, LNCS 13624, pp. 367–378, 2023.
https://doi.org/10.1007/978-3-031-30108-7_31

the sequence information and the location of the user's recent check-in records, which achieved good recommendation performance. But it still suffers from data sparsity that cannot accurately explore the impact of different spatial and temporal conditions on check-in records. In some following studies [3,4], they used network embedding to model the social influence of users, which alleviated the data sparsity problem to some extent. However, these studies only consider the similarity of users' check-in trajectories, making it difficult to effectively obtain users' social influence.

Although the above approaches have achieved encouraging results, they all relied heavily on the user's historical check-in records. The next POI recommendation still encounter with a challenge as follow: when users have few historical trajectory or users leave a familiar area to go to another area with few check-in records. If we only rely on their check-in records to make recommendations, it is difficult to obtain effective POI recommendations. Therefore, more information is needed to obtain better POI recommendations. In the paper, we use GNN to learn the potential non-linear relationships in users' historical trajectory to get effective neighbor information. The location popularity is also considered, which can reflect the user's location preferences. The more check-ins a location has, the more popular the location is. However, the impact of location popularity on users is dynamic. This impact varies for the user at different times and locations. For example, users are more likely to go to the cinema in the evening than in the morning. In addition, the user's preferences for location decrease with distance. Therefore, we design a dynamic location popularity method to evaluate the impact of location on user's check-ins. The main research contributions are as follows:

- We use GNN to capture high-order information of user-location in User-POI graph and location-location in POI-POI graph. Then we perform similarity calculations on users' information to obtain the neighbors with different influences. The more similar neighbors have more influence on users' check-ins.
- We design a dynamic popularity calculation method, which can get the dynamic change of location popularity with time and geographical factors.
- We conduct complex compared experiments on two real-world datasets to evaluate the performance. The results show that our approach outperforms several classical next POI recommendation approaches.

## 2    Related Work

### 2.1    General POI Recommendation

In the past few years, Collaborative Filtering (CF) techniques [5] have been widely used to evaluate the user's POI preferences. User-based CF techniques [6] usually recommend POI preferences for the target users. However, it still faced the data sparsity problem, which led to limited recommendation performance. To solve this problem, researchers have adopted different auxiliary information such as social influence [7,8], sequence influence [9], geographical influence [6], and temporal influence [10] in POI recommendations. In addition, the model [11] used various content-aware embeddings of context information to learn the features of POI.

## 2.2  Successive POI Recommendation

Unlike general POI recommendations, the successive POI recommendations focused on the historical trajectory of the user's recent check-ins. Markov Chain model learned the user's recent sequence behaviour. It simulated the user's movement trajectory and focused on the transfer relationship between related locations. Nowadays, RNN variant [12] can be well mined key information in sequences, which is widely used for successive POI recommendations. For example, ST-RNN [13] modeled the effects of local temporal and geographical influences. The effects of different time intervals and different location distances are represented by two matrices, which were the time-specific transfer matrix and the distance-specific transfer matrix, respectively. HST-LSTM [14] designed an encoder and a decoder to model historical trajectory by combining spatio-temporal influences. ST-CLSTM [15] mined the spatio-temporal relationships of the user's successive check-ins, which explored the long-term and short-term interests of the user. CatDM [16] divided user's check-in records into multiple time windows, which captured POI categories and POI preferences by using LSTM encoders. HSP [17] used item-level information about check-in records and social relationships to model sequence transitions. It was able to learn the location preferences of the user at different times.

However, the above approaches mainly performs the influence of user's check-in trajectory on the check-in behaviour, which rely heavily on the user's historical trajectory. On the one hand, the user's check-in information has the problem of data sparsity. On the other hand, reflecting preferences in the user's check-in records are not comprehensive. Therefore, we can combine more information to alleviate the data sparsity problem and describe the user's preferences more accurately.

## 3  Preliminaries

### 3.1  Problem Formulation

In this section, we will define the formulas and terms for the problem. Let $U=\{u_1,u_2,...,u_n\}$, $L=\{l_1,l_2,...,l_m\}$ and $T=\{t_1,t_2,...,t_m\}$ be the sets of users, locations, and times, respectively. Each POI is uniquely georeferenced with longitude and latitude. The check-in record for each user is a triple $r_i=\{u_i,l_i,t_i\}$, which means the location $l_i$ visited by the user $u_i$ in time $t_i$. The historical check-in trajectory for each user is represented by $tra(r_i)=\{r_1,r_2,...,r_{mi}\}$. We convert each user's check-in trajectory to the same length $seq(u_i)=\{r_1,r_2,...,r_k\}$, where $k$ is the maximum trajectory length considered.

**Definition 1** (POI-POI Graph). The POI-POI graph can be defined as a graph $G_l=\{L,E_l\}$. $L$ is the set of POIs and $E_l$ is the set of edges between POIs. $e_{i,j}$

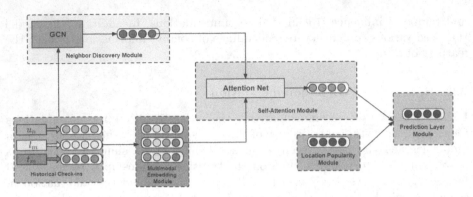

**Fig. 1.** The architecture of our approach.

belongs to the edge set $E_l$, it represents from location $l_i$ to location $l_j$. The weight $w_{i,j}^l$ of edge $e_{i,j}$ is defined as the number of times that location $l_j$ is checked in after location $l_i$.

**Definition 2** (User-POI Graph). The User-POI graph can be defined as a graph $G_u = \{U, L, E_u\}$. $U$ is the set of users and $E_u$ is the set of edges between users and POIs. $e_{i,j}$ belongs to the edge set $E_u$, it represents user $u_i$ check-in location $l_j$. The weight $w_{i,j}^u$ of edge $e_{i,j}$ is defined as the number of times that user $u_i$ check-in location $l_j$.

**Definition 3** (Next POI Recommendation). Given a sequence of users, the goal of the next POI is to predict the location that a user will visit at the following time.

## 4   Approach

In this section, we will introduce our approach in detail, which is divided into five modules. Figure 1 is the architecture of our approach.

### 4.1   Multimodal Embedding Module

The multimodal embedding module includes information about the user's historical trajectory and spatio-temporal intervals. As there is a cyclical pattern to the user's trajectory behaviour, we use a week to reflect the periodicity of the user's check-ins. The 168 h timestamp of a week can be converted into a 168-dimensional vector, which will help to understand the specific timestamp intervals of the user's check-in to a place. We use $e_u \in \mathrm{R}^d$, $e_l \in \mathrm{R}^d$ and $e_t \in \mathrm{R}^d$ to denote the user, location and time embedding vectors, respectively, where $d$ is the dimension of the embedding space. Inspired by literatures [18,19], in the embedding module, these vectors are converted into low-dimensional dense representations, which can model the user's check-in information more accurately

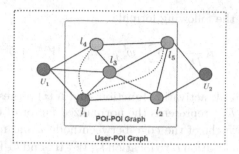

**Fig. 2.** The solid red boxes represent the POI-POI graph, where each node in the graph represents a location ($l_1, l_2, l_3, l_4, l_5$). The dashed boxes represent the User-POI graph, where the green nodes ($u_1, u_2$) represent the users.(Color figure online)

and reduce the amount of computation. We use the sum of these dense vectors to represent the user's trajectory. The embedding of each check-in record can be represented as $e_c = e_u + e_l + e_t \in R^d$. The user's history trajectory embedding matrix can be represented as $M(u) = [e_c^{r_1}, e_c^{r_2}, ..., e_c^{r_k}] \in R^{k \times d}$. The time and space intervals between the two locations are represented as $\triangle t \in R^{k \times d}$ and $\triangle s \in R^{k \times d}$, respectively. We represent the embedding matrix in time and the embedding matrix in space by $M(\triangle t) \in R^{k \times k \times d}$ and $M(\triangle s) \in R^{k \times k \times d}$, respectively. To represent the spatio-temporal relationship matrix $M(\triangle) \in R^{k \times k}$ of the user's historical trajectory, we use the weighted sum of the dimensions and sum their values, which are defined as follows:

$$M(\triangle) = Sum(M(\triangle t)) + Sum(M(\triangle s)) \tag{1}$$

## 4.2 Neighbor Discovery Module

The user's history of check-ins is often sparse. It is difficult to achieve good POI recommendations if you only rely on the historical check-in trajectory. In social networks, the check-in records of a user's neighbors can influence the user's behaviour. Therefore, finding effective neighbors can alleviate sparsity of the user's check-ins. The more similar the check-in trajectory information between users, the greater the social influence between them. The influence between users check-in records is often a complex non-linear relationship. We construct POI-POI graph and User-POI graph, from which we learn the embedding representation of locations and the embedding representation of users by GNN. Figure 2 consists of two parts: the POI-POI graph and the User-POI graph. In the POI-POI graph, each node represents a location, and there are both direct and indirect influences between locations. Node $l_1$ is influenced by its two first-order nodes ($l_2$ and $l_3$) and its higher-order neighbors ($l_4$ and $l_5$). The information of the locations is updated by gathering the information of the surrounding locations through the GNN. Formally, we update the information of locations at the

$k$ layer according to the following formula:

$$h_{l_i}^{(k)} = \sigma\left(\frac{1}{D_{(i)}} \sum_{l_j \in N(i)} w_{i,j}^l \times h_{l_j}^{(k-1)} W^{(k)} + b^{(k)}\right) \tag{2}$$

where $\sigma()$ is a LeakReLU activation function, and $N(i)$ represents the first-order neighbors of node $i$. $D_{(i)}$ represents the normalized factor of the sum of degrees. $w_{i,j}^l$ represents the weight of the edge between node $i$ and node $j$, which is the number of times from location $l_i$ to location $l_j$. $W \in R^d$ and $b \in R^d$ represent weight matrix and bias term, respectively. $h_{l_i}^{(k)}$ represents the feature representation of location $i$ at the $k$-th layer.

In order to obtain a more efficient latent representation of users in the User-POI graph, we consider not only the latent representation of users but also introduce the latent representation of location in Eq. 2. We use the inner product of the user's representations and the neighbor's representations to evaluate their similarity. Finally, all neighbors' influences are weighted, which represent by $e_n \in R^d$. The formula is defined as follows:

$$h_{u_i}^{(k)} = \sigma\left(h_{u_i}^{(k-1)} W_1^{(k)} + \sum_{l_j \in N(i)} w_{i,j}^u \times h_{l_j}^{(k)} W_2^{(k)} + b_1^{(k)}\right) \tag{3}$$

$$S_{i,j} = \phi(h_{u_i}^{(k)}, h_{u_j}^{(k)}) \tag{4}$$

$$e_n = \sum_{j=1}^{n} S_{i,j} h_{u_j}^{(k)} \tag{5}$$

where $W_1 \in R^d$ and $W_2 \in R^d$ represent weight matrix, $b_1 \in R^d$ represents bias term. $w_{i,j}^u$ represents the number of times that user $u_i$ check-in location $l_j$. $h_{u_i}^{(k)}$ and $h_{u_j}^{(k)}$ represent the feature representation of user $i$ and user $j$ at layer $k$, respectively. $S_{i,j}$ represents the similarity between user $i$ and user $j$.

### 4.3   Self-attention Module

In the next POI recommendation, different check-in trajectories affect the next check-in location in different degrees. The recommendation effect will be reduced if the same weight is given to the check-in location. Inspired by the self-attention mechanism [20], we expand on the self-attention mechanism. We add the neighbor information to the user's trajectory. We use $e_H = e_u + e_l + e_t + e_n \in R^d$ to represent the new embedding representation of check-in record. The trajectory embedding representation of the user is defined as a matrix $H \in R^{k \times d}$, which is used as input to the self-attention module. The self-attention module effectively captures long-term dependencies by taking into account spatio-temporal intervals and neighbor information, which enables to give different weights to locations in the historical check-in trajectory. Then we transform it into three matrices by linear projection and feed it to the attention layer. The formula is defined as follows:

$$W_s = softmax(\frac{HW_Q(HW^K)^T}{\sqrt{d_v}}) \qquad (6)$$

$$F = W_s(M(\triangle)W^V) \qquad (7)$$

where $Q \in R^{d \times d}$, $K \in R^{d \times d}$, and $V \in R^{d \times d}$ are projection matrices, $M(\triangle)$ is the spatio-temporal embedding vector matrix, $d_v$ is used to prevent excessive inner product in the self-attention module. $F \in R^{k \times d}$ is the preference matrix for the final output location.

### 4.4  Location Popularity Module

The number of POIs check-in by users reflects the popularity of POIs. Location popularity is influenced by different time and space, so we design a dynamic popularity method. We use each hour of the day as a time period and calculate the number of POI check-ins for and the sum of all POI check-ins during time period. We use their ratio as the probability that each location will be check-in. Since the popularity of POI is related to distance, in general, the popularity of POI decreases with increasing distance. To account for the effect on location popularity, we use a decay function that transforms the distance interval into an appropriate weight. Finally, the popularity of the user's check-in location can be represented by the matrix $P = [p_1, p_2, ..., p_k]$. The formula is defined as follows:

$$f_i = \frac{m_i}{M_i} \qquad (8)$$

$$p_i = g(d_i) \cdot f_i \qquad (9)$$

where $m_i$ is the number of the location check-in during this period. $M_i$ represents the number of all locations check-in within this time period. $g()$ is a decay function, which is denoted by $g(x) = \frac{1}{log(e+x)}$. $d_i$ represents the distance between the $i$-th location and the last location. $p_i$ represents the popularity of the current location $i$.

### 4.5  Prediction Layer Module

We will combine the user's location preference matrix and the location popularity matrix to calculate the score for each location. A higher score for a location indicates that users are more likely to visit that location. The system recommends the top $N$ locations based on the score. Next, we train our model by using a cross-entropy loss function and optimize the loss function by using stochastic gradient descent. The formula is defined as follows:

$$s_{l_i} = F \cdot e_{l_i}^T + P \cdot e_{l_i}^T \qquad (10)$$

$$loss = -\sum_i \sum_{m_i \in tra(r_i)} [log(\sigma(s_{l_i})) + \sum_{j=1, j \neq i}^{L} log(1 - \sigma(s_{l_j}))] \qquad (11)$$

**Table 1.** Characteristics of the different baselines.

| Approach | Category | Temporal influence | Geographical influence | Social influence |
|---|---|---|---|---|
| **GeoSoCa** | Traditional | ✗ | ✓ | ✓ |
| **ST-RNN** | Successive | ✓ | ✓ | ✗ |
| **HST-LSTM** | Successive | ✓ | ✓ | ✗ |
| **CatDM** | Successive | ✓ | ✓ | ✗ |
| **HSP** | Successive | ✓ | ✓ | ✓ |
| **OUR** | Successive | ✓ | ✓ | ✓ |

## 5    Experiments

### 5.1    Datasets and Parameter Setting

We conduct experiments on two real-world datasets. User's check-in records in New York and Tokyo are from April 12, 2012 to February 16, 2013. In the datasets, we randomly select 70% of the check-in data as training data and 30% as test data. We use the neural network to train the same hyperparameters. These parameters can then be applied to each module to reduce the experimental training time. In our experiments, the embedding dimension $d$ of each module is set to 50, the maximum length of the trajectory sequence is set to 100, and the values of learning rate and dropout rate are set to 0.001 and 0.2, respectively. We use the Adam optimizer to optimize our model, setting the epoch to 100 and the batch size to 128 for our experiments.

### 5.2    Evaluation Metrics

In order to effectively evaluate the experimental results, we use two metrics to evaluate Recall@$N$ and Precision@$N$, which are widely used in POI recommendation systems. We choose the values of $N=\{5, 10, 15\}$ on Recall@$N$ and Precision@$N$ as the experimental results. The formulae are as follows:

$$Recall = \frac{1}{|U|} \sum_{u \in U} \frac{|R(u) \cap T(u)|}{T(u)} \tag{12}$$

$$Precision = \frac{1}{|U|} \sum_{u \in U} \frac{|R(u) \cap T(u)|}{R(u)} \tag{13}$$

where $R(u)$ represents the recommendation list of POI and $T(u)$ represents the actual POI check-in list.

### 5.3    Baseline Approach

**GeoSoCa** [8]: It combined geographical, social, and POI classification to calculate the user's preference scores for POI. **ST-RNN** [13]: It used spatial-temporal recurrent neural network to simulate the local time and geographical influence.

**Table 2.** Recommendation performance comparison with baselines.

| | NYC | | | | | | TKY | | | | | |
|---|---|---|---|---|---|---|---|---|---|---|---|---|
| | Pre@5 | Pre@10 | Pre@15 | Rec@5 | Rec@10 | Rec@15 | Pre@5 | Pre@10 | Pre@15 | Rec@5 | Rec@10 | Rec@15 |
| GeoSoca | 0.0221 | 0.0183 | 0.0151 | 0.0734 | 0.1103 | 0.1422 | 0.0423 | 0.0288 | 0.0213 | 0.0953 | 0.1349 | 0.1504 |
| HST-LSTM | 0.0295 | 0.0228 | 0.0191 | 0.1073 | 0.1512 | 0.1937 | 0.0503 | 0.0335 | 0.0282 | 0.1153 | 0.1604 | 0.1807 |
| ST-RNN | 0.0482 | 0.0389 | 0.0282 | 0.1104 | 0.1596 | 0.2053 | 0.0605 | 0.0448 | 0.0331 | 0.1254 | 0.1649 | 0.1820 |
| CatDM | 0.0717 | 0.0456 | 0.0346 | 0.2403 | 0.3110 | 0.3465 | 0.0880 | 0.0588 | 0.0455 | 0.2147 | 0.2737 | 0.3382 |
| HSP | 0.0741 | 0.0490 | 0.0347 | 0.2447 | 0.3183 | 0.3566 | 0.0882 | 0.0596 | 0.0460 | 0.2169 | 0.2879 | 0.3494 |
| OUR | **0.0755** | **0.0502** | **0.0362** | **0.2524** | **0.3307** | **0.3715** | **0.0909** | **0.0617** | **0.0472** | **0.2296** | **0.3108** | **0.3692** |
| Improvement | 1.89% | 2.45% | 4.32% | 3.15% | 3.90% | 4.18% | 3.06% | 3.52% | 2.61% | 5.86% | 7.95% | 5.67% |

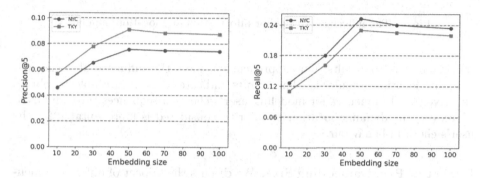

**Fig. 3.** Impact of POI embedding size

**HST-LSTM** [14]: It combined spatio-temporal influences to predict the next location. Encoders and decoders model access sequences to improve predictive performance. **CatDM** [16]: It divided the user's check-in record into multiple time windows. The time sequence data is modeled using an LSTM deep encoder, which is used to capture POI categories and the user's POI preferences. **HSP** [17]: It used item-level check-in sequences and area-level spatial information to model sequence transformations and learns user preferences for location at different times through recurrent neural networks.

### 5.4   Results and Analysis

**Performance Comparison.** Our approach is compared with multiple baselines for performance in two real-world datasets. According to Table 1, we can see the influencing factors considered in the different baselines. Table 2 shows the experimental results of the different approaches in NYC and TKY, which can be seen that our approach outperforms all baselines in evaluation metrics. In the NYC dataset, our approach improves over the best baseline HSP by 1.89% to 4.32% and 3.15% to 4.18% in Precision@$N$ and Recall@$N$, respectively. In the TKY dataset, our approach improves by 2.61% to 3.52% and 5.67% to 7.95% in the same metrics, respectively. We also observe that HSP combines spatio-temporal relationships with social influence and outperforms the rest of the baselines. It indicates that effective consideration of social impact can improve

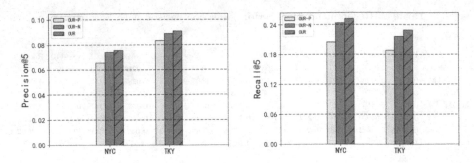

**Fig. 4.** The influence of neighbor information and location popularity

the recommended results. Our approach considers the influence of social and dynamic location popularity. The results indicate that we are able to obtain effective social influences for modeling user's check-in sequences through GNN. Moreover, the dynamic location popularity considered is more suitable for the user's check-in behaviour.

**Impact of POI Embedding Size.** We discuss the impact of different dimensional embedding vectors of POI on our approach, using Precision@5 and Recall@5 as our experimental evaluation metrics. Figure 3 shows the performance of our approach in different dimensions. The experimental results show that when the dimension of the POI vector is less than 50, the performance improves significantly, which indicates the ability to capture more POI information. When the dimension of the POI vector is more than 50, the performance gradually stabilizes, which indicates that more useful POI information cannot be captured. In our experiments, we set the dimension of the embedding to 50.

**Ablation Analysis.** We study the performance gains of the following modules: the neighbor discovery module and the location popularity module. Specifically, we use OUR-P and OUR-N to denote the removal of the neighbor discovery module and the location popularity module, respectively. Figure 4 shows the evaluation metrics of the different modules in Precision@5 and Recall@5. We can observe that two modules are able to improve the performance. The results show that we can effectively capture the relationship between users and locations to learn neighbor information through GNN. We consider the neighbor information in the user's check-in trajectory, which can effectively improve the performance. It also shows that users have different levels of preference for location at different times and check-in distances. We use dynamic location popularity to reflect the impact of location on user check-ins can also improve performance. Overall, we demonstrate the effectiveness of the neighbor discovery module and the location popularity module through ablation experiments.

# 6 Conclusion

In this paper, our approach improves the POI recommendation by exploring the neighbor relationship and location population of users. We use graph neural networks to find neighbors by learning the interactions between users and locations. At the same time, we optimize the influence of location popularity on the user's preference for check-in locations. Effectively combine neighbor information and location popularity in the user's historical check-in trajectory. Our approach is compared with the baseline approaches through two real-world datasets. The experiments show that our approach is better. In future research, it will be the focus of our study to use cross-domain data to alleviate the sparsity problem of POI data in the next POI recommendation.

**Acknowledgements.** This work is supported in part by the National Natural Science Foundation of China (Nos. U21A20474 and 62262003), the Guangxi Science and technology project (2021AA11006 and GuikeAD21220114), the Guangxi Collaborative Innovation Center of Multi-source Information Integration and Intelligent Processing, the Guangxi Talent Highland Project of Big Data Intelligence and Application, the Guangxi Natural Science Foundation (Nos. 2020GXNSFAA297075). The Innovation Project of Guangxi Graduate Education YCSW202-2162.

# References

1. He, R., McAuley, J.: Fusing similarity models with markov chains for sparse sequential recommendation. In: 2016 IEEE 16th International Conference on Data Mining (ICDM), pp. 191–200 (2016)
2. Schuster, M., Paliwal, K.K.: Bidirectional recurrent neural networks. IEEE Trans. Sig. Process. **45**(11), 2673–2681 (1997)
3. Huang, L., Ma, Y., Liu, Y., Sangaiah, A.K.: Multi-modal bayesian embedding for point-of-interest recommendation on location-based cyber-physical-social networks. Future Gener. Comput. Syst. **108**, 1119–1128 (2020)
4. Yang, C., Sun, M., Zhao, W.X., Liu, Z., Chang, E.Y.: A neural network approach to jointly modeling social networks and mobile trajectories. ACM Trans. Inf. Syst. (TOIS) **35**(4), 1–28 (2017)
5. Ye, M., Yin, P., Lee, W.C., Lee, D.L.: Exploiting geographical influence for collaborative point-of-interest recommendation. In: Proceedings of the 34th International ACM SIGIR Conference on Research and Development in Information Retrieval, pp. 325–334 (2011)
6. Liu, B., Fu, Y., Yao, Z., Xiong, H.: Learning geographical preferences for point-of-interest recommendation. In: Proceedings of the 19th ACM SIGKDD International Conference on Knowledge Discovery and Data Mining, pp. 1043–1051 (2013)
7. Huang, L., Ma, Y., Liu, Y.: Point-of-interest recommendation in location-based social networks with personalized geo-social influence. China Commun. **12**(12), 21–31 (2015)
8. Zhang, J.D., Chow, C.Y.: Geosoca: exploiting geographical, social and categorical correlations for point-of-interest recommendations. In: Proceedings of the 38th International ACM SIGIR Conference on Research and Development in Information Retrieval, pp. 443–452 (2015)

9. Zhang, J.D., Chow, C.Y.: Lore: Exploiting sequential influence for location recommendations. In: Proceedings of the 22nd ACM SIGSPATIAL International Conference on Advances in Geographic Information Systems, pp. 103–112 (2014)
10. Gao, H., Tang, J., Hu, X., Liu, H.: Exploring temporal effects for location recommendation on location-based social networks. In: Proceedings of the 7th ACM conference on Recommender Systems, pp. 93–100 (2013)
11. Chang, B., Park, Y., Park: Content-aware hierarchical point-of-interest embedding model for successive poi recommendation. In: IJCAI, vol. 2018, p. 27th (2018)
12. Zhang, Y., Dai, H., Xu, C., Feng, J., Wang, T., Bian, J.: Sequential click prediction for sponsored search with recurrent neural networks. In: Proceedings of the AAAI Conference on Artificial Intelligence, pp. 1369–1375 (2014)
13. Liu, Q., Wu, S., Wang, L., Tan, T.: Predicting the next location: a recurrent model with spatial and temporal contexts. In: Thirtieth AAAI Conference on Artificial Intelligence, pp. 194–200 (2016)
14. Kong, D., Wu, F.: Hst-lstm: A hierarchical spatial-temporal long-short term memory network for location prediction. In: IJCAI, vol. 18, pp. 2341–2347 (2018)
15. Zhao, P., Zhu, H., Liu, V.S.: Where to go next: a spatio-temporal LSTM model for next poi recommendation. arXiv preprint arXiv:1806.06671 (2018)
16. Yu, F., Cui, L., Guo: A category-aware deep model for successive poi recommendation on sparse check-in data. In: Proceedings of the Web Conference 2020, pp. 1264–1274 (2020)
17. Shi, M., Shen, D.: Next point-of-interest recommendation by sequential feature mining and public preference awareness. J. Intell. Fuzzy Syst. 40(3), 4075–4090 (2021)
18. Liu, Q., Liu, Z., Zhang, H.: An empirical study on feature discretization. arXiv preprint arXiv:2004.12602 (2020)
19. Liu, Q., Wu, S., Wang, L.: Multi-behavioral sequential prediction with recurrent log-bilinear model. IEEE Trans. Knowl. Data Eng. 29(6), 1254–1267 (2017)
20. Vaswani, A., Shazeer, N., Parmar, N., Uszkoreit, L., Polosukhin, I.: Attention is all you need. Adv. Neural Inf. Process. Syst. 30, 5998–6008 (2017)

# Trustworthiness and Confidence of Gait Phase Predictions in Changing Environments Using Interpretable Classifier Models

Danny Möbius[1], Jensun Ravichandran[2], Marika Kaden[2], and Thomas Villmann[2(✉)]

[1] ICM - Institute Chemnitzer Maschinen - und Anlagenbau e.v.,
Otto-Schmerbach-Street 19, 09117 Chemnitz, Germany
d.moebius@icm-chemnitz.de
[2] University of Applied Sciences Mittweida,
Technikumpatz 17, 09648 Mittweida, Germany
{jensun.ravichandran,marika.kaden,thomas.villmann}@hs-mittweida.de
https://www.ifm-chemnitz.de/

**Abstract.** The recognition of the different phases of human gait is valuable in areas such as rehabilitation and sports. Machine Learning models have been increasingly used for such recognition tasks. However, such models are usually trained on data obtained from participants in strictly controlled environments which—needless to say—might vary quite significantly from the environment in which the models are subsequently employed. Therefore, it is advisable to analyze the confidence of the model's predictions. To this end, we present an interpretable classifier for gait phase detection. Together with classification reliability estimation tools, classification predictions can be rejected in low confidence scenarios. Our classifier is based on a robust and distance-based Learning Vector Quantization classifier. Finally, we present our approach using a real-world application in gait phase detection, which consists of one learning scenario and two different prediction scenarios.

**Keywords:** gait phase recognition · interpretable machine learning · drift detection · classification certainty

## 1 Introduction

The analyses of the human gait is prerequisite for many motion analysis applications in rehabilitation and in professional and recreational sports [6,8]. A precise recording of the individual gait phases often serves as the basis for further analyses [12]. In general, the gait is partitioned into the *swing* and the *stance* phases, which can further be divided into eight sub-phases. For the development of a gait phase prediction, machine learning solutions have to contend with the fact that training data are usually recorded in strictly controlled environments whereas the

---

M.K is funded by the European Social Fund (ESF), ESF-SAB 100381749.

application environment might heavily differ from the training conditions. Moreover, precisely labeled test data from the application environment are often infeasible to obtain due to technical restrictions. Hence, a trained model should be able to decide in its recall mode, whether it is applicable for the given situation (data sample) or not. This problem is related to but not fully caused by data drift due to the changing environment. One possibility to tackle this problem is to use an interpretable classifier for gait phase recognition. Such classifiers provide valuable tools to estimate the classification certainty of the model for a given sample.

In the project reported here, the gait data for the training scenario (denoted as scenario A) was recorded via a marker-based motion capturing system and processed with the digital human model *alaska/Dynamicus*[1], which delivers the joint angles of the human body together with their velocity values in real time. For precise differentiation of the gait phases, measurements obtained from Force Measurement Plates (FMP) placed on the floor are used. However, this equipment is very expensive and only allows short strides to be studied. Furthermore, the FMP is sensitive to environmental conditions and can therefore only be used in rather limited scenarios as compared to the full range of effects of an exoskeleton on human gait in the work environment. A more realistic scenario in this sense would be a gait analysis on a treadmill for the investigation of the influence of exoskeleton on human gait. Throughout the rest of the paper, the treadmill scenario is denoted as B1 and the exoskeleton scenario is denoted as B2. Note that neither B1 nor B2 requires the expensive FMP-system for the detection of gait phases. Thus the machine learning task is to train a gait phase predictor for scenario A and to investigate the validity of their applications to scenarios B1 and B2. This validity has to be proven individually for each test person, because the influence of the changed environment is different for different persons. The transition from learning scenario A to the application scenarios B1 and B2 can be seen as a data drift in the context of machine leaning.

To tackle this task, an interpretable classifier is first trained for scenario A to predict the gait phases based on the labeled data obtained by the FMP-system for the angle and velocity trajectories provided by *alaska/Dynamicus*. For this purpose, Generalized Matrix Learning Vector Quantizer (GMLVQ) [17,19] is used as the classifier, which is known to be powerful, robust and interpretable. Furthermore, the robustness is based on implicit margin hypothesis optimization [16], which is a lower bound of the separation margin optimized by the popular Support Vector Machine (SVM). On the one hand, the evaluation of this margin in the working phase allows to estimate the certainty of a classification decision and on the other hand, it can be used to reject data points based on their proximity to the decision border. In other words, the properties of GMLVQ facilitate the detection of data drift violating the model validity. Moreover, GMLVQ internally applies an adaptive linear mapping of the data to achieve better classification performance. Hence, any data drift in the working phase, which only influences the nullspace of this linear mapping, does not disturb the classification

---

[1]    alaska/Dynamicus is a module for efficient, and comfortable generation and use of maker-based-system models of the human body provided by the ICM - Institute Chemnitzer Maschinen- und Anlagenbau e.V, Chemnitz, Germany.

behavior [21]. This key observation lies at the heart of the method we propose in this work. This observation is of particular interest, because label drift detection methods such as the one proposed in [23,24] cannot be applied because of the entirely different label distribution in the application phase. One reason why the label distributions are different is that humans walk differently when on a treadmill than they would walking on the floor.

## 1.1 Related Work

The unique problem described above demands a custom solution. To the best of our knowledge, we are not aware of any existing solutions that adequately solve the problem. Having said that, after training a regular classifier model on scenario A, subsequent *unsupervised* transfer learning methods could be applied in scenarios B1 and B2 in which no labels are available. This is referred to as transductive learning [23,24]. An appropriate model for such transductive learning is the aforementioned method [21] which explicitly evaluates the nullspace in GMLVQ.

## 2 Generalized Matrix Learning Vector Quantization as an Interpretable Machine Learning Classifier

For our approach, we require a classifier model that lends itself to making statements about its own validity (classification confidence) in scenarios B1 and B2 although only trained explicitly for scenario A. Thus, we focus on explainable or interpretable learning models, which allow for much better model inspection than black-box approaches do [11]. Having said that, interpretable models, e.g. nearest-prototype methods, should be favored over explainable methods because they are inherently interpretable (ante-hoc interpretability) [15]. For vector quantization methods, the training and application of models are both intuitive and transparent, subject to the (dis-)similarities used to compare data points and the trainable prototypes.

One such classifier is the Generalized Learning Vector Quantization method (GLVQ) [17] adapted from the heuristic Learning Vector Quantization (LVQ) introduced by Kohonen [10]. During training, GLVQ minimizes a cost function that approximates the classification error. Over the years, several extensions to GLVQ have been developed [1]. Among them, metric adaptation is one of the most successful improvements [7,19], which additionally performs an automatic input feature weighting and implicit classification correlation analysis to optimize the class discrimination.

For the training of a GLVQ model, data samples $\mathbf{x}_i \in \mathcal{X} \subset \mathbb{R}^n$ with corresponding class labels $c(\mathbf{x}_i) \in \mathcal{C}$, $|\mathcal{C}| = C$ are required. It is also assumed that prototype vectors $\mathbf{w}_k \in \mathcal{W} \subset \mathbb{R}^n$, which are equipped with class labels $c(\mathbf{w}_k) \in \mathcal{C}$, such that at least one prototype per class is available. Further, a dissimilarity measure $d(\mathbf{x}_i, \mathbf{w}_k)$ is supposed to judge the similarity between data and prototypes. Such a measure also has to be differentiable with respect to its second argument for Stochastic Gradient Descent (SGD) learning.

The model prediction for a data sample $\mathbf{x}$ with respect to the current prototype set $\mathcal{W}$ is realized via the Winner-Takes-All (WTA) rule

$$\mathbf{w}_{s(\mathbf{x},\mathcal{W})} = \arg\min_{\mathbf{w}\in\mathcal{W}} d(\mathbf{x}, \mathbf{w}) , \qquad (1)$$

yielding the class prediction $c\left(\mathbf{w}_{s(\mathbf{x})}\right)$. GLVQ training distributes the prototypes in the data space according to the class distribution given in $\mathcal{X}$ based on the cost function (classification loss)

$$L(\mathcal{X}, \mathcal{W}) = \sum_{i=1}^{|\mathcal{X}|} f\left(\mu(\mathbf{x}_i, \mathcal{W})\right) \qquad (2)$$

to be minimized usually by SGD or a variant thereof. In $L(\mathcal{X}, \mathcal{W})$, the so-called transfer function $f(\cdot)$ is a monotonously increasing function, often chosen to be a sigmoid squashing function. The classifier function

$$\mu(\mathbf{x}_i, \mathcal{W}) = \frac{d(\mathbf{x}_i, \mathbf{w}^+) - d(\mathbf{x}_i, \mathbf{w}^-)}{d(\mathbf{x}_i, \mathbf{w}^+) + d(\mathbf{x}_i, \mathbf{w}^-)} \qquad (3)$$

depends on the most similar positive prototype $\mathbf{w}^+$ to $\mathbf{x}_i$, which is equipped with a class label matching $c(\mathbf{x}_i)$, and the most similar negative prototype $\mathbf{w}^-$ with different class label. Thus $\mu(\mathbf{x}_i, \mathcal{W})$ delivers a negative value if the data is correctly classified for the current prototype configuration $\mathcal{W}$.

The SGD learning is performed with

$$\Delta\mathbf{w}^\pm \propto -\frac{\partial f\left(\mu(\mathbf{x}_i, \mathcal{W})\right)}{\partial\mathbf{w}^\pm}$$

using the local error $f\left(\mu(\mathbf{x}_i, \mathcal{W})\right)$. If $d$ is the squared Euclidean distance this adaptation realizes an attraction-repulsing-scheme (ARS) whereby $\mathbf{w}^+$ is shifted towards $\mathbf{x}_i$ whereas $\mathbf{w}^-$ is repelled.

In [3] it has been proven that GLVQ implicitly maximizes the local hypothesis margin

$$m\left(\mathbf{x}, \mathcal{W}\right) = d\left(\mathbf{x}, \mathbf{w}_{s_2(\mathbf{x},\mathcal{W}_2)}\right) - d\left(\mathbf{x}, \mathbf{w}_{s(\mathbf{x},\mathcal{W})}\right) , \qquad (4)$$

where $\mathcal{W}_2$ is the reduced prototype set containing all prototypes of $\mathcal{W}$ except those with label $c\left(\mathbf{w}_{s(\mathbf{x})}\right)$ is considered in (1) and the second class winner $s_2(\mathbf{x}, \mathcal{W}_2)$ is obtained. Hence, GLVQ delivers a robust classification decision.

Several factors contribute to the interpretability of the model overall. First of all, the prototypes are in the feature space and are directly responsible for the classification decision in a manner that is intuitive to understand. Further, distance values are generally easy to interpret as opposed to inner product values considered in (deep) neural networks (multilayer perceptrons) [5] or in SVMs [18]. In addition, distances are lower bounded whereas inner products and kernels are generally unbounded.

A classification certainty $(C)$ about a decision can be directly defined by

$$r_C(\mathbf{x}, \mathcal{W}) = \frac{m\left(\mathbf{x}, \mathcal{W}\right)}{d\left(\mathbf{x}, \mathbf{w}_{s_2(\mathbf{x},\mathcal{W}_2)}\right) + d\left(\mathbf{x}, \mathbf{w}_{s(\mathbf{x},\mathcal{W})}\right)} \qquad (5)$$

which is simply the normalized hypothesis margin. The same can be also interpreted as model-self-confidence [13]. Note that this quantity is the same as the one considered in [4] albeit without referring to the relationship to the hypothesis margin. Rather, the authors point out that this geometrically motivated certainty measurement is efficient for the use as classification reject option and, hence, could be used to optimize a respective reject threshold. The corresponding reject decision is denoted as *classification reject*. Here, we refrain from optimizing a classification reject threshold, because we do not specify a direct cost for misclassified or rejected data points. Instead, we apply the statistics of the classification certainty together with that of the rejected data points to detect variations in the decision process. Therefore, we use an appropriately chosen percentile of all $r_C$ values for the training data. In Fig. 1 on the left side, a respective example histogram of $r_C - values$ for the training in scenario A and a corresponding test on data of scenario B is depicted.

In addition to changes in the statistics near the decision boundary, data points can drift out of the *sight*-range of the prototypes. Data points with distance $d\left(\mathbf{x}, \mathbf{w}_{s(\mathbf{x}, \mathcal{W})}\right)$ significantly deviating from the majority in the training data are denoted as outliers. Accordingly, an *outlier reject* option can be integrated quite intuitively by specifying a suitable distance threshold [4,22]. As in the case of the classification rejection strategy, we do not learn the threshold. We obtain it after learning the model using the training data. For the training set a distribution of the distances $d\left(\mathbf{x}, \mathbf{w}_{s(\mathbf{x}, \mathcal{W})}\right)$ between the data points to the winning prototypes can be estimated such that a percentile based threshold is established, e. g. the above 99% percentile. An example histogram of the distance statistics is shown in Fig. 1 on the right side.

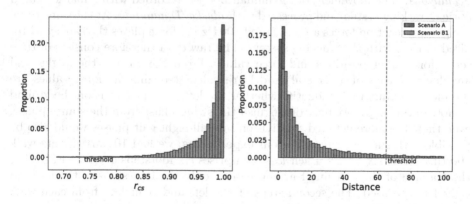

**Fig. 1.** Histogram of the $r_C$-values (left) and the distances (right) for the training data (scenario A) and the test data (scenario B).

As mentioned earlier GMLVQ differs from GLVQ in that an internal linear data mapping is incorporated to improve the flexibility and the subsequent classification performance [20]. For this purpose, the distance function is chosen as a quadratic form

$$d_{\Omega}(\mathbf{x}, \mathbf{w}) = (\mathbf{\Omega}(\mathbf{x} - \mathbf{w}))^2 \tag{6}$$

with $\mathbf{\Omega} \in \mathbb{R}^{m \times n}$ being the mapping matrix and $m \leq n$ the mapping dimension. The mapping matrix $\mathbf{\Omega}$ is learned in conjunction with the prototypes and improves the class separability [19]. When $m < n$ the so-called *limited rank* GMLVQ is obtained [2]. Both, full-rank as well as limited-rank GMLVQ have been shown to be margin optimizers [16] and thereby provide better robustness.

Additionally, the mapping matrix $\mathbf{\Omega}$ can be exploited to further improve interpretability. Rewriting the quadratic form (6) as

$$d_{\Omega}(\mathbf{x}, \mathbf{w}) = (\mathbf{x} - \mathbf{w})^T \mathbf{\Lambda} (\mathbf{x} - \mathbf{w}) \tag{7}$$

with $\mathbf{\Lambda} = \mathbf{\Omega}^T \mathbf{\Omega} \in \mathbb{R}^{n \times n}$, $\mathbf{\Lambda}$ can be interpreted as a classification correlation matrix (CCM) delivering information about those correlations of the data features, which contribute to a better class discrimination [9].

For limited-rank GMLVQ, frequently $m << n$ is chosen to reduce the complexity of the model significantly. However, this influences outlier detection as well as drift detection: if the respective change/deviation of the data occurs in the nullspace of the limited rank matrix $\mathbf{\Omega}$, these events have absolutely no impact on the prediction of the classifier [21].

## 3    Experimental Setup

### 3.1    The Collected Data

*Scenario A.* The movement of the human body is recorded with a marker-based optical tracking system and processed with *alaska/Dynamicus*. At the same time, the ground reaction forces are measured 100 Hz via force plates that are synchronised in time with the tracking system. The raw data therefore consists of time series for marker positions and force values. From the former, the angles and angular velocities of a skeletal model are calculated and the force values are used for labelling, i.e. to identify whether the left or right foot is on the ground or not. In this paper we only distinguish the swing phase from the stance phase, even though a more detailed subdivision into all eight gait phases would also be possible with such a system. For our experiments, we had 10 participants walk 100 times over six FMPs. Each *walk* comprises six footsteps over the FMP. For the training of the machine learning model, only the first steps of the left and right legs are used. The second steps of the left and right legs from each walk were used to validate the model.

*Scenario B.* In the second scenario, 10 participants walked on a treadmill for 15 min at 4.0 km/h, once without an exoskeleton (scenario **B1**) and once with an exoskeleton (scenario **B2**). As mentioned earlier, we do not have force plates available for the treadmill and thus these data were not labelled.

## 3.2   Feature Generation

The data features are generated from the joint angles and their angular velocities. Only the following 10 joints of the left and right side of the human body are used, namely the hips, knees, ankles, shoulders and elbows. The aim is to identify the two main gait phases for each time step and for each leg independently. Therefore, feature generation must be performed adequately. After extensive analyses, it was found that simple statistical values of a very short time interval are sufficient to obtain adequate classification results. The final feature extraction for each time step is the calculation of mean and standard deviation per sensor of a time window of the length of seven time steps. Including the skewness did not provide an advantage. This results in a feature vector of size 120 (mean and standard deviation of the absolute values of the angles and velocities in three directions at the 10 joints). The feature generation is done in the same way for all scenarios.

## 3.3   Model Training

We trained classification models for the left and right sides independently. For the sake of simplicity, we will only consider the model for the left side in the following discussion. For the right side, the consideration is completely analogous and both models are applied to the data at the end to obtain a picture of the entire gait phase. The GMLVQ model was trained on the features generated from scenario A. The mapping dimension of $\Omega$ was set to 15 and only one prototype per class was used, which means that the model can be interpreted as a linear classifier. For building and training our models, we used the ProtoTorch [14] Python package. To evaluate the generalization ability for the classifier in Scenario A, we apply 10-fold cross-validation using the walks of nine subjects for training and the one remaining for testing. We further used the data over the first and second full steps on the FMPs as training examples and testing examples respectively to exclude the influence of individual FMPs on the model.

## 3.4   Postprocessing and Evaluation

In this particular application, the performance of a classification model for a single data point, i.e., one time step, is not really meaningful. Neither is the average accuracy over a time interval. The most critical point is when a gait phase begins. Therefore, after classifying each time step, we extract the contiguous blocks for the swing and stance phases and compare the absolute difference between real and predicted start of the blocks. We denote this difference by $\Delta_{start}$. These blocks for multiple gait phases are visualized in Fig. 2. Here we assume that a block is more than ten contiguous time steps in size. We ignore individual misclassifications in a block, i.e., exclude them by logical reasoning in post-processing. We also compute classification confidence and confidence per time step (see (5)) and average them per block. We discard gait phases with small

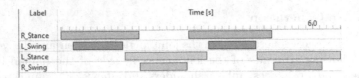

**Fig. 2.** Example of visualization of the result of the detected blocks of gait phases.

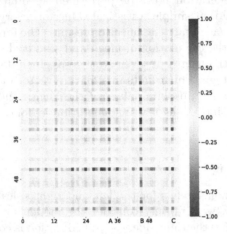

**Fig. 3.** Classification Correlation Matrix (CCM) for the mean values of the joint angles trained for scenario A to detect the gait phases for the left side. The feature A (rotational velocity of flexion/extension in the left hip joint), the feature B (left knee), and the feature C (left foot angle) are most important for discrimination.

values for the classification confidence, since it can be assumed here that the experiment was not performed properly. This avoids downgrading the model due to erroneous measurements or movements that were not performed accurately.

## 4    Results

To evaluate the learned classifier for Scenario A, we use several measures. The results are shown in Table 1. In Fig. 3, a part of the classification correlation matrix (7) is visualized. Despite the sparse model-complexity (limited-rank $\Omega$) and the simple feature extraction, the prediction of gait phases is good. It can be observed that the stance phases are slightly more difficult to detect than the swing phases. This is reflected in the $\Delta_{start}$-value and in the higher outlier rate. If the application requires higher accuracy, a more complex model, i.e., more prototypes per class or a more complex feature extraction can be performed. Having said that, we emphasize that there is no fundamental trade-off between model simplicity and accuracy.

For better illustration, we consider here only the mean values of the angles and their velocities. It can be observed that the rotational velocity of flex-

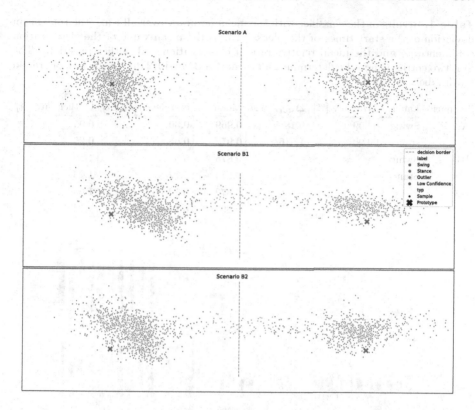

**Fig. 4.** Data points of one subject and prototypes mapped with the learned mapping matrix $\Omega$ of GMLVQ to dimension 15 and additionally PCA applied for all three scenarios.

ion/extension in the left hip joint (feature A), the angle of internal rotation/external rotation of the left knee (feature B), and the rotational velocity of the left ankle joint (feature C) and their positive correlation are most important to distinguish the swing and stance phases for the left foot. The body parts above the hip (head and shoulder), i.e., features No. 0 to No. 23, have little influence on the discrimination.

In the second step, we look at a 2D visualization of the data using principal component analysis (PCA) applied to the mapped data using the learned $\Omega$ matrix (4). In other words, the data and prototypes are mapped using the learned matrix and PCA is then applied to the resulting 15-dimensional latent data. First of all, we can see that the problem is linearly separable, which is also shown by the accuracy values in Table 1. In the second image of scenario B1, we can see a drift in the data, which can be observed by closely inspecting the prototype position of the model learned for scenario A. At this point, it should be noted that only the information in 2D is available here, but the decision takes place in a 15-dimensional space. The same is true for the third image of Fig. 4 for scenario

**Table 1.** Results of the classifier with the mean accuracy over all runs, the time mean deviation of the start times of the blocks $\Delta_{start}$, the mean value of the classification confidence $r_C$ and the mean relative rate of classification and outlier rejections. The first two entries cannot be determined for scenarios B1 and B2 due to the lack of ground truth labels.

| Scenario | Label | Accuracy [%] | $\Delta_{start}$ [ms] | Mean $r_C$ | cl. reject rate [%] | Outlier rate [%] |
|---|---|---|---|---|---|---|
| A | Swing | 99.99 | 9.18 | 0.969 | 0.30 | 0.05 |
| | stance | 99.90 | 11.16 | 0.972 | 0.68 | 0.25 |
| B1 | Swing | – | – | 0.875 | 5.35 | 1.20 |
| | Stance | – | – | 0.914 | 4.72 | 1.23 |
| B2 | Swing | – | – | 0.873 | 5.12 | 1.53 |
| | Stance | – | – | 0.897 | 6.13 | 1.74 |

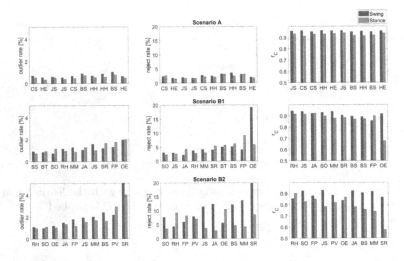

**Fig. 5.** Visualization of the percentage of outliers, rejected data, and classification confidence values for the three scenarios and individual subjects (acronyms).

B2. As tempting as it may be, these visualizations should not be over-interpreted as they are not faithful representations of the original model in their entirety.

We finally finish our analyses of the effect of data drift on model decisions by observing three key metrics: the mean classification certainty $r_C$, the outlier rejection rate and the classification rejection rate for all three scenarios A, B1 and B2. The outlier and classification rejection rates are elevated in scenarios B1 and B2 (see Table 1). In Fig. 5 we depict the key metrics for each participant. It can also be observed there that the metrics are relatively stable for all participants in scenario A. However, looking at the results of scenario B1 and B2, we find a high variance in the metrics. Thus, we need to analyze the participants individually to decide whether the model learned in scenario A can be meaningfully applied.

# 5 Conclusion

In this paper, we have shown how distance-based classifiers help to analyze the impact of data drift on predictive performance. Depending on the subsequent analyses and precision requirements of the downstream application, the reduction rate in classification certainty is still acceptable but must be defined on a problem-specific basis. In our case, the model can be applied to almost all subjects from scenario B1. For all but one test subject, the classification certainty was acceptable. However, the situation differs for scenario B2, where an exoskeleton was worn by the volunteers. In this scenario, the classification certainty values drop significantly and also fluctuate strongly. This is a clear indication that the drift in the data has a severe impact on predictive performance.

# References

1. Biehl, M., Hammer, B., Villmann, T.: Prototype-based models in machine learning. Wiley Interdisc. Rev. Cogn. Sci. **7**(2), 92–111 (2016). https://doi.org/10.1002/wcs.1378

2. Bunte, K., Schneider, P., Hammer, B., Schleif, F.M., Villmann, T., Biehl, M.: Limited rank matrix learning, discriminative dimension reduction and visualization. Neural Networks **26**(1), 159–173 (2012)

3. Crammer, K., Gilad-Bachrach, R., Navot, A., A.Tishby: Margin analysis of the LVQ algorithm. In: Becker, S., Thrun, S., Obermayer, K. (eds.) Advances in Neural Information Processing, vol. 15, pp. 462–469. MIT Press, Cambridge, MA (2003)

4. Fischer, L., Hammer, B., Wersing, H.: Optimal local rejection for classifiers. Neurocomputing **214**, 445–457 (2016)

5. Goodfellow, I., Bengio, Y., Courville, A.: Deep Learning. MIT Press, Cambridge (2016)

6. Güth, V., Klein, D., Rosenbaum, D.: Ganganalyse. In: Rehabilitation in Orthopädie und Unfallchirurgie, pp. 32–41. Springer, Cham (2005)

7. Hammer, B., Villmann, T.: Generalized relevance learning vector quantization. Neural Networks **15**(8–9), 1059–1068 (2002)

8. Jöllenbeck, T., Pietschmann, J.: Ganganalyse und gangtraining in der orthopädischen rehabilitation nach gelenkersatz-zurück zum normalen gang, aber wie? B&G Bewegungstherapie und Gesundheitssport **35**, 3–13 (2019)

9. Kaden, M., Lange, M., Nebel, D., Riedel, M., Geweniger, T., Villmann, T.: Aspects in classification learning - review of recent developments in learning vector quantization. Found. Comput. Decis. Sci. **39**(2), 79–105 (2014)

10. Kohonen, T.: Learning vector quantization. Neural Networks **1**(Supplement 1), 303 (1988)

11. Lisboa, P., Saralajew, S., Vellido, A., Villmann, T.: The coming of age of interpretable and explainable machine learning models. In: Verleysen, M. (ed.) Proceedings of the 29th European Symposium on Artificial Neural Networks, Computational Intelligence and Machine Learning, pp. 547–556. i6doc.com, Louvain-La-Neuve, Belgium (2021). https://doi.org/10.14428/esann/2021.ES2021-2

12. Perry, J., Schoneberger, B.: Gait Analysis: normal and pathological function. In: SLACK (1992)

13. Ravichandran, J., Kaden, M., Saralajew, S., Villmann, T.: Variants of dropconnect in learning vector quantization networks for evaluation of classification stability. Neurocomputing **403**, 121–132 (2020). https://doi.org/10.1016/j.neucom.2019.12. 131
14. Ravichandran, J.: Prototorch. https://github.com/si-cim/prototorch (2020)
15. Rudin, C., Chen, C., Chen, Z., Huang, H., Semenova, L., Zhong, C.: Interpretable machine learning: fundamental principles and 10 grand challenges. Stat. Surv. **16**, 1–85 (2022)
16. Saralajew, S., Holdijk, L., Villmann, T.: Fast adversarial robustness certification of nearest prototype classifiers for arbitrary seminorms. In: Larochelle, H., Ranzato, M., Hadsell, R., Balcan, M., Lin, H. (eds.) Proceedings of the 34th Conference on Neural Information Processing Systems, vol. 33, pp. 13635–13650. Curran Associates, Inc. (2020)
17. Sato, A., Yamada, K.: Generalized learning vector quantization. In: Touretzky, D.S., Mozer, M.C., Hasselmo, M.E. (eds.) Advances in Neural Information Processing Systems, pp. 423–9. MIT Press, Cambridge (1996)
18. Sch"olkopf, B., Smola, A.: Learning with kernels. MIT Press, Cambridge (2002)
19. Schneider, P., Hammer, B., Biehl, M.: Adaptive relevance matrices in learning vector quantization. Neural Comput. **21**, 3532–3561 (2009)
20. Schneider, P., Hammer, B., Biehl, M.: Distance learning in discriminative vector quantization. Neural Comput. **21**, 2942–2969 (2009)
21. Villmann, T., Staps, D., Ravichandran, J., Saralajew, S., Biehl, M., Kaden, M.: A learning vector quantization architecture for transfer learning based classification in case of multiple sources by means of null-space evaluation. In: Bouadi, T., Fromont, E., Hüllermeier, E. (eds.) IDA 2022. LNCS, vol. 13205, pp. 354–364. Springer, Cham (2022). https://doi.org/10.1007/978-3-031-01333-1_28
22. Villmann, T., Kaden, M., Nebel, D., Biehl, M.: Learning vector quantization with adaptive cost-based outlier-rejection. In: Azzopardi, G., Petkov, N. (eds.) CAIP 2015. LNCS, vol. 9257, pp. 772–782. Springer, Cham (2015). https://doi.org/10. 1007/978-3-319-23117-4_66
23. Yang, Q., Zhang, Y., Dai, W., Pan, J.: Transfer Learning. Cambridge University Press, Cambridge (2020)
24. Zhuang, F., et al.: A comprehensive survey on transfer learning. Proc. IEEE **109**(1), 43–76 (2021). https://doi.org/10.1109/JPROC.2020.3004555

# Enhance Gesture Recognition via Visual-Audio Modal Embedding

Yiting Cao, Yuchun Fang(✉), and Shiwei Xiao

School of Computer Engineering and Science, Shanghai University,
Shanghai 200444, China
{caoyiting12,ycfang,xiaoshiwei}@shu.edu.cn

**Abstract.** In recent years, gesture recognition has achieved remarkable advances, restrained from either the mainly limited attribute of the adopted single modality or the synchronous existence of multiple involved modalities. This paper proposes a novel visual-audio modal gesture embedding framework, aiming to absorb the information from other auxiliary modalities to enhance performance. The framework includes two main learning components, *i.e.*, multimodal joint training and visual-audio modal embedding training. Both are beneficial to exploring the fundamental semantic gesture information but with a shared recognition network or a shared gesture embedding space, respectively. The enhanced framework trained with this method can efficiently take advantage of the complementary information from other modalities. We experiment on a large-scale gesture recognition dataset. The obtained results demonstrate that the proposed framework is competitive or superior to other outstanding methods, emphasizing the importance of the proposed visual-audio learning for gesture recognition.

**Keywords:** Visual-audio modal learning · Gesture recognition · Gesture embedding

## 1 Introduction

Gestures, as a nonverbal body language, are a simple and natural way of communication. There is no doubt that it will become increasingly important in computer vision applications, such as human-computer interaction [25], human-robot interaction [15], virtual reality and sign language recognition. Gesture recognition aims to recognize and understand meaningful movements of human bodies. Over the past decade, enormous efforts have been made to improve the accuracy and robustness of gesture recognition in both unimodal and multimodal scenarios.

Unimodal gesture recognition generally explores the salient features of gestures independently by focusing on a specific modality, such as video and skeleton of body posture. With the advent of deep learning, such modality-specific approaches have continually achieved promising performance [6,23,27]. In contrast, multimodal approaches jointly utilize several modalities to learn discriminative representations from different modal sources. Due to this, such approaches

© The Author(s), under exclusive license to Springer Nature Switzerland AG 2023
M. Tanveer et al. (Eds.): ICONIP 2022, LNCS 13624, pp. 391–402, 2023.
https://doi.org/10.1007/978-3-031-30108-7_33

have been consistently evaluated to be more accurate than unimodal approaches in previous works [1,19,29].

The mainstream literature on gesture recognition focuses on the visual modality, such as color and depth modality [14,31], color and optic flow [4,8], but does not consider the relevant information contained in the audio modality. Wu *et al.* [31] proposed a deep dynamic neural network for simultaneous gesture segmentation and recognition and acquired high-level spatio-temporal representations using different neural networks suited to three kinds of input modality: skeleton sequences, RGB videos, and depth videos. Huang *et al.* [14] applied attention-based 3D-CNN for capturing spatio-temporal features. Moreover, Cui *et al.* [8] combined convolutional neural networks and bi-directional LSTM to capture features from RGB and optic flow modalities in an iterative optimization process. Chang *et al.* [4] leverage optical flow to understand human motion in gesture recognition. Their optical flow estimation method introduces four improvements: strong feature extractors, attention to contours, midway features, and a combination of these three. However, in reality, event perception and recognition are inherently multimodal as multiple sensory organs (*i.e.*, eyes, ears, skin, etc.) are involved simultaneously [12]. While eating an apple, we simultaneously perceive the color, taste, and sound of the bite.

Except for only using visual modality, combining audio with visual information can enhance the performance for different applications, such as automatic speech recognition [2,16] or emotion recognition [20,24]. Furthermore, audio information has been demonstrated to be helpful in human perception of objects by existing research [11]. However, although the visual-audio information can be utilized jointly by realizing the modalities fusion, the model cannot fully exploit the audio information due to the large heterogeneity between the visual and audio modality data. Furthermore, since the features learned from different modalities are initially located in different subspaces, the features linked with similar semantics would differ. This phenomenon would hamper the model from utilizing the multimodal data comprehensively.

In this paper, we propose a novel Visual-Audio modal gesture embedding framework to transfer the knowledge from audio modality to visual modality to enhance the performance of the visual modality gesture recognition model. Our method consists of two main processes: multimodal joint training and visual-audio modal embedding training. The former process does the joint training based on visual representation prediction and audio representation prediction and utilizes a shared recurrent neural network (RNN) to transfer or fuse the knowledge implicitly between modalities. Meanwhile, the latter one applies a triplet loss [5,26] on visual and audio features to minimize the distance of intra-class representations while maximizing the inter-class ones, regardless of their modality types. In doing this, it forces the extracted high-level representations between visual and audio modalities to share a similar space, where the intra-class representations have a close distance while the inter-class ones have a long distance.

The major contributions of this work include:

- We propose a novel learning framework to explore knowledge from visual and audio modalities for gesture recognition. We extract modality-invariant gesture embedding in a latent space via a triplet loss, further improving the fusion efficiency between heterogeneous modalities. Furthermore, utilizing this complementary information makes the model more discriminative and robust.
- We utilize a shared network to fuse visual-audio features instead of simple concatenation or average, which reduces the computation and complexity of the model and prevents the potential risk of overfitting. In addition, our method does not require alignment between audio and visual modalities and is concise.
- We assess the proposed method through qualitative and quantitative evaluations. The pre-experiments prove that large heterogeneous modalities can damage the accuracy. On the other hand, the t-SNE visualization and ablation experiment demonstrate that the proposed model can successfully integrate and utilize the complementary information between large heterogeneous modalities.

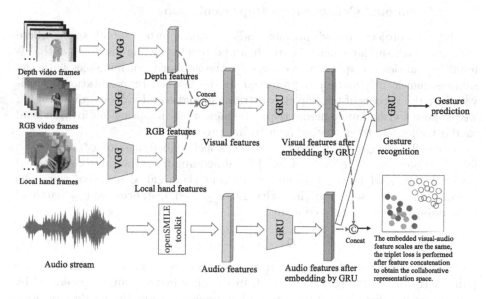

**Fig. 1.** The proposed visual-audio gesture embedding framework for gesture recognition. The upper and lower branches extract the visual and audio features respectively and use a triplet loss to improve the accuracy of gesture recognition by applying collaborative representation learning and joint training.

## 2  Proposed Approach

### 2.1  Model Architecture

In this paper, we aim to acquire a shared embedding space to explore the latent correlation between audio and visual signals for gesture recognition. The overall structure of our framework is depicted in Fig 1, visual modality is divided into three parts, including RGB videos, depth videos, and local gesture videos. We first extract features from visual and audio modalities. In order to capture the correlation between audio and visual modalities, this method embeds the visual features and the audio features into the common representation space through the embedding network. Mathematically, the two embedding functions $f_A(\cdot) : \mathbb{R}^M \to \mathbb{R}^E$ and $f_V(\cdot) : \mathbb{R}^N \to \mathbb{R}^E$, aim at mapping audio inputs in $\mathbb{R}^M$ and visual inputs in $\mathbb{R}^N$ onto a shared collaborative space $\mathbb{R}^E$. This step is the basis for creating a collaborative representation space in the latter method. After the embedding network, we jointly train these two modality-specific networks through a shared RNN while simultaneously performing visual-audio modal embedding training.

### 2.2  Multimodal Collaborative Representation

Triplet loss projects the visual and audio features into a latent space where instances with similar semantics are clustered together while instances with different semantics are split up. As a result, the similarity of instances with the same semantic information is preserved in the learned representations. Motivated by the success of the triplet constraint in audio and video studies [13] [9], we propose a visual-audio modal triplet framework by adopting audio and visual modal triplet loss to supervise the learning process.

For embedding a given instance $e$, we select embeddings of $e^+$ and $e^-$ to form a triplet $tr = \{e, e^+, e^-\}$. $\{e, e^+\}$ is embedding from the same class, called positive pair. And $\{e, e^-\}$ belongs to different class, called negative pair. The calculation of the semantic similarity $D$ of paired instances over a batch of embeddings can be calculated as in Eq. (1).

$$D_{e,e'} = ||e - e'||_2 \tag{1}$$

where $|| \cdot ||_2$ denotes the Euclidean distance between two embeddings in the pair. Accordingly, the pairwise Euclidean distance matrix can be generated by computing the distance between all embeddings. To form the hardest positive pair $\{e, e^+\}$ and the hardest negative pair $\{e, e^-\}$, for each $e$, $e^+$ is one embedding which has the maximum distance from $e$ in the batch, and $e^-$ is another embedding which has the minimum distance from $e$ in the batch. In this way, the obtained $e^+$ and $e^-$ applied together with $e$ to construct the hard triplet for each embedding, and the triplet loss constraint $\mathcal{L}_T$ can be estimated by all hard triplets as in Eq. (2).

$$\mathcal{L}_T = \sum(D_{e_i,e_i^+} - D_{e_i,e_i^-}) \tag{2}$$

In order to compute the visual-audio modal triplet loss $\mathcal{L}_T$, the audio embedding $e_A$ and the visual embedding $e_V$ are combined to form a double-sized batch of embeddings in the form of $\{e_A; e_V\}$. After that, a pairwise Euclidean distance matrix is obtained by computing the distance between all embedding pairs. Furthermore, each embedding (audio or visual) is combined with two others selected from the same batch to form a hard triplet. In doing this, the training process forces the model to narrow the distribution gap of embeddings derived from visual and audio modalities, while at the same time maintaining the specific gesture semantics.

## 2.3   Multimodal Joint Training

Apart from the color and texture information in the RGB videos, we also consider the spatial shape and geometric information in the depth videos. Besides those global hand locations/motions, the local hand gestures are taken into account. Hence, the visual modality in our work referred to RGB and depth videos, as well as the cropped hands image from the RGB videos. It is input to independent VGG networks and then concatenated as the visual feature vector. We denote an audio feature vector as $x_A \in \mathbb{R}^M$ and its corresponding visual feature vector as $x_V \in \mathbb{R}^N$, where M and N are the dimensions of the audio and visual vectors, respectively. The process of $x_A$ and $x_V$ are fed into two modality-specific subnetworks that can be formulated as in Eq. (3).

$$e_A = f_A(x_A), e_V = f_V(x_V) \tag{3}$$

where the function $f_A(\cdot) : \mathbb{R}^M \to \mathbb{R}^E$ and the function $f_V(\cdot) : \mathbb{R}^N \to \mathbb{R}^E$ map each input of audio and visual modalities into the same subspace, resulting in corresponding $E$-dimensional representations $e_A$ and $e_V$. After that, the following shared layer is used to estimate the final predictions. It can be given as in Eq. (4).

$$y_A = f(e_A), y_V = f(e_V) \tag{4}$$

where the function $f(\cdot) : \mathbb{R}^E \to \mathbb{R}$ estimates final predictions $y_A$ and $y_V$ separately. In order to take advantage of visual and audio modalities for gesture recognition, the model is trained with a set of visual-audio features $\{(x_A, x_V)\}$. The joint loss function $J(\theta)$ for gesture recognition is calculated by Eq. (5).

$$J(\theta) = \mathcal{L}_V + \alpha \cdot \mathcal{L}_A \tag{5}$$

where $\theta$ denotes the network parameters to be optimized, $\mathcal{L}_V$ and $\mathcal{L}_A$ represent the loss of visual data and audio data, respectively, and $\alpha$ stand for the weight of the audio prediction loss to balance its contribution to $J(\theta)$. The term $\alpha \cdot \mathcal{L}_A$ enforces the optimization to consider the auxiliary modality information. Moreover, the value of $\theta$ is optimized on the training set.

After the unimodal descriptors are extracted from the CNN and audio tool, embedding functions $f_A(\cdot)$ and $f_V(\cdot)$ are estimated by two RNNs. Subsequently, the visual and audio embeddings are fed into a shared gesture recognition neural network and a fully connected layer for classification. The total loss function of the framework for gesture recognition can be formatted as in Eq. (6).

$$J(\theta) = \mathcal{L}_V + \alpha \cdot \mathcal{L}_A + \beta \cdot \mathcal{L}_T + \lambda \cdot \mathcal{R}(\theta) \tag{6}$$

where $\mathcal{L}_V$ and $\mathcal{L}_A$ represent the discriminative loss function of visual and audio data, respectively. $\mathcal{L}_T$ represents the triplet loss function of both visual and audio data. The $\alpha$ and $\beta$ are introduced to balance the contribution of the audio data and the triplet loss. $\mathcal{R}(\theta)$ is the regularization and $\lambda$, $a$, $b$ are the hyperparameter.

## 3    Experiments

### 3.1    Dataset

We perform experiments on an Italian gesture dataset to evaluate our model. The dataset is provided by ChaLearn multimodal gesture recognition in 2013 and 2014, including RGB videos, depth videos, skeleton sequences, etc. There is audio modal data in ChaLearn13 but none in ChaLearn14. The gestures in the ChaLearn dataset are divided into 20 categories performed by different people and recorded by Kinect. ChaLearn13 is a dataset that includes 1,074 videos and about 13,900 gestures. ChaLearn14 is a dataset that consists of 940 videos and about 14,000 gestures. There are many same data between the video modal data, but a different metric is used for evaluation. In 2013, the Levenshtein distance was adopted, but it was replaced with the Jaccard index in 2014.

For isolated gesture recognition, we first find the same video data between ChaLearn13 and ChaLearn14. After that, we split these videos into video segments based on the labels of ChaLearn14, in which each video segment contains a gesture instance. Accordingly, skeleton sequences and ChaLearn13's audio data are divided in the same way. As a result, we obtain an augmented gesture dataset with 10,169 gesture instances, including RGB videos, depth videos, skeleton sequences, and audio data for each gesture instance. In this paper, we perform gesture recognition on this augmented gesture dataset.

### 3.2    Preliminary Experiments

In this section, we conduct a preliminary experiment to validate our ideas which heterogeneous modal data could influence the model learning the correlation between modalities. Based on the experimental requirements and time efficiency, this experiment selects a subset of the augmented gesture dataset containing 9,162 gesture instances, and the training and testing sets are split by 8:2.

**Experimental Setup.** Three experiments are designed for the preliminary experiment: gesture recognition based on RGB and depth modalities, RGB and skeleton modalities, and RGB and audio modalities. In these three experiments,

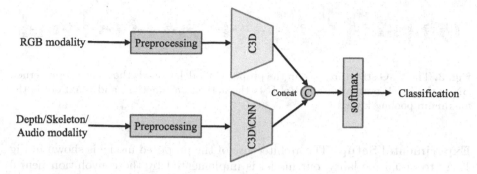

**Fig. 2.** The architecture of modal fusion in gesture recognition. It is an abstract architecture used in the preliminary experiment.

**Table 1.** Accuracy analysis in preliminary experiment.

| Methods | Modality | Accuracy |
|---------|----------|----------|
| C3D [30] + CNN [18] | RGB + audio | 58.5 |
| C3D [30] + CNN [17] | RGB + skeleton | 65.4 |
| C3D [30] | RGB + depth | 69.7 |

C3D [30] is used as the feature extractor for RGB and depth modalities, whereas AlexNet [17] is used for skeleton data. As to audio modality, first, use the audio processing library librosa [21] to extract the Mel-Frequency Cepstral Coefficients (MFCC) feature and then use CNN [18] to extract the discriminative feature. MFCC feature has been widely used in speech and speaker recognition. The multimodal fusion architecture of the preliminary experiment can be abstracted as shown in Fig 2. The Adam optimization algorithm with an initial learning rate of 0.0001 is adopted to train the network.

**Performance Analysis.** Table 1 lists the result of the preliminary experiment, the simple concatenation between RGB and audio modal features is worse than the other two. In general, simple fusion by concatenation in RGB modality with depth and skeleton modalities is reasonable because they all model the space. However, since audio and visual modalities are heterogeneous, the network cannot learn their correlation. On this basis, we propose a visual-audio modal embedding framework to capture the correlation between large heterogeneous modalities to improve the accuracy of gesture recognition.

### 3.3   Multimodal Embedding Experiments

The pre-experiments are performed with multimodal heterogeneous data that is hard to be handled with general joint representation methods. In this section, we validate our proposed model on the augmented dataset presented in Sect. 3.1.

**Fig. 3.** The VGG structure used in the proposed model. Conv is the convolution kernel, BN is the batch normalization, ReLU is the activation function, and MaxPool is the maximum pooling layer.

**Experimental Setup.** The architecture of the proposed model is shown in Fig 1. As to visual modality, our model is implemented by the convolution neural network (CNN) and recurrent neural network (RNN). CNN is a powerful feature extraction that leads to successful applications in classification tasks for image and temporal data. Specifically, we use VGG11 to extract features from visual modality, and its detailed structure is shown in Fig. 3. RNN can better capture temporal information. Specifically, the GRU is used as embedding networks and the shared network because GRU has fewer parameters than LSTM due to the lack of separate memory cells and output gates, which leads to a faster training process and increases generalization with less training data [7]. As to audio modality, OpenSMILE [10] is used to extract features. The modality-specific subnetworks and the modality-shared subnetwork have two hidden layers, and each hidden layer has 80 hidden nodes. The RGB and depth videos each contain 38 frames, and the local gesture videos contain 74 frames. The Adam optimization algorithm with an initial learning rate of 0.0001 is adopted to train the network. Furthermore, $l_2$ regularization is used to improve the generalization of the model. The mini-batch size is 12.

**Table 2.** Ablation analysis.

| Accuracy/Training Method | Visual | Audio |
|---|---|---|
| Unimodal Training | 84.2% | 55.1% |
| Visual-Audio Joint Training | 87.0% | 59.6% |
| Visual-Audio Embedding Training | 87.9% | 55.9% |
| Complete Model Training Method | **89.6%** | **60.1%** |

**Ablation Analysis.** The results of the ablation analysis are reported in Table 2. We perform unimodal training prediction, visual-audio modal joint training prediction, and visual-audio embedding training prediction, respectively. In unimodal training, we independently train and classify the gesture on visual and audio data. GRU-RNNs can capture the long-term temporal dependencies in visual data from the experimental result. In addition, the recognition accuracy of audio modality alone is much worse than that of visual modality, because the audio modality in the dataset is intentionally added with noise, and every person uses a different accent to convey the meaning of gestures, such as speed and tone.

Furthermore, jointly training audio and visual data can deliver higher accuracy than just one modality trained, which implies that the shared semantic information is somewhat transferred from other heterogeneous data to the target modality by implementing a shared subnetwork for joint training. Rather than the visual-audio joint training, when the triplet constraint across the audio and visual modalities is performed, the obtained accuracy (Line 3 in Table 2) shows that it can significantly improve the performance compared with the unimodal framework, which suggests that the implementation of the triplet constraint is beneficial to distill discriminative representations of gesture. In conclusion, the proposed method can essentially provide additional knowledge from audio signals to alleviate the shortage of video signals and vice versa.

**Table 3.** Comparison with benchmark models.

| Methods | Modality | Accuracy |
|---|---|---|
| VGG [28] | RGB-D | 69.2% |
| C3D [30] | RGB-D | 82.1% |
| 3D-CNN + LSTM [22] | RGB-D | 84.8% |
| Brousmiche et al. [3] | RGB-D + audio | 88.3% |
| Ours | RGB-D + audio | **89.6%** |

**Comparison Analysis.** The results on the comparison of our proposed model with other benchmark models in the ChaLearn dataset are reported in Table 3. Since we augmented the dataset, the benchmark models are also experimental on the augmented dataset. VGG and C3D are the typical feature extraction network for processing images and videos, respectively. As shown in Table 3, the accuracy of C3D is 82.1%, higher than the VGG network, because compared with the VGG, the C3D extracts not only the spatial feature of the video but also the temporal feature. Moreover, after adding LSTM to improve the ability to capture temporal features with the 3D network, the experimental result is improved by 2.7%. [3] proposed a feature-by-feature linear adjustment module, which makes multimodal information influence each other to change each other's feature learning. Compared with feature fusion methods, it has a strong ability to utilize multimodal information comprehensively. Our method independently learns visual-audio modal features based on CNN and GRU frameworks and establishes their correlation through triplet loss. Furthermore, after joint training with a shared feature extractor, its recognition accuracy is greatly improved. The results show that our visual-audio gesture embedding training method results in better discriminative ability and indicates that the method achieves good information complementarity, establishes a correlation between audio and visual modalities, and increases the robustness of the model.

## 3.4   Visualization of Gesture Embedding

To investigate how the proposed visual-audio modal embedding framework benefits gesture recognition, we extract the learned representations from the unimodal framework and the proposed gesture framework. By t-Distributed Stochastic

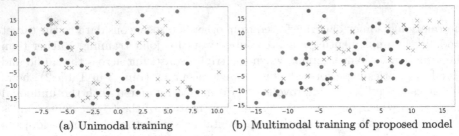

| (a) Unimodal training | (b) Multimodal training of proposed model |

**Fig. 4.** A visualization of the representation learned from the training set by the unimodal model and our proposed multimodal model. Red and blue points represent visual modality and audio modality, respectively, and circles and forks represent two different gestures: 'vieniqui' and 'chedupalle'. The abscissa and ordinate values represent the size of the two components of the two-dimensional vector after dimension reduction by t-SNE. (Color figure online)

Neighbour Embedding (t-SNE), we illustrate the distribution of the learned representations on the development set of the augmented version of gesture dataset of ChaLearn14 in Fig 4(a). It can be seen that in the unimodal framework, the learned representations can be easily distinguished into two parts according to the modal source. Specifically, the representations learned from different modalities almost have no overlap, although they belong to the same gesture. In contrast, the representations extracted from our proposed framework are visibly clustered together according to their gesture categories, as shown in Fig 4(b). Note that, for the sake of simplicity, we only chose two gesture categories for visualization. Likewise, one can find that the representations belonging to the same gesture category share almost the same latent space. These findings suggest that the representations learned by the proposed framework are somewhat invariant to the modalities. Furthermore, by taking advantage of the gesture embedding space, the gesture representations extracted from audio and visual signals can implicitly fuse the knowledge from each other. Hence, the exploitation of mutual information possibly enhances the performance of a unimodal framework.

## 4   Conclusion

In this paper, we propose a visual-audio modal gesture embedding framework to enhance gesture recognition. The proposed method uses concatenation to fuse different video features and obtain the joint representation of the video. In contrast, for audio features, triplet loss is applied to the audio and joint representation of the video to embed it in a shared space, thus establishing the correlation between audio and visual modalities. In addition, this paper uses a shared subnetwork for joint training to further transfer the audio information into the video information. We have evaluated the proposed method on an augmented hand gesture dataset consisting of ChaLearn13 and ChaLearn14. The experimental results on the augmented ChaLearn14 gesture dataset with audio signals demonstrate that the proposed method improves gesture recognition performance. Despite its success, our framework still has some limitations. First, although our method considers local hand gestures, the background information will make lots of redundant information. Therefore, proposing a method to predict the hand shape correctly is necessary.

**Acknowledgment.** The work is supported by the National Natural Science Foundation of China under Grant No.: 61976132, 61991411 and U1811461, and the Natural Science Foundation of Shanghai under Grant No.: 19ZR1419200.

We appreciate the High Performance Computing Center of Shanghai University and Shanghai Engineering Research Center of Intelligent Computing System No.: 19DZ2252600 for providing the computing resources.

# References

1. Abavisani, M., Joze, H.R.V., Patel, V.M.: Improving the performance of unimodal dynamic hand-gesture recognition with multimodal training. In: CVPR, pp. 1165–1174 (2019)
2. Afouras, T., Chung, J.S., Senior, A., Vinyals, O., Zisserman, A.: Deep audio-visual speech recognition. IEEE Trans. Pattern Anal. Mach. Intell. **44**(12), 8717–8727 (2018)
3. Brousmiche, M., Rouat, J., Dupont, S.: Audio-visual fusion and conditioning with neural networks for event recognition. In: MLSP, pp. 1–6 (2019)
4. Chang, J.Y., Tejero-de Pablos, A., Harada, T.: Improved optical flow for gesture-based human-robot interaction. In: ICRA, pp. 7983–7989 (2019)
5. Chechik, G.: Sharma, varun, Shalit, Uri, Bengio, Samy: large scale online learning of image similarity through ranking. J. Mach. Learn. Res. **11**, 1109–1135 (2010)
6. Chen, X., Guo, H., Wang, G., Zhang, L.: Motion feature augmented recurrent neural network for skeleton-based dynamic hand gesture recognition. In: ICIP, pp. 2881–2885 (2017)
7. Cho, K., Van Merriënboer, B., Bahdanau, D., Bengio, Y.: On the properties of neural machine translation: Encoder-decoder approaches. arXiv preprint arXiv:1409.1259 (2014)
8. Cui, R., Liu, H., Zhang, C.: A deep neural framework for continuous sign language recognition by iterative training. IEEE Trans. Multimedia **21**(7), 1880–1891 (2019)
9. Ding, C., Tao, D.: Trunk-branch ensemble convolutional neural networks for video-based face recognition. IEEE Trans. Pattern Anal. Mach. Intell. **40**(4), 1002–1014 (2016)
10. Eyben, F., Wöllmer, M., Schuller, B.: Opensmile: the munich versatile and fast open-source audio feature extractor. In: Proceedings of the 18th ACM International Conference on Multimedia, pp. 1459–1462 (2010)
11. Giard, P.: Auditory-visual integration during multimodal object recognition in humans: a behavioral and electrophysiological study. J. Cogn. Neurosci. **11**(5), 473–490 (1999)
12. Goldstein, E.B., Brockmole, J.: Sensation and perception. In: Cengage Learning (2016)
13. Han, J., Zhang, Z., Keren, G., Schuller, B.: Emotion recognition in speech with latent discriminative representations learning. Acta Acustica united with Acustica **104**(5), 737–740 (2018)

14. Huang, J., gang Zhou, W., Li, H., Li, W.: Attention-based 3d-cnns for large-vocabulary sign language recognition. IEEE Trans. Circ. Syst. Video Technol. **29**, 2822–2832 (2019)
15. Khan, A., et al.: Packerrobo: model-based robot vision self supervised learning in cart. Alexandria Eng. J. **61**(12), 12549–12566 (2022)
16. Kim, M., Hong, J., Park, S.J., Ro, Y.M.: Cromm-vsr: cross-modal memory augmented visual speech recognition. IEEE Trans. Multimedia **24**, 4342–4355 (2021)
17. Krizhevsky, A., Sutskever, I., Hinton, G.E.: ImageNet classification with deep convolutional neural networks. Commun. ACM **60**, 84–90 (2012)
18. Kumar, A., Khadkevich, M., Fügen, C.: Knowledge transfer from weakly labeled audio using convolutional neural network for sound events and scenes. In: ICASSP, pp. 326–330. IEEE (2018)
19. Liu, J., Furusawa, K., Tateyama, T., Iwamoto, Y., Chen, Y.W.: An improved hand gesture recognition with two-stage convolution neural networks using a hand color image and its pseudo-depth image. In: ICIP, pp. 375–379 (2019)
20. Maréchal, C., et al.: Survey on AI-based multimodal methods for emotion detection. In: High-Performance Modelling and Simulation for Big Data Applications (2019)
21. McFee, B., et al.: librosa: Audio and music signal analysis in python. In: Proceedings of the 14th Python in Science Conference, vol. 8, pp. 18–25 (2015)
22. Mullick, K., Namboodiri, A.M.: Learning deep and compact models for gesture recognition. In: ICIP (2017)
23. Nguyen, X.S., Brun, L., Lézoray, O., Bougleux, S.: A neural network based on SPD manifold learning for skeleton-based hand gesture recognition. In: CVPR, pp. 12036–12045 (2019)
24. Praveen, R.G., Granger, E., Cardinal, P.: Cross attentional audio-visual fusion for dimensional emotion recognition. In: FG 2021, pp. 1–8 (2021)
25. Rautaray, S.S., Agrawal, A.: Vision based hand gesture recognition for human computer interaction: a survey. Artif. Intell. Rev. **43**(1), 1–54 (2015)
26. Schroff, F., Kalenichenko, D., Philbin, J.: Facenet: a unified embedding for face recognition and clustering. In: CVPR, pp. 815–823 (2015)
27. Shi, L., Zhang, Y., Hu, J., Cheng, J., Lu, H.: Gesture recognition using spatiotemporal deformable convolutional representation. In: ICIP, pp. 1900–1904 (2019)
28. Simonyan, K., Zisserman, A.: Very deep convolutional networks for large-scale image recognition. In: ICLR (2015). http://arxiv.org/abs/1409.1556
29. Tang, J., Cheng, H., Zhao, Y., Guo, H.: Structured dynamic time warping for continuous hand trajectory gesture recognition. Pattern Recogn. **80**, 21–31 (2018)
30. Tran, D., Bourdev, L., Fergus, R., Torresani, L., Paluri, M.: Learning spatiotemporal features with 3d convolutional networks. In: ICCV, pp. 4489–4497 (2015)
31. Wu, D., et al.: Deep dynamic neural networks for multimodal gesture segmentation and recognition. IEEE Trans. Pattern Anal. Mach. Intell. **38**(8), 1583–1597 (2016)

# Learning from Fourier: Leveraging Frequency Transformation for Emotion Recognition

Binqiang Wang[1,2,3](✉), Gang Dong[2,3], Yaqian Zhao[2,3], and Rengang Li[2,3]

[1] Shandong Massive Information Technology Research Institute, Jinan, China
**binqiang2wang@qq.com**
[2] State Key Laboratory of High-end Server and Storage Technology, Jinan, China
[3] Inspur (Beijing) Electronic Information Industry Co., Ltd., Beijing, China

**Abstract.** With the help of feature fusion techniques, multi-modal emotion recognition has achieved great success, aiming to reach more naturally and human-likely communication during human-machine interaction. Existing methods focus on designing specific modules to generate better representations in the semantic space domain. However, we find that the frequency domain can enhance the emotion correlation among the same category, which is omitted by previous methods. To complement this feature, we design a novel feature fusion module based on the frequency domain to capture the information from both the space domain and frequency domain. Specifically, an attention-based mechanism is incorporated with Fourier transformation to inject the frequency information into the fused feature representation. Furthermore, analyzing features from the frequency domain may lose some normal semantic information such as appearance clues. Thus a residual connection is investigated during the feature representation and accomplishment of the final emotion recognition. Experimental results based on benchmark datasets demonstrate the effectiveness of the proposed module. In addition, we analyze the limitations and applicability of our method based on the existing datasets.

**Keywords:** Human-machine interaction · Emotion recognition · Fourier transformation

## 1 Introduction

Emotion recognition is a rising topic with prevalent applications related to human-machine interaction [1]. Aiming to employ emotional characteristics in artificial intelligence agents, emotion recognition plays a crucial role to make the machine feel the world like human beings. To capture emotion more robustly and accurately, multi-modal based emotion recognition has attracted many researchers' interests [2–5].

Supported by the Natural Science Foundation of Shandong Province (No. ZR2021QF145).

According to the pipeline position to perform emotion recognition, these methods can be divided into two categories: feature extraction and feature fusion. Feature extraction based methods concentrate on the representative feature extraction from individual modalities [6,7]. The sequential feature fusion is relatively naive because the point is to obtain discriminative representation from original input data. However, this kind of method omitted the fact that the complementary of different modalities may be lost during the individual feature extraction procedure. Thus, the feature fusion based method is proposed to design fusion strategies that can exploit the complementary of different modalities, achieving a more robust affective analysis [3]. It needs to be noted that the feature extraction in the feature fusion based methods is generally a carefully engineered representation.

To capture the emotional contribution information by combining the advantages of different information, many feature fusion methods have been proposed. In terms of types of techniques adopted to fuse features, these methods can summarize the tensor based method [2,8], attention-based method [5,9], and other methods [3]. The tensor based method utilizes tensor fusion strategies to combine different information, resulting in a high dimension representation [8]. The attention-based method introduces the attention mechanism into the feature fusion step to generate weighted representation [5], which enhanced the ability of discrimination. Other methods include the concatenation operation and quantum theory related operation [4]. However, all these methods analyze the fusion schedule of different features in the original domain, i.e. time domain or space domain. As the object analyzed is the feature, we utilized the semantic space domain herein to describe the specific domain. As a powerful tool in information processing, Fourier transformation has shown great success in many applications. We argue that the Fourier transformation has the potential to enhance the feature fusion performance, which is validated by a toy experiment.

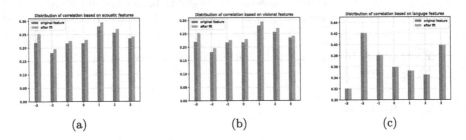

(a)                    (b)                    (c)

**Fig. 1.** A toy experiment conducted on CMU-MOSI to validate the enhanced similarities of samples coming from the same category after fast Fourier transformation.

The similarities of features coming from the same category are computed based on both original features and fast Fourier transformation of original features. The similarity is measured with Pearson product-moment correlation coefficients and the results of the dataset CMU-MOSI [10] are reported in Fig. 1.

It can be concluded that the fast Fourier transformation can enhance the similarities of features coming from the same category. Inspired by this phenomenon, we propose a Fourier transformation based attention mechanism to achieve better representation for emotion recognition. Specifically, the original input is encoded into different representations by the corresponding feature extraction module first. Then, a deliberately designed attention mechanism based on Fourier transformation is introduced to enhance the representation of original representations of different modalities. Furthermore, an optional residual connection is introduced to enhance the discrimination of the proposed attention mechanism. The main contributions of this paper can be summarized as follows:

- Firstly, inspired by the enhancement of Pearson product-moment correlation coefficients after fast Fourier transformation, an attention mechanism is incorporated with Fourier transformation to obtain better emotion recognition results.
- Secondly, residual connection based on the proposed attention mechanism is introduced to combine the original feature from the semantic space domain, aiming to utilize the advantages of the space domain and frequency domain at the same time.
- Finally, the experimental results validate the proposed method and the condition when the proposed method work is discussed.

## 2   The Proposed Method

Multi-modal based emotion recognition aims to recognize the emotion states given a short video containing three modalities: visual frames, acoustic sound, and textual language transcribed from the sound. Firstly, the pipeline of the proposed Fourier-based Attention Emotion Network (FATENet) is introduced. Then, the details of the Fourier-based Attention module are explained. Finally, an optional residual structure is introduced and the optimized loss function is constructed to train the parameters.

### 2.1   Pipeline of FATENet

The rough pipeline of FATENet can be divided into four modules: original video input, feature extraction, Fourier-based module, and decision phase. As illustrated in Fig. 2, the original input is a short video containing frame sequences, acoustic sound, and transcribed textual language.

Feature extraction: The data of different modalities are processed with different techniques. For visual feature extraction, the Facet[1] is utilized to encode emotion related information based on basic and advanced emotional clues. To extract acoustic representation from wave format files, VOCAREP [6] is exploited to

---

[1] https://pair-code.github.io/facets/.

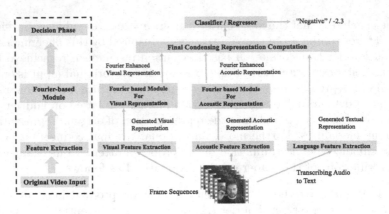

**Fig. 2.** The overall architecture of the proposed method.

obtain basic acoustic features: 12 Mel-frequency cepstral coefficients, pitch tracking, and so on (more details refer to [3]). As for language feature extraction, the pre-trained GloVe model [7] is employed which can embed a word into a 300 dimension semantic space.

As mentioned before, the focus of this paper is to study the fusion strategy, thus the feature extraction process is conducted based on the standard process in this domain. Because these features are extracted separately, P2FA [11] is utilized to match the frames and sound with words in the language. To format the input, these three modalities are represented by $v_i$, $a_i$, and $l_i$, where $i$ is the position of a word, and $v_i$ represents the short frame sequences corresponding to the word $l_i$. $a_i$ is the same as $v_i$.

Fourier-based Module: The video naturally is sequence related data, Long Short-Term Memory (LSTM) [12] is used to capture this sequence information, which is also the main structure of Fourier-based module. As there are three modalities of data, we utilize three LSTMs to model different data. It can be found from the right of Fig. 1 that fast Fourier transformation cannot enhance the emotional Pearson product-moment correlation coefficients of language modality. Thus, the Fourier-based Attention structure (introduced in Sect. 2.1) is only inserted into the visual modality and acoustic modality.

After encoded by Fourier based module, the visual representation is embedded into Fourier enhanced visual representation. The acoustic representation is encoded into Fourier enhanced acoustic representation.

Decision Phase: Enhanced representations of three different modalities are generated from the above Fourier-based module. To perform emotion recognition with a whole feature, these three representations need to be condensed to a final representation. This function can be implemented by the proposed final condensing representation computation structure (introduced in Sect. 2.2). The output of this module is a condensed vector representation, which can be input to the final classifier or regressor.

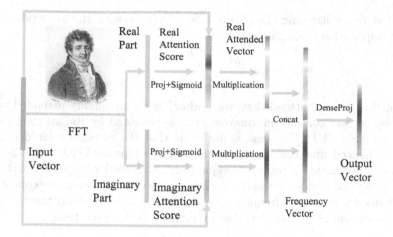

**Fig. 3.** Illustration of the Fourier based Attention structure.

**Fourier-Based Attention Structure.** The construction of the proposed Fourier-based Attention is introduced here. Note that the input of this structure is the semantic feature vector extracted by the previous feature extraction module, which is marked as $v_i$, $a_i$, and $l_i$. Figure 1 demonstrates that the fast Fourier transformation only works for visual and acoustic modalities on CMU-MOSI [10]. Thus, visual modality is chosen as an example to show the procedure of Fourier-based Attention. Figure 3 depicts the process of Fourier-based Attention structure.

Firstly, the input visual vector $v_i$ is transformed to the frequency domain by fast Fourier transformation:

$$v_i^R + iv_i^I = FFT(v_i) \tag{1}$$

where the $R$ in $v_i^R$ and $I$ in $iv_i^I$ represent the real part and imaginary part after the fast Fourier transformation (FFT). Note that $i$ in the left of $iv_i^I$ is the representation of complex numbers.

The point of the attention mechanism is how to obtain the attention score. Here, we process the real part and imaginary part with a projection layer with Sigmoid activation. Specifically, attention computation is formulated as follows:

$$s_i^R = \sigma(f_{proj}(v_i^R)) \tag{2}$$

$$s_i^I = \sigma(f_{proj}(v_i^I)) \tag{3}$$

where $s_i^R$ represents the attention score computed based on the real part after FFT. $\sigma$ is the Sigmoid activation function. $f_{proj}$ is a liner transformation whose parameters are trainable. The physical meaning of symbols in Eq. 3 is the same as that of Eq. 2. After attention scores of the real part and imaginary part are obtained, the real attended vector and imaginary attended vector can be calculated by multiplying the original input vector with the attention scores. This multiplication is shown as follows:

$$at_i^R = s_i^R * v_i \tag{4}$$

$$at_i^I = s_i^I * v_i \tag{5}$$

where $at_i^R$ is the real attended vector and $at_i^I$ is the imaginary attended vector, these two vectors contain information that is weighted by the attention scores computed from FFT. The ideal situation is that the emotion related information is enhanced and unrelated information is suppressed. At this stage, the frequency information is processed separately. To consider these two parts comprehensively and condense the dimension of the output vector, we concatenate two attended vectors to a frequency vector and project the concatenated vector to the dimension of the input vector by two fully connected layers:

$$v_i^o = f_{denseproj}[at_i^R; at_i^I] \tag{6}$$

where $v_i^o$ is the output of the Fourier-based Attention structure. The dimension of the output vector is the same as that of the input, which is controlled by the dense projection layer. $[;]$ represents the vector concatenation operation.

Note both visual modality vector and acoustic modality vector are processed by this structure. The output vector for acoustic modality is denoted by $a_i^o$.

## 2.2 Optional Structure and Loss Function

To model the sequence feature of the original video, three LSTMs are utilized to encode $v_i^o$, $a_i^o$, and $l_i$, $i = 1, 2, ..., N$, and $N$ is the length of the textual language. Before encoding by the LSTM, a residual connection is added to involve the original space semantic information. The attention mechanism proposed in Multi-Attention Recurrent Network (MARN) [9] is introduced to compress different information into an intermediate representation. This representation and output final hidden states of three LSTMs are concatenated to a whole vector, denoted with $x_{final}$. $x_{final}$ is input to a computation schema following Eq. 1-6 to obtain $x_{final}^o$. This vector is embedded by Final Condensing Representation Computation, which is a residual connection between $x_{final}$ and $x_{final}^o$. The output condensing vector is input to the classifier or regressor to perform the emotion recognition.

Due to the different formats of emotion labels, the L1 loss computed by Mean Absolute Error (MEA) is used to measure the predicted output and the ground truth label when the annotation is a float value. When the annotated label is discrete categories, the negative log likelihood loss function is adopted.

# 3    Experimental Results

## 3.1    Dataset

Three datasets are used to validate the effectiveness of the proposed method: CMU-MOSI, CMU-MOSEI [13], and IEMOCAP [14]. The original videos of

CMU-MOSI and CMU-MOSEI are generally self-recoded monologues. The annotation of these two datasets is a float value between -3 and 3, representing the most negative attitude to the most positive attitude. As for dataset IEMOCAP, the video content is the recording of two speakers in a conversation. And the annotation used herein is four different discrete emotion classes: neural, happy, sad, and angry. The data splits of these datasets are shown in Table 1.

**Table 1.** Data splits of samples in different datasets. # represents the number of samples.

|  | CMU-MOSI | CMU-MOSEI | IEMOCAP |
|---|---|---|---|
| Training # | 1284 | 16265 | 2717 |
| Validation # | 229 | 1869 | 798 |
| Testing # | 686 | 4643 | 938 |

## 3.2   Compared Methods

The compared methods in this paper are presented in this subsection. We compared the performance of the proposed methods with the EF-LSTM [3], LF-LSTM [3], TFN [8], LMF [2], NUAN [5], and MARN [9]. EF-LSTM concatenates features of three different modalities, and then an LSTM is used to model the sequence to obtain the final representation before the classifier. On the contrary, LF-LSTM encodes features coming from different modalities with three different LSTMs, and then the outputs of LSTMs are concatenated to form the final representation. TFN utilizes the tensor fusion strategy between different modalities to generate representative high-dimension fused features. LMF adopts the low-rank idea into TFN to reduce the amount of calculation. Among these methods, MARN is chosen as the baseline method due to the relatively same techniques.

## 3.3   Main Results

Firstly, we conduct experiments on the dataset CMU-MOSI and the results are summarized in the upper part of Table 2. For the meaning of the table's header, Acc-2 represents the classification performance after discretizing the float annotation according to zero point; F1 means the F1 score of the binary classification. Acc-5 and Acc-7 represent the classification of 5 discretizations and 7 discretizations. MAE is the mean absolute error and r is Pearson product-moment correlation coefficient. For MAE, the smaller is better. For others, the higher is better.

It can be observed that the proposed method is better in performance compared with other methods in terms of Acc-2, F1, and Acc-7. Then, the dataset CMU-MOSEI is used to validate the effectiveness of the proposed method. The performance is reported in the upper part of Table 3. The results show that the proposed method achieves the best performance in terms of F1, Acc-7, MAE,

**Table 2.** Performance on dataset CMU-MOSI. (- represents there is no reported results from the corresponding paper)

|  | Acc-2 | F1 | Acc-5 | Acc-7 | MAE | r |
|---|---|---|---|---|---|---|
| Comparison with other methods | | | | | | |
| EF-LSTM [3] | 75.8 | 75.6 | – | 32.7 | 1.000 | 0.630 |
| LF-LSTM [3] | 76.2 | 76.2 | – | 32.7 | 0.987 | 0.624 |
| TFN [8] | 75.6 | 75.5 | – | 34.9 | 1.009 | 0.605 |
| LMF [2] | 75.3 | 75.2 | – | 30.5 | 1.018 | 0.605 |
| MARN [9] | 76.4 | 76.2 | – | 31.8 | **0.984** | 0.625 |
| NUAN [5] | **78.3** | **77.9** | - | 26.8 | 1.034 | **0.654** |
| ours | 76.53 | 76.48 | **40.67** | **35.13** | 1.0254 | 0.6279 |
| Ablation study | | | | | | |
| MARN | 73.18 | 73.14 | 35.13 | 31.78 | 1.0420 | 0.5970 |
| Final-FFT-res | 72.30 | 72.16 | 32.36 | 29.15 | 1.0740 | 0.5654 |
| Final-FFT-wores | 72.74 | 72.71 | 35.28 | 30.32 | 1.0676 | 0.5945 |
| FFT-AV | **76.53** | **76.48** | **40.67** | **35.13** | **1.0254** | **0.6279** |
| FFT-AV-final-FFT-wores | 70.55 | 70.64 | 33.82 | 29.74 | 1.1026 | 0.5847 |
| FFT-AV-final-FFT-res | 73.18 | 73.18 | 32.94 | 29.59 | 1.0584 | 0.6056 |

**Table 3.** Performance on dataset CMU-MOSEI

|  | Acc-2 | F1 | Acc-5 | Acc-7 | MAE | r |
|---|---|---|---|---|---|---|
| Comparison with other methods | | | | | | |
| EF-LSTM [3] | 78.2 | 77.1 | – | 45.7 | 0.687 | 0.573 |
| LF-LSTM [3] | 79.2 | 78.5 | – | 47.1 | 0.655 | 0.614 |
| TFN [8] | 79.3 | 78.2 | – | 47.3 | 0.657 | 0.618 |
| LMF [2] | 78.2 | 77.6 | – | 47.6 | 0.660 | 0.623 |
| MARN [9] | **79.3** | 77.8 | – | 47.7 | 0.646 | 0.629 |
| ours | 79.09 | **78.82** | **50.51** | **49.36** | **0.6175** | **0.6513** |
| Ablation study | | | | | | |
| MARN | 78.18 | 78.53 | 49.21 | 48.33 | 0.6261 | 0.6579 |
| Final-FFT-res | 70.13 | 71.46 | 49.75 | 48.96 | 0.6321 | 0.6361 |
| Final-FFT-wores | 79.24 | 78.62 | 50.92 | 49.56 | 0.6274 | 0.6421 |
| FFT-AV | 79.09 | **78.82** | 50.51 | 49.36 | 0.6175 | 0.6513 |
| FFT-AV-final-FFT-wores | **79.32** | 78.62 | 50.51 | 49.13 | 0.6280 | 0.6390 |
| FFT-AV-final-FFT-res | 70.92 | 72.22 | **51.86** | **50.44** | **0.6105** | **0.6593** |

**Table 4.** Performance on dataset IEMOCAP (AVG Acc represents the average accuracy)

| | Neutral | | Happy | | Sad | | Angry | | AVG Acc |
|---|---|---|---|---|---|---|---|---|---|
| | Acc | F1 | Acc | F1 | Acc | F1 | Acc | F1 | |
| Comparison with other methods | | | | | | | | | |
| EF-LSTM [3] | – | 61.2 | – | 30.8 | – | 62.0 | – | 71.7 | – |
| LF-LSTM [3] | – | 60.0 | – | 40.0 | – | 56.0 | – | 69.6 | – |
| TFN [8] | – | 61.9 | – | 28.0 | – | 57.3 | – | 72.9 | – |
| LMF [2] | – | 54.7 | – | 40.6 | – | 54.3 | – | 72.9 | – |
| MARN [9] | – | 59.6 | – | 35.1 | – | 57.4 | – | 71.2 | – |
| ours | **68.34** | **67.16** | **84.54** | **82.65** | **83.26** | **82.70** | **84.65** | **85.25** | **80.20** |
| Ablation study | | | | | | | | | |
| MARN | 68.12 | 67.24 | 85.50 | 83.18 | **84.86** | **84.62** | 84.22 | 84.81 | **80.68** |
| Final-FFT-res | 68.12 | 68.39 | **87.21** | **83.47** | 82.20 | 80.60 | 84.33 | 83.71 | 80.46 |
| Final-FFT-wores | 68.66 | 67.24 | 85.61 | 81.73 | 80.28 | 80.37 | 83.90 | 83.45 | 79.62 |
| FFT-AV | 68.66 | 68.30 | 85.82 | 81.18 | 84.44 | 83.49 | 83.58 | 84.22 | 80.62 |
| FFT-AV-final-FFT-wores | **69.62** | **68.72** | 85.82 | 79.86 | 78.78 | 78.55 | 80.28 | 79.50 | 78.63 |
| FFT-AV-final-FFT-res | 68.34 | 67.16 | 84.54 | 82.65 | 83.26 | 82.70 | **84.65** | **85.25** | 80.20 |

and r. Finally, the experiments are conducted on the dataset IEMOCAP. According to the upper part of Table 4, the proposed method gives the best performance in terms of all metrics.

### 3.4   Ablation Studies

To validate the effectiveness of the proposed Fourier based attention structure and optional residual structure, ablation studies are conducted on three datasets. According to the ablation study reported in the lower part of Table 2 conducted on dataset CMU-MOSI, MARN is a self-run algorithm, which is considered as a baseline in the ablation study. FFT-AV represents the method in which Fourier based Attention is inserted into acoustic and visual modalities. Final-FFT-res represents the method that Fourier based Attention and residual connection are adopted (Final Condensing Representation Computation). As a comparison, Final-FFT-wores is the method that Fourier based Attention is used while the residual connection is not used. The meaning of FFT-AV-final-FFT-wores and FFT-AV-final-FFT-res can be inferred as the previous combination.

It can be observed from the lower part of Table 2 that FFT-AV achieves the best performance compared with other settings in terms of all metrics. This suggests that the residual connection is not necessary for dataset CMU-MOSI. Then, ablation experiments on dataset CMU-MOSEI are reported in the lower part of Table 3. The results illustrate that although FFT-AV can achieve better results compared with THE baseline method, i.e., MARN, FFT-AV-final-FFT-res gives better performance in terms of Acc-5, Acc-7, and r. This phenomenon suggests that the residual connection is necessary for dataset CMU-MOSEI. Finally, the ablation results in Table 4 exhibit an abnormal conclusion: the proposed method has no advantage compared with the baseline method. Even FFT-AV-final-FFT-res achieves a better score in terms of measurement under the angry category, the AVG ACC is lower than that of the baseline method, which will be discussed in Sect. 3.5.

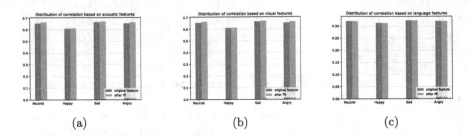

(a)                          (b)                          (c)

**Fig. 4.** Explanation of the performance on dataset IEMOCAP.

## 3.5   Analysis

According to the ablation study results in Tables 2, 3 and 4, the experimental results on datasets CMU-MOSI and CMU-MOSEI show that the proposed method illustrates a significant increase compared with the baseline method MARN. But for dataset IEMOCAP, there is no advantage of the proposed method compared with the baseline method. To find the reason, we conduct the same experiment that computing the Pearson product-moment correlation coefficients of the same category with or without FFT. As shown in Fig. 4, the results of language features are consistent with that of Fig. 1 while the results of acoustic features and visual features show different characteristics. For acoustic modality and visual modality, the correlation coefficients show a great increase after FFT when conducted on CMU-MOSI, but the increase is hardly obvious when conducted on IEMOCAP. Thus, we suspect the reason is the different distribution of feature vectors in different datasets. The condition the proposed method work is that the FFT can enhance the correlation coefficients coming from the same category.

## 4   Conclusion

Feature fusion considering different characteristics is essential for multi-modal emotion recognition. This paper proposes Fourier-based Attention Emotion Network inspired by the phenomenon that the correlation coefficients are improved after FFT on dataset CMU-MOSI. Unlike most previous methods, the frequency domain attention mechanism can achieve a better emotion recognition performance. In addition, the limitation of the proposed is analysed and we argue that the effective condition of the proposed method is when the correlation coefficient is improved after conducting FFT on the input data.

# References

1. Gao, K., Xu, H., Gao, C., Sun, X., Deng, J., Zhang, X.: Two-Stage Attention Network for Aspect-Level Sentiment Classification. In: Cheng, L., Leung, A.C.S., Ozawa, S. (eds.) ICONIP 2018, Part IV. LNCS, vol. 11304, pp. 316–325. Springer, Cham (2018). https://doi.org/10.1007/978-3-030-04212-7_27
2. Liu, Z., Shen, Y., Lakshminarasimhan, V.B., Liang, P.P., Zadeh, A., Morency, L.P.: Efficient low-rank multimodal fusion with modality-specific factors. arXiv preprint arXiv:1806.00064 (2018)
3. Gkoumas, D., Li, Q., Lioma, C., Yu, Y., Song, D.: What makes the difference? an empirical comparison of fusion strategies for multimodal language analysis. Inf. Fusion **66**, 184–197 (2021)
4. Gkoumas, D., Li, Q., Dehdashti, S., Melucci, M., Yu, Y., Song, D.: Quantum cognitively motivated decision fusion for video sentiment analysis. In: Proceedings of the AAAI Conference on Artificial Intelligence, vol. 35, pp. 827–835 (2021)
5. Wang, B., Dong, G., Zhao, Y., Li, R., Cao, Q., Chao, Y.: Non-uniform attention network for multi-modal sentiment analysis. In: THornór Jónsson, B., et al. (eds.) MMM 2022. LNCS, vol. 13141, pp. 612–623. Springer, Cham (2022). https://doi.org/10.1007/978-3-030-98358-1_48
6. Degottex, G., Kane, J., Drugman, T., Raitio, T., Scherer, S.: Covarep-a collaborative voice analysis repository for speech technologies. In: 2014 IEEE International Conference on Acoustics, Speech and Signal Processing (ICASSP), pp. 960–964. IEEE (2014)
7. Pennington, J., Socher, R., Manning, C.: Glove: global vectors for word representation. In: Proceedings of Empirical Methods Natural Language Process, pp. 1532–1543 (2014)
8. Zadeh, A., Chen, M., Poria, S., Cambria, E., Morency, L.P.: Tensor fusion network for multimodal sentiment analysis. In: Proceedings of the 2017 Conference on Empirical Methods in Natural Language Processing, pp. 1103–1114 (2017)
9. Zadeh, A., Liang, P.P., Poria, S., Vij, P., Cambria, E., Morency, L.P.: Multiattention recurrent network for human communication comprehension. In: Proceedings of the AAAI Conference on Artificial Intelligence, vol. 32, pp. 5642–5649 (2018)
10. Zadeh, A., Zellers, R., Pincus, E., Morency, L.P.: Mosi: multimodal corpus of sentiment intensity and subjectivity analysis in online opinion videos. arXiv preprint arXiv:1606.06259 (2016)
11. Yuan, J., Liberman, M., et al.: Speaker identification on the scotus corpus. J. Acoustical Soc. America **123**(5), 3878 (2008)
12. Hochreiter, S., Schmidhuber, J.: Long short-term memory. Neural Comput. **9**(8), 1735–1780 (1997)
13. Zadeh, A.B., Liang, P.P., Poria, S., Cambria, E., Morency, L.P.: Multimodal language analysis in the wild: Cmu-mosei dataset and interpretable dynamic fusion graph. In: Proceedings of the 56th Annual Meeting of the Association for Computational Linguistics (Volume 1: Long Papers), pp. 2236–2246 (2018)
14. Busso, C., et al.: Iemocap: interactive emotional dyadic motion capture database. Lang. Resour. Eval. **42**(4), 335–359 (2008)

# Efficient Double Oracle
# for Extensive-Form Two-Player
# Zero-Sum Games

Yihong Huang[1,2], Liansheng Zhuang[1(✉)], Cheng Zhao[1], and Haonan Liu[1]

[1] University of Science and Technology of China, Hefei 230027, China
hyh1109@mail.ustc.edu.cn, lszhuang@ustc.edu.cn
[2] Peng Cheng Laboratory, Shenzhen 518000, China

**Abstract.** Policy Space Response Oracles (PSRO) is a powerful tool for large two-player zero-sum games, which is based on the tabular Double Oracle (DO) method and has achieved state-of-the-art performance. Though having guarantee to converge to a Nash equilibrium, existing PSRO and its variants suffer from two drawbacks: (1) exponential growth of the number of iterations and (2) serious performance oscillation before convergence. To address these issues, this paper proposes Efficient Double Oracle (EDO), a tabular double oracle algorithm for extensive-form two-player zero-sum games, which is guaranteed to converge linearly in the number of infostates while decreasing exploitability every iteration. To this end, EDO first mixes best responses at every infostate so that it can make full use of current policy population and significantly reduce the number of iterations. Moreover, EDO finds the restricted policy for each player that minimizes its exploitability against an unrestricted opponent. Finally, we introduce Neural EDO (NEDO) to scale up EDO to large games, where the best response and the meta-NE are learned through deep reinforcement learning. Experiments on Leduc Poker and Kuhn Poker show that EDO achieves a lower exploitability than PSRO and XFP with the same amount of computation. We also find that NEDO outperforms PSRO and NXDO empirically on Leduc Poker and different versions of Tic Tac Toe.

**Keywords:** Two-player zero-sum games · Nash equilibrium · Deep reinforcement learning

## 1 Introduction

Two-player zero-sum games have been a long-standing interest of the development of artificial intelligence. In solving such games, an agent aims to minimize its exploitability, the performance of its opponent in the worst case. When both agents achieve zero exploitability, they reach a Nash equilibrium (NE) [13], a classical solution concept from Game Theory. Even though NE is a clear objective, developing a general algorithm capable of finding an approximate NE often requires tremendous human efforts.

M. Tanveer et al. (Eds.): ICONIP 2022, LNCS 13624, pp. 414–424, 2023.
https://doi.org/10.1007/978-3-031-30108-7_35

Policy Space Response Oracles (PSRO) [6] is a general multi-agent reinforcement learning algorithm which has been applied in many non-trivial tasks. Generally, PSRO aims to find an approximate NE by iteratively expanding a restricted game of a restricted policy population, which is ideally much smaller than the original game. Methods based on PSRO have achieved state-of-the-art performance on large two-player zero-sum games such as Starcraft [16] and Stratego [8]. Despite the empirical success achieved, PSRO and its variants still suffer from exponential growth of the number of iterations and potential serious performance oscillation. The reason for this is that PSRO is based on the tabular method Double Oracle (DO) [11], which makes an inefficient use of the policy population and has no guarantee to decrease exploitability from one iteration to the next.

In this work, we propose a new double oracle algorithm, Efficient Double Oracle (EDO), which is designed for extensive-form two-player zero-sum games. Like PSRO, each player in EDO maintains a population of pure strategies. However, EDO makes full use of the policy population by mixing best responses at every infostate instead of mixing them only at the root of the game, which is what DO does. EDO also removes the restriction on the opponent policy space and creates respective meta-games for each player. EDO is guaranteed to converge to a NE in a number of iterations that is linear in the number of infostates, while PSRO may require a number of iterations exponential in the number of infostates. EDO is also guaranteed to find the least-exploitable policy for the current policy population in each iteration, which avoids the problem of performance oscillation in PSRO.

We also introduce a neural version of EDO, called Neural EDO (NEDO). NEDO uses deep reinforcement learning (DRL) methods to compute a meta-NE in restricted games and compute best responses in each iteration for large games. The restricted games of NEDO contain meta-actions, each selects a corresponding population policy to play the next action. The meta-solver could be any neural extensive-form game solver, such as NFSP [4] and DREAM [15]. NEDO scales up EDO to large games using deep reinforcement learning.

To summarize, our contributions are as follows:

1. We present EDO, a tabular double oracle algorithm that converges linearly without performance oscillation.
2. We present NEDO, a deep reinforcement learning version of EDO that scales up EDO to large games and outperforms existing PSRO methods in all of our experiments.

## 2 Background

We consider extensive-form games with perfect recall. An extensive-form game progresses through a sequence of player actions, and has a *world state* $w \in \mathcal{W}$ at each step. In an $N$-player game, the space of joint actions for players is denoted as $\mathcal{A} = \mathcal{A}_1 \times \ldots \times \mathcal{A}_N$. $\mathcal{A}_i(w) \subseteq \mathcal{A}_i$ denotes the legal action set for player $i \in \{1, \ldots, N\}$ at world state $w$ and $a = (a_1, \ldots, a_N) \in \mathcal{A}$ denotes a joint action. At each world state, a transition function determines the probabilities of the next world state $w'$ after a joint action $a$ is chosen, which is denoted as $\mathcal{T}(w, a) \in \Delta^{\mathcal{W}}$. Player $i$ makes an *observation* $o_i = \mathcal{O}_i(w, a, w')$ upon each transition. The game ends when the players reach a terminal world state $w^T$ and player $i$ receives a reward $\mathcal{R}_i(w)$ in each world state $w$.

A *history* is a sequence of actions and world states representing a trace of the game, denoted $h = (w^0, a^0, w^1, a^1 \ldots, w^t)$, where $w^0$ is the initial world state of the game. $\mathcal{R}_i(h)$ and $\mathcal{A}_i(h)$ are the reward and the set of legal actions for player $i$ in the last world state of $h$. An *infostate* for player $i$, denoted by $s_i$, is a sequence of that player's observations and actions until that time, denoted $s_i(h) = (a_i^0, o_i^1, a_i^1, \ldots, o_i^t)$. It's also assumed that different world states corresponding to the same infostate share the same set of legal actions $\mathcal{A}_i(s_i(h)) = \mathcal{A}_i(h)$.

A player's *strategy* $\pi_i$ is a function mapping from an infostate to a distribution over legal actions. A *strategy profile* $\pi$ is a tuple $(\pi_1, \ldots, \pi_N)$. All players except $i$ are denoted $-i$ and their joint strategies are denoted $\pi_{-i}$. A strategy for a history is defined as $\pi_i(h) = \pi_i(s_i(h))$. We also refer to a strategy as a *policy* in the later parts of this article.

The *value* $v_i^\pi(h)$ is the expected sum of future rewards given that all players play the strategy profile $\pi$. The value of the entire game is denoted as $v_i(\pi)$. A *two-player zero-sum* game has $v_1(\pi) + v_2(\pi) = 0$ for all strategy profiles $\pi$. A *Nash equilibrium (NE)* is a strategy profile $\pi^*$ satisfying $v_i(\pi^*) = \max_{\pi_i}(\pi_i, \pi_{-i}^*)$ for each player $i$.

A *best response* to $\pi_{-i}$ is defined as $\mathbb{BR}_i(\pi_{-i}) = \arg\max_{\pi_i} v_i(\pi_i, \pi_{-i})$ and the *exploitability* of a strategy profile $\pi$ is defined as $e(\pi) = \sum_{i \in \mathcal{N}} \max_{\pi_i'} v_i(\pi_i', \pi_{-i})$. Lower exploitability means better approximation to Nash equilibrium for a strategy profile.

A *normal-form game* is a single-step extensive-form game, in which players choose their actions simultaneously. An extensive-form game can always induce a normal-form game in which legal actions for player $i$ are its deterministic strategies in the original game.

## 3    Related Work

In small two-player zero-sum games, Nash equilibrium can be found via linear programming and no-regret algorithms such as replicator dynamics, fictitious play (FP) and regret matching, which become infeasible when the size of game increases. In perfect information extensive-form games, algorithms based on minimax tree search have had success on games such as chess and Go [14]. Extensive-form fictitious play (XFP) [3] and counterfactual regret minimization (CFR) [18] extend FP and regret matching, respectively, to extensive-form games. Although CFR based on abstraction has been used in large imperfect-information extensive-form zero-sum games like heads-up no limit Texas Hold'em [2], this is not a general method for large games because finding efficient abstractions needs expert domain knowledge.

Recently, deep reinforcement learning (DRL) [7] has been proven effective on complex sequential decision-making problems like Atari games. Deep CFR [1] is a general method that trains a neural network on a buffer of counterfactual values. However, external sampling used by Deep CFR may be impractical for games with a large branching factor. AlphaStar based on self-play beat top human players at StarCraft using population-based reinforcement learning. Nevertheless, self-play methods are not guaranteed to converge to an approximate NE and can not handle some small games like Rock-Paper-Scissors. Neural Fictitious Self Play (NFSP) approximates XFP by progressively training a best response against an average of all past opponent policies via DRL. The average policy in NFSP is also represented by a neural network and trained by supervised learning using a reservoir-sampling buffer, which may become prohibitively large in large games.

Double Oracle (DO) is an algorithm for finding a NE in normal-form games. The algorithm works by maintaining a population of policies $\Pi^t$ at time $t$. A meta-Nash Equilibrium (meta-NE) is computed for the game restricted to policies in $\Pi^t$ in each iteration. Then, a best response to the meta-NE for each player is computed and added to the population. In the worst case, DO must expand all pure strategies. Policy Space Response Oracles (PSRO) approximates the DO algorithm. The meta-NE in PSRO is computed on the empirical game matrix, which is generated by sampling each pair of policies and tracking the average utility. The approximate best response in PSRO is computed via any reinforcement learning method. Extensive-form Double Oracle (XDO) [9] and its neural version, Neural XDO (NXDO) iteratively adds extensive-form BRs to a population and then computes a meta-NE on an entensive-form meta-game while this method is not guaranteed to decrease exploitability every iteration, which means potential performance oscillation.

The concept of finding a low-exploitability meta-policy in a restricted game has also been explored in recent years [10,17]. However, most of these work focus on normal-form restricted games. This paper proposes a method to compute the least-exploitable meta-policy in extensive-form restricted games.

# 4    Method

## 4.1    Efficient Double Oracle

In this paper, we propose Efficient Double Oracle (EDO), which is guaranteed to converge linearly and decrease the exploitability of meta-NE monotonically. Like other DO algorithms, EDO maintains a population of pure strategies and computes a meta-NE of a restricted game in each iteration. Also like other DO algorithms, EDO expands its population by best responses (BR) to the meta-NE. However, to overcome aforementioned drawbacks of PSRO, EDO creates different extensive-form restricted games instead of traditional normal-form ones for each player, and the restricted game $G^i$ for player $i$ is constructed while only this player is restricted to play actions suggested by any strategy in the population at each infostate, against the opponent able to play all legal actions in the full game.

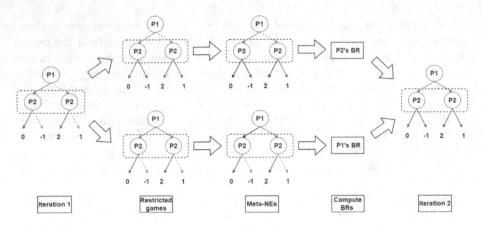

**Fig. 1.** In EDO, the following happens in one iteration: (1)Two extensive-form restricted games are created for each player. (2)For each restricted game, a meta-NE is computed. (3)For each player, a BR is trained against the opponent's restricted policy and this BR is added to the population. Dashed actions are outside the restricted games and the black solid actions are outside the meta-NE.

Formally, EDO creates a restricted extensive-form game for each player based on a pure strategy population $\Pi^t$ and computes a restricted NE at time $t$. Each restricted game $G^p$ for player $p$ has the same infostates as the full game with restricted actions:

$$\mathcal{A}_i^p(s_i) = \begin{cases} \{a \in \mathcal{A}_i(s_i) : \exists \pi_i \in \Pi_i^t \text{ s.t. } \pi_i(s_i, a) > 0\}, & i = p, \\ \mathcal{A}_i(s_i), & i \neq p. \end{cases} \tag{1}$$

The restricted game $G^i$ for player $i$ is then solved for both players to get a meta-NE $(\pi_1^i, \pi_2^i)$. EDO uses a tabular method like XFP or CFR as the restricted game solver. We refer to the restricted player's strategy as the restricted NE $\pi_i^r = \pi_i^i$. The restricted NE $\pi^r$ is the strategy profile with the least exploitability supported by current population. Then, a novel best response with some unseen action against the meta-NE of the restricted opponent is computed and added to the strategy population for each player. Therefore, at each non-final iteration of EDO, at least one new action at some non-terminal infostate is added to the extensive-form restricted game by some player, otherwise the algorithm terminates. As a result, the upper bound of the number of iterations is linear to the number of infostates while the number of iterations DO takes to terminate is exponential to the number of infostates in the worst case, which may make computing meta-NE intractable in later iterations. To illustrate how EDO works, we demonstrate a simple game in Fig. 1. Note that in large games this restricted game may become prohibitively large to solve, which requires NEDO, introduced later in this paper. EDO is described in Algorithm 1.

**Algorithm 1: EDO**

---

1  Input: Initial Population $\Pi^0$, $t = 0$
2  **repeat**
3  $\quad$ Define restricted game $G^i$ for $\Pi^t$ via eq. 1, for $i \in \{1,2\}$
4  $\quad$ $\pi_i^r \leftarrow$ NE policy of player $i$ in $G^i$
5  $\quad$ **for** $i \in \{1,2\}$ **do**
6  $\quad\quad$ Compute a novel best response $\beta_i \leftarrow \mathbb{BR}_i(\pi_{-i}^r)$
7  $\quad\quad$ $\Pi_i^{t+1} = \Pi_i^t \cup \beta_i$
8  $\quad$ **end**
9  $\quad$ $t = t + 1$
10  **until** No action of best responses is outside $G^i$;
11  **return** $\pi^r$

---

EDO is guaranteed to terminate because the number of actions of every infostate in a extensive-form game is finite. The returned retricted NE $\pi^r$ is also a NE in the original game (Proposition 1), as shown below.

**Proposition 1.** *The restricted NE of both players is a Nash equilibrium in the original game, when EDO terminates.*

*Proof.* Suppose that $(\pi_1^r, \pi_2^{ur})$ and $(\pi_1^{ur}, \pi_2^r)$ are the meta-NE of the restricted games for player 1 and 2, respectively, when the algorithm is terminated. Let $\beta_1$ and $\beta_2$ be the novel best responses to $\pi_2^r$ and $\pi_1^r$. If $\beta_1$ or $\beta_2$ on some infostate has support outside the respective restricted game, EDO would not terminate. Therefore, the support of $\beta_1$ and $\beta_2$ on all infostates is guaranteed to be inside the respective restricted game, which means $\pi_1^{ur}$ and $\pi_2^{ur}$ also have this property. Refer to the policy space of the restricted players as $\Lambda^r$. Then:

$$v_2(\pi_1^r, \pi_2^r) \leq v_2(\pi_1^r, \pi_2^{ur})$$
$$= \min_{\pi_1 \in \Lambda_1^r} v_2(\pi_1, \pi_2^{ur})$$
$$\leq v_2(\pi_1^{ur}, \pi_2^{ur})$$
$$\leq \max_{\pi_2 \in \Lambda_2^r} v_2(\pi_1^{ur}, \pi_2) \tag{2}$$
$$= v_2(\pi_1^{ur}, \pi_2^r)$$
$$= \min_{\pi_1} v_2(\pi_1, \pi_2^r)$$
$$\leq v_2(\pi_1^r, \pi_2^r).$$

$$v_2(\pi_1^r, \pi_2^r) = v_2(\pi_1^r, \pi_2^{ur}) = \max_{\pi_2} v_2(\pi_1^r, \pi_2). \tag{3}$$

Therefore, player 2 has no motivation to deviate from $\pi_2^r$, which is same for player 1. Hence, $\pi^r$ is a NE in the original game.

While DO and EDO both return a NE upon termination, EDO has the guarantee of non-increasing monotonicity on exploitability (Propostition 2), which doesn't hold for DO.

**Proposition 2.** *The exploitability of EDO is monotonically non-increasing.*

*Proof.* Suppose that $\pi^t$ is the meta-NE at iteration $t$, and $\Lambda_i^t$ is the policy space of player $i$ in the restricted game $G^i$. It's obvious that $\Lambda_i^t \subseteq \Lambda_i^{t+1}$ because $\Pi_i^t \subseteq \Pi_i^{t+1}$. Then,

$$
\begin{aligned}
e_i(\pi_i^t) &= - \min_{\pi_{-i}} v_i(\pi_i^t, \pi_{-i}) \\
&= - \max_{\pi_i \in \Lambda_i^t} \min_{\pi_{-i}} v_i(\pi_i, \pi_{-i}) \\
&\geq - \max_{\pi_i \in \Lambda_i^{t+1}} \min_{\pi_{-i}} v_i(\pi_i, \pi_{-i}) \\
&= e_i(\pi_i^{t+1}).
\end{aligned}
\tag{4}
$$

$$
e(\pi^t) = \sum_i e_i(\pi_i^t) \geq \sum_i e_i(\pi_i^{t+1}) = e(\pi^{t+1}).
\tag{5}
$$

### 4.2    Neural Efficient Double Oracle

As mentioned above, the tabular EDO cannot handle large games because the restricted games may become prohibitively large to solve. Hence, we propose Neural Efficient Double Oracle (NEDO) algorithm, which extends EDO to large games by deep reinforcement learning (DRL) methods. NEDO uses approximate best responses trained by DRL and also uses DRL methods as the meta-solver. However, approximate best responses usually have many actions with positive probability at every infostate while a oracle best response usually has a single action with probability 1, which means restricted action space defined in EDO may be quite large and even equal to the original game with approximate BRs. To avoid this problem, NEDO defines a different meta-action space for the restricted game $G^p$ for each player $p$ based on the player's DRL policy population $\Pi_p$:

$$
\mathcal{A}_i^p(s_i) = \begin{cases} \{1, 2, ..., |\Pi_i|\}, & i = p, \\ \mathcal{A}_i(s_i), & i \neq p. \end{cases}
\tag{6}
$$

NEDO is described in Algorithm 2. Although the action space differs, the restricted game states, histories and infostates are still the same as in the original game. After the restricted player $p$ chooses an action indicating a DRL population policy, it would actually sample an action in the orginal game according to this DRL policy. The transition function of $G^p$ then becomes:

$$
T^p(h, a^p, w) = \sum_a \pi_p^{a_p^p}(s_p(h), a_p) \mathbb{1}(a_{-p}^p = a_{-p}) T(h, a, w).
\tag{7}
$$

After the restricted game for each player $i$ is defined, an approximate meta-NE $(\pi_1^{i*}, \pi_2^{i*})$ is computed via a DRL method like NFSP. We refer to the restricted player's strategy as the restricted NE $\pi_i^{r*} = \pi_i^{i*}$. Approximate BRs to $\pi_2^{r*}$ and $\pi_1^{r*}$ are computed via a DRL method like DQN [12] and added to the corresponding policy polulation: $\Pi_i^{t+1} = \Pi_i^t \cup \mathbb{BR}_i(\pi_{-i}^{r*})$, $i \in \{1, 2\}$.

---

**Algorithm 2: NEDO**

---

1  Input: Initial Population $\Pi^0$, $t = 0$
2  **repeat**
3       Define restricted game $G^i$ for $\Pi^t$ via eq. 2, for $i \in \{1, 2\}$
4       $\pi_i^{r*} \leftarrow$ approximate NE policy of player $i$ in $G^i$ via DRL
5       **for** $i \in \{1, 2\}$ **do**
6           Compute an approximate best response $\beta_i \leftarrow \mathbb{BR}_i(\pi_{-i}^{r*})$ via DRL
7           $\Pi_i^{t+1} = \Pi_i^t \cup \beta_i$
8       **end**
9       $t = t + 1$
10 **until** *termination conditions are satisfied*;
11 **return** $\pi^{r*}$

---

Although contemporary DRL methods lack guarantee of the approximate solutions, we show that NEDO outperforms PSRO and some other methods experimentally in later sections. A potential drawback of NEDO is that the meta-action space of the restricted player in the corresponding restricted game grows linearly with the number of iterations, which may cause the restricted game hard to solve in later iterations, even harder than the original game. Nevertheless, it's shown that the algorithm achieves significant performance with a small number of iterations in our experiments, which means this issue does not become an real obstacle.

## 5  Experiments

In this section, we report experiments upon EDO and NEDO. For the tabular experiments, we use EDO with an oracle best response and CFR as the meta-solver. For the neural experiments with deep reinforcement learning methods, we use NEDO taking DQN as the best response solver and NFSP as the meta-solver. In both experiments, we use some extensive-form zero-sum games in Openspiel [5] as our environments and report results of different algorithms on each environment respectively.

### 5.1  Experiments with Tabular Methods

As mentioned above, DO often requires more pure strategies to achieve an approximate NE of the original game and the exploitability is not guaranteed to decrease every iteration. Figure 2(a) presents a bad case of DO, in which DO increases exploitability every iteration except the last one, while EDO always decreases it. The game is the equivalent extensive-form game of a zero-sum normal-form game, in which all values are 0, except if the row $r$ is one more than the column $c$ or the column $c$ is one more than the column $r$. The values of these are $\sum_{i=0}^{r-1} 4^i + 3i$ and $\sum_{i=0}^{c-1} -4^i + 3i$ respectively. We plot the performance of EDO and DO in this game with 10 actions.

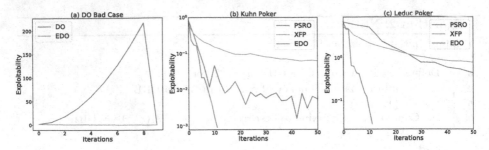

**Fig. 2.** (a) DO Bad Case; (b) Exploitability in Kuhn Poker of PSRO, XFP and EDO; (c) Exploitability in Leduc Poker of PSRO, XFP and EDO.

We also compare EDO with XFP and PSRO in Kuhn Poker and Leduc Poker in Fig. 2(b) and Fig. 2(c). XFP and PSRO here also use oracle BRs and PSRO uses fictitious play (FP) as the meta-solver. We plot the exploitability of these algorithms as a function of iterations. It's shown that EDO achieves lower exploitability in much fewer iterations than PSRO and XFP. Although EDO requires more computation in a single iteration because the policy space of an extensive-form restricted game is often larger than the one of a normal-form restricted game for the same population, much less iterations to achieve a low exploitability means less amount of computation in total. In larger games, the expensive cost of computing BRs so many times would also make PSRO infeasible.

### 5.2   Experiments with Deep Reinforcement Learning

For all neural expeiments, we use the same DRL best response hyperparameters in NEDO, PSRO and NXDO as well as in the measure of approximate exploitability. In DQN, the learning rate of DQN is 0.01 and the size of replay buffer is 2e5. In NFSP, the anticipatory factor is set to be 0.1 and the learning rate of average network in NFSP is 0.1. For PSRO, the entries of empirical payoff tables are computed by sampling 1000 episodes for each new pair of populations strategies. FP is used as the meta-NE solver in PSRO and the number of inner-loop iterations for FP is 2000.

In Fig. 3(a), we compare the exploitability of NEDO and PSRO and NXDO on Leduc Poker. DQN is used to train BRs in these methods. Although NEDO uses DRL methods to solve the restricted games and train BRs, we find that the number of iterations it takes to achieve a low exploitability is much smaller than PSRO and similar to NXDO, which means NEDO inherits the property of EDO in exntensive-form games empirically.

We also compare NEDO, PSRO and NXDO on Tic Tac Toe and Phantom Tic Tac Toe. We plot the exploitability of these methods as a funtion of episodes. However, it's shown that NXDO does not outperform other methods as much as tabular EDO. One of these reasons could be training episodes of NFSP takes a large proportion compared with training BRs via DQN. Although episodes

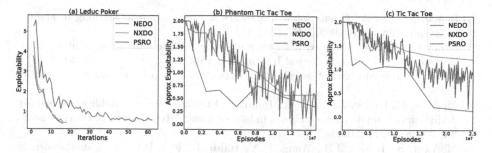

**Fig. 3.** (a) Exploitability in Leduc of NEDO, NXDO and PSRO as a function of iterations; (b and c) Approximate Exploitability in (Phantom) Tic Tac Toe of NEDO, NXDO and PSRO as a function of episodes gathered.

required by a single iteration in NEDO is more than PSRO, NEDO outperforms PSRO and NXDO in both Tic Tac Toe and Phantom Tic Tac Toe which we conjecture is due to the unrestricted opponent policy space and more effective use of population strategies.

# 6   Conclusion

In this paper, we propose EDO, a modification of DO that converges to a Nash equilibrium linearly decreasing exploitability monotonically. We also propose NEDO, an algorithm that scales up EDO to large games via deep reinforcement learning. As shown in our tabular experiments, EDO outperforms PSRO and XFP on different poker games. In the neural experiments, it's shown that NEDO outperforms PSRO and NXDO on Leduc Poker and different Tic Tac Toe games, which we conjecture is due to the unrestricted opponent policy space in restricted games and more effective use of population strategies.

**Acknowledgements.** This work was supported in part to Dr. Liansheng Zhuang by NSFC under contract No. U20B2070 and No. 6197619, and in part to Dr. Houqiang Li by NSFC under contract No. 61836011.

# References

1. Brown, N., Lerer, A., Gross, S., Sandholm, T.: Deep counterfactual regret minimization. In: International Conference on Machine Learning, pp. 793–802. PMLR (2019)
2. Brown, N., Sandholm, T.: Superhuman AI for heads-up no-limit poker: libratus beats top professionals. Science **359**(6374), 418–424 (2018)
3. Heinrich, J., Lanctot, M., Silver, D.: Fictitious self-play in extensive-form games. In: International Conference on Machine Learning, pp. 805–813. PMLR (2015)
4. Heinrich, J., Silver, D.: Deep reinforcement learning from self-play in imperfect-information games. arXiv preprint arXiv:1603.01121 (2016)

5. Lanctot, M., et al.: Openspiel: a framework for reinforcement learning in games. arXiv preprint arXiv:1908.09453 (2019)

6. Lanctot, M., et al.: A unified game-theoretic approach to multiagent reinforcement learning. In: Advances in Neural Information Processing Systems, vol. 30 (2017)

7. Li, Y.: Deep reinforcement learning: an overview. arXiv preprint arXiv:1701.07274 (2017)

8. McAleer, S., Lanier, J.B., Fox, R., Baldi, P.: Pipeline psro: a scalable approach for finding approximate nash equilibria in large games. Adv. Neural Inf. Process. Syst. **33**, 20238–20248 (2020)

9. McAleer, S., Lanier, J.B., Wang, K.A., Baldi, P., Fox, R.: Xdo: a double oracle algorithm for extensive-form games. Adv. Neural Inf. Process. Syst. **34**, 23128–23139 (2021)

10. McAleer, S., Wang, K., Lanctot, M., Lanier, J., Baldi, P., Fox, R.: Anytime optimal psro for two-player zero-sum games. arXiv preprint arXiv:2201.07700 (2022)

11. McMahan, H.B., Gordon, G.J., Blum, A.: Planning in the presence of cost functions controlled by an adversary. In: Proceedings of the 20th International Conference on Machine Learning (ICML-03), pp. 536–543 (2003)

12. Mnih, V., et al.: Human-level control through deep reinforcement learning. Nature **518**(7540), 529–533 (2015)

13. Nash, J.F., Jr.: Equilibrium points in n-person games. Proc. Natl. Acad. Sci. **36**(1), 48–49 (1950)

14. Silver, D., et al.: Mastering the game of go without human knowledge. Nature **550**(7676), 354–359 (2017)

15. Steinberger, E., Lerer, A., Brown, N.: Dream: deep regret minimization with advantage baselines and model-free learning. arXiv preprint arXiv:2006.10410 (2020)

16. Vinyals, O., et al.: Grandmaster level in starcraft ii using multi-agent reinforcement learning. Nature **575**(7782), 350–354 (2019)

17. Wang, Y., Ma, Q., Wellman, M.P.: Evaluating strategy exploration in empirical game-theoretic analysis. arXiv preprint arXiv:2105.10423 (2021)

18. Zinkevich, M., Johanson, M., Bowling, M., Piccione, C.: Regret minimization in games with incomplete information. In: Advances in Neural Information Processing Systems, vol. 20 (2007)

# FastThaiCaps: A Transformer Based Capsule Network for Hate Speech Detection in Thai Language

Krishanu Maity[1]([✉]), Shaubhik Bhattacharya[1], Sriparna Saha[1],
Suwika Janoai[2], and Kitsuchart Pasupa[2]

[1] Department of Computer Science and Engineering,
Indian Institute of Technology Patna, Patna, India
{krishanu_2021cs19,shaubhik_2111cs19,sriparna}@iitp.ac.in
[2] School of Information Technology, King Mongkut's Institute of Technology,
Ladkrabang, Bangkok 10520, Thailand
{64607044,kitsuchart}@it.kmitl.ac.th

**Abstract.** The advent of technology has led to people sharing their views openly like never before. Parallelly, cyberbullying and hate speech content have also increased as a side effect that is potentially hazardous to society. While plenty of research is going on to detect online hate speech in English, there is very little research on the Thai language. To investigate how noisy Thai posts can be handled effectively, in this work, we have developed a two-channel deep learning model *FastThaiCaps* based on BERT and FastText embedding along with a capsule network. The input to one channel is the BERT language model, and that to the other is the pre-trained FastText embedding. Our model has been evaluated on a benchmark Thai dataset categorized into four categories, i.e., peace speech, neutral speech, level-1 hate speech, and level-2 hate speech. Experiments show that *FastThaiCaps* outperforms state-of-the-art methods by up to 3.11% in terms F1 score.

**Keywords:** Hate Speech · Thai · Transformer · Capsule Network · FastText

## 1 Introduction

Social media interactions are becoming more popular due to advancements in online communication technology. Though Social media helps spread knowledge more effectively, but it also stimulates the propagation of online abuse and harassment, including hate speech. Hate speech [18] is any communication that disparages a person or group on the basis of a characteristic such as color, gender, race, sexual orientation, ethnicity, nationality, religion, or other features. These unpleasant incidents can have a measurable detrimental effect on users. Therefore, it is crucial to identify these at the right time and stop them from spreading to a broader group.

© The Author(s), under exclusive license to Springer Nature Switzerland AG 2023
M. Tanveer et al. (Eds.): ICONIP 2022, LNCS 13624, pp. 425–437, 2023.
https://doi.org/10.1007/978-3-031-30108-7_36

According to the Pew Research Center, 40% of social media users have experienced some sort of online harassment[1] [1]. Over the last decade, plenty of research has been conducted to develop datasets and models for automatic online hate speech detection in English language [3,20,26]. There are very few works in other languages like Italian [4], Indonesian [9], code-mixed [13,14,17] and Thai [19,25]. According to a recent report from Reuters, the incident of hate speech increased rapidly in Thailand during the COVID-19 outbreak[2]. Many Myanmar workers at a fish market in Samut Sakhon were reported to be infected with COVID-19. In response, hate speech against them began circulating on social media, particularly on YouTube, Facebook, and Twitter. Due to this, migrant and immigrant workers from Myanmar were terrified for their lives. This motivated us to develop a better model for online hate speech detection in the Thai language with the hope that these automatic hate speech detection systems will automatically flag the hate messages. This in turn will help the law and enforcement departments for taking some actions against people who are spreading hate speech.

We have chosen the ThaiText [19] dataset as, unlike other Thai hate speech datasets tagged with only hate or non-hate, it has a severity level. The vocabulary utilized on these social media sites deviates from the accepted language used in literature [2]. One of the most challenging problems with social media Thai data is noisiness (spelling variations, short-form). Social media users frequently purposefully obfuscate terms by using short words, acronyms, and misspelled words to avoid automatic inspection. The pre-trained word embedding model does not include such words, so morphological information is lost. Furthermore, tokenization of Thai text is not that easy compared to other languages as there is no space or any special symbol between words or sentences.

To overcome the challenges mentioned above, in this work, we have proposed a two-channel *FastThaiCaps* framework to represent Thai data efficiently. The first channel uses WangchanBERTa, a variant of RoBERTa followed by Capsule networks. WangchanBERTa model was pre-trained on the largest Thai language dataset of size 78.5 GB. The role of capsule networks is to learn hierarchical relationships between successive layers by employing an iterative dynamic routing approach. In the second channel, FastText [7] embedding with Bi-LSTM has been employed. Contrary to word2vec [11] and Glove [21], which employ word-level representations, FastText takes advantage of the character level when putting words into the vectors.

The following are the primary contributions of this work:

1. We have examined the BERT language model and FastText pre-trained embedding to investigate how effectively they can handle Thai text data.
2. We have proposed the two-channel deep neural network model, *FastThaiCaps*, where one channel's input is the BERT+Capsule, and another is FastText with LSTM.

---

[1] https://www.pewresearch.org/internet/2017/07/11/online-harassment-2017/.
[2] https://www.reuters.com/article/us-health-coronavirus-thailand-myanmar-idUSKBN28Y0KS.

3. Experimental results illustrate that using BERT and FastText together significantly enhances the performance of hate speech detection and outperforms SOTA by 3.11% in terms of F1-score.

## 2    Related Works

Text mining and NLP paradigms have been used to investigate numerous subjects linked to hate speech detection, including identifying online sexual predators, detection of internet abuse and cyberterrorism [24]. The associated research described below demonstrates that hate speech detection in some low-resource languages should get more attention as most of the current work is conducted in the English language.

### 2.1    Works on English Data

Authors in [26] proposed an approach based on unigrams and patterns automatically collected from training set to detect hate expressions on Twitter. They achieved an accuracy of 87.4% on classifying hate vs non hate tweets. Based on some specific keywords, authors in [3] collected tweets and labeled them as hate, offensive, and none using crowd-souring. They have developed a multiclass classifier to detect hate and offensive tweets. Authors in [6] investigated cyberbullying detection using a corpus of 4500 YouTube comments and various binary and multiclass classifiers. The SVM classifier attained an overall accuracy of 66.70%, while the Naive Bayes classifier attained an accuracy of 63%. Authors in [22] developed a Cyberbullying dataset by collecting data from Formspring.me and finally achieved 78.5% accuracy by applying C4.5 decision tree algorithm using Weka tool kit. CyberBERT, a BERT based framework developed by [20] achieved state-of-the-art results on Formspring (12k posts), Twitter (16k posts), and Wikipedia (100k posts) datasets.

### 2.2    Works on Thai Data

In 2021, Wanasukapunt et al. [25] developed binomial (SVM, RF) and multinomial (LSTM, DistilBERT) models to detect abusive speech from social media in the Thai language. They found that the performance of deep learning models is remarkably better than the machine learning models and achieved the best F1 score of 90.67% using DistilBERT. Authors in [19] created a benchmark Thai hate speech dataset from Facebook, Twitter, and YouTube posts. Each post is annotated with four labels, i.e., peace speech, neutral speech, level-1 hate speech, and level-2 hate speech. They fine-tuned the WangchanBERTa using Ordinal regression loss function and achieved state-of-the-art performance.

| Peace Speech Message – Label (0) | |
|---|---|
| **Thai:** | จิตใจแสนประเสริฐ หม่าก็เป็นมนุษย์ ยามดีใช้แรงงานเขา ยามร้ายจะผลักไสได้ยังไง |
| **English:** | Such a good-hearted person. Burmese is human. We hire Burmese workers in a normal situation. How can we expel them in this dire situation? |

| Neutral Message – Label (1) | |
|---|---|
| **Thai:** | คนไทยคนพม่าก็พี่พี่น้องกันครับ |
| **English:** | Thais and Burmese are brothers and sisters. |

| Level 1 Hate Speech Message – Label (2) | |
|---|---|
| **Thai:** | นี่กูจะต้องตายเพราะพม่าไช่ป้อ |
| **English:** | Am I going to die just because of Myanmar? |

| Level 2 Hate Speech Messages – Label (3) | |
|---|---|
| **Thai:** | เกลียดหม่าส่งพวกมันกลับไปตายบ้านมัน |
| **English:** | Hate Burmese, send them back to die in their homeland. |

**Fig. 1.** Examples of each category in the Thai hate speech dataset

## 3 Dataset Description

The dataset we used to evaluate our proposed model was collected from the three widely used social media platforms: Facebook, Twitter, and YouTube. They were collected between 18/12/2020 to 23/12/2020 after getting the news that a merchandiser at a market in Samut Sakhon, Thailand, got infected by the COVID-19 virus and was admitted to a hospital [19]. Many Myanmar migrant workers were working in the market, including illegal ones and illegal immigration of Burmese was behind the outbreak of COVID-19 in Thailand. This incident triggered the spread of hate speech towards Myanmar migrant workers on social media platforms and made them hardly live their life as usual. There were 7,597 messages in the dataset. The length of each message was between 1 and 428 words. The dataset was labeled into four classes: 3,198 positive or peace speech messages, 2,246 neutral messages, 1,441 Level 1 hate speech messages, and 712 Level 2 hate speech messages. Notice that there were two levels of hate speech messages in this dataset: Level 1 indicates hate or fear of Myanmar for no reason, while level 2 indicates all types of violence. Examples of each category are shown in Fig. 1.

## 4 Methodology

In this section, we have described the proposed methodology for hate speech detection in Thai languages. We have developed the two-channel deep neural network model, namely *FastThaiCaps*, where one channel's input is the BERT language model, and another is pre-trained FastText embedding. Figure 2 depicts the overall architecture of our proposed *FastThaiCaps* model.

**Problem Statement.** We formulate our problem definition as follows: Let $I = <X_t, b_t>_{t=1}^N$ be a set of $N$ instances where $b_t$ represent the corresponding hate

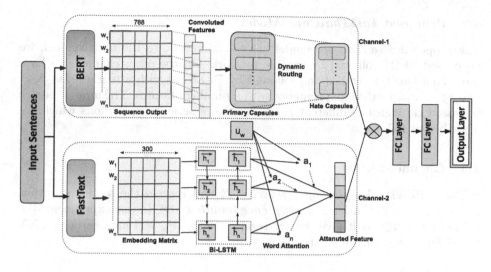

**Fig. 2.** FastThaiCaps architecture

labels for $X^{th}$ instance. Here, $X_t \in N$, $b_t \in B$ (hate classes). The objective of our proposed framework is to maximize the function (1) that maps each instance $X_t$ to its fitting hate label, $b_t$.

$$\arg\max_{\theta} \left( \prod_{t=1}^{N} P(b_t|X_t, \theta) \right) \tag{1}$$

where $X_t$ is the input sentence whose hate label ($b_t$) is to be predicted. $\theta$ denotes the model's parameters we aim to optimize.

### 4.1    Text Embedding Generation

To generate the embedding of input sentence $X$ (say) containing $N$ number of tokens, we have experimented with BERT and FastText.

i) **BERT** [5] is a language model based on bidirectional transformer encoder with a multi-head self-attention mechanism. The sentences in our dataset are written in Thai language, so we choose WangchanBERTa [12], pre-trained in Thai languages. We have considered the sequence output from BERT, where each word of the input sentence has a 768 dimensional vector representation.

ii) **FastText** [7] was created by the Facebook Research Team for effective word embedding of more than 157 different languages. The FastText model provides a 300-dimensional dense vector for each token after being trained using the CBOW approach. In our model, we have used the pre-trained FastText Thai embedding.

## 4.2    Proposed *FastThaiCaps* Model

The proposed *FastThaiCaps* model has two channels. In the first channel, we have used BERT followed by capsule network. On the other channel, we have used Thai FastText embedding followed by Bi-LSTM with attention. Let $X = \{x_1, x_2, \ldots x_n\}$ be the input sentence with $n$ number of words. The input sentence is passed through two different channels with a series of operations which are described as follows:

## 4.3    Channel-1

**BERT.** BERT takes the input sentence X and returns the sequence output $W_B \in \mathbb{R}^{n \times d}$ of dimension $max\_sequence\_length \times 768$. In our experiment, the value of $n = 128$ as it gives better results. Next, $W_B$ will pass through CNN layers for abstract feature extraction.

**CNN** [10] effectively captures abstract representations that reflect semantic meaning at various positions in a text. To obtain N-gram feature map, $\mathbf{c} \in \mathbb{R}^{n-k_1+1}$ using filter $F \in \mathbb{R}^{k_1 \times d}$, we perform convolution operation, an element-wise dot product over each possible word-window, $W_{j:j+k_1-1}$. Each element $c_j$ of feature map $\mathbf{c}$ is generated after convolution by

$$c_j = f(w_{j:j+k_1-1} * F_a + b), \tag{2}$$

where $f$ is a non-linear activation function and $b$ is the bias. After applying $t$ distinct filters of the same N-gram size, $t$ feature maps will be generated, which can then be rearranged as

$$\mathbf{C} = [\mathbf{c_1}, \mathbf{c_2}, \mathbf{c_3}, \ldots, \mathbf{c_t}] \tag{3}$$

After convolution, instead of applying a pooling operation, we have used a capsule network [23] to keep the special features, which are generally lost due to poling operation.

**Primary Capsule Layer.** This initial layer of capsule networks incorporates CNN-generated convoluted features into the primary capsules. In order to maintain the instantiation parameters, such as the local order of words and semantic representations of words, primary capsules maintain a group of neurons to represent each element in the feature maps as opposed to a scalar. We get a series of capsules, $p_i \in \mathbb{R}^d$, by sliding each kernel, $K_i$, across the $C$, where $d$ represents the number of neurons in a capsule. In the primary capsule layer, a channel $PC_i$ consisting of a list of capsules, described as

$$PC_i = g(C * K_i + b), \tag{4}$$

where $g$ is a squashing function with bias $b$.

**Dynamic Routing Between Capsules.** The core concept of dynamic routing is to devise a non-linear map iteratively while ensuring that the lower label capsule has a solid connection to its appropriate upper label capsule. Let $a_i$ be a capsule in layer $L$. A capsule $b_j$ in layer $L + 1$ is calculated as:

$$b_j = g(\sum_i P_{ij}\hat{a}_{j|i}) \ and \ \hat{a}_{j|i} = W_{ij}a_i, \tag{5}$$

where weight matrix is $W_{ij}$ and from the capsule $a_i$, a predicted vector $\hat{a}_{j|i} \in \mathbb{R}^d$ is calculated. The capsule length was constrained by the nonlinear squashing function $g$ to the range of $[0, 1]$, and the coupling coefficient $P_{ij}$ was iteratively updated using the dynamic routing method [23].

**Hate Capsule Layer.** The hate capsule layer is the final capsule layer consisting of $k$ capsules with 16-dimensional ($d = 16$) instantiated parameters. Hate capsules are flattened into a 1d vector with dimension ($k \times 16$) to concatenate with attenuated features generated by channel-2.

### 4.4  Channel-2

**Bi-LSTM.** [8] learns long term context-dependent semantic features into hidden states by sequentially encoding the embedding vectors, $e$, generated by Fast-Text model as

$$\overrightarrow{h}_t = \overrightarrow{LSTM}_{fd}(e_t, h_{t-1}), \ \overleftarrow{h}_t = \overleftarrow{LSTM}_{bd}(e_t, h_{t+1}), \tag{6}$$

where $\overleftarrow{h}_t$ and $\overrightarrow{h}_t$ and are the backward and forward hidden states, respectively. The final hidden state representation for the input sentence is obtained as,

$$H_e = [h_1, h_2, h_3, \dots, h_N], \tag{7}$$

where $h_t = \overrightarrow{h}_t, \overleftarrow{h}_t$ and $H_e \in \mathbb{R}^{N \times 2D_h}$. The number of hidden units in LSTM is $D_h$.

**Attention Layer.** The underlying idea behind the attention mechanism is to assign more weight to the words that contribute the most to the phrase's meaning. We use the word label attention [27] on the Bi-LSTM layer's output to create an attended sentence vector. Specifically,

$$u_t^i = \tanh(W_w h_t^i + b_w), \tag{8}$$

$$\sigma_t^i = \frac{\exp(u_t^i{}^T u_w)}{\sum_t \exp(u_t^i{}^T u_w))}, \tag{9}$$

$$S_i = \sum_t (\sigma_t^i * h_t^i), \tag{10}$$

where $u_w$ is the context vector. Using a single layer MLP, we calculate the hidden representation $u_t^i$ from the word vector $h_t^i$. The output of the attention layer is $S_i$, and the attention weight for a given word is $\sigma_t^i$.

**FC Layers.** The outputs returned by the BERT+Capsule and FastText+ LSTM+ Attention are concatenated to make a combined representation, $J$, of the input sentence, $X$. Next, $J$ is passed through two fully connected layers, ($FC_1$(200 *neurons*) + $FC_2$(100 *neurons*)) followed by a softmax output layer to predict the probabilities of a sample belonging to target classes.

### 4.5  Loss Function

To train the parameter and back-propagating the loss, categorical cross-entropy $L(\hat{Y}, Y)$ has been employed as

$$L_{CE}(\hat{Y}, Y) = -\frac{1}{N} \sum_{j=1}^{M} \sum_{i=1}^{N} Y_i^j log(\hat{Y}_i^j), \tag{11}$$

where $\hat{Y}_i^j$ is predicted label and $Y_i^j$ is true label. $N$ represent the number of tweets in the dataset and $M$ represents the number of classes.

## 5  Experimental Results and Analysis

This section describes the outcomes of various baseline models and our proposed model, tested on the *Thai Hate speech* dataset. We split our dataset into 80% train, 10% validation and 10% test sets. During validation, we experimented with different network configurations and attained optimal performance with batch size = 32, activation function = ReLU, dropout rate= 0.25, learning rate = 1e–4, epoch = 20. We used Adam optimizer with a weight_decay = 1e–3 (for avoiding overfitting) for training. We performed all our experiments on a hybrid cluster of multiple GPUs comprised of RTX 2080Ti.

### 5.1  Baseline Setup

We have experimented with standard machine learning baselines like Naive Bayes, Support Vector Machine (SVM), Random forest as mentioned in [25]. Some advanced deep learning models like BERT+Capsule [15] and LSTM+Attn+ FC [16] are also included in baselines for a strong comparison with our proposed model. For machine learning-based baselines, we have considered the pooled output of dimension 768 returned by WangchanBERTa as input. Whereas for Fast-Text embedding, firstly, we tokenized the sentence using PyThaiNLP[3], extracted the embedding of each token from the pre-trained Thai FastText model, and took the average to represent the entire sentence by a 300 dimension vector. **Capsule**: The input is sent through a 1D CNN with 64 filters of window size 2. Then the convoluted feature is passed through the capsule network and the final output is generated by the hate capsule layer. **Capsule+FC**: Here hate capsule layer flattened output again passed through two FC layers (100 neurons per layer), followed

---

[3] https://pythainlp.github.io/docs/2.2/.

by a soft-max output layer. **LSTM+Attn+FC**: We pass the input features to BiLSTM. The obtained hidden representation from BiLSTM is sent to the word attention layer followed by fully connected layers to obtain the output.

**Table 1.** Experimental results of different baselines and proposed *FastThaiCaps* evaluated on Thai Hate speech dataset

| Embedding | Model | Accuracy | Precision | Recall | F1 Score |
|-----------|-------|----------|-----------|--------|----------|
| **Machine Learning Baselines** | | | | | |
| WangchanBERTa | Naive Bayes | 59.76 | 61.25 | 59.45 | 60.34 |
| | SVM | 66.56 | 66.71 | 66.53 | **66.61** |
| | Random forest | 60.33 | 62.51 | 60.17 | 61.32 |
| FastText | Naive Bayes | 54.56 | 56.43 | 54.63 | 55.52 |
| | SVM | 67.04 | 68.32 | 67.11 | **67.71** |
| | Random forest | 62.35 | 64.47 | 62.05 | 63.24 |
| **Deep learning Baselines** | | | | | |
| WangchanBERTa | Capsule | 74.60 | 74.56 | 72.35 | 73.43 |
| | Capsule+FC | 76.13 | 76.48 | 75.86 | **76.17** |
| | LSTM+Attn+FC | 73.86 | 73.63 | 73.86 | 73.74 |
| FastText | Capsule | 73.25 | 73.45 | 70.42 | 71.90 |
| | Capsule+FC | 74.19 | 74.53 | 73.22 | 73.87 |
| | LSTM+Attn+FC | 75.33 | 75.23 | 76.15 | **75.69** |
| **SOTA [19]** | | | | | |
| WangchanBERTa | CE + JT | 75.12 | 74.96 | 75.46 | 75.21 |
| | OR + JT | 75.52 | 75.36 | 75.86 | 75.61 |
| **Proposed Approach** | | | | | |
| BERT+FastText | FastThaiCaps | 78.65 | 77.35 | 80.13 | **78.72** |
| Improvements | | 3.13 | | | 3.11 |

## 5.2  Findings from Experiments

Table 1 shows the results of our proposed model, *FastThaiCaps*, and other baselines and SOTA in terms of accuracy, precision, recall and macro F1 score.

From Table 1, we can conclude the following:

(1) The proposed *FastThaiCaps* model outperforms all the baselines and SOTA significantly, improving 2.52% and 3.13% accuracy over the best baseline (WangchanBERTa+Capsule+FC) and SOTA, respectively. The joint optimization of channel-1 (BERT+Capsule+FC) and channel-2 (Fast-Text+LSTM+Attn) in the proposed model lead to the classifier's better performance and the gain in accuracy.

(2) When comparing three machine learning baselines, SVM always attains the best results with both the embeddings.

(3) Capsule+FC has consistently performed better than Capsule in both embedding strategies (BERT/FastText). Like, *Capsule+FC* performs better than *Capsule* with improvements in the F1 score of 2.74% and 1.97% for BERT and FastText embeddings, respectively. This finding supports the idea of keeping Capsule+FC as channel-1 in the proposed model.

(4) We have also observed that excluding the Naive Bayes classifier, other machine learning models perform better when utilizing FastText embedding instead of WangchanBERTa.

(5) We can notice that Capsule and Capsule+FC perform better than LSTM+Attn+FC when embedded with WangchanBERTa, and the reverse scenario occurs for the FastText embedding. That is why in channel -1, we keep WangchanBERTa, and for channel 2, FastText embedding is utilized.

(6) The individual performances of channel 1 (WangchanBERTa+Capsule+FC) and channel 2 (FastText+LSTM+Attn+FC) are 76.17% and 75.69% in terms of F1 score, respectively. But when we combined both channels, we achieved an F1 score of 78.7%. This significant improvement suggests that combinations of BERT and FastText embedding can efficiently handle the noisy text.

We have conducted a statistical t-test on the results of five different runs of our proposed model and other baselines and obtained a p-value less than 0.05.

### 5.3   Error Analysis

The most confused classes of this study are peace speech and neutral speech classes. The model predicted peace speech messages as neutral and vice versa, with 39.0% of the number of misclassified samples. The semantics of both categories are close together in our case. Individually considering a word or phase may give us one meaning, but it changes the meaning when considering the whole sentence. We usually find this in long sentences, metaphors, and sentences to encourage or express feelings. For example, "คนไทยไม่แล้งน้ำใจหรอกครับ," it means" Thais are very kind. The Thai government is helping the Burmese. Let's work together to prevent the disease. Please do not hide." The model classified this message as peace speech instead of neutral. According to this sentence, two words can interpret and dominate the model to predict peace speech: "ทางการไทยช่วยเหลือพี่น้องชาวพม่าอยู่แล้ว ขอให้ทุกคนร่วมมือกันป้องกันโรคอย่าหลบหนี" (kindness) and "น้ำใจ" (helpful). However, considering the whole message, it is found that the speaker mentioned the fact and did not convey it in any direction of showing pity or encouragement to the Burmese.

## 6   Conclusion and Future Work

The recent increase in online hate speech and trolling on various social media networks has become a critical problem. In this paper, we have proposed a novel two channel framework *FastThaiCaps* for hate speech detection in Thai, a complex language that is structurally different from other Asian languages. We have examined the performance of channel-1 (BERT+Capsule+FC) and channel-2 (FastText+LSTM+Attention+FC) separately and noticed that they achieved overall accuracies of 76.13% and 75.33%, respectively, on a benchmark Thai hate speech dataset. We have proposed the *FastThaiCaps* model with the intuition that the joint optimization of channel-1 and channel-2 leads to the development of a better classifier. The overall accuracy of 78.65% on the proposed dataset establishes that BERT and FastText together can handle noisy social media text more efficiently. Furthermore, Our proposed model outperforms all the baselines and beats state-of-the-art with an F1 score of 3.11%.

In future, we would like to enrich the existing Thai hate speech dataset with sentiment and emotion labels and develop a deep multitask framework to investigate how sentiment and emotion information enhance the performance of the main task, i.e., hate speech detection.

**Acknowledgement.** This work was supported by the Ministry of External Affairs (MEA) and the Department of Science & Technology (DST), India, under the ASEAN-India Collaborative R&D Scheme. The Authors also would like to acknowledge the support of Ministry of Home Affairs (MHA), India for conducting this research.

# References

1. Chan, T.K., Cheung, C.M., Wong, R.Y.: Cyberbullying on social networking sites: the crime opportunity and affordance perspectives. J. Manage. Inf. Syst. **36**(2), 574–609 (2019)
2. Choudhury, M., Saraf, R., Jain, V., Mukherjee, A., Sarkar, S., Basu, A.: Investigation and modeling of the structure of texting language. Int. J. Doc. Anal. Recogn. (IJDAR) **10**(3–4), 157–174 (2007)
3. Davidson, T., Warmsley, D., Macy, M., Weber, I.: Automated hate speech detection and the problem of offensive language. In: Proceedings of the International AAAI Conference on Web and Social Media, vol. 11, pp. 512–515 (2017)
4. Del Vigna12, F., Cimino23, A., Dell'Orletta, F., Petrocchi, M., Tesconi, M.: Hate me, hate me not: hate speech detection on facebook. In: Proceedings of the First Italian Conference on Cybersecurity (ITASEC17), pp. 86–95 (2017)
5. Devlin, J., Chang, M.W., Lee, K., Toutanova, K.: Bert: pre-training of deep bidirectional transformers for language understanding. arXiv preprint arXiv:1810.04805 (2018)
6. Dinakar, K., Reichart, R., Lieberman, H.: Modeling the detection of textual cyberbullying. In: Proceedings of the International Conference on Weblog and Social Media 2011. Citeseer (2011)
7. Grave, E., Bojanowski, P., Gupta, P., Joulin, A., Mikolov, T.: Learning word vectors for 157 languages. arXiv preprint arXiv:1802.06893 (2018)
8. Hochreiter, S., Schmidhuber, J.: Long short-term memory. Neural Comput. **9**(8), 1735–1780 (1997)
9. Ibrohim, M.O., Budi, I.: Multi-label hate speech and abusive language detection in Indonesian twitter. In: Proceedings of the Third Workshop on Abusive Language Online, pp. 46–57 (2019)

10. Kim, Y.: Convolutional neural networks for sentence classification. arXiv preprint arXiv:1408.5882 (2014)
11. Le, Q., Mikolov, T.: Distributed representations of sentences and documents. In: International Conference on Machine Learning, pp. 1188–1196. PMLR (2014)
12. Lowphansirikul, L., Polpanumas, C., Jantrakulchai, N., Nutanong, S.: Wangchanberta: pretraining transformer-based Thai language models. arXiv preprint arXiv:2101.09635 (2021)
13. Maity, K., Jha, P., Saha, S., Bhattacharyya, P.: A multitask framework for sentiment, emotion and sarcasm aware cyberbullying detection from multi-modal code-mixed memes. In: Amigó, E., Castells, P., Gonzalo, J., Carterette, B., Culpepper, J.S., Kazai, G. (eds.) SIGIR 2022: The 45th International ACM SIGIR Conference on Research and Development in Information Retrieval, Madrid, Spain, 11–15 July 2022, pp. 1739–1749. ACM (2022). https://doi.org/10.1145/3477495.3531925
14. Maity, K., Kumar, A., Saha, S.: A multi-task multi-modal framework for sentiment and emotion aided cyberbully detection. In: IEEE Internet Computing (2022)
15. Maity, K., Saha, S.: BERT-capsule model for cyberbullying detection in code-mixed Indian languages. In: Métais, E., Meziane, F., Horacek, H., Kapetanios, E. (eds.) NLDB 2021. LNCS, vol. 12801, pp. 147–155. Springer, Cham (2021). https://doi.org/10.1007/978-3-030-80599-9_13
16. Maity, K., Saha, S.: A multi-task model for sentiment aided cyberbullying detection in code-mixed Indian languages. In: Mantoro, T., Lee, M., Ayu, M.A., Wong, K.W., Hidayanto, A.N. (eds.) ICONIP 2021. LNCS, vol. 13111, pp. 440–451. Springer, Cham (2021). https://doi.org/10.1007/978-3-030-92273-3_36
17. Maity, K., Saha, S., Bhattacharyya, P.: Emoji, sentiment and emotion aided cyberbullying detection in hinglish. In: IEEE Transactions on Computational Social Systems (2022)
18. Nockleby, J.T.: Hate speech in context: the case of verbal threats. Buff. L. Rev. **42**, 653 (1994)
19. Pasupa, K., Karnbanjob, W., Aksornsiri, M.: Hate speech detection in Thai social media with ordinal-imbalanced text classification. In: Proceedings of the 19th International Joint Conference on Computer Science and Software Engineering (JCSSE 2022), 22–25 June 2022, Bangkok, Thailand, pp. 1–6 (2022)
20. Paul, S., Saha, S.: Cyberbert: Bert for cyberbullying identification. Multimedia Syst. 1–8 (2020)
21. Pennington, J., Socher, R., Manning, C.D.: Glove: Global vectors for word representation. In: Proceedings of the 2014 Conference on Empirical Methods in Natural Language Processing (EMNLP), pp. 1532–1543 (2014)
22. Reynolds, K., Kontostathis, A., Edwards, L.: Using machine learning to detect cyberbullying. In: 2011 10th International Conference on Machine Learning and Applications and Workshops, vol. 2, pp. 241–244. IEEE (2011)
23. Sabour, S., Frosst, N., Hinton, G.E.: Dynamic routing between capsules. arXiv preprint arXiv:1710.09829 (2017)
24. Simanjuntak, D.A., Ipung, H.P., Nugroho, A.S., et al.: Text classification techniques used to facilate cyber terrorism investigation. In: 2010 Second International Conference on Advances in Computing, Control, and Telecommunication Technologies, pp. 198–200. IEEE (2010)
25. Wanasukapunt, R., Phimoltares, S.: Classification of abusive thai language content in social media using deep learning. In: 2021 18th International Joint Conference on Computer Science and Software Engineering (JCSSE), pp. 1–6. IEEE (2021)

26. Watanabe, H., Bouazizi, M., Ohtsuki, T.: Hate speech on twitter: a pragmatic approach to collect hateful and offensive expressions and perform hate speech detection. IEEE Access **6**, 13825–13835 (2018)
27. Yang, Z., Yang, D., Dyer, C., He, X., Smola, A., Hovy, E.: Hierarchical attention networks for document classification. In: Proceedings of the 2016 conference of the North American Chapter of the Association for Computational Linguistics: Human Language Technologies, pp. 1480–1489 (2016)

# Author Index